VOLUME ONE HUNDRED AND SIXTEEN

VITAMINS AND HORMONES

Hormones and Stem Cells

VOLUME ONE HUNDRED AND SIXTEEN

VITAMINS AND HORMONES

Hormones and Stem Cells

Series Editor

GERALD LITWACK, PhD

Toluca Lake, North Hollywood, California

ELSEVIER

ACADEMIC PRESS

An imprint of Elsevier

Academic Press is an imprint of Elsevier
50 Hampshire Street, 5th Floor, Cambridge, MA 02139, United States
525 B Street, Suite 1650, San Diego, CA 92101, United States
The Boulevard, Langford Lane, Kidlington, Oxford OX5 1GB, United Kingdom
125 London Wall, London, EC2Y 5AS, United Kingdom

First edition 2021

Notices
Knowledge and best practice in this field are constantly changing. As new research and experience broaden our understanding, changes in research methods, professional practices, or medical treatment may become necessary.

Practitioners and researchers must always rely on their own experience and knowledge in evaluating and using any information, methods, compounds, or experiments described herein. In using such information or methods they should be mindful of their own safety and the safety of others, including parties for whom they have a professional responsibility.

To the fullest extent of the law, neither the Publisher nor the authors, contributors, or editors, assume any liability for any injury and/or damage to persons or property as a matter of products liability, negligence or otherwise, or from any use or operation of any methods, products, instructions, or ideas contained in the material herein.

ISBN: 978-0-323-85550-1
ISSN: 0083-6729

For information on all Academic Press publications
visit our website at https://www.elsevier.com/books-and-journals

Publisher: Zoe Kruze
Acquisitions Editor: Ashlie M. Jackman
Editorial Project Manager: Chris Hockaday
Production Project Manager: Denny Mansingh
Cover Designer: Alan Studholme

Typeset by SPi Global, India

Contents

Contributors

Mohammad Foad Abazari
Research Center for Clinical Virology, Tehran University of Medical Sciences, Tehran, Iran

Elizabeth T. Ables
Department of Biology, East Carolina University, Greenville, NC, United States

Seyedeh Fatemeh Ahmadi
Student Research Committee; Department of Medical Biotechnology, School of Advanced Technologies in Medicine, Mazandaran University of Medical Sciences, Sari, Iran

Caroline M. Alexander
McArdle Laboratory for Cancer Research, University of Wisconsin-Madison School of Medicine and Public Health, Madison, WI, United States

Vito Antonio Baldassarro
Interdepartmental Center for Industrial Research in Health Sciences and Technologies, University of Bologna, Bologna, Italy

J.S. Beltrame
Center of Pharmacological and Botanical Studies (CEFyBO), School of Medicine University of Buenos Aires (UBA)—National Research Council (CONICET), Buenos Aires, Argentina

E. Bogetti
Center of Pharmacological and Botanical Studies (CEFyBO), School of Medicine University of Buenos Aires (UBA)—National Research Council (CONICET), Buenos Aires, Argentina

Lucile Butruille
UMR 7221 Phyma, CNRS/Muséum National d'Histoire Naturelle, Paris, France

Sérgio Paulo Bydlowski
Lipids, Oxidation and Cell Biology Team, Laboratory of Immunology (LIM19), Heart Institute (InCor), School of Medicine, University of São Paulo, São Paulo, Brazil

Laura Calzà
Montecatone Rehabilitation Institute, Imola; IRET Foundation, Ozzano Emilia; Department of Pharmacy and Biotechnology, University of Bologna, Bologna, Italy

V.A. Cañumil
Center of Pharmacological and Botanical Studies (CEFyBO), School of Medicine University of Buenos Aires (UBA)—National Research Council (CONICET), Buenos Aires, Argentina

Hugues Dardente
PRC, INRAE, CNRS, IFCE, Université de Tours, Nouzilly, France

Fabio Alessandro de Freitas
Lipids, Oxidation and Cell Biology Team, Laboratory of Immunology (LIM19), Heart Institute (InCor), School of Medicine, University of São Paulo, São Paulo, Brazil

F.L. de la Cruz Borthiry
Center of Pharmacological and Botanical Studies (CEFyBO), School of Medicine University of Buenos Aires (UBA)—National Research Council (CONICET), Buenos Aires, Argentina

Barbara A. Demeneix
UMR 7221 Phyma, CNRS/Muséum National d'Histoire Naturelle, Paris, France

Shirley Suet Lee Ding
Stem Cells and Diabetes Laboratory, Institute of Molecular and Cell Biology (IMCB), A*STAR, Proteos, Singapore, Singapore

Seyed Ehsan Enderami
Department of Medical Biotechnology, School of Advanced Technologies in Medicine; Molecular and Cell biology Research Center, Faculty of Medicine, Mazandaran University of Medical Sciences, Sari, Iran

Naomi Even-Zohar
Pituitary Center, Department of Medicine, Cedars-Sinai Medical Center, Los Angeles, CA, United States

Danielle S. Finger
Department of Biology, East Carolina University, Greenville, NC, United States

Alessandra Flagelli
Interdepartmental Center for Industrial Research in Health Sciences and Technologies, University of Bologna, Bologna, Italy

Liezhen Fu
Section on Molecular Morphogenesis, Eunice Kennedy Shriver National Institute of Child Health and Human Development (NICHD), National Institutes of Health (NIH), Bethesda, MD, United States

Jan-Åke Gustafsson
Department of Biosciences and Nutrition, Karolinska Institutet, Huddinge, Sweden; Center for Nuclear Receptors and Cell Signaling, University of Houston, Houston, TX, United States

Jose Inzunza
Department of Biosciences and Nutrition, Karolinska Institutet, Huddinge, Sweden

Robert J. Johnston Jr
Department of Biology, Johns Hopkins University, Baltimore, MD, United States

Samaneh Khazaei
Department of Hematology and Cell Therapy, Faculty of Medical Sciences, Tarbiat Modares University, Tehran; Department of Cell and Molecular Biology, Faculty of Biological Science and Technology, Isfahan University, Isfahan, Iran

Débora Levy
Lipids, Oxidation and Cell Biology Team, Laboratory of Immunology (LIM19), Heart Institute (InCor), School of Medicine, University of São Paulo, São Paulo, Brazil

Reyhaneh Nassiri Mansour
Student Research Committee; Department of Tissue Engineering, School of Advanced Technologies in Medicine, Mazandaran University of Medical Sciences, Sari, Iran

Christina McNerney
Department of Biology, Johns Hopkins University, Baltimore, MD, United States

Shlomo Melmed
Pituitary Center, Department of Medicine, Cedars-Sinai Medical Center, Los Angeles, CA, United States

Derya Metin Armagan
Pituitary Center, Department of Medicine, Cedars-Sinai Medical Center, Los Angeles, CA, United States

Martine Migaud
PRC, INRAE, CNRS, IFCE, Université de Tours, Nouzilly, France

Seyedeh Mahsoumeh Mirtaghi
Student Research Committee; Department of Medical Biotechnology, School of Advanced Technologies in Medicine, Mazandaran University of Medical Sciences, Sari, Iran

Ivan Nalvarte
Department of Biosciences and Nutrition, Karolinska Institutet, Huddinge, Sweden

Claire Wen Ying Neo
Stem Cells and Diabetes Laboratory, Institute of Molecular and Cell Biology (IMCB), A*STAR, Proteos; Department of Medicine, Yong Loo Lin School of Medicine, National University of Singapore, Singapore, Singapore

Natasha Hui Jin Ng
Stem Cells and Diabetes Laboratory, Institute of Molecular and Cell Biology (IMCB), A*STAR, Proteos, Singapore, Singapore

Daniel N. Phipps
Department of Biology, East Carolina University, Greenville, NC, United States

Mohammad Rahmati
Department of Medical Biotechnology, Faculty of Paramedicine, Guilan University of Medical Sciences, Rasht, Iran

Cadiele Oliana Reichert
Lipids, Oxidation and Cell Biology Team, Laboratory of Immunology (LIM19), Heart Institute (InCor), School of Medicine, University of São Paulo, São Paulo, Brazil

Sylvie Remaud
UMR 7221 Phyma, CNRS/Muséum National d'Histoire Naturelle, Paris, France

M.L. Ribeiro
Center of Pharmacological and Botanical Studies (CEFyBO), School of Medicine University of Buenos Aires (UBA)—National Research Council (CONICET), Buenos Aires, Argentina

Michele Sannia
Interdepartmental Center for Industrial Research in Health Sciences and Technologies, University of Bologna, Bologna, Italy

Yun-Bo Shi
Section on Molecular Morphogenesis, Eunice Kennedy Shriver National Institute of Child Health and Human Development (NICHD), National Institutes of Health (NIH), Bethesda, MD, United States

Yuki Shibata
Section on Molecular Morphogenesis, Eunice Kennedy Shriver National Institute of Child Health and Human Development (NICHD), National Institutes of Health (NIH), Bethesda, MD, United States

Adele Soltani
Department of Clinical Biochemistry, Faculty of Medical Sciences, Tarbiat Modares University, Tehran, Iran

Yuta Tanizaki
Section on Molecular Morphogenesis, Eunice Kennedy Shriver National Institute of Child Health and Human Development (NICHD), National Institutes of Health (NIH), Bethesda, MD, United States

Adrian Kee Keong Teo
Stem Cells and Diabetes Laboratory, Institute of Molecular and Cell Biology (IMCB), A*STAR, Proteos; Department of Medicine; Department of Biochemistry, Yong Loo Lin School of Medicine, National University of Singapore, Singapore, Singapore

Pieter Vancamp
UMR 7221 Phyma, CNRS/Muséum National d'Histoire Naturelle, Paris, France

Mukesh Varshney
Department of Biosciences and Nutrition, Karolinska Institutet, Huddinge, Sweden

Julia von Maltzahn
Leibniz-Institute on Aging—Fritz-Lipmann-Institute, Jena, Germany

Kaitlin M. Whitehead
Department of Biology, East Carolina University, Greenville, NC, United States

Preface

In early development the inner cell mass of the blastocyst contains totipotent cells that, in the appropriate specific environment, can form all the tissues of the human body. Following fertilization, after the first and second cell divisions, embryonic stem cells are totipotent. In the mature organism, most bodily tissues (even tumors) have a store of stem cells (caches) that are capable of replacing tissue cells after they die off. These caches of stem cells seem to diminish during aging and this reduction in stem cell stores may be a major factor in the aging process. Although stem cells can differentiate and proliferate in the process of replacing tissue cells, they require specific molecules to be present in their environment. Many of these molecules are hormones.

In this volume, the roles of hormones are described as they affect differentiation as well as other activities of stem cells. The following list includes the titles of chapters and authors in the sequence as they appear in this book:

N. Even-Zohar, D.M. Armagan and S. Melmed

Pituitary stem cells

C.M. Alexander

Wnt signaling in mammary stem cells

C. McNerney and R.J. Johnston Jr.

Thyroid hormone signaling specifies cone photoreceptor subtypes during eye development: Insights from model organisms and human stem cell-derived retinal organoids

H. Dardente and M. Migaud

Thyroid hormone and hypothalamic stem cells in seasonal functions

L. Butruille, P. Vancamp, B.A. Demeneix and S. Remaud

Thyroid hormone regulation of adult neural stem cell fate: A comparative analysis between rodents and primates

N.H.J. Ng, C.W.Y. Neo, S. Ding and A.K.K. Teo

Insights from single cell studies of human pancreatic islet cells and stem cell-derived islet cells to guide functional beta cell maturation in vitro

A. Soltani, S. Khazaei, S.M. Mirtaghi, S.F. Ahmadi, M. Reyhaneh, N. Mansour, M.F.A. Abazari and S.F. Enderami

Generation of high yield insulin-producing cells (IPCs) from various sources of stem cells

Y.-B. Shi, Y. Shibata, Y. Tamizaki and L. Fu

The development of adult intestinal stem cells: Insights from studies on thyroid hormone-dependent anuran metamorphosis

J. von Matlzahn

Regulation of muscle stem cell function

I. Nalvarte, M. Varshney, J. Inzunza and J.-A. Gustafsson

Estrogen receptor beta and neural development

D.S. Finger, K.M. Whitehead, D.N. Phipps and E.T. Ables

Nuclear receptors linking physiology and germ line stem cells in Drosophila

V.A. Canumil, E. Bogetti, F.L. De la Cruz, M.L. Ribeiro and J.S. Beltrame

Steroid hormones and first trimester vascular remodeling

V.A. Baldassarro, A. Flegelli, M. Samnia and L. Calza

Nuclear receptors and differentiation of oligodendrocyte precursor cells

C.O. Reichert, F.A. de Freitas, D. Levy and S.P. Bydlowski

Oxysterols and mesenchymal stem cell biology

The cover is one that perpetuates from one book to the next in this Series, except that the background color varies with each new volume.

I am grateful to the group in the Oxford, United Kingdom, office of Elsevier, especially to Ashlie Jackman, who have helped in the early phases of the production of this book and to the production group in Chennai, India, for their precise development of the printed volume.

GERALD LITWACK
Toluca Lake, CA, United States

CHAPTER ONE

Pituitary stem cells

Naomi Even-Zohar[†], Derya Metin Armagan[†], and Shlomo Melmed*
Pituitary Center, Department of Medicine, Cedars-Sinai Medical Center, Los Angeles, CA, United States
*Corresponding author: e-mail address: melmed@csmc.edu

Contents

Abstract

The anterior pituitary is derived from Rathke's pouch precursors, which differentiate into specific hormone-secreting cell lineages. Sustained low postnatal and adult pituitary cell turnover is governed by stem/progenitor cells that undergo slow mitotic activity and give rise to hormone-secreting cells in response to physiological demands and feedback loops. Pituitary cell populations exhibit stem cell properties, which include stem cell marker expression, non-hormone expression, and the ability to self-renew and to potentially differentiate into any of five hormone-secreting cell lineages. Specific signaling pathways underlie differentiated pituitary cell development and regulation. Several validated pituitary stem cell models have been reported and have the potential for functional regeneration of pituitary hormone-secreting cell functions.

[†] Equal contribution.

Vitamins and Hormones, Volume 116
ISSN 0083-6729
https://doi.org/10.1016/bs.vh.2021.02.008

1. Introduction

The developmental tissue origin of the adenohypophysis (anterior pituitary) is Rathke's pouch, which is formed by invagination of the oral epithelium, specified in the mouse embryo as early as 9 days post coitum (Kelberman, Rizzoti, Lovell-Badge, Robinson, & Dattani, 2009). Cell precursors in Rathke's pouch proliferate and differentiate into specific cell lineages, and eventually into one of the major hormone-secreting anterior pituitary cell types (Bilodeau, Roussel-Gervais, & Drouin, 2009). Pituitary organogenesis takes place prenatally and involves migration and differentiation of multiple stem/progenitor cells. After birth, in the neonatal period, there is a second proliferative growth wave with progenitor cell amplification and differentiation and increased numbers of hormone-producing cells (Zhu, Gleiberman, & Rosenfeld, 2007).

Only relatively recently, it has become clear that, even later in life, there is a sustained low pituitary turnover activity, governed by stem/progenitor cells within the gland (Vankelecom & Gremeaux, 2010). These cells undergo slow mitotic activity and give rise to hormone-secreting cells in response to physiological demands and complex feedback loops (Castinetti, Davis, Brue, & Camper, 2011). Several pituitary cell populations exhibit stem cell properties, which fulfill the following criteria: non-hormone-secreting, ability to self-renew, and capacity to differentiate into any of the five hormone-secreting cell lineages. Stem cell markers, and specifically SOX2, are expressed in these cell populations (Caffarini, Orciani, Trementino, Di Primio, & Arnaldi, 2018).

This chapter addresses signaling pathways underlying differentiated pituitary cell development and stem cell regulation. We also review the current state of the art for pituitary stem cell models, addressing the potential use of validated and functional pituitary stem cells in regenerative medicine.

2. Signaling pathways in pituitary development and stem cell regulation

Several temporally regulated signaling pathways determine pituitary organogenesis, respective cell type proliferation, and ultimately mature cell-specific hormone differentiation. This cellular cascade culminates in unique, highly differentiated pituitary trophic hormone synthesis and secretion for maintenance of healthy developmental and adult endocrine and metabolic

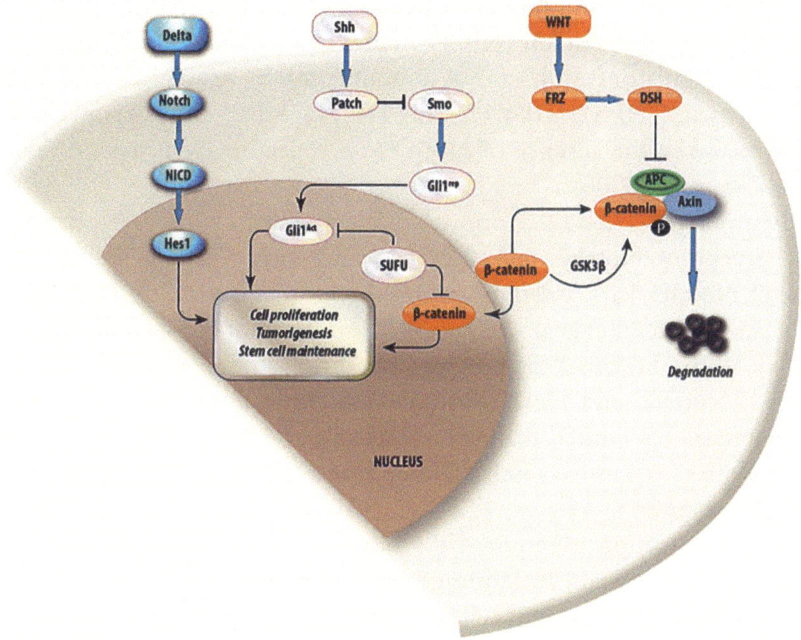

Fig. 1 Notch, WNT, and SHH signaling in stem cell regulation and development. These pathways play critical roles in stem cell maintenance and cell proliferation, involved in both pituitary organogenesis and tumorigenesis. *Reprinted by permission from Yavrapoulou, M. P., et al. (2015). The role of Notch and Hedgehog signaling pathways in pituitary development and pathogenesis of pituitary adenomas.* Hormones (Athens), 14*(1), 5–18. © 2015, Springer Nature.*

homeostasis. Pituitary stem cells have been proposed as driving progenitor differentiation and proliferation, and a subset of such cells may also be sustained throughout adulthood to mount a differentiated pituitary cell response to provoked stress or extrinsic and intrinsic signals. Disrupted progenitor cell function has also been implicated in pathogenesis of pituitary adenomas.

Notch, WNT, SHH, and Hippo (YAP/TAZ) signaling pathways play essential roles in maintaining pituitary gland homeostasis during embryonic development and postnatally by regulating regeneration and differentiation of stem or progenitor cells (Fig. 1).

2.1 Notch signaling

Notch pathway ligands and receptors are expressed early in pituitary development and serve to regulate pituitary cell terminal differentiation. Notch signaling requires cell-to-cell contact between the ligand and receptor.

Upon ligand binding, Notch receptors promote proteolytic cleavage of the Notch intracellular domain (NICD), enter the nucleus, and bind the transcriptional regulator RBPJ, a DNA-binding protein, to activate transcription of target genes HES1, HEY1,2 and HEYL. RBPJ binds transcriptional co-repressors in the absence of NICD (Yavropoulou, Maladaki, & Yovos, 2015; Zhu et al., 2006).

By embryonic day E9.5, mouse oral ectoderm expresses Notch ligand Jag1, receptors Notch2 and Notch3, and target protein Hes1, while Notch2, Notch3, Jag1, and the ligand Dll1 appear in Rathke's pouch between E10.5 and E12.5 (Zhu et al., 2006). At E12.5, Hes1 is restricted to the ventral region. With the onset of cell-specific Pit1 expression on E13.5, Dll1, Notch2, Notch3, and Hes1 expression continue in periluminal cells, and Jag1 expression is limited mainly to Rathke's pouch mesenchymal cells. By E14.5 and E17.5, Hes1, Notch2, Notch3, and Dll1 are only detected in cells near the lumen.

Notch activation is required for pituitary-specific Prop1 induction at E12.5, with subsequent Pit1 lineage, underscoring the Notch requirement in determining differentiated cell fate (Zhu et al., 2006). Notch signaling also regulates proliferation of cells expressing Sca1, Notch1, and Hes1 *via* several growth factors including bFGF, EGF, and LIF (Chen, Crabbe, Van Duppen, & Vankelecom, 2006; Chen et al., 2005).

E-cadherin is required for the appearance of Notch2 and Jag1 in the marginal layer of adult pituitary and non-endocrine folliculostellate (FS) cells, as well as for activation of the Notch signal. E-cadherin is expressed in FS cells, which also contain both Notch receptor and ligand, suggesting that neighboring cells interact to activate Notch signaling (Batchuluun, Azuma, Yashiro, & Kikuchi, 2017; Kikuchi et al., 2006, 2007).

Enhanced cell cycle exit occurring between E10.5 and E14.5 was observed in the Hesl mutant pituitary due to expansion of p27 and p57 cyclin-dependent kinase inhibitors (Monahan, Rybak, & Raetzman, 2009), suggesting a role for Hes1 in the control of cell cycle exit and timely development of differentiated hormone-synthesizing cells. Thus, these lines of evidence support the essential role of Notch signaling in cell specification by regulating a balance between pituitary cell proliferation and differentiation (Monahan et al., 2009).

Periluminal stem cells and stem cell clusters continue to increase postnatally and represent more than half of Ki67-positive anterior pituitary proliferating cells (Zhu, Tollkuhn, Taylor, & Rosenfeld, 2015). As Notch signaling protects Sox2+ cells and regulates timing of Pit1+ cell re-entry into the cell cycle, Notch may also induce Sox2 expression (Zhu et al., 2015).

Genetic inhibition of Notch2 reveals that, despite its important role in regulating postnatal location and number of pituitary progenitors, the molecule has little effect on early embryonic pituitary proliferation. Rather, study of Notch2 knockout mice shows decreased Prop1 expression at both E12.5 and wild-type postnatal P1, with increased Tpit and Pomc expression. DAPT inhibits Notch signaling with decreased Prop1 mRNA levels. Several lines of evidence support the important function of Notch signaling in regulating pituitary size and cell specification during both embryonic and postnatal periods, suppressing the corticotroph line while triggering Prop1 expression and the Pit1 cell lineage (Nantie, Himes, Getz, & Raetzman, 2014).

2.2 Wnt signaling

Binding of Wnt proteins to the Fzd receptor family activates and stabilizes β-catenin, which enters the nucleus and interacts with TCF/LEF transcription factors regulating cell fate and morphogenesis (Nusse, 2012; van Amerongen & Nusse, 2009).

Studies with Wnt knockout mouse models demonstrate the importance of pituitary Wnt signaling *via* the ventral diencephalon, and respective roles of these ligands and receptors differ based on the developmental stage (Potok et al., 2008). Wnt5a mutants exhibit increased Fgf10 and Bmp4 expression in the ventral diencephalon as well as reduced Rathke's pouch Lhx3 expression, while Wnt4/Wnt5a double mutants show dysmorphology and mild pituitary hypoplasia. Pituitary Wnt3 is detected at E12.5, Wnt2b at E14.5, and Wnt10b in adults, while Fzd6 and Fzd8 are detected in the developing pituitary at both E12.5 and E14.5, but not in adult tissue (Potok et al., 2008).

Wnt/β-catenin signaling is critical to pituitary gland organogenesis. Both loss and gain of β-catenin causes pituitary progenitors to exit the cell cycle and undergo EMT-like transition, leading to early developmental hypopituitarism (Youngblood, Coleman, & Davis, 2018). Interestingly, excess Wnt/β-catenin attenuates Shh, Fgf10, and Bmp4 expression and enhances Fgf8, while loss of β-catenin decreases Fgf8 but not Fgf10 or Bmp4, leading to a smaller Rathke's pouch (Osmundsen, Keisler, Taketo, & Davis, 2017). Wnt/β-catenin also promotes proliferation of cells expressing Sox2, enhancing proliferation of pituitary adenoma cells. Thus, Wnt/β-catenin signaling plays an important role both during pituitary embryonic development and in mechanisms underlying pituitary adenoma pathogenesis (Tang, Chen, Wang, Huang, & Hu, 2019).

2.3 SHH signaling

Binding of the glycoprotein SHH inactivates PTCH receptors in the cell membrane; as the SHH ligand-deficient SMO receptor is inhibited by PTCH, SMO induces translocation of zinc finger transcription factor GLI proteins to initiate transcription of target genes (Carballo, Honorato, de Lopes, & Spohr, 2018; Yavropoulou et al., 2016).

In the adult pituitary, induced Shh signals trigger proliferation of Sox2+ and Sox9+ mouse pituitary stem cells and increases immunopositive expression of ACTH, GH, and PRL in adenomas (Pyczek et al., 2016).

By contrast, during pituitary organogenesis, Shh affects both proliferation and cell-type determination. Mouse Shh is expressed in the oral ectoderm until E12.0 and along the ventral diencephalon adjacent to Rathke's pouch until E14.0. Although not detected in the developing Rathke's pouch, Shh signaling to the developing pituitary occurs *via* the Ptch1 receptor and through transcriptional targets highly expressed throughout Rathke's pouch (Treier et al., 2001). Shh/Gli signaling promotes Rathke's pouch progenitor proliferation, as well as Bmp4/Fgf8 expression in the ventral diencephalon to control early Rathke's pouch modeling (Wang, Martin, & Bai, 2010; Yavropoulou et al., 2015).

2.4 Hippo signaling

The Hippo pathway regulates pituitary size through stem cell function and regeneration, proliferation, and control of apoptosis. Upstream components of the Hippo pathway, namely MST1/2 and LATS1/2, have tumor suppressor activity, while downstream elements YAP, TAZ, and TEAD act as oncogenes (Johnson & Halder, 2014).

Nuclear localization of Yap and Taz in mouse Rathke's pouch and anterior postnatal pituitary is restricted to Sox2+ stem cells (Lodge, Russell, Patist, Francis-West, & Andoniadou, 2016). Accordingly, postnatal deletion of Lats kinases induces pituitary Yap/Taz, resulting in increased Sox2+ cells and formation of aggressive pituitary tumors, indicating the requirement of the Hippo pathway for both regulation of normal pituitary physiology and cell proliferation as well as tumor formation (Lodge et al., 2019).

3. Pituitary stem cell models

3.1 Immortalized pituitary cell lines

Several immortalized pituitary cell lines have been created for modeling pituitary adenomas (Ooi, Tawadros, & Escalona, 2004; Zhu, Cui, Zhu, Gao, & Zhu, 2020). As these may exhibit characteristics of immature pituitary cells, suggesting their source could be a primitive/stem cell within anterior pituitary tissue, they could potentially serve as models of pituitary cell differentiation, as well as disrupted pituitary functions.

3.1.1 MtT/E (rat)

In 1990, an estrogen-induced rat mammotropic tumor was used to derive four different cell lines (Inoue et al., 1990), three of which secreted GH and/or PRL, morphologically resembled pituitary cells, and were round/oval in appearance. Expression of the cell-specific Pit-1 gene strongly suggested that these cells were committed mammosomatotroph progenitors with the potential to differentiate into hormone-secreting cells, as demonstrated by differentiation into GH-secreting cells under the influence of ATRA (Mogi et al., 2005; Nogami, Soya, Hiraoka, Aiso, & Hisano, 2012).

3.1.2 TtT/GF (mouse)

This line was derived from a mouse pituitary tumor induced by thyroid gland radioablation (Inoue et al., 1992; Matsumoto et al., 1993). Source cells were pituitary FS cells, which do not produce hormones but express the FS markers S100 and GFAP, and morphologically display long cytoplasmic protrusions, forming follicles in culture and showing some phagocytic activity. This model enabled study of paracrine regulation of anterior pituitary hormones (Brokken, Leendertse, Bakker, Wiersinga, & Prummel, 2004; Chapman, Nishimura, Buckingham, Morris, & Christian, 2002; Koike et al., 1997; Vlotides, Chen, Eigler, Ren, & Melmed, 2009; Vlotides et al., 2006). Furthermore, the same qualities that support use of FS cells as candidates for stem/progenitor cells (Vankelecom, 2007) are also evident in the TtT/GF cell line. For example, stem cell markers including ABC transporters Abcb1a, Abcb1b, Abcg2 and Sca1 are expressed in TtT/GF but not in Tpit/F1 and LβT2 pituitary cell lines (Mitsuishi et al., 2013), and expression of SOX2 as well as the transcription factor Isl1 were also demonstrated (Yoshida et al., 2014). Although the endocrine origin of these cells has been questioned (Tsukada et al., 2018; Yoshida et al., 2014), they are considered a valuable tool for studying pituitary organogenesis and tumorigenesis and serve as a model for pituitary stem cells (Tsukada et al., 2019, 2018; Zhu et al., 2020).

3.1.3 Tpit/F1, Tpit/E (mouse)

These two lines were created (Chen et al., 2000) using cells derived from the anterior pituitary of a temperature-sensitive large-T-antigen-transgenic mouse. Both lines express Sox2, Isl1, and Sca1 pituitary stem cells markers. Interestingly, there is also expression of PitX1 and PitX2, important for early pituitary development (Yoshida et al., 2014). With bFGF and B27 supplemented in serum-free medium, Tpit/F1 cells form small floating clumps that express stem cell markers and exhibit proliferative ability (Higuchi et al., 2017). Some cells "migrated" from the clump and were immune-positive for GH without being positive for PIT-1, suggesting validated pituitary cell lineage differentiation. Differentiation of Tpit/F1 cells into skeletal muscle cells is believed to reflect their FS-like properties (Mogi et al., 2004).

3.1.4 αT1-1, αT3-1 (mouse)

Several immortalized cell lines representing different stages of differentiation in the anterior pituitary were generated (Alarid, Windle, Whyte, & Mellon, 1996; Windle, Weiner, & Mellon, 1990) using targeted mutagenesis in transgenic mice. A transgene with 1.8 kb of the 5′-regulatory region of the human α-subunit gene linked to the coding region for the oncogene SV40 T-antigen was inserted, thus immortalizing immature pituitary cells during the early stages of pituitary embryonic development. Expression of -early-appearing pituitary transcription factors such as LHX3 and HESX1 was demonstrated (Alarid et al., 1996). Cells from the αT1-1 line, corresponding to E11.5, express α-subunit, which responds to hypothalamic GnRH; FSH and LH β-subunits are not expressed. It thus represents a multipotent precursor to thyrotroph and gonadotroph lineages. The more mature αT3-1 corresponds to E14.5 and represents a committed immature gonadotroph, expressing GnRH receptor and SF-1. TαT-1 and LβT2 correspond to mature thyrotroph and gonadotroph cells at E14.5 and E17.5, respectively. These cells have been used to determine expression of genes that might play a role in anterior pituitary development and differentiation (Aikawa, Sato, Ono, Kato, & Kato, 2006; Barnhart & Mellon, 1994; Holley, Hall, & Mellon, 2002; Horn, Windle, Barnhart, & Mellon, 1992; Laverrière et al., 2016; Xie et al., 2017).

3.1.5 *Pit1/0, Pit1/Prl, Pit1/Triple (mouse)*

SV40 T-antigen induced cell transformation established three cell lines representing distinct stages in Pit-1-dependent cell differentiation in the mouse pituitary (Sizova, Ho, Cooke, & Liebhaber, 2010). Pit1/0 represents an early step in anterior pituitary development and correlates with E12.5-13.5, at which point Pit-1 is already expressed but hormone gene activation cannot yet be detected. By contrast, Pit1/Prl represents differentiated lactotroph cells, and Pit1/Triple represents a triple-positive intermediate precursor positive for GH, PRL, and TSHβ.

3.2 Somatic stem cell colonies

The adult pituitary contains a side population of stem cells involved with slow-but-steady turnover of pituitary cells as well as with pituitary regenerative responses to extrinsic or intrinsic insults (Vankelecom & Chen, 2014). Primary anterior pituitary cells cultured from animals and humans (Melmed, Odenheimer, Carlson, & Hershman, 1982; Zhu et al., 2020) are a challenging model, as the pituitary is a relatively small organ with a limited number of cells, and, in general, these cultures are short-lived. Furthermore, preparing primary cultures from adult pituitary stem cells is difficult, as only a small percentage (3–5%) of pituitary cells display stem cell qualities (Fauquier, Rizzoti, Dattani, Lovell-Badge, & Robinson, 2008).

Upon culturing adult mouse pituitary cells in serum-free, chemically defined culture medium supplemented with B27 and bFGF (Chen et al., 2005), free-floating spheres appeared and enlarged during the first 6 days. Around 40 spheres were observed per 200,000 anterior pituitary cells, with an average of 42 cells per sphere. After 6 days, spheres formed large aggregates, with the original spheres continuing to enlarge inside the aggregate; BrdU labeling demonstrated that cell division was responsible for sphere enlargement. Using eGFP-expressing and WT mice, sphere cell clonal origin was validated. Accordingly, spheres could be dissociated into single cells and re-seeded and secondary spheres formed in culture medium. A third generation of spheres could not be attempted as the number of cells did not permit further seeding.

Pluripotent stem cell markers and pituitary differentiation related transcription factors are expressed in the sphere-forming cells, including SOX2, SOX9, Nanog, Pitx1, Pitx2, Isl1, Lhx3, Lhx4, Pax6, Six1, Six6, Msx1, HesX1, Prop1, and Otx2 (Chen et al., 2009). Interestingly, expression of several of these transcription factors, including Hesx1, Prop1, Lhx4, and Pax6, are down-regulated after embryonic development is completed. More than 70% showed no hormone gene expression, and others were restricted to α-subunit mRNA and/or POMC mRNA expression. When spheres cultured from Sca1-negative cells were dissociated into single cells and plated on Matrigel in differentiation conditions, all endocrine cell types of the anterior pituitary including GH, PRL, ACTH, TSHβ, LHβ, FSHβ and α-GSU were detected, whereas expression of the stem cell marker SOX2 was lost, thus confirming the pluripotency of this cell population.

Pituitary colony-forming cells were also identified in 2D primary murine pituitary cultures that have the potential to differentiate into GH-positive cells (Lepore et al., 2005), and SOX2+ E-cadherin–positive "pituispheres" (Fauquier et al., 2008) contained progenitor cells that differentiated into all endocrine hormone expressing cells. Transgenic mice expressing GFP driven by regulatory elements of the nestin gene were used to derive colony-forming stem cells in primary pituitary culture (Gleiberman et al., 2008). Exponentially growing colonies survived at least 25 divisions until they differentiated spontaneously into hormone-secreting anterior pituitary cells. Sphere-forming GFRa2+ cells isolated from mouse pituitary (Garcia-Lavandeira et al., 2009) were maintained in attached/floating colonies for more than seven passages; cells were positive for Prop1, SOX2, SOX9, and Oct, but negative for nanog and nestin, demonstrating that these cells can differentiate into hormone forming cells under highly specific conditions. The sphere-forming trait of these cells was employed to create similar *in vitro* models (Andoniadou et al., 2013; Rizzoti, Akiyama, & Lovell-Badge, 2013; Yoshida et al., 2009), including human spheres in culture (Weiss, Siebzehnrubl, Kreutzer, Blumcke, & Buslei, 2009) that were differentiated into PRL-secreting cells under forskolin treatment.

Colony-forming cells isolated in these studies using very different methodologies were not homogenous, and likely represent different cell populations with heterogenous progenitor capacities, or even similar populations at different stages of differentiation (Vankelecom, 2012). Nevertheless, creating colonies *in vitro* from adult pituitary stem cells remains a promising approach for studying stem cell qualities.

3.3 Differentiation of mouse and human multipotent cells

Mouse models of embryonal stem-cell differentiation are a key source of multipotent embryonal cells. The embryonal period in mice is considerably shorter than in humans, 20 days vs 300 days, allowing for more results and easier establishment of protocol modifications, while the considerable similarity between human and mouse development renders research with mouse stem cells highly relevant for extrapolation to human diseases (Kelberman et al., 2009).

A 3D model for pituitary differentiation was established and referred to as "serum-free floating culture of embryoid body-like aggregates" (Suga et al., 2011). Culturing cells in high-density conditions in low-adhesion plates induced aggregates to self-form, and high-density seeding or addition of BMP4 to these aggregates on days 6–18 induced PITX-1/2 expression, marking the formation of rostral head ectoderm. Addition of an SHH agonist on days 6–30 induced expression of Lhx3, a Rathke's pouch marker, and morphological changes were also demonstrated, including invagination of ectoderm and formation of hollowed vesicles, recapitulating those occurring in the embryo during Rathke's pouch formation. Aggregates self-organized into a hybrid hypothalamus-pituitary organoid, with an outer layer positive for PITx1 and LHX3 and resembling surface ectoderm differentiated into pituitary cells, while an inner layer positive for neuroepithelial markers nestin, Rx, and Nkx2.1 resembled hypothalamic neural ectoderm. Co-induction of both hypothalamus and pituitary properties enabled interactions critical for pituitary differentiation between the embryonic hypothalamus and the developing pituitary. Cells in the respective pouches then differentiated into functional hormone-producing cells. Modulations of notch, Wnt, glucocorticoid, and estradiol signaling influenced differentiation. For example, treatment with the notch inhibitor DAPT increased TBX-19 and ACTH positive cells, and those same aggregates were also responsive to CRH, with increased ACTH secretion *in vitro* observed after 10 min of CRH stimulation, while addition of WNT agonist BIO increased Pou1f1 (PIT-1) expression and GH and PRL immunopositive cells. Furthermore, DAPT-treated aggregates ectopically transplanted into hypophysectomized mice showed partial recovery of ACTH levels and improved survival compared to non-transplanted hypophysectomized mice (Suga et al., 2011).

This protocol was adapted to human embryonic stem cells (Ozone et al., 2016) with 3D generation of a hybrid hypothalamus-pituitary culture, recapitulating interactions between the hypothalamus and oral ectoderm during differentiation. Both high cell density conditions and exogenous BMP4 were required for induction of the hybrid culture. However, in contrast to mouse embryonic stem cells, morphological invagination of human oral ectoderm did not consistently occur without influencing tissue functionality.

Also, DAPT did not induce corticotrophs; in fact, Notch pathway inhibition by DAPT led to gonadotroph differentiation. Responsiveness to positive and negative regulatory signals was also demonstrated. For example, human embryonic stem-cell–derived corticotrophs responded to CRH with increased ACTH secretion *in vitro* and *in vivo* when transplanted to hypophysectomized mice, while pre-exposure of aggregates to hydrocortisone *in vitro* led to significantly suppressed ACTH secretion. Furthermore, addition of dexamethasone to the culture medium favored induction of somatotrophs over corticotrophs, while GH secretion by terminally differentiated somatotrophs was induced by exogenous GHRH and inhibited by somatostatin.

Differentiated pituitary 2D cultures were established for mouse and human embryonic stem cells (Dincer et al., 2013; Zimmer et al., 2016), first using human embryonic stem cells and later using induced pluripotent stem cells (iPSCs). Pituitary placode cells were established (Dincer et al., 2013) using a universal "placode induction protocol" with human embryonic stem cells as well as with iPSCs, in which the BMP inhibitor Noggin was used in combination with SB-431542, an inhibitor of TGFβ-R1, ALK4, and ALK7; Noggin was then withdrawn on day 3. Using this protocol, more than 70% of cells expressed the placodal precursor cell marker SIX1 on day 11 of differentiation. Addition of SHH agonist on days 7–11 induced expression of PITX1, SIX6, and LHX3, while addition of the Notch inhibitor DAPT on days 13–18 induced GATA2 and Pou1f1 positive cells. By differentiation day 30, ACTH-, FSH-, and GH-expressing cells were evident; ACTH production was detectable by ELISA. Cells transplanted into rats survived *in vivo* for at least 6 weeks, and human-specific circulating GH and ACTH levels were detected.

Recently, both models were adjusted to iPSCs (Kasai et al., 2020; Matsumoto et al., 2020; Zimmer et al., 2016), and key components of each model are presented in Table 1.

3.4 Potential role of iPSCs in regenerative pituitary medicine

Unique iPSC models offer an option for clinical application and patient benefit. As no human cell lines are available, use of iPSC-derived functional, validated human hormone-secreting cells offers a potential novel therapeutic horizon. Establishment of validated human iPSCs differentiated to respective hormone-secreting proliferating lineages can be used for studying mechanisms underlying human pituitary adenoma signaling and pathogenesis, thereby identifying novel therapeutic targets for these adenomas, which

Table 1 2D and 3D models for pituitary differentiation from iPSCs.

	2D model (Dincer et al., 2013; Zimmer et al., 2016)	**3D model (Kasai et al., 2020; Matsumoto et al., 2020)**
Type of culture	2D, monolayer of pituitary cells only	3D, floating aggregates of hybrid hypothalamus-pituitary culture
Reagents used for pituitary differentiation induction	• Days 0–30: SB-431542 (TGFβRI, ALK4/7 inhibitor) • Days 0–3: BMP4 • Days 4–30: FGF10, FGF8, SHH • Days 30–60: BMP2 ± FGF8	• Days 0–3: LDN 193189 (ALK 2/3 inhibitor) • Days 6–18: BMP4 • Days 6–30: SHH agonist
Source of FGF signals	Exogenous	Endogenous
O_2 concentration	21% throughout	40% from day 18
Days until hormone-secreting cells apparent	~30–60 days	~100 days
Responsiveness to ACTH stimulatory signals (e.g., CRH, low glucose)	+	+
Responsiveness to ACTH inhibitory signals (e.g., dexamethasone)	–	+
Responsiveness to GHRH	+	+

are often resistant to current medical therapies. Screening these subcellular targets for novel therapeutics by employing these cell models also offers an approach for testing pituitary cell-targeted therapeutic efficacy.

For example, hypopituitarism is currently treated with hormone replacement (Fleseriu et al., 2016), which is costly, life-long, and requires daily or twice daily use of oral or injectable drugs or their combinations. These medications, although replacing deficient hormones, do not mimic intrinsic circadian or stress-related changes in hormone requirements. By contrast, explanted pituitary organoids differentiated *in vitro* and immunologically identical to the patient's own tissue offer hope for attaining a cell-based approach to improve efficiency of hormone substitution and responsiveness to physiologic feedback signals. Orthotopic placement of regenerating, functional pituitary cells into the native pituitary niche is technically challenging but could facilitate the close interrelationship between hypothalamic signaling through the pituitary portal circulation to impinge on differentiated anterior pituitary cell types. Therefore, establishment of functional hypothalamic-pituitary cell units to recapitulate neuroendocrine hormonal axes offers a novel discovery tool for gaining further insights into both normal and dysfunctional human pituitary cell biology.

References

Aikawa, S., Sato, T., Ono, T., Kato, T., & Kato, Y. (2006). High level expression of Prop-1 gene in gonadotropic cell lines. *The Journal of Reproduction and Development*, *52*(2), 195–201. https://doi.org/10.1262/jrd.17050.

Alarid, E. T., Windle, J. J., Whyte, D. B., & Mellon, P. L. (1996). Immortalization of pituitary cells at discrete stages of development by directed oncogenesis in transgenic mice. *Development*, *122*(10), 3319–3329.

Andoniadou, C. L., Matsushima, D., Mousavy Gharavy, S. N., Signore, M., Mackintosh, A. I., Schaeffer, M., et al. (2013). Sox2(+) stem/progenitor cells in the adult mouse pituitary support organ homeostasis and have tumor-inducing potential. *Cell Stem Cell*, *13*(4), 433–445. https://doi.org/10.1016/j.stem.2013.07.004.

Barnhart, K. M., & Mellon, P. L. (1994). The orphan nuclear receptor, steroidogenic factor-1, regulates the glycoprotein hormone alpha-subunit gene in pituitary gonadotropes. *Molecular Endocrinology*, *8*(7), 878–885. https://doi.org/10.1210/mend.8.7.7527122.

Batchuluun, K., Azuma, M., Yashiro, T., & Kikuchi, M. (2017). Notch signaling-mediated cell-to-cell interaction is dependent on E-cadherin adhesion in adult rat anterior pituitary. *Cell and Tissue Research*, *368*(1), 125–133. https://doi.org/10.1007/s00441-016-2540-5.

Bilodeau, S., Roussel-Gervais, A., & Drouin, J. (2009). Distinct developmental roles of cell cycle inhibitors p57Kip2 and p27Kip1 distinguish pituitary progenitor cell cycle exit from cell cycle reentry of differentiated cells. *Molecular and Cellular Biology*, *29*(7), 1895–1908. https://doi.org/10.1128/mcb.01885-08.

Brokken, L. J., Leendertse, M., Bakker, O., Wiersinga, W. M., & Prummel, M. F. (2004). Expression of adenohypophyseal-hormone receptors in a murine folliculo-stellate cell line. *Hormone and Metabolic Research*, *36*(8), 538–541. https://doi.org/10.1055/s-2004-825758.

Caffarini, M., Orciani, M., Trementino, L., Di Primio, R., & Arnaldi, G. (2018). Pituitary adenomas, stem cells, and cancer stem cells: What's new? *Journal of Endocrinological Investigation*, *41*(7), 745–753. https://doi.org/10.1007/s40618-017-0803-y.

Carballo, G. B., Honorato, J. R., de Lopes, G. P. F., & Spohr, T. (2018). A highlight on Sonic hedgehog pathway. *Cell Communication and Signaling: CCS*, *16*(1), 11. https://doi.org/10.1186/s12964-018-0220-7.

Castinetti, F., Davis, S. W., Brue, T., & Camper, S. A. (2011). Pituitary stem cell update and potential implications for treating hypopituitarism. *Endocrine Reviews*, *32*(4), 453–471. https://doi.org/10.1210/er.2010-0011.

Chapman, L., Nishimura, A., Buckingham, J. C., Morris, J. F., & Christian, H. C. (2002). Externalization of annexin I from a folliculo-stellate-like cell line. *Endocrinology, 143*(11), 4330–4338. https://doi.org/10.1210/en.2002-220529.

Chen, J., Crabbe, A., Van Duppen, V., & Vankelecom, H. (2006). The notch signaling system is present in the postnatal pituitary: Marked expression and regulatory activity in the newly discovered side population. *Molecular Endocrinology, 20*(12), 3293–3307. https://doi.org/10.1210/me.2006-0293.

Chen, J., Gremeaux, L., Fu, Q., Liekens, D., Van Laere, S., & Vankelecom, H. (2009). Pituitary progenitor cells tracked down by side population dissection. *Stem Cells, 27*(5), 1182–1195. https://doi.org/10.1002/stem.51.

Chen, J., Hersmus, N., Van Duppen, V., Caesens, P., Denef, C., & Vankelecom, H. (2005). The adult pituitary contains a cell population displaying stem/progenitor cell and early embryonic characteristics. *Endocrinology, 146*(9), 3985–3998. https://doi.org/10.1210/en.2005-0185.

Chen, L., Maruyama, D., Sugiyama, M., Sakai, T., Mogi, C., Kato, M., et al. (2000). Cytological characterization of a pituitary folliculo-stellate-like cell line, Tpit/F1, with special reference to adenosine triphosphate-mediated neuronal nitric oxide synthase expression and nitric oxide secretion. *Endocrinology, 141*(10), 3603–3610. https://doi.org/10.1210/endo.141.10.7710.

Dincer, Z., Piao, J., Niu, L., Ganat, Y., Kriks, S., Zimmer, B., et al. (2013). Specification of functional cranial placode derivatives from human pluripotent stem cells. *Cell Reports, 5*(5), 1387–1402. https://doi.org/10.1016/j.celrep.2013.10.048.

Fauquier, T., Rizzoti, K., Dattani, M., Lovell-Badge, R., & Robinson, I. C. (2008). SOX2-expressing progenitor cells generate all of the major cell types in the adult mouse pituitary gland. *Proceedings of the National Academy of Sciences of the United States of America, 105*(8), 2907–2912. https://doi.org/10.1073/pnas.0707886105.

Fleseriu, M., Hashim, I. A., Karavitaki, N., Melmed, S., Murad, M. H., Salvatori, R., et al. (2016). Hormonal replacement in hypopituitarism in adults: An Endocrine Society clinical practice guideline. *The Journal of Clinical Endocrinology and Metabolism, 101*(11), 3888–3921. https://doi.org/10.1210/jc.2016-2118.

Garcia-Lavandeira, M., Quereda, V., Flores, I., Saez, C., Diaz-Rodriguez, E., Japon, M. A., et al. (2009). A GRFa2/Prop1/stem (GPS) cell niche in the pituitary. *PLoS One, 4*(3), e4815. https://doi.org/10.1371/journal.pone.0004815.

Gleiberman, A. S., Michurina, T., Encinas, J. M., Roig, J. L., Krasnov, P., Balordi, F., et al. (2008). Genetic approaches identify adult pituitary stem cells. *Proceedings of the National Academy of Sciences of the United States of America, 105*(17), 6332–6337. https://doi.org/10.1073/pnas.0801644105.

Higuchi, M., Yoshida, S., Kanno, N., Mitsuishi, H., Ueharu, H., Chen, M., et al. (2017). Clump formation in mouse pituitary-derived non-endocrine cell line Tpit/F1 promotes differentiation into growth-hormone-producing cells. *Cell and Tissue Research, 369*(2), 353–368. https://doi.org/10.1007/s00441-017-2603-2.

Holley, S. J., Hall, S. B., & Mellon, P. L. (2002). Complementary expression of IGF-II and IGFBP-5 during anterior pituitary development. *Developmental Biology, 244*(2), 319–328. https://doi.org/10.1006/dbio.2002.0608.

Horn, F., Windle, J. J., Barnhart, K. M., & Mellon, P. L. (1992). Tissue-specific gene expression in the pituitary: The glycoprotein hormone alpha-subunit gene is regulated by a gonadotrope-specific protein. *Molecular and Cellular Biology, 12*(5), 2143–2153. https://doi.org/10.1128/mcb.12.5.2143.

Inoue, K., Hattori, M., Sakai, T., Inukai, S., Fujimoto, N., & Ito, A. (1990). Establishment of a series of pituitary clonal cell lines differing in morphology, hormone secretion, and response to estrogen. *Endocrinology, 126*(5), 2313–2320. https://doi.org/10.1210/endo-126-5-2313.

Inoue, K., Matsumoto, H., Koyama, C., Shibata, K., Nakazato, Y., & Ito, A. (1992). Establishment of a folliculo-stellate-like cell line from a murine thyrotropic pituitary tumor. *Endocrinology*, *131*(6), 3110–3116. https://doi.org/10.1210/endo.131.6.1446645.

Johnson, R., & Halder, G. (2014). The two faces of Hippo: Targeting the Hippo pathway for regenerative medicine and cancer treatment. *Nature Reviews. Drug Discovery*, *13*(1), 63–79. https://doi.org/10.1038/nrd4161.

Kasai, T., Suga, H., Sakakibara, M., Ozone, C., Matsumoto, R., Kano, M., et al. (2020). Hypothalamic contribution to pituitary functions is recapitulated in vitro using 3D-cultured human iPS cells. *Cell Reports*, *30*(1), 18–24.e15. https://doi.org/10.1016/j.celrep.2019.12.009.

Kelberman, D., Rizzoti, K., Lovell-Badge, R., Robinson, I. C., & Dattani, M. T. (2009). Genetic regulation of pituitary gland development in human and mouse. *Endocrine Reviews*, *30*(7), 790–829. https://doi.org/10.1210/er.2009-0008.

Kikuchi, M., Yatabe, M., Fujiwara, K., Takigami, S., Sakamoto, A., Soji, T., et al. (2006). Distinctive localization of N- and E-cadherins in rat anterior pituitary gland. *The Anatomical Record. Part A, Discoveries in Molecular, Cellular, and Evolutionary Biology*, *288*(11), 1183–1189. https://doi.org/10.1002/ar.a.20384.

Kikuchi, M., Yatabe, M., Kouki, T., Fujiwara, K., Takigami, S., Sakamoto, A., et al. (2007). Changes in E- and N-cadherin expression in developing rat adenohypophysis. *The Anatomical Record: Advances in Integrative Anatomy and Evolutionary Biology*, *290*(5), 486–490. https://doi.org/10.1002/ar.20516.

Koike, K., Zhang, Z. X., Sakamoto, Y., Kanda, Y., Murakami, K., Miyake, A., et al. (1997). The pituitary folliculo-stellate cell line TtT/GF augments basal and TRH-induced prolactin secretion by GH3 cell. *Life Sciences*, *61*(25), 2491–2497. https://doi.org/10.1016/s0024-3205(97)00984-3.

Laverrière, J. N., L'Hôte, D., Tabouy, L., Schang, A. L., Quérat, B., & Cohen-Tannoudji, J. (2016). Epigenetic regulation of alternative promoters and enhancers in progenitor, immature, and mature gonadotrope cell lines. *Molecular and Cellular Endocrinology*, *434*, 250–265. https://doi.org/10.1016/j.mce.2016.07.010.

Lepore, D. A., Roeszler, K., Wagner, J., Ross, S. A., Bauer, K., & Thomas, P. Q. (2005). Identification and enrichment of colony-forming cells from the adult murine pituitary. *Experimental Cell Research*, *308*(1), 166–176. https://doi.org/10.1016/j.yexcr.2005.04.023.

Lodge, E. J., Russell, J. P., Patist, A. L., Francis-West, P., & Andoniadou, C. L. (2016). Expression analysis of the Hippo cascade indicates a role in pituitary stem cell development. *Frontiers in Physiology*, 7, 114. https://doi.org/10.3389/fphys.2016.00114.

Lodge, E. J., Santambrogio, A., Russell, J. P., Xekouki, P., Jacques, T. S., Johnson, R. L., et al. (2019). Homeostatic and tumourigenic activity of SOX2+ pituitary stem cells is controlled by the LATS/YAP/TAZ cascade. *eLife*, *8*, e43996. https://doi.org/10.7554/eLife.43996.

Matsumoto, H., Koyama, C., Sawada, T., Koike, K., Hirota, K., Miyake, A., et al. (1993). Pituitary folliculo-stellate-like cell line (TtT/GF) responds to novel hypophysiotropic peptide (pituitary adenylate cyclase-activating peptide), showing increased adenosine 3′,5′-monophosphate and interleukin-6 secretion and cell proliferation. *Endocrinology*, *133*(5), 2150–2155. https://doi.org/10.1210/endo.133.5.8404665.

Matsumoto, R., Suga, H., Aoi, T., Bando, H., Fukuoka, H., Iguchi, G., et al. (2020). Congenital pituitary hypoplasia model demonstrates hypothalamic OTX2 regulation of pituitary progenitor cells. *The Journal of Clinical Investigation*, *130*(2), 641–654. https://doi.org/10.1172/jci127378.

Melmed, S., Odenheimer, D., Carlson, H. E., & Hershman, J. M. (1982). Establishment of functional human pituitary tumor cell cultures. *In Vitro*, *18*(1), 35–42. https://doi.org/10.1007/bf02796383.

Mitsuishi, H., Kato, T., Chen, M., Cai, L. Y., Yako, H., Higuchi, M., et al. (2013). Characterization of a pituitary-tumor-derived cell line, TtT/GF, that expresses Hoechst efflux ABC transporter subfamily G2 and stem cell antigen 1. *Cell and Tissue Research, 354*(2), 563–572. https://doi.org/10.1007/s00441-013-1686-7.

Mogi, C., Goda, H., Mogi, K., Takaki, A., Yokoyama, K., Tomida, M., et al. (2005). Multistep differentiation of GH-producing cells from their immature cells. *The Journal of Endocrinology, 184*(1), 41–50. https://doi.org/10.1677/joe.1.05870.

Mogi, C., Miyai, S., Nishimura, Y., Fukuro, H., Yokoyama, K., Takaki, A., et al. (2004). Differentiation of skeletal muscle from pituitary folliculo-stellate cells and endocrine progenitor cells. *Experimental Cell Research, 292*(2), 288–294. https://doi.org/10.1016/j.yexcr.2003.09.018.

Monahan, P., Rybak, S., & Raetzman, L. T. (2009). The notch target gene HES1 regulates cell cycle inhibitor expression in the developing pituitary. *Endocrinology, 150*(9), 4386–4394. https://doi.org/10.1210/en.2009-0206.

Nantie, L. B., Himes, A. D., Getz, D. R., & Raetzman, L. T. (2014). Notch signaling in postnatal pituitary expansion: Proliferation, progenitors, and cell specification. *Molecular Endocrinology, 28*(5), 731–744. https://doi.org/10.1210/me.2013-1425.

Nogami, H., Soya, H., Hiraoka, Y., Aiso, S., & Hisano, S. (2012). Epidermal growth factor-activated extracellular signal-regulated kinase suppresses growth hormone expression and stimulates proliferation in MtT/E cells. *Journal of Neuroendocrinology, 24*(2), 357–365. https://doi.org/10.1111/j.1365-2826.2011.02247.x.

Nusse, R. (2012). Wnt signaling. *Cold Spring Harbor Perspectives in Biology, 4*(5), a011163. https://doi.org/10.1101/cshperspect.a011163.

Ooi, G. T., Tawadros, N., & Escalona, R. M. (2004). Pituitary cell lines and their endocrine applications. *Molecular and Cellular Endocrinology, 228*(1–2), 1–21. https://doi.org/10.1016/j.mce.2004.07.018.

Osmundsen, A. M., Keisler, J. L., Taketo, M. M., & Davis, S. W. (2017). Canonical WNT signaling regulates the pituitary organizer and pituitary gland formation. *Endocrinology, 158*(10), 3339–3353. https://doi.org/10.1210/en.2017-00581.

Ozone, C., Suga, H., Eiraku, M., Kadoshima, T., Yonemura, S., Takata, N., et al. (2016). Functional anterior pituitary generated in self-organizing culture of human embryonic stem cells. *Nature Communications*, 7, 10351. https://doi.org/10.1038/ncomms10351.

Potok, M. A., Cha, K. B., Hunt, A., Brinkmeier, M. L., Leitges, M., Kispert, A., et al. (2008). WNT signaling affects gene expression in the ventral diencephalon and pituitary gland growth. *Developmental Dynamics, 237*(4), 1006–1020. https://doi.org/10.1002/dvdy.21511.

Pyczek, J., Buslei, R., Schult, D., Holsken, A., Buchfelder, M., Hess, I., et al. (2016). Hedgehog signaling activation induces stem cell proliferation and hormone release in the adult pituitary gland. *Scientific Reports, 6*, 24928. https://doi.org/10.1038/srep24928.

Rizzoti, K., Akiyama, H., & Lovell-Badge, R. (2013). Mobilized adult pituitary stem cells contribute to endocrine regeneration in response to physiological demand. *Cell Stem Cell, 13*(4), 419–432. https://doi.org/10.1016/j.stem.2013.07.006.

Sizova, D., Ho, Y., Cooke, N. E., & Liebhaber, S. A. (2010). Research resource: T-antigen transformation of pituitary cells captures three novel cell lines in the Pit-1 lineage. *Molecular Endocrinology, 24*(11), 2232–2240. https://doi.org/10.1210/me.2010-0235.

Suga, H., Kadoshima, T., Minaguchi, M., Ohgushi, M., Soen, M., Nakano, T., et al. (2011). Self-formation of functional adenohypophysis in three-dimensional culture. *Nature, 480*(7375), 57–62. https://doi.org/10.1038/nature10637.

Tang, J., Chen, L., Wang, Z., Huang, G., & Hu, X. (2019). SOX2 mediates crosstalk between Sonic Hedgehog and the Wnt/beta-catenin signaling pathway to promote proliferation of pituitary adenoma cells. *Oncology Letters, 18*(1), 81–86. https://doi.org/10.3892/ol.2019.10311.

Treier, M., O'Connell, S., Gleiberman, A., Price, J., Szeto, D. P., Burgess, R., et al. (2001). Hedgehog signaling is required for pituitary gland development. *Development*, *128*(3), 377–386.

Tsukada, T., Isowa, Y., Kito, K., Yoshida, S., Toneri, S., Horiguchi, K., et al. (2019). Identification of TGFβ-induced proteins in non-endocrine mouse pituitary cell line TtT/GF by SILAC-assisted quantitative mass spectrometry. *Cell and Tissue Research*, *376*(2), 281–293. https://doi.org/10.1007/s00441-018-02989-2.

Tsukada, T., Yoshida, S., Kito, K., Fujiwara, K., Yako, H., Horiguchi, K., et al. (2018). TGFβ signaling reinforces pericyte properties of the non-endocrine mouse pituitary cell line TtT/GF. *Cell and Tissue Research*, *371*(2), 339–350. https://doi.org/10.1007/s00441-017-2758-x.

van Amerongen, R., & Nusse, R. (2009). Towards an integrated view of Wnt signaling in development. *Development*, *136*(19), 3205–3214. https://doi.org/10.1242/dev.033910.

Vankelecom, H. (2007). Non-hormonal cell types in the pituitary candidating for stem cell. *Seminars in Cell & Developmental Biology*, *18*(4), 559–570. https://doi.org/10.1016/j.semcdb.2007.04.006.

Vankelecom, H. (2012). Pituitary stem cells drop their mask. *Current Stem Cell Research & Therapy*, 7(1), 36–71. https://doi.org/10.2174/157488812798483467.

Vankelecom, H., & Chen, J. (2014). Pituitary stem cells: Where do we stand? *Molecular and Cellular Endocrinology*, *385*(1–2), 2–17. https://doi.org/10.1016/j.mce.2013.08.018.

Vankelecom, H., & Gremeaux, L. (2010). Stem cells in the pituitary gland: A burgeoning field. *General and Comparative Endocrinology*, *166*(3), 478–488. https://doi.org/10.1016/j.ygcen.2009.11.007.

Vlotides, G., Chen, Y. H., Eigler, T., Ren, S. G., & Melmed, S. (2009). Fibroblast growth factor-2 autofeedback regulation in pituitary folliculostellate TtT/GF cells. *Endocrinology*, *150*(7), 3252–3258. https://doi.org/10.1210/en.2008-1625.

Vlotides, G., Cruz-Soto, M., Rubinek, T., Eigler, T., Auernhammer, C. J., & Melmed, S. (2006). Mechanisms for growth factor-induced pituitary tumor transforming gene-1 expression in pituitary folliculostellate TtT/GF cells. *Molecular Endocrinology*, *20*(12), 3321–3335. https://doi.org/10.1210/me.2006-0280.

Wang, Y., Martin, J. F., & Bai, C. B. (2010). Direct and indirect requirements of Shh/Gli signaling in early pituitary development. *Developmental Biology*, *348*(2), 199–209. https://doi.org/10.1016/j.ydbio.2010.09.024.

Weiss, S., Siebzehnrubl, F. A., Kreutzer, J., Blumcke, I., & Buslei, R. (2009). Evidence for a progenitor cell population in the human pituitary. *Clinical Neuropathology*, *28*(4), 309–318. https://doi.org/10.5414/npp28309.

Windle, J. J., Weiner, R. I., & Mellon, P. L. (1990). Cell lines of the pituitary gonadotrope lineage derived by targeted oncogenesis in transgenic mice. *Molecular Endocrinology*, *4*(4), 597–603. https://doi.org/10.1210/mend-4-4-597.

Xie, H., Hoffmann, H. M., Iyer, A. K., Brayman, M. J., Ngo, C., Sunshine, M. J., et al. (2017). Chromatin status and transcription factor binding to gonadotropin promoters in gonadotrope cell lines. *Reproductive Biology and Endocrinology*, *15*(1), 86. https://doi.org/10.1186/s12958-017-0304-z.

Yavropoulou, M. P., Maladaki, A., Topouridou, K., Kotoula, V., Poulios, C., Daskalaki, E., et al. (2016). Expression pattern of the Hedgehog signaling pathway in pituitary adenomas. *Neuroscience Letters*, *611*, 94–100. https://doi.org/10.1016/j.neulet.2015.10.076.

Yavropoulou, M. P., Maladaki, A., & Yovos, J. G. (2015). The role of Notch and Hedgehog signaling pathways in pituitary development and pathogenesis of pituitary adenomas. *Hormones (Athens, Greece)*, *14*(1), 5–18. https://doi.org/10.1007/BF03401377.

Yoshida, S., Higuchi, M., Ueharu, H., Nishimura, N., Tsuda, M., Yako, H., et al. (2014). Characterization of murine pituitary-derived cell lines Tpit/F1, Tpit/E and TtT/GF.

The Journal of Reproduction and Development, *60*(4), 295–303. https://doi.org/10.1262/jrd.2014-031.

Yoshida, S., Kato, T., Susa, T., Cai, L. Y., Nakayama, M., & Kato, Y. (2009). PROP1 coexists with SOX2 and induces PIT1-commitment cells. *Biochemical and Biophysical Research Communications*, *385*(1), 11–15. https://doi.org/10.1016/j.bbrc.2009.05.027.

Youngblood, J. L., Coleman, T. F., & Davis, S. W. (2018). Regulation of pituitary progenitor differentiation by beta-catenin. *Endocrinology*, *159*(9), 3287–3305. https://doi.org/10.1210/en.2018-00563.

Zhu, Z., Cui, W., Zhu, D., Gao, N., & Zhu, Y. (2020). Common tools for pituitary adenomas research: Cell lines and primary cells. *Pituitary*, *23*(2), 182–188. https://doi.org/10.1007/s11102-019-01003-4.

Zhu, X., Gleiberman, A. S., & Rosenfeld, M. G. (2007). Molecular physiology of pituitary development: Signaling and transcriptional networks. *Physiological Reviews*, *87*(3), 933–963. https://doi.org/10.1152/physrev.00006.2006.

Zhu, X., Tollkuhn, J., Taylor, H., & Rosenfeld, M. G. (2015). Notch-dependent pituitary SOX2(+) stem cells exhibit a timed functional extinction in regulation of the postnatal gland. *Stem Cell Reports*, *5*(6), 1196–1209. https://doi.org/10.1016/j.stemcr.2015.11.001.

Zhu, X., Zhang, J., Tollkuhn, J., Ohsawa, R., Bresnick, E. H., Guillemot, F., et al. (2006). Sustained Notch signaling in progenitors is required for sequential emergence of distinct cell lineages during organogenesis. *Genes & Development*, *20*(19), 2739–2753. https://doi.org/10.1101/gad.1444706.

Zimmer, B., Piao, J., Ramnarine, K., Tomishima, M. J., Tabar, V., & Studer, L. (2016). Derivation of diverse hormone-releasing pituitary cells from human pluripotent stem cells. *Stem Cell Reports*, *6*(6), 858–872. https://doi.org/10.1016/j.stemcr.2016.05.005.

CHAPTER TWO

Wnt signaling and mammary stem cells

Caroline M. Alexander*
McArdle Laboratory for Cancer Research, University of Wisconsin-Madison School of Medicine and Public Health, Madison, WI, United States
*Corresponding author: e-mail address: cmalexander@wisc.edu

Contents

Abstract

Wnt signaling is an important morphogenetic signaling pathway best known for its essential role in determining embryonic cell fates; it is often activated to re-specify fetal cells or to maintain the lineage flexibility of somatic stem cells. In this review, we consider the role of this pathway in the remarkable process of differentiation, growth and morphogenesis of the mammary gland during embryogenesis, ductal outgrowth and

Vitamins and Hormones, Volume 116
ISSN 0083-6729
https://doi.org/10.1016/bs.vh.2021.02.001

pregnancy. Specifically, mammary stem cells are compared with stem cells from other tissues, to identify commonalities and differences. Wnt signaling is known to be required to maintain the bipotent basal stem cell present in adult mammary ductal trees, however, the absence of this stem cell has little effect on growth or morphogenesis, and Wnt signaling is not induced during the ductal/alveolar expansion during pregnancy. The evidence for pre-determined hierarchies of mammary epithelial cells is reviewed, together with the role of signaling between mixtures of specified mammary epithelial cells in the maintenance of Wnt-dependent clonagenic stem cells. The dazzling variety of Wnt signaling components expressed by mammary epithelial cells is presented, along with some potential stromal sources of Wnt proteins that may be important starting points for the induction of plasticity in the epithelium.

1. Expression of Wnt signaling components in mammary epithelial cells

1.1 Diverse Wnt signaling activities

This review will be biased towards the so-called "canonical" Wnt signaling pathway, which is defined by a β-catenin-TCF transcriptional response. There are other signaling outputs from distinct Wnt binding reactions at the cell surface, notably the Wnt-Fzd or Wnt-Ror "non-canonical" pathways; these often compete or interact with the canonical binding reactions (Kessenbrock et al., 2013; Roarty & Serra, 2007). These are also likely to be important for the regulation of growth and morphogenesis of mammary tissues, however from a molecular point of view, they are still obscure. Although the transcriptional response of the canonical pathway serves as a defining readout, it is very unlikely that this is the only significant output of Wnt signaling; there are other, acute, proteomic reactions which function as a component of the total cell response. For example, the metabolic checkpoint mTOR is activated by Wnt ligand exposure in some cell types, including the classic intestinal epithelial stem cell (Inoki et al., 2006; Valvezan, Huang, Lengner, Pack, & Klein, 2014). This intestinal example is perhaps unusual, because these cells divide in response to a gain of function for Wnt signaling; typically, a Wnt signal induces a morphogenetic reaction but does not initiate a growth response. Perhaps a dedicated connection between Wnt and mTOR activation may explain the different responses of different tissues (Ly, Lynch, & Ryall, 2020; Valvezan & Manning, 2019).

1.2 The mammary Wnt signaling repertoire

The regulation of Wnt signaling is known to have tissue-specific elements. Fig. 1A illustrates the expression of specific Wnt pathway-associated components, the "Wnt landscape," expressed by mouse mammary epithelial cells

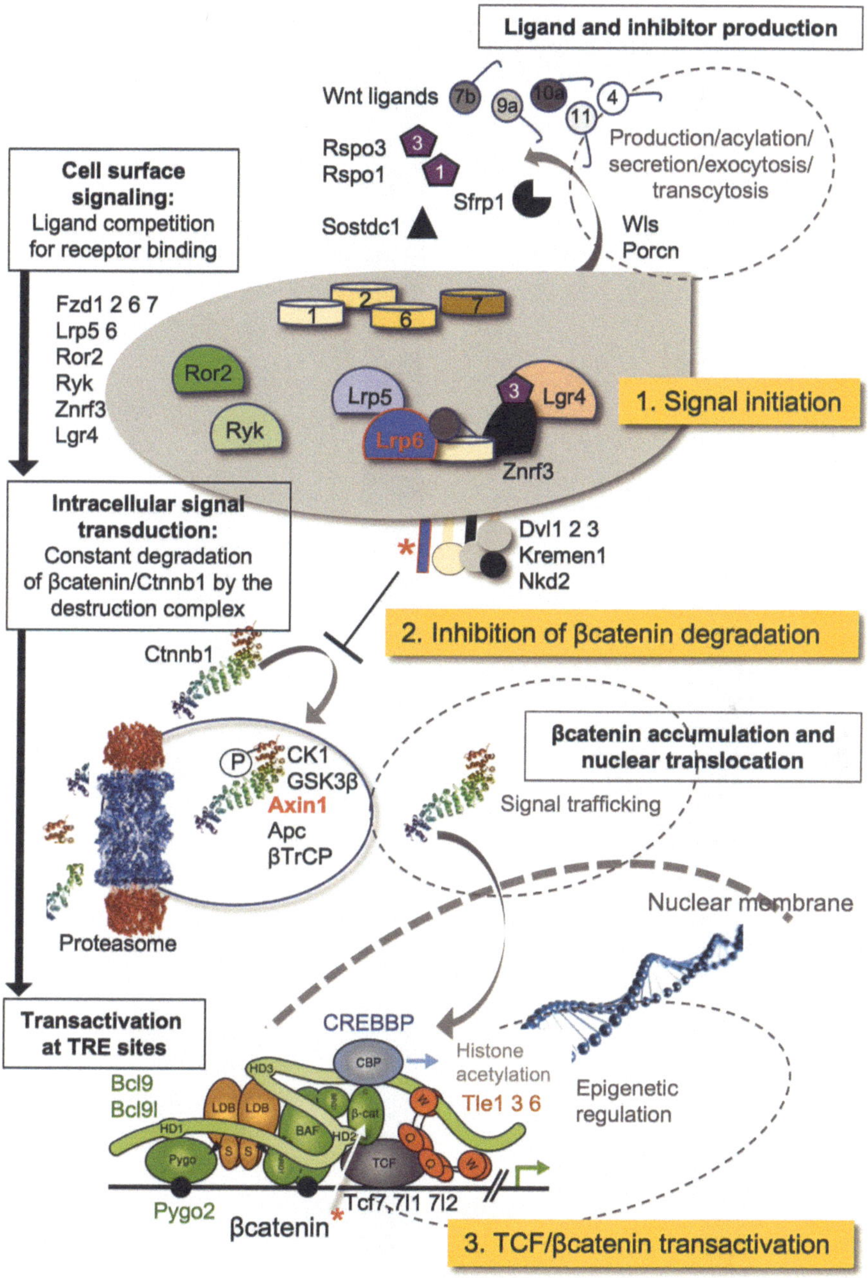

Fig. 1 The Wnt landscape in mouse mammary epithelial cells. (A) The components of the Wnt signaling pathway are assembled in a summary schematic; these data are derived from a non-tumorigenic mouse mammary epithelial cell line (EP cells) in regular

(Continued)

	RPKM	mRNA
Biosynthesis	38	Wls
	4	Porcn
	2	Notum
Ligands and inhibitors	1	Wnt4
	18	Wnt7B
	1	Wnt9a
	6	Wnt10a
	1	Wnt11
	1	Rspo3
	14	Sostdc1
	21	Sfrp1
	2	Wif1
Receptors/cell surface regulators	42	Fzd1
	12	Fzd2
	2	Fzd3
	1	Fzd5
	21	Fzd6
	37	Fzd7
	2	Fzd8
	2	Fzd9
	24	Lrp5
	6	Lrp6
	13	Lgr4
	2	Ror2
	30	Ryk
	3	Znfr3
	31	Fat1

	RPKM	mRNA
Catenins	51	Ctnna1
	134	Ctnnb1
	11	Ctnnd1
Signal transduction	15	Dvl1
	1. 8	Dvl2
	7	Dvl3
	4	Kremen1
	107	Nkd2
	28	Dab2
Destruction complex	33	Axin1
	4	Apc
	4	Btrc
	66	Gsk3a
	10	Gsk3b
Transcription complex	13	Tcf7
	11	Tcf7l1
	3	Tcf7l2
	6	Tle1
	14	Tle3
	1	Tle6
	28	Pygo2
	3	Bcl9
	45	Bcl9l
	5	Crebbp

Fig. 1—Cont'd growth culture conditions (Chin, Martin, Kim, Fakhraldeen, & Alexander, 2015). Assuming that mRNA expression results in protein expression, the landscape is described as follows: the extracellular signaling environment shows five Wnt ligands (colored in tones of grey with acyl tails). The translation and export of acylated Wnt proteins requires the chaperone **Wls** and the acyl-transferase, **Porcn** (Yu & Virshup, 2014):

Fig. 1—Cont'd Acyl tails mediate Fzd binding and oligomerization (Nile, Mukund, Stanger, Wang, & Hannoush, 2017). Also shown is the Wnt signaling-related RSpondin ligand (**Rspo3**; purple) and two secreted inhibitors, **Sfrp1** (that binds competitively to Fzd Wnt binding domain) and **Sostdc1** (also known as WISE), known to interact with Lrp4 to inhibit Lrp-dependent Wnt signaling (Ahn, Sanderson, Klein, & Krumlauf, 2010; Ahn, Sims, Logue, Weatherbee, & Krumlauf, 2013; Ohazama, Porntaveetus, Ota, Herz, & Sharpe, 2010). The array of receptors at the cell surface (represented by a grey disc) could include any of four Fzd species, together with either/both Lrp proteins, **Lrp5** and/or **Lrp6**. Other Wnt-binding receptors have different signaling outputs: not covered in this review are the effects of non-canonical Fzd- and **Ror/Ryk**-dependent signaling events, though these are clearly important (Kessenbrock et al., 2017; Roarty, Shore, Creighton, & Rosen, 2015). Molecules known to be present in limiting amounts for Wnt signaling are indicated in red (Lrp6 and axin1). 1. Signaling is initiated in this example by a **hypothetical** complex comprising a ternary Lrp6/Wnt10a/Fzd1 complex in combination with an **Lgr4/Rspo3/Znrf3** ubiquitin ligase "amplifier". This is discussed more in Section 1. 2. Levels of the Wnt signal transducer, βcatenin, are low due to continuous proteolysis by the destruction complex, shown here to include the scaffolding proteins **Apc** and **Axin1**, the ubiquitin ligase **βTrCP** (Btrc) and two kinases, casein kinase-1α (**CK1**) and **GSK3β**. These kinases promote the processive phosphorylation of the N-terminal domain of βcatenin, creating a substrate for βTrCP-induced ligation and processing by the proteasome. Ligand/receptor complexation results in recruitment of axin1 and GSK3β away from the destruction complex, inducing the phosphorylation and binding of the carboxy terminal PPPSPxP domains of Lrp6 (red starred). This serves to inhibit the degradation of **βcatenin** (gene name is Ctnnb1), allowing this signal transducer to accumulate and translocate to the nucleus. 3. Recruitment of βcatenin to TCF factors bound in Wnt enhanceosomes at TCF-responsive elements (TRE) initiates the Wnt-induced changes of gene transcription (van Tienen, Mieszczanek, Fiedler, Rutherford, & Bienz, 2017). Core components of the Wnt enhanceosome include three TCF DNA binding transcriptional activator species, Tcf1 (TCF7) Tcf3 (TCF7L1) and Tcf4 (TCF7L2) and three Graucho-related transcription repressors, Tle1, -3 and -6, along with other DNA binding proteins (including Pygopus2; Pygo2) and Bcl proteins Bcl9 and Bcl9l, serving to scaffold the complex. If there are specific properties for each family member, they are not understood. Several aspects of the signaling pathway are incompletely understood, especially transfer of Wnt and Rspo proteins from cell to cell and through cell layers, the role of trafficking in signal activation (Langton, Kakugawa, & Vincent, 2016), and the determination of Wnt transactivation targets by epigenetic cell type-specific regulation (these are shown circled in grey dotted lines). (B) The expression levels (RPKM, reads per kB of transcript per million mapped reads) of each component were derived by RNASeq analysis. Genes that were not significantly expressed in EP cells (defined here as expression level <1) include the non-canonical Wnt agonists, Wnt5a and Wnt5b, other lesser characterized Wnt ligands such as Wnt6, Wnt7a, Wnt9b, Wnt10b and the better known RSpondin species, Rspo1 and -2. None of the three Dkk inhibitor species are significantly expressed (Dkk1, -2 and -3). Other components notably absent include Norrin, Fzd4, Lrp4, Lgr5, Lgr6, Ror1, Apc2, Lef1, Tle2 and Tle4; compare these data to those collected from many studies of dissociated mammary epithelial cells, from both mouse and human, reviewed earlier (Alexander, Goel, Fakhraldeen, & Kim, 2012). Several species are commonly known by other names: Sostdc1 (WISE), Fat1 (atypical cadherin), Ctnna1 (αcatenin), Ctnnb1 (βcatenin), Ctnnd1 (p120), Btrc (βTRCP Ubiquitin ligase), Tcf7 (TCF1), Tcf7l1 (TCF3), Tcf7l2 (TCF4), Tle1/3/6 (Graucho repressors). As with any pathway, this one is intertwined with others, but those components are not included on this diagram.

growing in typical tissue culture conditions. Note that primary mouse mammary epithelial cells de-differentiate and/or die quickly in culture; the data of Fig. 1A are drawn from one of the very few immortal, multi-potent, differentiation-competent cell lines available. The mRNA expression (RPKM) of each component is shown in Fig. 1B, and may imply the relative abundance of components of this pathway. Not all of the components known to be functional *in vivo* are expressed *in vitro* in cultured cells, either because they are not in the correct inductive environment, or because those components are not epithelial in origin. For example, the Rspo1 ligand is not commonly expressed *in vitro* (discussed in Sections 1.6 and 3.1) Wnt5a and Wnt5b are expressed by some luminal cells in mammary gland, but not in this cell line. None of the Wnt ligands known to be oncogenic for mouse mammary epithelia (Wnt1, Wnt3 and Wnt10b) are expressed in adult mammary epithelial cells (Callahan & Smith, 2008), although Wnt10b is developmentally important and shows canonical βcatenin/TCF transactivation activity (Veltmaat, Van Veelen, Thiery, & Bellusci, 2004). The detailed expression patterns of Wnt signaling components in separated mammary epithelial populations of mouse and human origin has been reviewed and reported previously (Alexander et al., 2012).

1.3 The Lrp co-receptors

Wnt signaling is initiated by the binding of Wnt ligands to both Frizzled (Fzd) and Low Density Lipoprotein-Receptor Related Protein-5 or -6 (Lrp5/6). These Lrp co-receptor species are especially important to the analysis of Wnt signaling, since they represent a rate-limiting and non-redundant component of the signaling pathway (Saito-Diaz et al., 2013). Notably, although the Lrp receptors are commonly co-expressed, they are not functionally redundant. For mammary epithelial cells, Lrp6 directly transmits a canonical Wnt signal; Lrp5 is not phosphorylated efficiently at the PPSPxP motif in the carboxy-terminal domain, and is therefore a poor Wnt signaling receptor (Macdonald, Semenov, Huang, & He, 2011). However, Lrp5 is required to respond to endogenously expressed Wnt ligands, including Wnt 9a and Wnt10a (Chin et al., 2015; Goel et al., 2012). The basis for that is not yet clear but may reflect the participation of Lrp5 in broader trafficking functions (Chin et al., 2015).

1.4 Range and secretion of highly hydrophobic Wnt ligands

Wnt ligands are palmitoleoylated by a highly specific acyl transferase, Porcn (*porcupine*) and transported to the cell surface via a dedicated multipass transmembrane protein called Wntless (WLS) (Langton et al., 2016). The palmitoyl lipid moiety is essential for activity of these ligands (Janda & Garcia, 2015; Langton et al., 2016; McGough et al., 2020). The range of action of paracrine Wnt ligands is therefore likely to be limited by the distribution mechanisms available for these hydrophobic ligands, which (may) include their incorporation into lipoprotein particles, or exosomes, their presentation on cytonemal cell extensions, or partnering with chaperones able to maintain their solubility. For example, the expression of the cell surface heparan sulfate glypican serves to enable the secretion of Wnt ligands from Drosophila cells (McGough et al., 2020).

1.5 Competition for receptor occupancy

The basic Wnt signaling receptor complex includes Fzd and Lrp cell surface proteins, where several Fzd proteins are typically expressed by any specific cell type. The population of mammary epithelial cells expresses four Fzd mRNAs at mid to high levels (Fzd1, -2, -6 and -7), the location of Fzd and Lrp proteins on the polarized surfaces of mammary cells is not yet known.

Similarly, there is often redundant expression of Wnt ligands. In order of abundance, the Wnt ligands produced by a cultured mammary epithelial cell line are **Wnt7b**, **Wnt10a**, **Wnt9a**, **Wnt4** and **Wnt11** (Fig. 1B). Biochemically, it is difficult to demonstrate specific pairing reactions for ligands and Fzd species. However, it is likely that within the competitive environment of multiple ligand/Fzd binding reactions at the cell surface, specific ligands predominantly bind specific Fzd receptors (van Amerongen & Nusse, 2009). An alternative idea suggests that it is the cholesterol content of the cell membranes that determines the activity of any given ligand/receptor signaling complex (Sheng et al., 2014).

Of the ligands expressed by mammary epithelial cells, Wnt10a is considered to be a canonical pathway activator (Xu et al., 2017), where Wnt4 has a paradoxical activity that typically does not result in canonical pathway activation *in vitro* or *in vivo* (see Section 3.1). The other ligands are likely to have non-canonical activities. Classic studies of the oncogenicity of Wnt ligands for mammary epithelial cells showed that Wnt7b was non-transforming *in vitro* or *in vivo*, unlike Wnt1 and Wnt3 (Naylor et al., 2000; Shimizu et al., 1997; Wong, Gavin, & McMahon, 1994). Wnt9a is a known agonist for the non-canonical receptor tyrosine kinases, Ror1 and Ror2, during the development of bone and cartilage (Weissenbock et al., 2019). Wnt11 has more hybrid activities: it is a classic non-canonical ligand, vital for convergent extension during embryonic development (Williams & Solnica-Krezel, 2020), but also activates the canonical pathway *in vivo* (Halt & Vainio, 2014; Lebreton & Jones, 2006).

Note that the vast majority of experimental data is derived from the addition of Wnt3a *in vitro,* due to its unique solubility properties. However, this ligand is rarely expressed *in vivo* and tends to be the exception in many other ways besides its convenient solubility. For example, although endogenous Wnt signaling reactions in mammary epithelial cells require Lrp5 both *in vivo* and *in vitro*, the response to Wnt3 does not (Goel et al., 2012).

1.6 The RSpo amplifier complex

More recently, the "basic" Wnt signaling cell surface components have been joined by an "amplifier" complex, which comprises an RSpondin (RSpo) ligand together with a cell surface associated Leucine Rich Repeat Containing G Protein-Coupled Receptor (Lgr) family member. This ligand-receptor pair serve to inhibit the Znrf3 and Rnf43 (**ZR**) ubiquitin ligases to decrease Fzd degradation rates, and amplify the Wnt signal by at least an order of magnitude (Cai et al., 2014; Carmon, Gong, Yi, Thomas, & Liu, 2014; Kumar, Burgess, & Gulbis, 2014; Yan et al., 2017). This complex is key to the amped-up Wnt signaling observed during embryonic specification, and perhaps to somatic stem cell activity.

1.7 Difficulties in the assay of Wnt signaling activation

Binding of a Wnt ligand to the two cell surface receptors initiates the signal transduction cascade (Fig. 1). The allosteric interaction of the ecto-domain of Lrp with Fzd/Wnt promotes binding of key elements of the βcatenin destruction complex to the PPPSPxP domains of the carboxy-tail of Lrp6 (Macdonald et al., 2011). This phosphorylation is a definitive mark of the activation of canonical Wnt signaling. Lrp activation suppresses proteasomal targeting and destruction of βcatenin, allowing the accumulation and nuclear translocation of this primary signal transducer. Nuclear βcatenin is a good marker of Wnt signaling, although it is not always present, therefore a negative signal is difficult to interpret. By binding to the TCF group of transcription factors in the Wnt "enhanceosome," this pathway initiates a highly cell type-specific transcriptional response. Contrary to common practice, almost no gene can serve as a common signature or reporter between different cell types; perhaps the exception is Axin2 mRNA. Unfortunately, axin2 induction serves only poorly as a protein marker, since the conservation of Axin1 and Axin2 is high and currently available antibodies cannot reliably distinguish them. However, in any given context, Wnt reporters can be identified using a positive control of the Wnt-treated cells or tissues of interest, over a short timeline (2–6 h), to specifically exclude all but the acute response to Wnt ligand interaction, and eliminate the inhibitory phase (Section 1.8). Alternatively, it may be practical to harness a knock-in Wnt reporter such as Axin2-lacZ or Axin2-GFP, or use a viral vector containing a series of TCF-repeat elements, TREs.

1.8 Feedback inhibition

The induction of the destruction complex component, Axin2, in response to a Wnt signal reveals a key element of this pathway, which is that activation induces feedback inhibitor expression at multiple levels (Cruciat & Niehrs, 2013). Indeed, most developmentally important pathways show this feedback, with the effect that the response is limited to a pulse of morphogenetic activity. For example, as observed in Section 1.2 and Fig. 1A, there are significant levels of Wnt ligand expression in normal mammary epithelial cells growing in 2D culture conditions, however there is insignificant activation of Wnt-dependent gene expression, despite a fully functional Wnt signaling response (Chin et al., 2015).

Other inhibitors besides Axin2 are specifically induced by activation of the cascade, including Naked (here Nkd2) (Larraguibel et al., 2015) and the ZR cell surface ubiquitin ligases, Znrf3 and Rnf43 (Saito-Diaz et al., 2013). In other cell types, these ligases have been shown to be induced in response to Wnt activation and to inhibit further signaling by targeting Fzd cell surface receptors for degradation (Xie et al., 2013). The significance of these ubiquitin ligases for mammary gland development and tumorigenesis is not yet known.

1.9 Co-expressed inhibitors

There are high levels of the mRNA for a soluble inhibitor of Wnt signaling, Sfrp1, which competes for Fzd binding by Wnt ligands, amongst other activities. Loss of Sfrp1 has been observed for more than 80% of invasive breast carcinomas (Ugolini et al., 2001) though the mechanism of action is not yet clear. Other Wnt signaling inhibitors expressed in mammary gland include Sostdc1 (Wise; (Ahn et al., 2013)) and high levels of Naked2, known to interact with Dvl (1, 2, 3) to inhibit productive signal transduction (). The transactivation response is regulated by several cross-talking signaling partners of the βcatenin/TCF transcription complex (Valenta, Hausmann, & Basler, 2012), both inhibitory and stimulatory. For example, the Wnt "enhanceosome" is shown in Figure 1, to illustrate the variants of this complex that exist in mammary epithelium. Thus, cultured mammary epithelial cells express three TCF factors TCF1 > TCF3 > TCF4 and three Groucho repressors, Tle1, -3 and -6, together with the chromatin modifier, Pygopus2 (Roarty & Rosen, 2013). Activation of transcriptional targets is achieved by the integration of β-catenin into the enhanceosomes, pre-assembled on the cell type-specific Wnt target genes (van Tienen et al., 2017).

1.10 Reasons for assaying Wnt responses in 3D cultures

One of the principal developmental roles of Wnt signaling is to organize the axes of division and tissue expansion. However, the majority of data describing Wnt responses in mammary epithelial cells is derived from 2D plastic dishes of serum-exposed cells. This is notoriously inaccurate at recapitulating the signaling reactions observed for confluent epithelia on matrices with signaling activities. There are reasons to suggest that the Wnt signaling pathway could be particularly sensitive to intercellular cues and matrix interactions: it regulates, and is regulated by, mechano-transduction cues via integrated pathways such as YAP/TAZ (Aragona et al., 2013; Dupont et al., 2011; Zanconato, Cordenonsi, & Piccolo, 2016). With the introduction of cell polarization, components of the Wnt signaling pathway become organized across the apical to basolateral axis, separating components that are otherwise mixed together in cell membranes in 2D cultures (Butler & Wallingford, 2017). Therefore, accurate Wnt responses will be better reconstructed *in vitro* by maintaining cell-cell interactions, for example by studies of explanted mammary organoids (Jamieson et al., 2017; Linnemann, Meixner, Miura, & Scheel, 2017; Nguyen-Ngoc et al., 2015). The mammary epithelial "neighborhood" is described in the next section.

2. Growth and development of mammary gland

2.1 Pre- and post-natal growth of the mammary gland

Most of the data available to describe mammary gland development comes from mouse (and occasionally from rat or rabbit). Thus mammary placodes form from the ectodermal sheet of female mouse skins at E13.5d pc, as the result of boundary patterning by Wnt ligands such as Wnt10b and Wnt6, in concert with other morphogens (Rivetti, Chen, Chen, & Bellusci, 2020;

Veltmaat, 2017). This specification event is associated with a pulse of active Wnt signaling, as reported by a TOP-GAL Wnt transactivation reporter (Chu et al., 2004). At this stage of development, all the cells allocated to the placode could be defined as stem cells; they are bipotent, plastic and actively dividing. Fetal cells often co-express luminal and basal cytokeratins (so called "double positive" cells), whereas adult cells do not. Using an elegant clonal analysis technique (Confetti labeling), Blanpain and colleagues showed that by post-natal day 1 there is a restriction of the bipotency of embryonic mammary cells to the two dedicated lineages of the mammary gland, basal and luminal cells. This requires the p53 super family member and basal cell marker, p63 (Blanpain & Fuchs, 2007; Wuidart et al., 2018). The role and arrangement of these two principal mammary epithelial cells types is shown in Fig. 2.

Wnt signaling is necessary for mammogenesis; without Wnt signaling, mammary placodes are not specified from ectoderm. However, it is not known whether a Wnt signal is sufficient to induce mammogenesis (see for example the work of Fallon and colleagues for a remarkable example of the inductive power of Wnt signaling (Harris, Hasso, Ferguson, & Fallon, 2006)). After birth, Wnt/Rspo signaling declines (assayed by the same reporter), leaving far fewer Wnt reporter-positive cells in the adult gland, with a much decreased expression signal (Makarem et al., 2013; van Amerongen, Bowman, & Nusse, 2012). The ductal tree colonizes all the way to the edge of the host fat pad (Fig. 2B), allowing room for pregnancy-associated expansion of the alveoli into the interdigitating spaces. All post-natal mammary structures comprise specified basal and luminal epithelial cells (Fig. 2C); at least for the ductal lineage, the basal cells are characterized by high expression of most of the Wnt signaling pathway components, including high cell surface presentation of Lrp5 (Fig. 2D) and Lrp6. Note that any assay returning a gene ontology (KEGG)-based Wnt signature is likely to be reflect an accumulation of cells with a basal phenotype.

Within hours of the induction of pregnancy (or indeed hours of initiation of estrus), the ductal tree is induced to divide, to create side-branches. For mice in estrus, these will regress after a day, but for pregnant mice, the full hormonal program induces the mammary epithelial cell population to expand and differentiate to create a complex anastomosing network of cells (Brisken & Ataca, 2015; Brisken & O'Malley, 2010). During pregnancy, there is no increase in Wnt signaling; instead the growth of a distinct, alveolar population (both luminal and basal) is mediated by estrogen stimulation of an estrogen receptor (ER)-positive cell minority (Fig. 2E); this alveolar

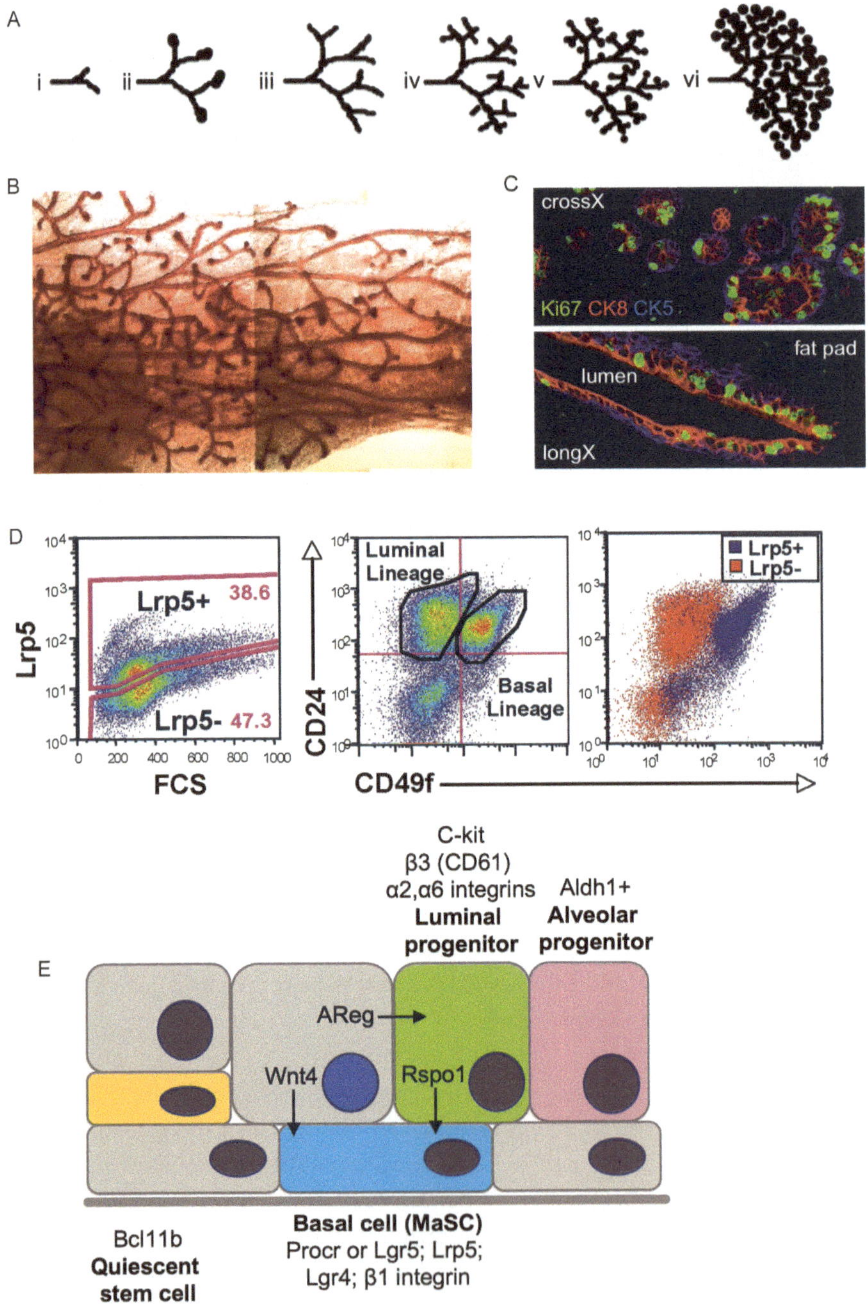

Fig. 2 Development of mouse mammary epithelial cell populations. (A) *Morphogenesis of the mammary ductal tree.* During post-natal development, the embryonic rudiment invaginates from the skin nipple to colonize the growing mammary fat pad underneath

Fig. 2—Cont'd (i). Ductal outgrowth is led by the proliferative engine provided by the dilated terminal end buds (ii). Full ductal branching is complete after puberty (iii). Pregnancy initiates alveolar development, starting from side-branches of the ductal tree (iv, v) and upon parturition, these alveoli stop dividing and differentiate to become truly remarkable secretory lactating cells (vi). (B) *Whole mount evaluation of morphogenesis.* The ductal mammary epithelial cell population is typically visualized using Carmine stain (pale background is the fat pad); ducts are hollow. (C) *Identification of distinct cell types.* Immunostains of sections of this structure show the two main cell types, arranged into apposed sheets; the top image is shown in cross-section (crossX) and the lower image as a longitudinal section (longX). Luminal cells (differentiating into the major milk producers during lactation), face the lumen (stained red with anti-cytokeratin 8), and are sheathed by the myoepithelial basal cells (blue stained with anti-cytokeratin 5). Myoepithelial cells are contractile and lay down basement membrane as a coordinated effort with the stroma. This provides a mechanical substructure and a signaling matrix (Xu, Spencer, Groesser, & Bissell, 2010), together providing the basis for oxytocin-induced milk ejection. Almost all cells of both lineages are replication competent, here cells in cycle are stained green (Ki67). (D) *Flow cytometric analysis of mammary epithelial cells.* The luminal and basal cell types can be dissociated by enzymatic digestion, and after "singlizing," they can be sorted into two populations by flow cytometry, here using forward scatter (FSC), and anti-Lrp5, CD24 or CD49f antisera (Badders et al., 2009). Basal cells are Lrp5-positive and the basal cell population contains all the *in vivo* stem cell activity, as defined by the establishment of new ductal outgrowths after transfer to cleared fat pads. Some of the luminal cells with the most CD24/CD49f can proliferate in culture and are therefore designated "progenitors"; they have no clonogenic activity *in vivo*. Note there are now three definitions of basal – location, cytokeratin 5/14 expression and cell surface immunophenotype. These do not always correspond to one another. (E) *A model of mammary epithelial cell interactions that govern stem-ness.* Colored cells indicate replication-competent cells, summarized as stem and progenitor fractions. Nuclei are painted blue if they express estrogen and progesterone receptors; these have been described as "mature" or terminally differentiated; they process the hormone signal that governs hormone-dependent growth of the entire population (Brisken & O'Malley, 2010). Most of the markers noted here for stem/progenitor cell types have been shown to be functionally important, as summarized by Visvader and Stingl (2014). The Bcl11b-positive quiescent stem cell fraction is colored yellow, and was described by Clarke and colleagues (Cai et al., 2017). A triad of cells that maintain the stem cell unit was deduced from Zeng and colleagues (Cai et al., 2014; Cai et al., 2020; Wang, Cai, et al., 2015), including an ER-positive mature luminal cell expressing Wnt4 and amphiregulin (AReg) and an ER-negative luminal progenitor expressing RSpo1, which together signal to the Lrp5-positive MaSC basal cell. Note that the cell surface phenotypes that discriminate basal and luminal cells *in vivo* do not apply to cells transferred to culture, despite persistent basal/luminal marker expression; typically these two (or more) populations merge into one with intermediate expression. Note also that the function and phenotype of human mammary stem and progenitor cell populations is not necessarily directly comparable to the same populations in mouse (Lim et al., 2009). *Shown diagrammatically, adapted from Brisken, C., & O'Malley, B. (2010). Hormone action in the mammary gland.* Cold Spring Harbor Perspectives in Biology 2, *a003178.*

population is entirely dependent upon hormones for survival, and regresses after weaning, leaving only the ductal tree in place (so called involution, one of the most controlled apoptotic processes in the adult mammal). These alveolar cells are rarely studied but have been shown to undergo a lysosomal-mediated cell death upon weaning (Kreuzaler et al., 2011). In contrast, a gain of function for Wnt signaling induces a ductal hyper-proliferation which provides a large, dysregulated, continuously growing cell population that is a substrate for tumor development (Liu, McDermott, Khwaja, & Alexander, 2004; Mastroianni et al., 2010).

During this time, the two types of mammary epithelial cells, basal and luminal, preserve their ratio (approximately 1:1.4, respectively) together with their geometry and cohesion. These two cells are essential partners and coordinate the outgrowth and function of both ductal and alveolar lineages (Alexander, Joshi, & Khokha, 2014). The identification of molecular players that mediate paracrine interactions demands the design of ingenious assays. Studies have demonstrated the recruitment of developmentally important pathways, such as BMPs and their inhibitors, RANK/RANKL, Notch, FGF and EGF/amphiregulin signals (Ataca et al., 2020; Brisken & Ataca, 2015; Fillmore et al., 2010; Forster et al., 2014; Rivetti et al., 2020; Robinson, 2007; Zabala et al., 2020).

This mammary tree will ultimately respond to both local and distant cues to synthesize and expel milk. During lactation, luminal cells absorb nutrients from the blood, synthesizing milk fat and proteins and other specialized components; basal cells differentiate into fully differentiated myoepithelial cells, which contract in response to oxytocin to expel milk. Ultimately, this post-natal development represents a highly interactive and coordinated population expansion, approximately 10 × increase in cell number over the course of 21 days (Paguirigan, Beebe, & Alexander, 2007).

2.2 Mammary stem cells

The first proof of a mammary stem cell activity came from the functional analysis of dissociated and flow sorted mammary epithelial cells (Shackleton et al., 2006; Stingl et al., 2006). Remarkably, a single basal cell was enough to recreate an entire, functional mammary ductal tree within weeks of transplantation. The cells were separated by their immunophenotype using antibodies to CD49f (α6 integrin), EpCAM (epithelial cell adhesion molecule) and/or CD24 (a GPI-linked protein core with high N- and O- glycosylation) (Altevogt, Sammar, Huser, & Kristiansen, 2021). None of these cell surface molecules can separate basal and luminal cells alone, and the difference between the basal and luminal cell signal is typically less than fivefold. These markers have not been tested for their functional role in stemness, although α6 integrin/CD49f dimerizes with β1 integrin, which is known to be functionally important for stem cell activity (Taddei et al., 2008). Given that CD49f is the most widely used cell surface marker for separating stem cells, it is perhaps surprising that its activity has not been established (Klinowska et al., 2001; Krebsbach & Villa-Diaz, 2017).

Basal cell-specific cytokeratin expression provided a strategy for lineage tracing of mammary stem cell activity during growth. Given the prevailing dogma, surprisingly, the daughters of basal cells were not found in significant numbers in the luminal cell fractions (Van Keymeulen et al., 2011; Van Keymeulen et al., 2017; Wuidart et al., 2016). These investigators found that when tagged luminal and basal cells were separated and purified, basal cells expressed their bipotent stem-ness when re-implanted, except when they are co-implanted with luminal cells (Van Keymeulen et al., 2011). Thus, the expression of bipotency may be actively suppressed by an inhibitor from luminal cells.

Lineage tracing tends to rely upon daughter cells remaining contiguous with parents. However, the studies of Watson and colleagues have shown that the progeny of dividing cells do not stay contiguous in patches, instead the daughter cells become dispersed through the population (Davis et al., 2016 and Alexander, unpublished). This could be unusual for an epithelial tissue and makes at least two predictions—(1) that unlike the intestinal stem cell crypt, tissue geometry is unlikely to create the usual cues for asymmetric division and the organization of a hierarchical lineage, and (2) tumor precursor cells with specific DNA mutations could become widely dispersed in the mammary gland and seed several different tumors.

2.3 Mammary stem cells—Are they entities or activities?

Most of the stem cell "rules" arising from the regenerative lineages either do not apply to the mammary gland or they do not (yet) have data that support their application to mammary gland growth. First, it is not accurate to state that mammary stem cell activation accounts for the expansion of the mammary epithelial populations, indeed it is unlikely.

So far, we do not have genetic loci capable of tagging and tracing daughters of "rare" stem cells; therefore our understanding of mammary gland development has been based on the expression of lineage-specific markers (keratin loci—keratin5, keratin8), or tracing offspring of randomly tagged single cells (Davis et al., 2016). This has shown that for normal development, once luminal and basal cells differentiate (post-natal day1), they are committed (Van Keymeulen et al., 2011; Van Keymeulen et al., 2017), thus basal cells stay basal, and luminal cells stay luminal.

The concept of hierarchical arrangements of more and more differentiated cells comes to us by way of the single cell hematopoietic lineages, and from lineages strictly constrained along a conveyor-belt style regenerative pattern, such as intestine. However, there are few data to suggest that this principle applies to mammary gland, and plenty to suggest that it does not. Thus in the intestine, there is a cohesive patch of daughter cells of various intestinal lineages that creep up the crypt wall after intestinal stem cell activation (Barker et al., 2007; Koo & Clevers, 2014). In contrast, daughters of the accredited stem cells in the mammary gland, the basal cells, do not express their bipotency during normal growth cycles.

It is even unlikely that stem cells are a discreet and determined entity, instead stem-ness is more likely to be flexible, induced in the majority basal cell community by local factors and environment. Note that the mammary neighborhood itself is able to provide a dominant cue that reprograms cells to allow their participation in the mammary epithelium. Thus tumor cells, or cells with dramatically different phenotype, can be transplanted into mammary glands to integrate into the ductal unit and look for all intents and purposes like normal mammary epithelial cells (Bruno & Smith, 2012; Frank-Kamenetskii et al., 2020)

Consider these points:

1) The majority of mammary epithelial cells in a ductal tree are fully replication-competent, and will undergo several rounds of cell division; modeling shows that only five divisions/cell will account for the extensive cell growth associated with pregnancy (and involution) (Paguirigan et al., 2007). Almost all epithelial cells, basal or luminal, incorporate mitotic markers such as EdU during pregnancy-associated expansion of the ductal tree, and will become re-labeled when another growth cycle is induced (Van Keymeulen et al., 2011).
2) During normal development, the stem cell activity of basal cells is not expressed. Thus lineage tracking shows that basal cells do not differentiate to generate luminal cells during ductal elongation or pregnancy; instead, the vast majority of growth and expansion of the mammary ductal tree follows lineage-specific specification; luminal cells divide to give luminal cells, and basal epithelial cells generate basal cells

(Van Keymeulen et al., 2011). Indeed, even the ER-positive and ER-negative lineages of luminal cells that were previously assumed to be related turn out to be independent. Thus, when they divide in the adult gland, they recreate themselves (Giraddi et al., 2015; Van Keymeulen et al., 2017; Wang, Christin, Oktay, & Guo, 2017). This ER-positive subpopulation comprises approximately 5% of the total population and were previously described as the terminally differentiated luminal cell type at the end of the hierarchy tree. How they distribute and space themselves across the total epithelial sheet will be fascinating to understand. Likewise, a specified set of alveolar progenitors are distinguished by their Notch1-expressing parent cell (Rodilla et al., 2015).

3) Mammary glands with no somatic clonogenic activity perform the same as wild type glands during any phase of regeneration. Thus mammary epithelial populations deficient in β1 integrin, Lgr4 or Lrp5, show no/little clonogenic activity (Badders et al., 2009; Taddei et al., 2008; Wang et al., 2013) and yet show almost normal development of ductal trees, and totally normal pregnancy-associated division, lactation and involution, through >3 cycles.

4) The majority of mammospheres established in suspension cultures from healthy, singlized cells from mouse epithelia derive from luminal cells. None of these mammosphere-associated "stem cells" are competent to establish mammary gland outgrowths *in vivo*, so there is no direct connection between increased matrix-independent clonogens *in vitro* and cells with stem cell activity *in vivo*.

5) Stem-ness in mammary basal epithelial cells can appear and disappear within hours during the mouse estrus cycle (Joshi et al., 2010); a pulse of progesterone induces a transient increase in the number of cells that serve as clonogens *in vivo*, without significant cell division and death. Progesterone induces expression of Wnt4, which is required as part of a paracrine set of interactions between (at least) a triad of mammary epithelial cell types that confers stemness on a minority of basal cells (Fig. 2E). It is likely that the plasticity of the system is such that any replication-competent basal cell provided with a Wnt/Rspo source, or even recently exposed to Wnt/Rspo, could serve as a stem cell, if it remains undamaged by dissociation protocols (Magee, Piskounova, & Morrison, 2012).

6) Other broad generalizations do not apply; the basal stem cells with stem cell activity are not slow-dividing (they do not retain label), and other markers able to enrich for stem cell activities do not apply to mammary epithelial cells (such as CD133, Bmi1, Aldh activity, prominin).

This situation is therefore very different from that of the continuously regenerating lineages that include colorectal, skin and hematopoietic tissue; for these lineages, stem cell deficiencies are represented as organ failure after a few weeks (Visvader & Clevers, 2016). Perhaps a better organ homolog is liver: thus, growth of hepatocytes is focused around a peri-central Wnt-secreting endothelial niche (Wang, Zhao, Fish, Logan, & Nusse, 2015), and diploid hepatocytes within the Wnt-exposed field are mostly quiescent but replication-competent. The hepatocyte majority shows high plasticity, and the organ is peppered with "stem" cells of different potency and origins, with variable impact on regeneration and repair (Kopp, Grompe, & Sander, 2016).

Note also that the functional characterization of tissue-associated somatic stem cells is affected by the technical difficulties and limitations of this technique (Rosen, 2012); thus tissue dissociation takes hours, allowing the physiology of the cells to change (including progression of the cell cycle), the sorting relies on potentially function perturbing (or activating) binding to cell surface antibodies, the dissociation enzymes cleave and damage the cell surface proteins, the flow sorting process is rarely the same from lab to lab, and the time and physical stressors typically take a toll on cell viability, often more than ten-fold, and affecting some cell types more than others (Alexander et al., 2009).

A current inventory of stem/progenitor fractions that score on different assays is shown in Fig. 2E and are described in more detail in the legend. These include basally located mammary stem cells, luminal and alveolar progenitors and a quiescent Bcl11b+ stem cell that lies between luminal and basal cells.

3. How does Wnt signaling affect mammary growth and morphogenesis?

3.1 Which of the stem/progenitor cells is Wnt-dependent?

Basal mammary epithelial cells are enriched in Wnt signaling components and are Wnt-responsive. More stem cell activity is induced by the expression of auto-active βcatenin or Wnt1, and basal stem cell activity is absent in mammary glands from *Lrp5-/-* mice. (Badders et al., 2009; Li et al., 2003; Liu et al., 2004; Teuliere et al., 2005). However, although it is clear that Wnt signaling is required for embryonic specification of mammary glands, it has yet to be proven that Wnt signaling is important to the maintenance of adult gland. Given that Lrp5 has moonlighting functions in other signaling pathways (Chin et al., 2015), the stem cell activity of mammary glands will need to be tested using a conditional approach loss of function for a non-redundant Wnt component (such as Lrp6).

Note that there are two similarly named, and therefore confusing, cell surface receptors, Lgr5 and Lrp5, which may both be involved in Wnt signaling in mammary gland. Approximately half the basal cells express Lgr5 (Plaks et al., 2013; Zhang et al., 2017), and all express Lgr4 and Lrp5 (Badders et al., 2009; Wang et al., 2013). The phenotype of Lgr4 mutant mammary glands resembles that of Lrp5 mutants: mice show somewhat hypomorphic ductal outgrowth, and little adult regenerative activity. Furthermore, cultured *Lgr4-/-* organoids proliferate in response to Wnt3A but not Rspo1 (Wang et al., 2013). Thus, Lrp5 and Lgr4 cell surface receptors are required for the expression of stem cell activity.

Zeng and colleagues reported that the basal stem cell activity could be maintained by a combination of Rspo1 and Wnt4, produced by PR-negative and PR–positive luminal cells, respectively, in the presence of an estrogenic inducer (Cai et al., 2014; Wang, Cai, et al., 2015) (Fig. 2E). Rspo ligands interact with Lgr receptors to promote canonical Wnt signals by preventing the degradation of Fzd receptors (via inhibition of cell surface ubiquitin ligases) (Carmon et al., 2014; Kumar et al., 2014). Thus, although Rspo ligands collaborate in generating a Wnt signal, they are not sufficient to induce Wnt signaling alone. This could align mammary stem cell regulation with the dual Rspo/Wnt interactions that maintain intestinal stem cells (Yan et al., 2017). The Wnt4 ligand itself did not robustly generate a canonical signal *in vitro* or induce hyperplasia when over-expressed (Alexander et al., 2012 and references therein); however, Zeng and colleagues claimed they could observe the expression of canonical signaling in response to Wnt4 (Cai et al., 2014).

The requirement for two groups of ligands and receptors (Rspo/Wnt and Lrp/Lgr), linked by an inhibitory ubiquitin ligase, may explain why the regenerative capacity of Lgr5-positive mammary epithelial cells has been so controversial, as recently summarized by Kolesnick and colleagues (Zhang et al., 2017). Further controversy arises from the description of a basal stem cell activity that is Lgr5-negative, and positive instead for the Wnt target gene, Procr (protein C receptor), also from Zeng and colleagues (Wang, Cai, et al., 2015). Perhaps Procr-positive cells represent a population newly exposed to Wnt signaling, with down-regulated cell surface Lgr5 (Lgr5-negative during flow sorting).

3.2 How similar is the human mammary epithelium?

The data described so far refer to mouse mammary epithelial cell populations. On the whole, experimental approaches are much more limited for the human tissue homolog; there are challenges associated with obtaining and analyzing human breast tissues, and cell lines *in vitro* rarely recapitulate the developmental processes accurately. Reassuringly, the flow cytometric separation of human mammary epithelial cells from *Brca* +/- subjects showed that there were important changes in the luminal cell phenotypes that predicated tumor formation (Lim et al., 2009); this matched the results of studies in mice (Molyneux et al., 2010; Smalley, Reis-Filho, & Ashworth, 2008).

For the human mammary epithelial population, this progenitor cell with the ability to switch from a luminal to a basal phenotype is important to understand. Sartorius and colleagues have described a basal-type cytokeratin5-positive cell population that exists in luminal, ER-positive human breast tumors that has clonogenic properties *in vitro* (forming mammospheres) (Haughian et al., 2012). Indeed, this cell minority exists in normal glands, and could reflect the *Brca1* +/- phenotype. The co-existence of a minority of CK+ cells in an otherwise luminal cell population are mimicked in a classic human breast cancer cell line, T47D (Horwitz, Dye, Harrell, Kabos, & Sartorius, 2008). Even more interesting, cytokeratin5 expression is essential to the functional stemness; it is increased by treatment with Wnt ligands or progestin, and these cells are resistant to killing by anti-estrogenic drugs (Axlund et al., 2013; Kabos et al., 2011). The CK5 intermediate filament protein binds to βcatenin and induces a novel gene program that includes the expression of Wnt1 (McGinn et al., 2020). When these CK5-positive cells divide, they recreate the original mixture of cells; this signaling mechanism will be fascinating to explore.

3.3 Other sources of Wnt signals?

The intestinal stem cell niche includes several sources of Wnt ligands, secreted either by the interdigitated Paneth epithelial cells, or by the near-adjacent stroma (Clevers, 2013). In this review, several paracrine loops of interactive mammary epithelial cells have been described, but there are other potential sources of Wnt ligands in the stromal population. The stroma regulates the development and morphogenesis for mouse mammary gland, from the specification of the mammary placode onward through ductal development (Howard & Lu, 2014). Using heterologous recombination, the pattern of ductal outgrowth can be shown to depend upon the properties of the host fat pad and systemic endocrine factors; indeed, human breast epithelium grows only in fat pads transplanted with human stroma.

For mouse mammary gland, macrophages are likely to be important point sources of regulated Wnt ligand secretion (Cosin-Roger, Ortiz-Masia, & Barrachina, 2019; Malsin, Kim, Lam, & Gottardi, 2019). These have been known to participate in the morphogenesis of the developing mammary ductal tree for several decades (Gouon-Evans, Rothenberg, & Pollard, 2000; Ingman, Wyckoff, Gouon-Evans, Condeelis, & Pollard, 2006). Resident macrophages in mammary gland have been shown to be enriched sources of Wnt ligands (Wnt10A and Wnt16) when induced by proximal mammary epithelial cells expressing the Notch ligand, Dll1 (Chakrabarti et al., 2018). This interaction was demonstrated to promote the expression of clonogenic activity both *in vitro* and *in vivo*, so that depleting macrophages, or ablating the Notch transcriptional response, reduced clonogenicity.

3.4 Are there other functions for Wnt ligands that could increase growth potential of mammary glands?

Wnt signaling is typically associated with developmental re-specification; perhaps not surprisingly it has been shown to increase plasticity of mammary epithelial cells. Interestingly, little is known about how this works, beyond a rudimentary appreciation of the associated gene signatures. Although over-simplified, plasticity is often summarized as an epithelial mesenchymal transition (EMT) reaction; reassuringly, this signature exists at the highly plastic embryonic stages of normal mammary gland development (Wuidart et al., 2018). EMT has been shown to be the common endpoint of many stimuli, such as dysregulated expression of master transcriptional programmers (e.g., p63, Sox,Twist, Snail or Slug), matrix stiffness, hormone exposure, metabolic shock or circadian rhythms *in vitro*, or by more global stressors such as telomerase inhibitors (Gross et al., 2019; Gupta, Pastushenko, Skibinski, Blanpain, & Kuperwasser, 2019; Sun et al., 2020; Wuidart et al., 2018).

One of the big surprises of the last two decades was the discovery that so-called "basal" breast tumors did not derive from basal mammary epithelial stem cells (Lim et al., 2009; Molyneux et al., 2010; Proia et al., 2011). Instead, using lineage tracing, the origin of these tumors was shown to be luminal cells, and the oncogenic event was the acquisition of plasticity. Plasticity is inherent in lineages with high demands for wound repair or regeneration (Ge & Fuchs, 2018). This plasticity can be induced in luminal mammary epithelial cells by the expression of kingpin regulators of differentiation, such as Sox9 (Christin et al., 2020; Domenici et al., 2019), and bipotency can be activated in basal cells by gain of function mutations in PIK3CA (Van Keymeulen et al., 2015).

Wnt signaling has been implicated in this EMT reaction. A potentially novel molecular basis for this effect involves the post-transcriptional modulation of one of the zinc finger transcription factors, Slug (Snail2) by Wnt signaling; thus Wnt-activated GSK3 phosphorylates and stabilizes Slug, allowing it to accumulate in the nucleus to promote EMT (Wu et al., 2012). Other mechanisms have been implicated in the adoption of transient mammary epithelial cell states that arise during metastasis; for example an elegant *in vivo* genetic fate mapping strategy identified N-cadherin accumulation as important (Li et al., 2020). In general, the induction of plasticity is considered a powerful tumor initiation event, fusing previously separated paracrine activities into one cell type that evades the normal controls (Gupta et al., 2019). The role of Wnt signaling in the induction of plasticity will therefore be important to the study of breast cancer initiation.

4. Summary

We have emphasized that terminology is important; all cells that divide are not stem cells. Expansion of cell populations in adult tissues does not necessarily implicate a stem cell, classically defined as the origin of all terminally differentiated cell types in a given lineage. Thus, unlike the replication demands of the regenerative blood or intestine lineages, the expansion and involution of the mammary gland during ductal outgrowth and pregnancy is minor. Instead, the challenge for mammary gland growth is to maintain all the participating cell types in balance with one another, so that the proportion of basal cells, luminal cells, sensory ER-positive luminal cells, and alveolar progenitors support functional demands. Molecular dissection has begun to show the complexity of paracrine intercellular communications that maintain these communities. The result is the development of a biologically responsive and adaptable architecture, much more flexible than the division of a stem cell into pre-determined cell types.

Wnt signaling is recruited for the essential phase of embryonic specification from the ectodermal precursor; canonical Wnt signaling acts in concert with the RSpo/Lgr amplifier complex, to promote the plasticity of the embryonic placode. As development proceeds, this pulse of Wnt signaling recedes so that the two principal lineages, basal and luminal, can become specified and self-organize. Residual mammary stem cell activity left dormant in the ductal trees only re-activates when suppressor signals from luminal cells are absent; this could rescue an otherwise apocalyptic scenario for mammals.

The interaction of the two principal lineages of mammary epithelial cell to induce and organize mammary gland morphogenesis is the perfect playground for biologists, and includes a number of unanswered questions in Wnt biology, including, how does the paradoxical Wnt4 work? Wnt signaling is highly oncogenic for other epithelia such as intestine, but does not appear on the list of tumor drivers for breast cancer (Alexander, 2018). Yet it is highly bioactive in mouse mammary gland and induces a hyper-proliferative precursor state which is susceptible to mammary tumor development (Li et al., 2003; Liu et al., 2004). Is there a missing link in mammary epithelia that suppresses the oncogenicity of a gain of function for Wnt signaling?

We have made remarkable progress over the last decades, especially with the testing of the cancer stem cell hypothesis. Perhaps surprisingly, it turns out that somatic mammary basal stem cells are rarely recruited as tumor precursors. Even though basaloid breast tumors share their phenotype and heterogeneity of fates with basal stem cells and their daughters, there are no data to support these cells as the origin of these tumors. Perhaps the susceptibility of basal stem cells to regulation by their neighbors, and their reliance on incoming signals, prevents them from expressing their inherent longevity and plasticity, the qualities that most of us anticipated would be a dangerously tumorigenic combination.

Acknowledgments

This work was supported by RO1GM113142.

References

Ahn, Y., Sanderson, B. W., Klein, O. D., & Krumlauf, R. (2010). Inhibition of Wnt signaling by Wise (Sostdc1) and negative feedback from Shh controls tooth number and patterning. *Development*, *137*, 3221–3231.

Ahn, Y., Sims, C., Logue, J. M., Weatherbee, S. D., & Krumlauf, R. (2013). Lrp4 and Wise interplay controls the formation and patterning of mammary and other skin appendage placodes by modulating Wnt signaling. *Development*, *140*, 583–593.

Alexander, C. M. (2018). The wnt signaling landscape of mammary stem cells and breast tumors. *Progress in Molecular Biology and Translational Science*, *153*, 271–298.

Alexander, C. M., Goel, S., Fakhraldeen, S. A., & Kim, S. (2012). Wnt signaling in mammary glands: Plastic cell fates and combinatorial signaling. *Cold Spring Harbor Perspectives in Biology*, *4*(10), 1–28, a008037.

Alexander, C. M., Joshi, P. A., & Khokha, R. (2014). Fully interlocking: a story of teamwork among breast epithelial cells. *Developmental Cell*, *28*, 114–115.

Alexander, C. M., Puchalski, J., Klos, K. S., Badders, N., Ailles, L., Kim, C. F., et al. (2009). Separating stem cells by flow cytometry: reducing variability for solid tissues. *Cell Stem Cell*, *5*, 579–583.

Altevogt, P., Sammar, M., Huser, L., & Kristiansen, G. (2021). Novel insights into the function of CD24: A driving force in cancer. *International Journal of Cancer*, *148*(3), 546–559, Epub 2020 Sep 3. https://doi.org/10.1002/ijc.33249.

Aragona, M., Panciera, T., Manfrin, A., Giulitti, S., Michielin, F., Elvassore, N., et al. (2013). A mechanical checkpoint controls multicellular growth through YAP/TAZ regulation by actin-processing factors. *Cell*, *154*, 1047–1059.

Ataca, D., Aouad, P., Constantin, C., Laszlo, C., Beleut, M., Shamseddin, M., et al. (2020). The secreted protease Adamts18 links hormone action to activation of the mammary stem cell niche. *Nature Communications*, *11*, 1571.

Axlund, S. D., Yoo, B. H., Rosen, R. B., Schaack, J., Kabos, P., Labarbera, D. V., et al. (2013). Progesterone-inducible cytokeratin 5-positive cells in luminal breast cancer exhibit progenitor properties. *Hormones and Cancer*, *4*, 36–49.

Badders, N. M., Goel, S., Clark, R. J., Klos, K. S., Kim, S., Bafico, A., et al. (2009). The Wnt receptor, Lrp5, is expressed by mouse mammary stem cells and is required to maintain the basal lineage. *PLoS One*, *4*, e6594.

Barker, N., van Es, J. H., Kuipers, J., Kujala, P., van den Born, M., Cozijnsen, M., et al. (2007). Identification of stem cells in small intestine and colon by marker gene Lgr5. *Nature*, *449*, 1003–1007.

Blanpain, C., & Fuchs, E. (2007). p63: revving up epithelial stem-cell potential. *Nature Cell Biology*, *9*, 731–733.

Brisken, C., & Ataca, D. (2015). Endocrine hormones and local signals during the development of the mouse mammary gland. *Wiley Interdisciplinary Reviews: Developmental Biology*, *4*, 181–195.

Brisken, C., & O'Malley, B. (2010). Hormone action in the mammary gland. *Cold Spring Harbor Perspectives in Biology*, *2*, a003178.

Bruno, R. D., & Smith, G. H. (2012). Reprogramming non-mammary and cancer cells in the developing mouse mammary gland. *Seminars in Cell & Developmental Biology*, *23*, 591–598.

Butler, M. T., & Wallingford, J. B. (2017). Planar cell polarity in development and disease. *Nature Reviews. Molecular Cell Biology*, *18*, 375–388.

Cai, C., Geng, A., Wang, M., Yang, L., Yu, Q. C., & Zeng, Y. A. (2020). Amphiregulin mediates the hormonal regulation on Rspondin-1 expression in the mammary gland. *Developmental Biology*, *458*, 43–51.

Cai, S., Kalisky, T., Sahoo, D., Dalerba, P., Feng, W., Lin, Y., et al. (2017). A quiescent Bcl11b high stem cell population is required for maintenance of the mammary gland. *Cell Stem Cell*, *20*(247–260), e245.

Cai, C., Yu, Q. C., Jiang, W., Liu, W., Song, W., Yu, H., et al. (2014). R-spondin1 is a novel hormone mediator for mammary stem cell self-renewal. *Genes & Development*, *28*, 2205–2218.

Callahan, R., & Smith, G. H. (2008). Common integration sites for MMTV in viral induced mouse mammary tumors. *Journal of Mammary Gland Biology and Neoplasia*, *13*, 309–321.

Carmon, K. S., Gong, X., Yi, J., Thomas, A., & Liu, Q. (2014). RSPO-LGR4 functions via IQGAP1 to potentiate Wnt signaling. *Proceedings of the National Academy of Sciences of the United States of America*, *111*, E1221–E1229.

Chakrabarti, R., Celia-Terrassa, T., Kumar, S., Hang, X., Wei, Y., Choudhury, A., et al. (2018). Notch ligand Dll1 mediates cross-talk between mammary stem cells and the macrophageal niche. *Science*, *360*(6396), eaan4153.

Chin, E. N., Martin, J. A., Kim, S., Fakhraldeen, S. A., & Alexander, C. M. (2015). Lrp5 has a wnt-independent role in glucose uptake and growth for mammary epithelial cells. *Molecular and Cellular Biology*, *36*, 871–885.

Christin, J. R., Wang, C., Chung, C. Y., Liu, Y., Dravis, C., Tang, W., et al. (2020). Stem cell determinant SOX9 promotes lineage plasticity and progression in basal-like breast cancer. *Cell Reports*, *31*, 107742.

Chu, E. Y., Hens, J., Andl, T., Kairo, A., Yamaguchi, T. P., Brisken, C., et al. (2004). Canonical WNT signaling promotes mammary placode development and is essential for initiation of mammary gland morphogenesis. *Development*, *131*, 4819–4829.

Clevers, H. (2013). The intestinal crypt, a prototype stem cell compartment. *Cell*, *154*, 274–284.

Cosin-Roger, J., Ortiz-Masia, M. D., & Barrachina, M. D. (2019). Macrophages as an emerging source of Wnt ligands: Relevance in mucosal integrity. *Frontiers in Immunology*, *10*, 2297.

Cruciat, C. M., & Niehrs, C. (2013). Secreted and transmembrane wnt inhibitors and activators. *Cold Spring Harbor Perspectives in Biology*, *5*, a015081.

Davis, F. M., Lloyd-Lewis, B., Harris, O. B., Kozar, S., Winton, D. J., Muresan, L., et al. (2016). Single-cell lineage tracing in the mammary gland reveals stochastic clonal dispersion of stem/progenitor cell progeny. *Nature Communications*, *7*, 13053.

Domenici, G., Aurrekoetxea-Rodriguez, I., Simoes, B. M., Rabano, M., Lee, S. Y., Millan, J. S., et al. (2019). A Sox2-Sox9 signalling axis maintains human breast luminal progenitor and breast cancer stem cells. *Oncogene*, *38*, 3151–3169.

Dupont, S., Morsut, L., Aragona, M., Enzo, E., Giulitti, S., Cordenonsi, M., et al. (2011). Role of YAP/TAZ in mechanotransduction. *Nature*, *474*, 179–183.

Fillmore, C. M., Gupta, P. B., Rudnick, J. A., Caballero, S., Keller, P. J., Lander, E. S., et al. (2010). Estrogen expands breast cancer stem-like cells through paracrine FGF/Tbx3 signaling. *Proceedings of the National Academy of Sciences of the United States of America*, *107*, 21737–21742.

Forster, N., Saladi, S. V., van Bragt, M., Sfondouris, S. E., Jones, F. E., & Ellisen, L. W. (2014). Basal cell signaling by p63 controls luminal progenitor function and lactation via NRG1. *Developmental Cell*, *28*(2), 147–160.

Frank-Kamenetskii, A., Mook, J., Reeves, M., Boulanger, C. A., Meyer, T. J., Ragle, L., et al. (2020). Induction of phenotypic changes in HER2-postive breast cancer cells in vivo and in vitro. *Oncotarget*, *11*, 2919–2929.

Ge, Y., & Fuchs, E. (2018). Stretching the limits: from homeostasis to stem cell plasticity in wound healing and cancer. *Nature Reviews. Genetics*, *19*, 311–325.

Giraddi, R. R., Shehata, M., Gallardo, M., Blasco, M. A., Simons, B. D., & Stingl, J. (2015). Stem and progenitor cell division kinetics during postnatal mouse mammary gland development. *Nature Communications*, *6*, 8487.

Goel, S., Chin, E. N., Fakhraldeen, S. A., Berry, S. M., Beebe, D. J., & Alexander, C. M. (2012). Both LRP5 and LRP6 receptors are required to respond to physiological Wnt ligands in mammary epithelial cells and fibroblasts. *The Journal of Biological Chemistry*, *287*, 16454–16466.

Gouon-Evans, V., Rothenberg, M. E., & Pollard, J. W. (2000). Postnatal mammary gland development requires macrophages and eosinophils. *Development*, *127*, 2269–2282.

Gross, K. M., Zhou, W., Breindel, J. L., Ouyang, J., Jin, D. X., Sokol, E. S., et al. (2019). Loss of slug compromises DNA damage repair and accelerates stem cell aging in mammary epithelium. *Cell Reports*, *28*(394–407), e396.

Gupta, P. B., Pastushenko, I., Skibinski, A., Blanpain, C., & Kuperwasser, C. (2019). Phenotypic plasticity: Driver of cancer initiation, progression, and therapy resistance. *Cell Stem Cell*, *24*, 65–78.

Halt, K., & Vainio, S. (2014). Coordination of kidney organogenesis by Wnt signaling. *Pediatric Nephrology*, *29*, 737–744.

Harris, M. P., Hasso, S. M., Ferguson, M. W., & Fallon, J. F. (2006). The development of archosaurian first-generation teeth in a chicken mutant. *Current Biology*, *16*, 371–377.

Haughian, J. M., Pinto, M. P., Harrell, J. C., Bliesner, B. S., Joensuu, K. M., Dye, W. W., et al. (2012). Maintenance of hormone responsiveness in luminal breast cancers by suppression of Notch. *Proceedings of the National Academy of Sciences of the United States of America*, *109*, 2742–2747.

Horwitz, K. B., Dye, W. W., Harrell, J. C., Kabos, P., & Sartorius, C. A. (2008). Rare steroid receptor-negative basal-like tumorigenic cells in luminal subtype human breast cancer xenografts. *Proceedings of the National Academy of Sciences of the United States of America*, *105*, 5774–5779.

Howard, B. A., & Lu, P. (2014). Stromal regulation of embryonic and postnatal mammary epithelial development and differentiation. *Seminars in Cell & Developmental Biology*, *25-26*, 43–51.

Ingman, W. V., Wyckoff, J., Gouon-Evans, V., Condeelis, J., & Pollard, J. W. (2006). Macrophages promote collagen fibrillogenesis around terminal end buds of the developing mammary gland. *Developmental Dynamics*, *235*, 3222–3229.

Inoki, K., Ouyang, H., Zhu, T., Lindvall, C., Wang, Y., Zhang, X., et al. (2006). TSC2 integrates Wnt and energy signals via a coordinated phosphorylation by AMPK and GSK3 to regulate cell growth. *Cell*, *126*, 955–968.

Jamieson, P. R., Dekkers, J. F., Rios, A. C., Fu, N. Y., Lindeman, G. J., & Visvader, J. E. (2017). Derivation of a robust mouse mammary organoid system for studying tissue dynamics. *Development*, *144*, 1065–1071.

Janda, C. Y., & Garcia, K. C. (2015). Wnt acylation and its functional implication in Wnt signalling regulation. *Biochemical Society Transactions*, *43*, 211–216.

Joshi, P. A., Jackson, H. W., Beristain, A. G., Di Grappa, M. A., Mote, P. A., Clarke, C. L., et al. (2010). Progesterone induces adult mammary stem cell expansion. *Nature*, *465*, 803–807.

Kabos, P., Haughian, J. M., Wang, X., Dye, W. W., Finlayson, C., Elias, A., et al. (2011). Cytokeratin 5 positive cells represent a steroid receptor negative and therapy resistant subpopulation in luminal breast cancers. *Breast Cancer Research and Treatment*, *128*, 45–55.

Kessenbrock, K., Dijkgraaf, G. J., Lawson, D. A., Littlepage, L. E., Shahi, P., Pieper, U., et al. (2013). A role for matrix metalloproteinases in regulating mammary stem cell function via the Wnt signaling pathway. *Cell Stem Cell*, *13*, 300–313.

Kessenbrock, K., Smith, P., Steenbeek, S. C., Pervolarakis, N., Kumar, R., Minami, Y., et al. (2017). Diverse regulation of mammary epithelial growth and branching morphogenesis through noncanonical Wnt signaling. *Proceedings of the National Academy of Sciences of the United States of America*, *114*, 3121–3126.

Klinowska, T. C., Alexander, C. M., Georges-Labouesse, E., Van der Neut, R., Kreidberg, J. A., Jones, C. J., et al. (2001). Epithelial development and differentiation in the mammary gland is not dependent on alpha 3 or alpha 6 integrin subunits. *Developmental Biology*, *233*, 449–467.

Koo, B. K., & Clevers, H. (2014). Stem cells marked by the R-spondin receptor LGR5. *Gastroenterology*, *147*, 289–302.

Kopp, J. L., Grompe, M., & Sander, M. (2016). Stem cells versus plasticity in liver and pancreas regeneration. *Nature Cell Biology*, *18*, 238–245.

Krebsbach, P. H., & Villa-Diaz, L. G. (2017). The role of integrin alpha6 (CD49f) in stem cells: More than a conserved biomarker. *Stem Cells and Development*, *26*, 1090–1099.

Kreuzaler, P. A., Staniszewska, A. D., Li, W., Omidvar, N., Kedjouar, B., Turkson, J., et al. (2011). Stat3 controls lysosomal-mediated cell death in vivo. *Nature Cell Biology*, *13*, 303–309.

Kumar, K. K., Burgess, A. W., & Gulbis, J. M. (2014). Structure and function of LGR5: an enigmatic G-protein coupled receptor marking stem cells. *Protein Science*, *23*, 551–565.

Langton, P. F., Kakugawa, S., & Vincent, J. P. (2016). Making, exporting, and modulating Wnts. *Trends in Cell Biology*, *26*, 756–765.

Larraguibel, J., Weiss, A. R., Pasula, D. J., Dhaliwal, R. S., Kondra, R., & Van Raay, T. J. (2015). Wnt ligand-dependent activation of the negative feedback regulator Nkd1. *Molecular Biology of the Cell, 26*, 2375–2384.

Lebreton, S., & Jones, C. M. (2006). Emergence of organizer function: a lot of "stuff" involved. *Seminars in Cell & Developmental Biology, 17*, 110–116.

Li, Y., Lv, Z., Zhang, S., Wang, Z., He, L., Tang, M., et al. (2020). Genetic fate mapping of transient cell fate reveals N-cadherin activity and function in tumor metastasis. *Developmental Cell, 54*(593–607), e595.

Li, Y., Welm, B., Podsypanina, K., Huang, S., Chamorro, M., Zhang, X., et al. (2003). Evidence that transgenes encoding components of the Wnt signaling pathway preferentially induce mammary cancers from progenitor cells. *Proceedings of the National Academy of Sciences of the United States of America, 100*, 15853–15858.

Lim, E., Vaillant, F., Wu, D., Forrest, N. C., Pal, B., Hart, A. H., et al. (2009). Aberrant luminal progenitors as the candidate target population for basal tumor development in BRCA1 mutation carriers. *Nature Medicine, 15*, 907–913.

Linnemann, J. R., Meixner, L. K., Miura, H., & Scheel, C. H. (2017). An organotypic 3D assay for primary human mammary epithelial cells that recapitulates branching morphogenesis. *Methods in Molecular Biology, 1612*, 125–137.

Liu, B. Y., McDermott, S. P., Khwaja, S. S., & Alexander, C. M. (2004). The transforming activity of Wnt effectors correlates with their ability to induce the accumulation of mammary progenitor cells. *Proceedings of the National Academy of Sciences of the United States of America, 101*, 4158–4163.

Ly, C. H., Lynch, G. S., & Ryall, J. G. (2020). A metabolic roadmap for somatic stem cell fate. *Cell Metabolism, 31*, 1052–1067.

Macdonald, B. T., Semenov, M. V., Huang, H., & He, X. (2011). Dissecting molecular differences between Wnt coreceptors LRP5 and LRP6. *PLoS One, 6*, e23537.

Magee, J. A., Piskounova, E., & Morrison, S. J. (2012). Cancer stem cells: impact, heterogeneity, and uncertainty. *Cancer Cell, 21*, 283–296.

Makarem, M., Spike, B. T., Dravis, C., Kannan, N., Wahl, G. M., & Eaves, C. J. (2013). Stem cells and the developing mammary gland. *Journal of Mammary Gland Biology and Neoplasia, 18*, 209–219.

Malsin, E. S., Kim, S., Lam, A. P., & Gottardi, C. J. (2019). Macrophages as a source and recipient of Wnt signals. *Frontiers in Immunology, 10*, 1813.

Mastroianni, M., Kim, S., Kim, Y. C., Esch, A., Wagner, C., & Alexander, C. M. (2010). Wnt signaling can substitute for estrogen to induce division of ERalpha-positive cells in a mouse mammary tumor model. *Cancer Letters, 289*, 23–31.

McGinn, O., Ward, A. V., Fettig, L. M., Riley, D., Ivie, J., Paul, K. V., et al. (2020). Cytokeratin 5 alters beta-catenin dynamics in breast cancer cells. *Oncogene, 39*, 2478–2492.

McGough, I. J., Vecchia, L., Bishop, B., Malinauskas, T., Beckett, K., Joshi, D., et al. (2020). Glypicans shield the Wnt lipid moiety to enable signalling at a distance. *Nature, 585*, 85–90.

Molyneux, G., Geyer, F. C., Magnay, F. A., McCarthy, A., Kendrick, H., Natrajan, R., et al. (2010). BRCA1 basal-like breast cancers originate from luminal epithelial progenitors and not from basal stem cells. *Cell Stem Cell, 7*, 403–417.

Naylor, S., Smalley, M. J., Robertson, D., Gusterson, B. A., Edwards, P. A., & Dale, T. C. (2000). Retroviral expression of Wnt-1 and Wnt-7b produces different effects in mouse mammary epithelium. *Journal of Cell Science, 113*(Pt 12), 2129–2138.

Nguyen-Ngoc, K. V., Shamir, E. R., Huebner, R. J., Beck, J. N., Cheung, K. J., & Ewald, A. J. (2015). 3D culture assays of murine mammary branching morphogenesis and epithelial invasion. *Methods in Molecular Biology, 1189*, 135–162.

Nile, A. H., Mukund, S., Stanger, K., Wang, W., & Hannoush, R. N. (2017). Unsaturated fatty acyl recognition by Frizzled receptors mediates dimerization upon Wnt ligand binding. *Proceedings of the National Academy of Sciences of the United States of America*, *114*, 4147–4152.

Ohazama, A., Porntaveetus, T., Ota, M. S., Herz, J., & Sharpe, P. T. (2010). Lrp4: A novel modulator of extracellular signaling in craniofacial organogenesis. *American Journal of Medical Genetics. Part A*, *152A*, 2974–2983.

Paguirigan, A., Beebe, D. J., & Alexander, C. M. (2007). Simulating mouse mammary gland development: Cell ageing and its relation to stem and progenitor activity. *Cell Proliferation*, *40*, 106–124.

Plaks, V., Brenot, A., Lawson, D. A., Linnemann, J. R., Van Kappel, E. C., Wong, K. C., et al. (2013). Lgr5-expressing cells are sufficient and necessary for postnatal mammary gland organogenesis. *Cell Reports*, *3*, 70–78.

Proia, T. A., Keller, P. J., Gupta, P. B., Klebba, I., Jones, A. D., Sedic, M., et al. (2011). Genetic predisposition directs breast cancer phenotype by dictating progenitor cell fate. *Cell Stem Cell*, *8*, 149–163.

Rivetti, S., Chen, C., Chen, C., & Bellusci, S. (2020). Fgf10/Fgfr2b signaling in mammary gland development, homeostasis, and cancer. *Frontiers in Cell and Development Biology*, *8*, 415.

Roarty, K., & Rosen, J. M. (2013). Pygopus 2: tilting the Wnt-Notch balance in mammary epithelial lineage determination. *Breast Cancer Research*, *15*, 322.

Roarty, K., & Serra, R. (2007). Wnt5a is required for proper mammary gland development and TGF-beta-mediated inhibition of ductal growth. *Development*, *134*, 3929–3939.

Roarty, K., Shore, A. N., Creighton, C. J., & Rosen, J. M. (2015). Ror2 regulates branching, differentiation, and actin-cytoskeletal dynamics within the mammary epithelium. *The Journal of Cell Biology*, *208*, 351–366.

Robinson, G. W. (2007). Cooperation of signalling pathways in embryonic mammary gland development. *Nature Reviews. Genetics*, *8*, 963–972.

Rodilla, V., Dasti, A., Huyghe, M., Lafkas, D., Laurent, C., Reyal, F., et al. (2015). Luminal progenitors restrict their lineage potential during mammary gland development. *PLoS Biology*, *13*, e1002069.

Rosen, J. M. (2012). On murine mammary epithelial stem cells: discovery, function, and current status. *Cold Spring Harbor Perspectives in Biology*, *4*, a013268.

Saito-Diaz, K., Chen, T. W., Wang, X., Thorne, C. A., Wallace, H. A., Page-McCaw, A., et al. (2013). The way Wnt works: components and mechanism. *Growth Factors*, *31*, 1–31.

Shackleton, M., Vaillant, F., Simpson, K. J., Stingl, J., Smyth, G. K., Asselin-Labat, M. L., et al. (2006). Generation of a functional mammary gland from a single stem cell. *Nature*, *439*, 84–88.

Sheng, R., Kim, H., Lee, H., Xin, Y., Chen, Y., Tian, W., et al. (2014). Cholesterol selectively activates canonical Wnt signalling over non-canonical Wnt signalling. *Nature Communications*, *5*, 4393.

Shimizu, H., Julius, M. A., Giarre, M., Zheng, Z., Brown, A. M., & Kitajewski, J. (1997). Transformation by Wnt family proteins correlates with regulation of b-catenin. *Cell Growth & Differentiation*, *8*, 1349–1358.

Smalley, M. J., Reis-Filho, J. S., & Ashworth, A. (2008). BRCA1 and stem cells: tumour typecasting. *Nature Cell Biology*, *10*, 377–379.

Stingl, J., Eirew, P., Ricketson, I., Shackleton, M., Vaillant, F., Choi, D., et al. (2006). Purification and unique properties of mammary epithelial stem cells. *Nature*, *439*, 993–997.

Sun, X., Wang, M., Wang, M., Yao, L., Li, X., Dong, H., et al. (2020). Exploring the metabolic vulnerabilities of epithelial-mesenchymal transition in breast cancer. *Frontiers in Cell and Development Biology*, *8*, 655.

Taddei, I., Deugnier, M. A., Faraldo, M. M., Petit, V., Bouvard, D., Medina, D., et al. (2008). Beta1 integrin deletion from the basal compartment of the mammary epithelium affects stem cells. *Nature Cell Biology, 10*, 716–722.
Teuliere, J., Faraldo, M. M., Deugnier, M. A., Shtutman, M., Ben-Ze'ev, A., Thiery, J. P., et al. (2005). Targeted activation of beta-catenin signaling in basal mammary epithelial cells affects mammary development and leads to hyperplasia. *Development, 132*, 267–277.
Ugolini, F., Charafe-Jauffret, E., Bardou, V. J., Geneix, J., Adelaide, J., Labat-Moleur, F., et al. (2001). WNT pathway and mammary carcinogenesis: loss of expression of candidate tumor suppressor gene SFRP1 in most invasive carcinomas except of the medullary type. *Oncogene, 20*, 5810–5817.
Valenta, T., Hausmann, G., & Basler, K. (2012). The many faces and functions ofbeta-catenin. *The EMBO Journal, 31*, 2714–2736.
Valvezan, A. J., Huang, J., Lengner, C. J., Pack, M., & Klein, P. S. (2014). Oncogenic mutations in adenomatous polyposis coli (Apc) activate mechanistic target of rapamycin complex 1 (mTORC1) in mice and zebrafish. *Disease Models & Mechanisms, 7*, 63–71.
Valvezan, A. J., & Manning, B. D. (2019). Molecular logic of mTORC1 signalling as a metabolic rheostat. *Nature Metabolism, 1*, 321–333.
van Amerongen, R., Bowman, A. N., & Nusse, R. (2012). Developmental stage and time dictate the fate of wnt/beta-catenin-responsive stem cells in the mammary gland. *Cell Stem Cell, 11*, 387–400.
van Amerongen, R., & Nusse, R. (2009). Towards an integrated view of Wnt signaling in development. *Development, 136*, 3205–3214.
Van Keymeulen, A., Fioramonti, M., Centonze, A., Bouvencourt, G., Achouri, Y., & Blanpain, C. (2017). Lineage-restricted mammary stem cells sustain the development, homeostasis, and regeneration of the estrogen receptor positive lineage. *Cell Reports, 20*, 1525–1532.
Van Keymeulen, A., Lee, M. Y., Ousset, M., Brohee, S., Rorive, S., Giraddi, R. R., et al. (2015). Reactivation of multipotency by oncogenic PIK3CA induces breast tumour heterogeneity. *Nature, 525*, 119–123.
Van Keymeulen, A., Rocha, A. S., Ousset, M., Beck, B., Bouvencourt, G., Rock, J., et al. (2011). Distinct stem cells contribute to mammary gland development and maintenance. *Nature, 479*, 189–193.
van Tienen, L. M., Mieszczanek, J., Fiedler, M., Rutherford, T. J., & Bienz, M. (2017). Constitutive scaffolding of multiple Wnt enhanceosome components by Legless/BCL9. *Elife, 6*, e20882.
Veltmaat, J. M. (2017). Prenatal mammary gland development in the mouse: Research models and techniques for its study from past to present. *Methods in Molecular Biology, 1501*, 21–76.
Veltmaat, J. M., Van Veelen, W., Thiery, J. P., & Bellusci, S. (2004). Identification of the mammary line in mouse by Wnt10b expression. *Developmental Dynamics, 229*, 349–356.
Visvader, J. E., & Clevers, H. (2016). Tissue-specific designs of stem cell hierarchies. *Nature Cell Biology, 18*, 349–355.
Visvader, J. E., & Stingl, J. (2014). Mammary stem cells and the differentiation hierarchy: current status and perspectives. *Genes & Development, 28*, 1143–1158.
Wang, D., Cai, C., Dong, X., Yu, Q. C., Zhang, X. O., Yang, L., et al. (2015). Identification of multipotent mammary stem cells by protein C receptor expression. *Nature, 517*, 81–84.
Wang, C., Christin, J. R., Oktay, M. H., & Guo, W. (2017). Lineage-biased stem cells maintain estrogen-receptor-positive and -negative mouse mammary luminal lineages. *Cell Reports, 18*, 2825–2835.
Wang, Y., Dong, J., Li, D., Lai, L., Siwko, S., Li, Y., et al. (2013). Lgr4 regulates mammary gland development and stem cell activity through the pluripotency transcription factor Sox2. *Stem Cells, 31*, 1921–1931.

Wang, B., Zhao, L., Fish, M., Logan, C. Y., & Nusse, R. (2015). Self-renewing diploid Axin2(+) cells fuel homeostatic renewal of the liver. *Nature*, *524*, 180–185.
Weissenbock, M., Latham, R., Nishita, M., Wolff, L. I., Ho, H. H., Minami, Y., et al. (2019). Genetic interactions between Ror2 and Wnt9a, Ror1 and Wnt9a and Ror2 and Ror1: Phenotypic analysis of the limb skeleton and palate in compound mutants. *Genes to Cells*, *24*, 307–317.
Williams, M. L. K., & Solnica-Krezel, L. (2020). Cellular and molecular mechanisms of convergence and extension in zebrafish. *Current Topics in Developmental Biology*, *136*, 377–407.
Wong, G. T., Gavin, B. J., & McMahon, A. P. (1994). Differential transformation of mammary epithelial cells by Wnt genes. *Molecular and Cellular Biology*, *14*, 6278–6286.
Wu, Z. Q., Li, X. Y., Hu, C. Y., Ford, M., Kleer, C. G., & Weiss, S. J. (2012). Canonical Wnt signaling regulates Slug activity and links epithelial-mesenchymal transition with epigenetic Breast Cancer 1, Early Onset (BRCA1) repression. *Proceedings of the National Academy of Sciences of the United States of America*, *109*, 16654–16659.
Wuidart, A., Ousset, M., Rulands, S., Simons, B. D., Van Keymeulen, A., & Blanpain, C. (2016). Quantitative lineage tracing strategies to resolve multipotency in tissue-specific stem cells. *Genes & Development*, *30*, 1261–1277.
Wuidart, A., Sifrim, A., Fioramonti, M., Matsumura, S., Brisebarre, A., Brown, D., et al. (2018). Early lineage segregation of multipotent embryonic mammary gland progenitors. *Nature Cell Biology*, *20*, 666–676.
Xie, Y., Zamponi, R., Charlat, O., Ramones, M., Swalley, S., Jiang, X., et al. (2013). Interaction with both ZNRF3 and LGR4 is required for the signalling activity of R-spondin. *EMBO Reports*, *14*, 1120–1126.
Xu, M., Horrell, J., Snitow, M., Cui, J., Gochnauer, H., Syrett, C. M., et al. (2017). WNT10A mutation causes ectodermal dysplasia by impairing progenitor cell proliferation and KLF4-mediated differentiation. *Nature Communications*, *8*, 15397.
Xu, R., Spencer, V. A., Groesser, D. L., & Bissell, M. J. (2010). Laminin regulates PI3K basal localization and activation to sustain STAT5 activation. *Cell Cycle*, *9*, 4315–4322.
Yan, K. S., Janda, C. Y., Chang, J., Zheng, G. X. Y., Larkin, K. A., Luca, V. C., et al. (2017). Non-equivalence of Wnt and R-spondin ligands during Lgr5+ intestinal stem-cell self-renewal. *Nature*, *545*, 238–242.
Yu, J., & Virshup, D. M. (2014). Updating the Wnt pathways. *Bioscience Reports*, *34*, 593–607.
Zabala, M., Lobo, N. A., Antony, J., Heitink, L. S., Gulati, G. S., Lam, J., et al. (2020). LEFTY1 Is a Dual-SMAD inhibitor that promotes mammary progenitor growth and tumorigenesis. *Cell Stem Cell*, *27*(284–299), e288.
Zanconato, F., Cordenonsi, M., & Piccolo, S. (2016). YAP/TAZ at the roots of cancer. *Cancer Cell*, *29*, 783–803.
Zhang, L., Adileh, M., Martin, M. L., Klingler, S., White, J., Ma, X., et al. (2017). Establishing estrogen-responsive mouse mammary organoids from single Lgr5+ cells. *Cellular Signalling*, *29*, 41–51.

CHAPTER THREE

Thyroid hormone signaling specifies cone photoreceptor subtypes during eye development: Insights from model organisms and human stem cell-derived retinal organoids

Christina McNerney and Robert J. Johnston Jr*
Department of Biology, Johns Hopkins University, Baltimore, MD, United States
*Corresponding author: e-mail address: robertjohnston@jhu.edu

Contents

Abstract

Cones are the color-detecting photoreceptors of the vertebrate eye. Cones are specialized into subtypes whose functions are determined by the expression of color-sensitive opsin proteins. Organisms differ greatly in the number and patterning of cone subtypes.

Vitamins and Hormones, Volume 116
ISSN 0083-6729
https://doi.org/10.1016/bs.vh.2021.03.001

Despite these differences, thyroid hormone is an important regulator of opsin expression in most vertebrates. In this chapter, we outline how the timing of thyroid hormone signaling controls cone subtype fates during retinal development. We first examine our current understanding of cone subtype specification in model organisms and then describe advances in human stem cell-derived organoid technology that identified mechanisms controlling development of the human retina.

1. Introduction: The thyroid hormone signaling pathway

Thyroid hormones are a class of nuclear hormones that are iodinated tyrosine derivatives that have diverse functions in metabolism, growth, and development. *Thyroxine* (T4) is the main thyroid hormone circulating throughout an organism, carried by carrier proteins such as thyroxine binding globulin (TBG). T4 is transported throughout the body to target tissues. T4 is modified via inner or outer ring deiodination, or removal of an iodine moiety, to produce other thyroid hormone derivatives. Inner ring deiodination produces rT3, which is degraded (Gereben, Zavacki, et al., 2008). Outer ring deiodination of T4 produces *triiodothyronine* (T3). Inner ring deiodination of T3 produces T2, which is degraded along with rT3. T3 and T4 are both capable of binding thyroid hormone receptors, although T3 has greater affinity and is referred to as the "active form" of thyroid hormone despite the greater abundance of T4 in circulation (Samuels, Tsai, Casanova, & Stanley, 1974). This balance of abundance and affinity allows for local regulation of thyroid hormone signaling as well as global regulation through the hypothalamus/anterior pituitary gland/thyroid gland/target axis.

The generation and secretion of thyroid hormone is largely controlled by a cascade of signals through the hypothalamus/anterior pituitary gland/thyroid gland/target axis (Fig. 1). Tight regulation of the release of thyroid hormones is important for maintenance of global metabolic processes (Gelfand, Hutchinson-Williams, Bonde, Castellino, & Sherwin, 1987). The hypothalamus secretes thyrotropin-releasing hormone (TRH), which promotes thyroid stimulating hormone (TSH) release by the anterior pituitary gland (Fig. 1A). TSH acts on the thyroid to stimulate the uptake of iodide (Fig. 1B). T4 is generated when tyrosine residues in the protein thyroglobulin are iodinated and subsequently cleaved to release T4. The thyroid secretes T4, and to a lesser extent T3, which circulates throughout the organism (Fig. 1B and C). Local deiodination occurs in target tissues to either produce active T3 from T4 or degrade thyroid hormones (Chiamolera & Wondisford, 2009) (Fig. 1D–G).

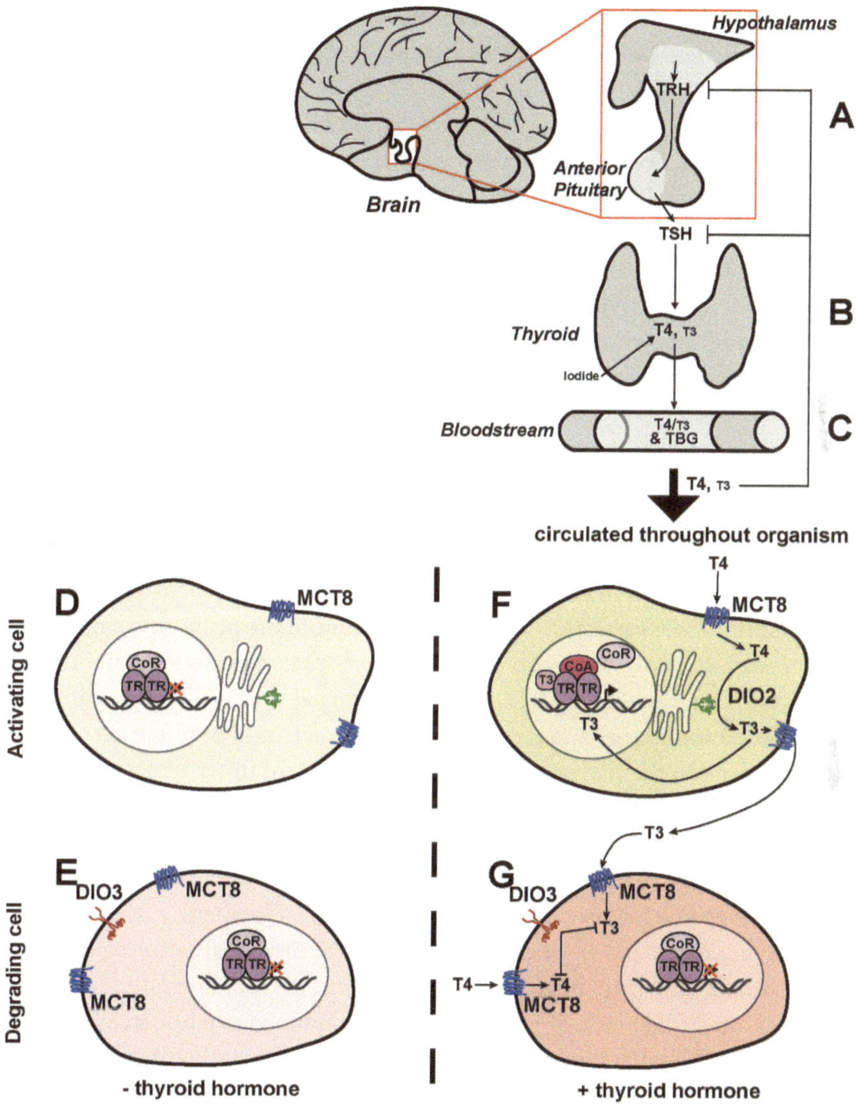

Fig. 1 Thyroid hormone synthesis, release, circulation, and local regulation/action. (A) Activation of the hypothalamus releases TRH, which promotes release of TSH from the anterior pituitary. (B) TSH stimulates the thyroid to produce thyroid hormone (T4 and T3), which is stored or released. (C) Thyroid hormones are carried through the bloodstream by TBG. (D and E) In the absence of thyroid hormone, thyroid hormone receptors (TR) bind to TREs as dimers (here, represented as a homodimer), which can also bind to a corepressor (CoR). (F and G) In the presence of thyroid hormone, T4 and T3 enter cells through membrane transporters (here, represented by MCT8). Internally, thyroid hormone can either be activated (F) or degraded (G). (F) DIO2 can convert T4 into T3, which can be released from the cell or brought to the nucleus. T3 binding TR in the nucleus promotes CoR release and coactivator (CoA) binding to promote transcription of response genes. (G) If DIO3 is present, it degrades both T4 and T3. CoR remains bound to TR, inhibiting transcription of response genes.

Thyroid hormone is transported from the thyroid and distributed to peripheral tissues. While circulating in plasma, thyroid hormone is carried by thyroxine binding globulin (TBG) (Fig. 1C) or transthyretin (TTR). Though thyroid hormones are nuclear hormones, they cannot freely diffuse into cells. Membrane transport proteins including monocarboxylate transporters 8 and 10 (MCT8 and MCT10, respectively) bind thyroid hormones and transport them into the cytoplasm (Fig. 1D–G). MCT8 and MCT10 transport both T4 and T3, but MCT8 has a greater affinity for T4 whereas MCT10 has a greater affinity for T3. In addition, the large neutral amino acid transporters (LAT) have weak affinities for T3 and T4 but do transport thyroid hormones across cell membranes. The organic anion transporter polypeptides (OATPs) bind and transport T4 in the brain. Different tissues express different transporters for regulatory purposes (Benvenga & Robbins, 1996; Bernal, Guadano-Ferraz, & Morte, 2015; Friesema et al., 2003; Friesema, Jansen, & Visser, 2005; Korcek & Tabachnick, 1976; Palha, 2002).

Within a cell, thyroid hormones are activated or deactivated through deiodination. *Deiodinases (DIOs)* are integral membrane proteins characterized by cytosolic selenocysteine-containing active sites that deiodinate thyroid hormones. Deiodinase 1 (DIO1) and Deiodinase 2 (DIO2) are embedded in the ER membrane and convert T4 into active T3 (Fig. 1D and F). DIO3 is found in the plasma membrane and performs inner ring deiodination to convert T3 into T2 or convert T4 into rT3, thereby enabling their degradation (Fig. 1E and G). Deiodination regulates local levels of thyroid hormone in a careful balance between active T3 and inactive T4 (Bianco, Salvatore, Gereben, Berry, & Larsen, 2002; Galton, 2017; Gereben, Zavacki, et al., 2008; Muller et al., 2014). Feedback on the organismal scale also regulates circulating levels of thyroid hormone, as high serum thyroid hormone negatively feeds back on TRH and TSH expression to maintain global homeostatic levels (Fig. 1A–C) (Alkemade et al., 2005; Fliers, Unmehopa, & Alkemade, 2006; Larsen, 1982).

Once modified within the cell, thyroid hormone travels to the nucleus to interact with its receptors, generally referred to as *thyroid hormone receptors (TRs)* (Fig. 1D–G). Many vertebrates have two TR genes: THRα and THRβ, and these often have splice isoforms that are differentially expressed across tissues (Brent, 2012). Nomenclature varies between species: in model organisms, the genes are named TRα and TRβ, while in humans, these same genes are called THRα and THRβ. TRs bind to DNA at sites called *thyroid hormone response elements (TREs)*. In the absence of thyroid hormone, TRs are often bound by *corepressors* (Fig. 1D). Upon binding thyroid hormone (of which T3 has a much higher affinity than T4), corepressor binding partners are exchanged for *coactivators* and transcription is activated (Fig. 1F).

In the absence of T3, corepressors, including nuclear receptor corepressors 1 and 2 (NCOR1, NCOR2/SMRT, respectively), repress expression at target genes (Fig. 1D and E). These proteins repress transcription by recruiting histone deacetylases to render the transcription start site inaccessible to transcriptional machinery (Sun et al., 2013; You et al., 2013). T3 binding to TRs disrupts this repressive complex and recruits local coactivators which in turn recruit histone acetylases. Histone acetylases function to increase accessibility of the start site and promote transcription (Astapova, 2016). Many of these coactivators have overlapping response elements with TREs (Oppenheimer & Schwartz, 1997; Williams, 2008), and local concentrations of the cofactors and their ligands affect how they interact with TRs (Brent, 2012). While TRs can homodimerize, most examples of transcriptional activation involve a heterodimer binding partner such as COUP TF-II or a member of the RXR class of transcriptional regulators (Butler & Parker, 1995; Koenig, 1998).

Summary: Thyroid hormones are produced by the thyroid via a signaling cascade and exist mainly as circulating T4 (inactive) and T3 (active). Thyroid hormones enter the cell via transport proteins and are modified by deiodinases to regulate local levels of thyroid hormone by either activating (DIO1, DIO2) or degrading (DIO1, DIO3) thyroid hormone. Upon entering the nucleus, T3 binds to DNA-bound TRs causing the TR to exchange its binding partner from a corepressor to a coactivator, leading to transcription initiation.

2. Thyroid hormone regulates cell fates during development

Thyroid hormone has important roles in the development of several cell types and tissues. Roles for thyroid hormone signaling have been found in oligodendrocyte differentiation, cochlear development, brain development, muscle stem cell development, and retinal development. In both cochlear and brain development, thyroid hormone signaling is suppressed early and is promoted late to temporally modulate differentiation (Billon, Tokumoto, Forrest, & Raff, 2001; Campos-Barros et al., 1999; Crantz, Silva, & Larsen, 1982; Guadano-Ferraz, Obregon, St. Germain, & Bernal, 1997; Ng et al., 2004, 2009; Oppenheimer & Schwartz, 1997; Tu et al., 1997, 1999). Dynamic thyroid hormone signaling also plays a critical role in cone subtype specification during the development of the retina, the light-detecting tissue in the eye.

3. Cones are the color-detecting photoreceptors of the retina

Vision begins in the *retina*, the inner layer of tissue that lines the back of the eye (Fig. 2A). The development of retinal cells occurs in a temporal progression. All seven broad cell types (six neurons and one glia; Fig. 2B) arise from a pool of *retinal progenitor cells* (RPCs) and are born in a series of overlapping temporal birth windows (Fig. 2C). How these fates are driven through signaling, lineage, and stochastic mechanisms remains unclear. Though RPCs are capable of differentiating into any cell type, cell fates are restricted based on developmental timing (Cepko, 2014).

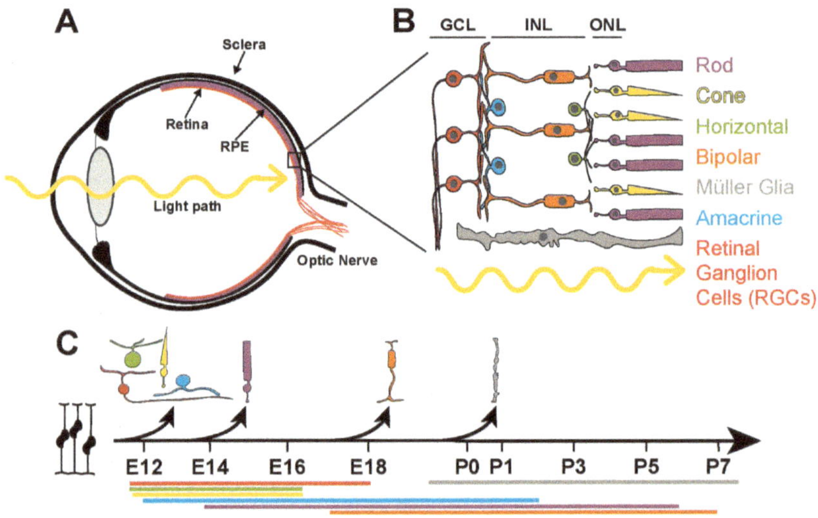

Fig. 2 The anatomy and development of the retina. (A) The eye is comprised of a series of layers. (B) Retinal layers include: the ganglion cell layer (GCL), the inner nuclear layer (INL), and the outer nuclear layer (ONL). Müller glia span the length of the retina. Light passes through the GCL and INL before hitting the ONL and the photoreceptors. (C) The cell types of the retina differentiate from retinal precursor cells (RPCs) in a specific temporal order. The black arrows represent the onset of differentiation for each cell type. The colored bars beneath the developmental timeline depict the full birth window during which these cell types are born, color-coded to the cell type. The days listed beneath are for mouse. While the precise days and rates of development differ among organisms, the birth order is conserved. *Adapted from Cepko, C. (2014). Intrinsically different retinal progenitor cells produce specific types of progeny.* Nature Reviews. Neuroscience, 15*(9), 615–627. https://doi.org/10.1038/nrn3767.*

The immature retina has two early layers: the outer neuroblastic layer (ONBL) and inner neuroblastic layer (INBL) which contain RPCs. As RPCs become fate-restricted and start to differentiate, the immature layers further stratify into the three layers of the fully developed retina: the outer nuclear layer (ONL), the inner nuclear layer (INL), and the ganglion cell layer (GCL) (Fig. 2B). The retinal ganglion cells inhabit the GCL, the nuclei of the interneurons (amacrine cells, horizontal cells, and bipolar cells) lie in the INL, and the *photoreceptors* (rods and cones) are found in the ONL. The Müller glia span the retina with nuclei in the inner nuclear layer (Fig. 2B) (Masland, 2012). Additionally, astrocytes migrate from the brain to the retina through the optic nerve, a structure comprised of the axons of retinal ganglion cells. Microglia enter the retina through the CNS and blood vessels (Langmann, 2007; Ramírez, Triviño, Ramírez, & Salazar, 1998; Rashid, Akhtar-Schaefer, & Langmann, 2019; Vecino, Rodriguez, Ruzafa, Pereiro, & Sharma, 2016).

There are two main types of light-sensitive photoreceptor cells: rods and cones. *Rods* are responsible for low-light vision whereas *cones* detect color and are our primary daytime light sensors (Kawamura & Tachibanaki, 2008; Nathans & Hogness, 1984; Nathans, Thomas, & Hogness, 1986; Yokoyama, 2000). *Opsin proteins* in the photoreceptors are sensitive to specific wavelengths of light. Opsin expression defines three subtypes of cones in the human retina. Blue/S cones express short wavelength-sensitive S-opsin, green/M cones express medium wavelength-sensitive M-opsin, and red/L cones express long wavelength-sensitive L-opsin (Deeb, 2005; Hunt, 2001; Kainz, Neitz, & Neitz, 1998; Nathans et al., 1989, 1986). Upon sensation of light by the opsin proteins, the signal is relayed through the interneurons to the ganglion cells, which connect to and transmit visual information to the brain via the optic nerve.

Progress has been made in elucidating mechanisms of retinal development, especially for the specification of photoreceptors. Over the past three decades, work in model organisms and human retinal organoids has uncovered a role for thyroid hormone receptor-driven signaling in regulating cell fate choices during cone subtype specification. As cone subtypes are different between species, the choices are species-dependent, but all show a common theme: a tightly regulated, temporally specific change in cone subtype specification occurs upon the initiation of thyroid hormone signaling. Here, we outline our understanding of how thyroid hormone signaling specifies cone subtypes across model organisms and how human retinal organoids have been used to study this choice in cone subtype fate during human development.

Summary: The retina is the light-detecting tissue of the eye. Cone and rod photoreceptors receive this visual information through their opsins. Different opsin wavelength sensitivity defines cone subtypes.

4. Thyroid hormone specifies cone subtypes in mouse, chicken, and fish

4.1 Mouse

4.1.1 Mouse opsins are expressed in opposing dorsal/ventral gradients

In mice, expression of two cone opsins (S/blue and M/green) defines three cone subtypes: S, M, and S/M co-expressing cones. The subtypes are arranged in a dorsal-ventral gradient with S cones being concentrated ventrally and M cones being found dorsally, and a gradient of co-expression throughout the central retina (Fig. 3A) (Applebury et al., 2000; Baden et al., 2013; Calderone & Jacobs, 1995; Eldred, Avelis, Johnston, & Roberts, 2020; Haverkamp et al., 2005; Rohlich, van Veen, & Szel, 1994; Szel et al., 1994). Cone precursors become postmitotic around embryonic day (E) 11.5. S-opsin mRNA expression is first detected 4 days later at E15.5 and S-opsin protein is detectable at birth (P0), ~5.5 days later. M-opsin expression occurs later, with mRNA first detected at P7 and protein detectable approximately at P14 (Applebury et al., 2007; Fei, 2003; Fujieda, Bremner, Mears, & Sasaki, 2009; Glaschke, Glosmann, & Peichl, 2010; Katoh et al., 2010; Szel, Rohlich, Mieziewska, Aguirre, & van Veen, 1993; Wikler, Szel, & Jacobsen, 1996). Thus, the gradient of cone subtypes begins to form around P0 and is fully established by P14 (Fig. 3I and K) (Roberts, Srinivas, Forrest, Morreale de Escobar, & Reh, 2006).

4.1.2 TRβ2—A key regulator of cone opsin expression

How are cone subtypes in the developing retina patterned? Pioneering work in mouse identified TRβ2 as an important regulator of S- and M-opsin expression. *Trβ2* mutant mice show a complete loss of M-opsin and an increase in S-opsin expression. Additionally, the gradient of S-opsin expression is lost (Fig. 3B) (Applebury et al., 2007; Eldred et al., 2020; Ng et al., 2001, 2011). These phenotypes show that TRβ2 is necessary for M-opsin expression and S-opsin repression.

While several thyroid hormone receptors are present in the retina, expression of *Trβ2* is spatiotemporally dynamic during development. *Trβ2* expression is detectable starting at embryonic day 13 (E13) and *Trβ2*-expressing

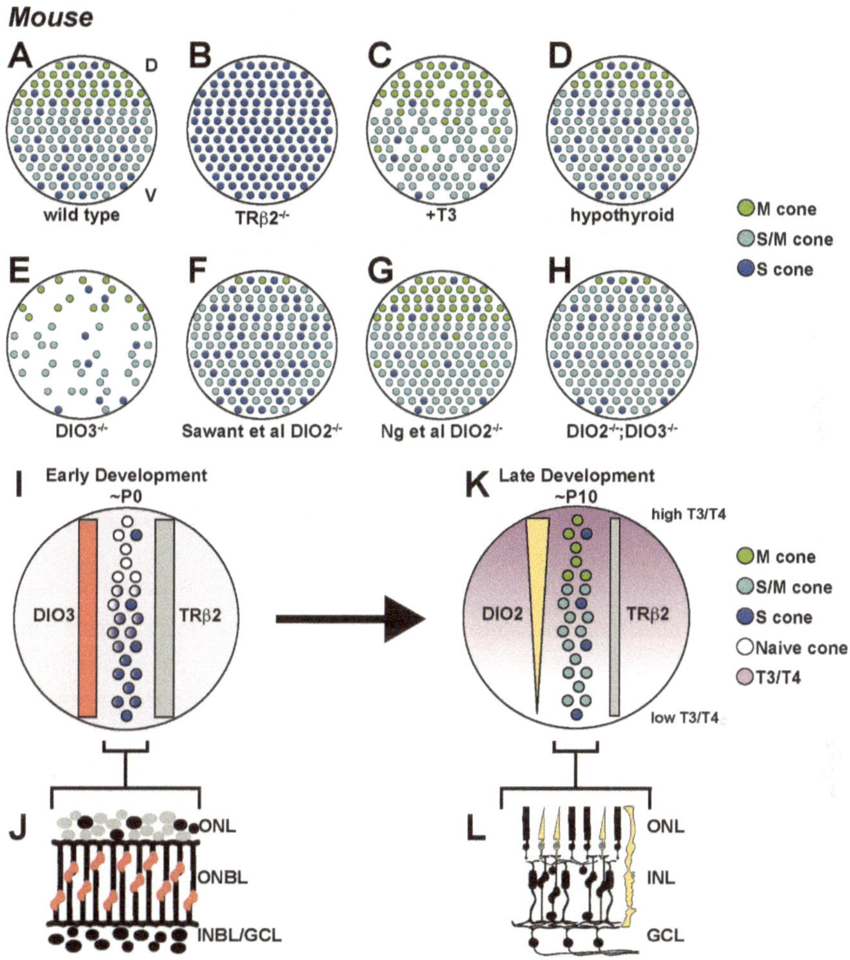

Fig. 3 Thyroid hormone signaling regulates M and S opsin expression in mouse cones. (A–H) Phenotypes of cone subtypes in the mouse retina upon changes in thyroid hormone signaling. (A) The wild type mouse retina has a dorsal/ventral gradient of cones. The dorsal region is primarily M cones, the ventral region is primarily S cones, and the transition region contains cones that co-express M and S. (B) *Trβ2* mutants display a loss of M cones and an expansion of S cones throughout the retina. (C) T3 treated retinas display a loss of cones and an expansion of M-opsin expression in the ventral region. (D) Hypothyroid retinas display an increase in S-opsin and a decrease in M-opsin expression. (E) *Dio3* mutant retinas display a cone death phenotype similar to retinas treated with T3. (F and G) *Dio2* mutant retinas from two studies display: (F) an increase in S-opsin and a decrease in M-opsin (Sawant et al., 2017) and (G) an increase in M-opsin and a decrease in S-opsin (Ng, Liu, St Germain, Hernandez, & Forrest, 2017). (H) Double knockout of DIO2 and DIO3 restores the cone gradient, with M only and S only cones restricted to narrower regions of the dorsal and ventral retina, respectively. (I–L) Expression

(Continued)

cells are uniformly distributed throughout the retina in the ONBL, where developing photoreceptors are found (Applebury et al., 2007). Expression peaks around E17 and then decreases sharply at birth (P0). After birth, expression persists at a low level until P10, and then decreases further in adulthood (Fig. 3I and J) (Applebury et al., 2007; Ng et al., 2009, 2001). Together, there are two temporal phases of *Trβ2* expression. The first phase includes the peak, which coincides with the end of cone precursor generation and suggests a role in the cone vs rod decision. The second phase of expression occurs during the shift from S- to M-opsin activation (Ng et al., 2009). In summary, *Trβ2* is expressed in developing cone precursors and terminal cones (Fig. 3K and L) (Roberts et al., 2006).

TRβ2 requires binding to DNA, thyroid hormone, and cofactors to regulate gene expression (Brent, 2012). TRβ2 mutations that block coactivator binding have no effect on S-opsin expression but decrease M-opsin levels across the retina. This result suggests an M-opsin-specific response to T3 when corepressors cannot be exchanged with coactivators. TRβ2 mutations causing either constitutive binding of a corepressor or disruption of T3 binding, resulted in an increase in S-opsin expression in the dorsal retina and a loss of the dorsal-ventral gradient through loss of M-opsin expression in the dorsal retina (Pessoa et al., 2008). Other TRβ2 mutations that prevented ligand binding but were otherwise functional, lost all M cones and induced S cones throughout the retina (Roberts et al., 2006). These results suggest that the S- vs M-opsin decision is controlled primarily by the corepressor's binding activity: first, when corepressor binds TRβ2 in the absence of T3, and later, upon T3 binding to TRβ2 and release of the corepressor suppression (Pessoa et al., 2008; Roberts et al., 2006).

Fig. 3—Cont'd patterns of thyroid hormone regulators through development inform the formation of the thyroid hormone gradient. (I and J) In early development, DIO3 and TRβ2 are expressed uniformly across the retina, with DIO3 in RPCs and TRβ2 in cones. Thyroid hormone levels are uniformly low. Here, it is assumed that the cone distribution does not change over time and that first, S-opsin is activated in S cones and future co-expressing cones in a gradient pattern, and later, M-opsin is activated in a reciprocal gradient pattern. (K and L) In late development, DIO2 is expressed in cones and Müller glia in a gradient pattern with high expression in the dorsal retina and low expression in the ventral retina. This results in a gradient of T3/T4 levels, with high T3/T4 in the dorsal region and low T3/T4 in the ventral region, promoting M-opsin expression in M cones and co-expressing cones. TRβ2 is expressed in cones but at lower levels to mediate thyroid hormone signaling.

4.1.3 Regulation of S- and M-opsin expression by thyroid hormone signaling

Expression and function of TRβ2 in cones suggest that endogenous thyroid hormone signaling regulates S- and M-opsin expression. The distribution of endogenous thyroid hormone in the mouse retina is dynamic throughout development. At the onset of S cone specification around P0, thyroid hormone is evenly distributed throughout the retina (Fig. 3I) (Roberts et al., 2006). This suggests that the gradient of S cones is generated through another mechanism that inhibits S-opsin dorsally and/or activates S-opsin ventrally. After birth, differences in thyroid hormone levels are observed between the dorsal and ventral regions of the retina. In the dorsal retina, thyroid hormone levels increase. In the ventral retina, thyroid hormone levels increase slightly until P3 and then sharply decline. By P10, a gradient of thyroid hormone is observed with high levels in the dorsal retina and low levels in the ventral retina (Fig. 3K). At this time, M cones are specified in the dorsal retina, likely due to the local increase in thyroid hormone signaling (Fig. 3A and K) (Roberts et al., 2006). This dorsal-ventral gradient of thyroid hormone levels is consistent with a role of thyroid hormone signaling in promoting M-opsin expression in the dorsal retina and restricting S-opsin expression to the ventral retina.

Mouse retinas that were treated with exogenous T3 (active thyroid hormone) to increase thyroid hormone signaling displayed fewer S-opsin-positive cones and more M-opsin-positive cones (Fig. 3C) (Roberts et al., 2006). In either genetically or pharmacologically induced *hypothyroid* mice (i.e., with low levels of thyroid hormone in their serum), the opposite phenotype was observed: mice showed an increase in S-opsin and a decrease in M-opsin with no change in the expression of *Trβ2* (Fig. 3D) (Glaschke et al., 2010, 2011; Lu et al., 2009). M cone specification experienced a delay of approximately 10 days, but was eventually able to form the gradient (Lu et al., 2009). Treatment with T3 in hypothyroid mice rescued M-opsin distribution and slightly decreased S-opsin in the ventral region (Lu et al., 2009). Thus, thyroid hormone signaling through TRβ2 induces M-opsin and represses S-opsin, and these processes are separable.

Summary: Mouse cone opsins are expressed in opposing gradients, with S-opsin high in the ventral retina and low in the dorsal retina, and M-opsin expressed in the opposite pattern. S-opsin expressed precedes M-opsin expression during development. Thyroid hormone signaling through TRβ2 in cones and cone precursors is necessary and sufficient for M-opsin expression and S-opsin repression.

4.1.4 Thyroid hormone signaling is regulated to ensure cone survival

In addition to its role in cone opsin specification, thyroid hormone also regulates cone survival. Very high levels of T3 signaling through TRβ2 cause a loss of cones through apoptosis, with S cones being particularly susceptible (Fig. 3C) (Lu et al., 2009; Ng et al., 2010). This role of thyroid hormone signaling in cone death is independent of its role in cone subtype specification (Ma et al., 2014). *Dio3*$^{-/-}$ mutant mice, which have increased thyroid hormone levels in their serum, display a drastic loss of all cone subtypes (Fig. 3E) (Ng et al., 2017). Increasing T3 levels in mice with retinal degeneration exacerbated cone degeneration. Opposite phenotypes were observed in mice with decreased T3. When mice with retinal degeneration were exposed to thyroid hormone antagonists, cone survival improved, with higher S-opsin and lower M-opsin expression (Ma et al., 2014). Taken together, very high T3 signaling results in cone death, predominantly in S cones, whereas low T3 signaling results in a decrease in M-opsin expression.

Summary: Excess thyroid hormone induces cone apoptosis. Early DIO3 activity likely protects against cell death.

4.1.5 Deiodinases pattern thyroid hormone signaling in the mouse retina

Because TRβ2 levels are relatively uniform throughout the retina (Ng et al., 2001; Roberts, Hendrickson, McGuire, & Reh, 2005; Roberts et al., 2006), a gradient of thyroid hormone signaling appears to be generated to regulate the dorsal-ventral patterning of cone subtypes. Local thyroid hormone levels in other settings are often regulated by deiodinases (Bianco et al., 2002; Galton, 2017; Gereben, Zeold, Dentice, Salvatore, & Bianco, 2008; Muller et al., 2014), suggesting that the spatiotemporal dynamics of DIO3 and DIO2 expression may also play a role in retinal cone patterning. DIO3, which degrades T3 and T4, is highly expressed from E13 through birth, and then maintains low levels of expression (Ng et al., 2017, 2010). DIO3 is expressed in a uniform pattern across the retina and throughout all retinal layers, likely in undifferentiated RPCs and immature differentiating cells (Fig. 3I and J) (Ng et al., 2010).

DIO2, which converts T4 to active T3, displays different expression dynamics. *Dio2* is very lowly expressed until about P7, at which time its expression increases rapidly (Ng et al., 2017; Sawant et al., 2017). *Dio2* expression is reciprocal to *Dio3*: as *Dio3* decreases, *Dio2* increases (Ng et al., 2017, 2010). DIO2 is expressed in cones (Sawant et al., 2017) and Müller glia (Ng et al., 2017) (Fig. 3L), which contact and exchange materials with photoreceptors (Reichenbach & Bringmann, 2013).

Importantly, this transition from *Dio3* expression to *Dio2* expression is coincident with the expression of M-opsin in the retina (Fig. 3K).

DIO2 and DIO3 are functionally important for cone subtype specification and survival. Loss of DIO3 causes an increase in thyroid hormone levels early in development, resulting in a dramatic loss of both S and M cones (Ng et al., 2017, 2010), consistent with excess T3 causing cell death (Fig. 3E). In the remaining cones, the opsin proteins are mislocalized to the cell body, pedicle, and axons, instead of the outer segments (Ng et al., 2010).

The role of DIO2 in cone subtype specification and survival remains unclear. Two groups observed different phenotypes in *Dio2* knockout mice. *Dio2* mutants have low levels of T3, and therefore would be expected to show a phenotype similar to hypothyroid mice, with increased S-opsin and decreased M-opsin expression. Sawant et al. reported an increase in S-opsin expression, an expanded region of co-expressing cones, and a decrease in cones expressing only M-opsin in *Dio2* knockout mice (Fig. 3F) (Sawant et al., 2017). In contrast, Ng et al. observed the opposite phenotype in *Dio2* knockout mice, reporting a small reduction in S-opsin expression and an increase in M-opsin expression (Fig. 3G), similar to the effect on cone opsin expression observed upon increasing T3 levels (Ng et al., 2017; Roberts et al., 2006). When double knockout *Dio2; Dio3* mice were generated, the opsin expression gradient was restored (Fig. 3H) (Ng et al., 2017). The conflicting and weak phenotypes observed in *Dio2* mutant mice suggest that other factors likely act with the deiodinases to determine cone opsin expression in the mouse retina. Moreover, the expression gradient observed in *Dio2, Dio3* double mutant mice suggests an additional, parallel mechanism acts redundantly with thyroid hormone signaling to regulate opsin expression.

Summary: *Dio3* is expressed in RPCs early in development. *Dio2* is expressed later in cones and Müller glia. The switch from *Dio3* to *Dio2* expression correlates with the switch from the initiation of S- to M-opsin expression. *Dio3* mutants display apoptosis of cones, whereas *Dio2* knockouts exhibit weak, conflicting phenotypes. Early DIO3 expression may protect against cone death from excess T3 and allow for S-opsin expression, whereas late expression of DIO2 may increase local thyroid hormone signaling to repress S-opsin and promote M-opsin expression.

4.2 Chicken

4.2.1 Thyroid hormone regulators are expressed in spatiotemporal waves in chicken

Chickens have four subtypes of single cones, which are sensitive to red, green, blue, or UV light, as well as two sets of double cones, which are sensitive to long wavelengths of light and likely have a role in motion detection (Bruhn & Cepko, 1996; Govardovskii & Zueva, 1977). Each cone subtype is regularly distributed but the relative distance is different for each cone, generating a semi-random pattern (Kram, Mantey, & Corbo, 2010). The chicken also has a high acuity region at the center of the retina with dense cone photoreceptor packing, analogous to the human fovea (Bruhn & Cepko, 1996; Morris, 1982; Viets, Eldred, & Johnston, 2016).

Studies in chickens revealed temporal patterns of thyroid hormone regulation and signaling. Several thyroid hormone receptors are expressed in the chicken retina. TRα is expressed in the progenitor layer. The two isoforms of TRβ, TRβ0 and TRβ2, have different expression patterns. TRβ0 is weakly expressed later in development in the INL, while TRβ2 is expressed early in the ONBL, likely in developing photoreceptors (Fig. 4B) (Sjoberg, Vennström, & Forrest, 1992; Trimarchi, Harpavat, Billings, & Cepko, 2008). Although TRα expression likely has a role in retinal development, expression of TRβ2 in the photoreceptor layer suggests that it is the primary thyroid hormone receptor responsible for cone subtype specification in chickens (Forrest, Sjoberg, & Vennstrom, 1990; Sjoberg et al., 1992).

In addition to the dynamic expression of the thyroid hormone receptors, the deiodinases are expressed in three distinct spatiotemporal waves corresponding to landmark events in retinal development. The first wave, during early development in Hamburger-Hamilton (HH) stages 20–26, corresponds with early retina neurogenesis. The wave moves from the central retina to the peripheral retina and involves *TRα*, *DIO3* and *TRβ2* (Fig. 4A and B). Expression of *DIO3* is initiated in the center and expands to the peripheral retina in precursor cells, slightly preceding the start of neurogenesis (Fig. 4A, red). *TRβ* expression coincides with the wave of neurogenesis, beginning after *DIO3* expression, and is limited to a subset of postmitotic cells. After a period of time where *TRβ* and *DIO3* are both expressed (Fig. 4A, purple), *DIO3* expression ceases, leaving *TRβ* expression in a subset of progenitor cells (Fig. 4A, blue) (Trimarchi et al., 2008).

The second wave at mid development in HH stages 29–33, includes photoreceptor differentiation and migration (Fig. 4C and D). This phase begins with high *DIO2* expression in a subset of developing photoreceptors in the ventral region and low expression in the dorsal region (Fig. 4C and D, yellow). A stripe expressing *DIO3* and lacking *DIO2* expression runs nasal-temporal (Fig. 4C). During the second wave, *DIO3* expression expands out from this central stripe to the periphery as *DIO2* expression is lost (Fig. 4E and F). *TRβ* is expressed in developing photoreceptors

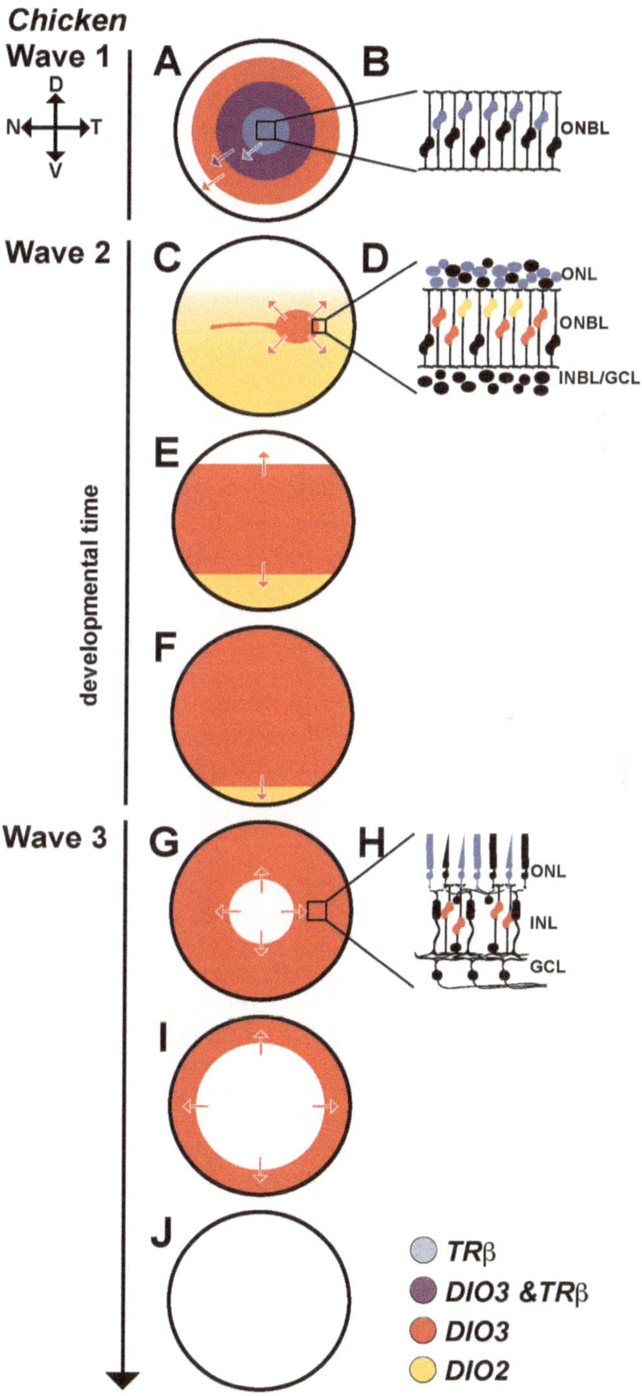

Fig. 4 See figure legend on next page.

during this stage (Fig. 4D, blue). *DIO2* is expressed in a subset of photoreceptors, likely increasing local thyroid hormone levels (Trimarchi et al., 2008).

The final phase of deiodinase expression occurs as Müller glia are generated during late development in HH stages 35–39 (Fig. 4G–I). *DIO3* expression is lost in a wave from the center to the periphery, corresponding with the differentiation of the last RPCs into Müller glia (Fig. 4G–J, red) (Trimarchi et al., 2008). As Müller glia are the final retinal cell type to differentiate, it is likely that the last wave of *DIO3* expression maintains the RPC population and that loss of *DIO3* expression allows for Müller glia specification. These findings in chicken were fundamental in establishing the wave-based differentiation dynamics of cone subtype specification during retinal development.

In addition to these spatiotemporal waves, roles for thyroid hormone signaling regulators have been characterized in chickens. Knockdown of MCT8, a thyroid hormone transporter, in the early chicken retina results in a decrease in L/M cones and an increase in UV/S cones, suggesting that thyroid hormone availability influences cone subtype specification (Vancamp, Bourgeois, Houbrechts, & Darras, 2019).

Summary: Expression of thyroid hormone regulators in the chicken retina follows a series of spatiotemporal waves, moving from the central retina to the periphery. Early in development, DIO3 in RPCs maintains a low thyroid hormone environment. RPCs differentiate into photoreceptor precursors, which express TRβ. A subset of TRβ positive photoreceptors

Fig. 4 Thyroid hormone regulators are expressed in waves in the developing chicken retina. (A–J) All three waves move from center to periphery during development. (A) In wave 1, the front of the wave is led by *DIO3* (red). Slightly delayed/behind *DIO3* expression is *TRβ* (blue). At the transition, there is a region of overlap (purple) between the *DIO3* and *TRβ* waves. (B) *TRβ* is expressed in a subset of progenitors, mostly in those closest to the developing photoreceptor layer. *DIO3* is expressed in all cells in the wave. (C–F) In wave 2, *DIO3* expression is limited to a strip running across the central retina (red), and *DIO2* is expressed in the ventral retina and lowly expressed in the dorsal retina (yellow). During wave 2, this strip of *DIO3* expands out toward the periphery as *DIO2* expression becomes limited to the ventral periphery. (D) *TRβ* is expressed in developing photoreceptors (blue). *DIO2* is expressed in a subset of cells in the upper ONBL (yellow). *DIO3* is expressed in RPCs (red). (G, I and J) In wave 3, *DIO3* expression (red) is lost from the center to the periphery, corresponding with the birth of Müller glia. (H) *TRβ* (blue) remains active in a subset of photoreceptors. *DIO3* expression (red) is limited to the remaining RPC population. *Adapted from Trimarchi, J. M., Harpavat, S., Billings, N. A., & Cepko, C. L. (2008). Thyroid hormone components are expressed in three sequential waves during development of the chick retina.* BMC Developmental Biology, 8, *101. https://doi.org/10.1186/1471-213X-8-101.*

also express DIO2. DIO2 expression likely generates a locally high thyroid hormone signaling environment, which may have a role in photoreceptor patterning. As DIO2 expression decreases, it is replaced by DIO3 expression, which creates a low thyroid hormone environment to maintain the remaining RPCs. In the final wave, DIO3 expression is lost, allowing Müller glia to differentiate.

4.3 Zebrafish

4.3.1 Thyroid hormone signaling controls cone fates in zebrafish

Zebrafish have four classes of cones: LWS, RH2, SWS2, and SWS1, which are homologous to red-, green-, blue-, and UV-sensitive cones of other species. There are four subtypes of RH2/green cones and two subtypes of SWS2/blue cones. LWS/red cones and RH2/green cones fuse into a cone pair (LWS/RH2 pair) (Chinen, Hamaoka, Yamada, & Kawamura, 2003; Ng et al., 2009; Raymond, Barthel, Rounsifer, Sullivan, & Knight, 1993; Vihtelic, Doro, & Hyde, 1999). The cones are arranged in a regular pattern, with two types of rows: the first is a row of the LWS/RH2 cone pair, and the second is an alternating pattern of blue/SWS2 and UV/SWS1 cones. These two rows run in parallel and alternate across the retina (Branchek & Bremiller, 1984; Raymond, Barthel, & Curran, 1995).

The differentiation of cones occurs in a "fan gradient" in which cells exit mitosis radially, from the ventral to the nasal to the dorsal and finally, to the temporal region of the retina (Easter & Malicki, 2002). Differentiation occurs asynchronously among cell types, with photoreceptors differentiating last (Easter & Malicki, 2002; Hu & Easter, 1999; Schmitt & Dowling, 1996). Cones differentiate into their subtypes in a temporal order: LWS/red cones first, RH2/green cones, SWS1/UV cones, and finally SWS2/S cones (Stenkamp, Barthel, & Raymond, 1997).

In zebrafish, *trβ2* is expressed in LWS/red cones and LWS/red cone precursors. *trβ2* expression is necessary and sufficient for LWS/red cone specification and repression of the SWS1/UV cone fate (Mackin et al., 2019; Suzuki et al., 2013). LWS/red and SWS1/UV cones are similar to human L/M and S cones, respectively (Ebrey & Koutalos, 2001; Vihtelic et al., 1999). Thyroid hormone signaling through *trβ2* also plays a role in promoting the developmental switch in opsin expression from *lws2* to *lws1*, two tandemly arranged long wavelength sensitive genes (Mackin et al., 2019).

Summary: In zebrafish, thyroid hormone signaling through Trβ2 regulates two choices in cone subtype specification. Thyroid hormone signaling specifies LWS/red cone subtypes while suppressing SWS1/UV cone fate, and promotes the developmental switch from *lws2* to *lws1* expression.

4.4 Salmonid

4.4.1 Dramatic remodeling of the cone mosaic occurs during salmonid development

Cone subtype fates in salmonids, including salmon and trout, are regulated by changes in thyroid hormone signaling that accompany several major life stage transitions. As salmonids progress through development, the retina is significantly modified during the alevin, freshwater-dwelling parr, saltwater-dwelling smolt, and adult stages (Fig. 5G). In particular, changes occur during *smoltification*, the transition from parr to smolt (Barron, 1986), and during the final maturation to adulthood (Beaudet, Novales Flamarique, & Hawryshyn, 1997; Novales Flamarique, 2000). Both of these events coincide with marked changes in the thyroid gland and thyroid hormone levels (Hoar, 1988; Specker, Eales, Tagawa, & Tyler, 2000). These retinal alterations involve rearrangements of cone positions and changes in opsin expression. Depending on retinal region and stage of development, salmonid retinas exhibit different permutations of cone mosaics that can be classified into three major patterns including: (1) a square mosaic, in which each unit contains an L/M double cone per side with a single cone in the center (Fig. 5A and B); (2) a row mosaic, in which the L/M double cones are turned 90 degrees relative to the square mosaic (Fig. 5C and D); and (3) a mix of square and row mosaics in a transition zone (Fig. 5E and F) (Beaudet et al., 1997; Novales Flamarique, 2000). Classification of these patterns is further subdivided based on the presence or absence of UV cones at the four corners of each unit, giving rise to six total cone mosaic patterns (Fig. 5A–F).

The cone mosaics of salmonids are highly dynamic during development. Alevin retinas have square mosaics with UV corner cones throughout much of the retina (Fig. 5A and H) (Novales Flamarique, 2000; Novales Flamarique, Sayed Ahmed, Cheng, Molday, & Devlin, 2019). This pattern deviates in the ventral-peripheral regions, which contain square mosaics without UV corner cones (Fig. 5B and H), and in the centrotemporal and nasal peripheral regions, which contain row mosaics (Fig. 5C and H). Upon smoltification, the UV-positive mosaics become restricted to the central retina (Fig. 5A and I), and row mosaics become more regularly distributed in the temporal region (Fig. 5D and I). At adulthood, retinas have row mosaics in the dorsal region, a mix of row and square mosaics in the transition zone between regions, and square mosaics in the ventral region (Fig. 5B, D, F and J). Additionally, corner UV cones reappear in the central region (Fig. 5C, E and J) (Novales Flamarique, 2000).

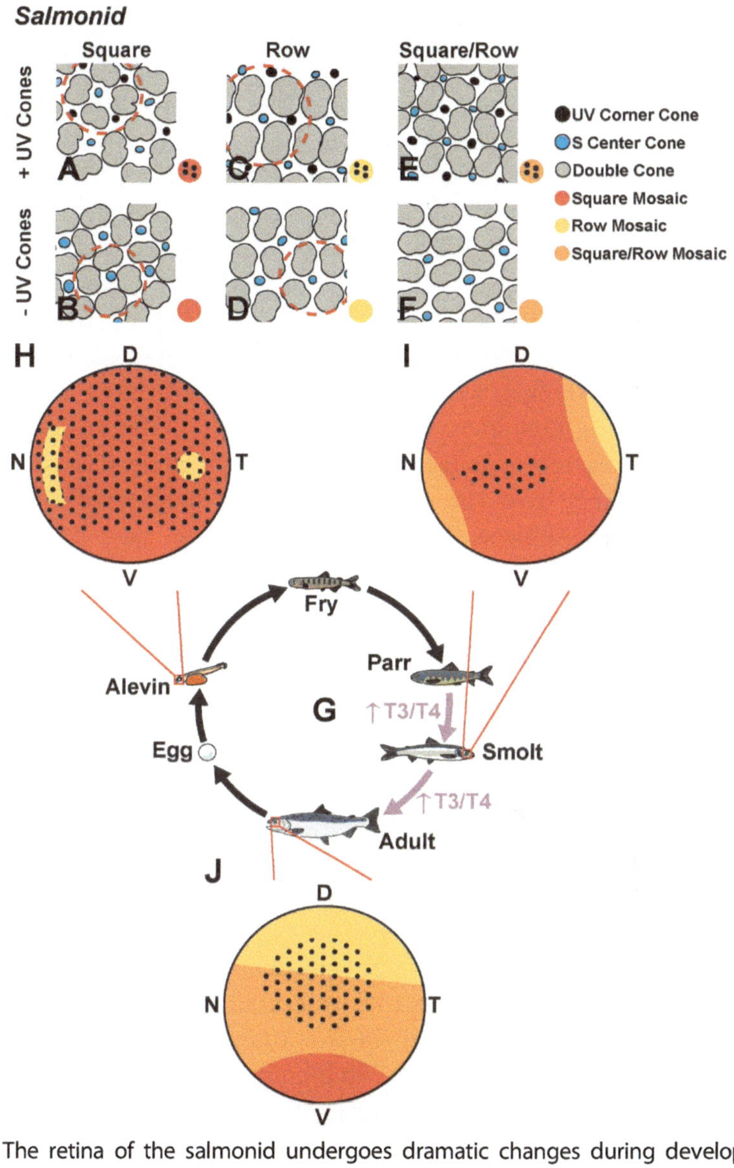

Fig. 5 The retina of the salmonid undergoes dramatic changes during development. (A–F) Six types of retinal mosaics in the salmonid. Red dashed circles outline the unit of the mosaic pattern. (G) The lifecycle of the salmonid. (H) The alevin retina is almost entirely square mosaic with UV corner cones except for the ventral region. (I) The smolt retina displays an expansion of the row mosaic in the dorso-temporal region and only has UV corner cones in the central retina. (J) At the adult stage, there is a major expansion of the row mosaic from the dorsal into the ventral region, and the UV corner cones are limited to the central retina. *Adapted from Beaudet, L., Novales Flamarique, I., & Hawryshyn, C. W. (1997). Cone photoreceptor topography in the retina of sexually mature Pacific salmonid fishes.* The Journal of Comparative Neurology, 383*(1), 49–59; Hawryshyn, C. W., Martens, G., Allison, W. T., & Anholt, B. R. (2003). Regeneration of ultraviolet-sensitive cones in the retinal cone mosaic of thyroxin-challenged post-juvenile rainbow trout (Oncorhynchus mykiss).* The Journal of Experimental Biology, 206*(Pt. 15), 2665–2673. https://doi.org/10.1242/jeb.00470; Novales Flamarique, I. (2000). The ontogeny of ultraviolet sensitivity, cone disappearance and regeneration in the sockeye salmon Oncorhynchus nerka.* The Journal of Experimental Biology, 203, *1161–1172.*

While several remodeling events occur in the salmonid retina, the patterning of UV corner cones has been related to changes in thyroid hormone levels during development. Consistent with the changes in UV cone constituency during development, UV sensitivity decreases during smoltification and recovers during adulthood. The loss of UV cones during smoltification was originally thought to be due to thyroid hormone-dependent apoptosis (Allison et al., 2003; Beaudet, Browman, & Hawryshyn, 1993; Bowmaker & Kunz, 1987; Flamarique & Hawryshyn, 1996; Hawryshyn, Arnold, Chaisson, & Martin, 1989; Hawryshyn, Martens, Allison, & Anholt, 2003; Kunz, 1987; Kunz, Wildenburg, Goodrich, & Callaghan, 1994; Loew & Wahl, 1991; Lyall, 1957). More recent studies identified two processes that control this change in patterning: thyroid hormone-independent apoptosis of corner cones and thyroid hormone-dependent switching of opsin expression (Cheng, Gan, & Flamarique, 2009).

4.4.2 Thyroid hormone signaling controls opsin switching in single cones

During development, thyroid hormone levels and thyroid hormone receptor expression increase at the alevin stage through smoltification, coinciding with the conversion of single cones from UV to blue fate (Cheng et al., 2009; Greenblatt, Brown, Lee, Dauder, & Bern, 1989; Jones, Rogers, Kille, & Sweeney, 2002). In the single cones, thyroid hormone signaling represses UV opsin and induces blue opsin (Cheng et al., 2009; Gan & Novales Flamarique, 2010).

Thyroid hormone signaling in the retina likely occurs through activation of *trα*, which is expressed in a dynamic pattern. Prior to opsin switching and smoltification, *trα* is expressed in all single cones across the retina. As thyroid hormone levels increase and opsin switching begins in the ventral region, a gradient of *trα* is established, with high expression in the ventral region and low expression in the dorsal region. As the wave of opsin switching progresses into the dorsal region, the gradient reverses, with high *trα* in the dorsal region and low *trα* in the ventral region. Once opsin switching is completed after smoltification, *trα* expression is turned off in all single cones (Gan & Novales Flamarique, 2010; Raine & Hawryshyn, 2009). These observations suggest that Trα mediates the switch from UV to blue opsin.

Exogenous thyroid hormone accelerates the onset of opsin expression and the expansion of cone specification in development. Additionally, after differentiation, thyroid hormone promotes switching of UV cones to blue single cones and induces *trα* and *trβ* expression. These are likely two separate processes, with the initial opsin expression occurring through an unknown thyroid hormone receptor (possibly Trβ) or another factor, and the opsin switching being mediated by Trα signaling (Gan & Novales Flamarique, 2010; Raine & Hawryshyn, 2009).

Summary: Thyroid hormone signaling drives opsin expression switching during salmonid smoltification. Thyroid hormone signaling, likely mediated by Trα, induces blue opsin expression and represses UV opsin expression in single cones.

5. Human retinal organoids are a model to study how thyroid hormone signaling specifies human cone subtypes

Extensive work in model systems laid the groundwork for understanding how thyroid hormone signaling specifies cone subtypes during retinal development. Studies in mouse identified how differential, local regulation of thyroid hormone signaling through TRβ2 controls cone opsin expression. Work in chicken revealed how regulators of thyroid hormone signaling are expressed in waves during retinal development. Experiments in fish revealed new roles for thyroid hormone signaling in cone development, including switching of opsin expression.

Thyroid hormone signaling plays a conserved role in determining opsin expression in cones despite dramatic differences in cone subtype patterning, roles of thyroid hormone receptor isoforms, developmental timing, and cell type-specific expression of regulatory proteins. Considering these differences, it is important to understand how thyroid hormone signaling specifies cone subtypes in the human retina. Our understanding of human cone subtype patterning and development largely comes from studies of human retinal tissue and clinical connections between color vision and changes in thyroid hormone level status. Recent advances in retinal organoid technology have enabled the visualization and manipulation of cone subtype specification to gain mechanistic insights. Here, we discuss our current understanding of how thyroid hormone signaling specifies cone subtypes in human retinas and retinal organoids and explore the challenges that need to be addressed in future studies.

5.1 Human cone subtypes are patterned in a series of waves

Humans have three subtypes of cones, each expressing either L-opsin (red-sensitive), M-opsin (green-sensitive), or S-opsin (blue-sensitive) (Nathans et al., 1986). L- and M-opsin proteins are extremely similar and are currently indistinguishable by immunohistochemistry, making them difficult to study independently. As a result, these subtypes are often grouped together during analysis as "L/M" cones (Gowdy & Cicerone, 1998; Hofer, Carroll, Neitz, Neitz, & Williams, 2005; Li & Roorda, 2007; Otake, Gowdy, & Cicerone, 2000; Roorda & David, 1999; Roorda, Metha, Lennie, & Williams, 2001; Rossi et al., 2011).

In the adult human retina, cone density is generally higher in the central retina and decreases toward the periphery, with a slight increase again at the outer rim of the retina. The central retina contains a structure called the macula, which is responsible for high acuity vision. The macula can be broken down into concentric sub-regions that are defined in part by photoreceptor constituency, photoreceptor density, and a physical pit structure. Moving from the center to the periphery, the central foveola contains L/M cones exclusively. Immediately outside of the foveola, the fovea contains S cones along with L/M cones. The outer edge of the macula is denoted by high cone density with sparse rods. In the peripheral retina, rods are densely packed, whereas S cones and L/M cones are distributed at lower density, with L/M cones in a greater proportion than S cones. The far peripheral outer rim of the retina is again cone-rich (Cornish, Hendrickson, & Provis, 2004; Curcio et al., 1991; Curcio, Sloan, Kalina, & Hendrickson, 1990; Provis, Dubis, Maddess, & Carroll, 2013; Xiao & Hendrickson, 2000).

In the developing human retina, cones are generated and specified in a series of waves starting from the foveola and moving out to the periphery (Fig. 6). Cones are first specified during week 11 of fetal development as S cones are generated in the central retina. S cones are then specified in a wave that spreads out to the periphery. Starting around week 14, L/M cones begin to develop at the foveola and continue out to the periphery, following the initial wave of S cone specification. At the front between developing S and L/M cones, there are cones that co-express S and L/M-opsin, and a small percentage of these co-expressing cones persist through adulthood (Cornish et al., 2004; Xiao & Hendrickson, 2000).

Summary: Humans have three types of cones that are patterned in two waves: first, a wave of S cone specification, followed by a wave of L/M cone specification.

5.2 Clinical thyroid dysfunction correlates with color vision defects

Thyroid hormone dysfunction has been linked to changes in the proportions of cone subtypes and color perception. In one case, an infant with severe thyroid hormone resistance was found to have increased S cone function

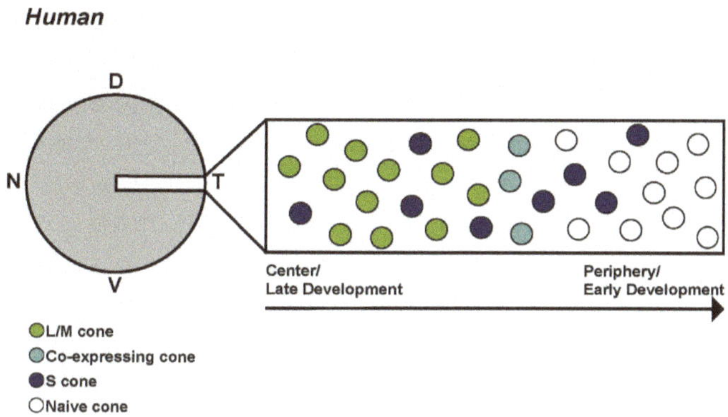

Fig. 6 Human cones are specified in waves. A section of the developing fetal retina representing both the progression of developmental time and distance from central to peripheral retina. Note that there is significant retinal growth over time which is not described in this depiction (Xiao & Hendrickson, 2000).

and decreased L/M cone function. Sequencing of the THRβ locus revealed compound heterozygous mutations in the ligand binding domain, suggesting a disruption of T3 binding affinity (Weiss, Kelly, Bisset, & Deeb, 2012). This was originally thought to be specific to THRβ2, but re-evaluation of the mutations revealed that the mutations affected both THRβ1 and THRβ2 isoforms (Eldred et al., 2018). These findings suggested that thyroid hormone signaling induces L and M cones and suppresses S cone specification, providing a link between thyroid hormone signaling and cone subtype fate in humans.

The first step of local thyroid hormone signaling is entry into the cell through transporters (Fig. 1F and G). The thyroid hormone membrane transporter MCT8 plays a primary role in the retina. A mutation in the *SLC16A2* gene, which encodes MCT8, causes Allan-Herndon-Dudley syndrome. This syndrome is characterized by high serum T3, low serum T4, and slightly upregulated TSH. Individuals with Allan-Herndon-Dudley syndrome have several visual impairments, although it is currently unclear if these are directly due to retinal dysfunction or another non-retinal neurological defect (Schwartz & Stevenson, 2007).

While connections between disruption of thyroid hormone signaling during development and defects in color vision are just beginning to be identified, many adult humans, from 1% to 10% depending on the population, have some degree of thyroid disorder (Canaris, Manowitz, Mayor, & Ridgway, 2000; Sawin, Castelli, Hershman, McNamara, & Bacharach, 1985; Tunbridge et al., 1977). Several studies have been conducted to identify color vision defects related to these disorders. Adults with *hypothyroidism* (i.e., low thyroid hormone levels) have impaired color contrast sensitivity compared to *euthyroid* (i.e., normal thyroid hormone levels) adults. After L-thyroxine/T4 treatment returned some hypothyroid patients to euthyroid status, the color sensitivities of these patients were significantly improved. However, comparison to the euthyroid control group revealed that, while the red/green sensitivity was restored, the blue/yellow sensitivity had improved but had not been completely restored (Cakir et al., 2015; Racheva et al., 2020). These findings hint at a plasticity in opsin expression and cone function upon changes in thyroid hormone levels even after specification.

Summary: Dysfunctions in the thyroid gland and alterations in serum thyroid hormone levels have been linked to color vision defects in humans.

5.3 Human retinal organoids are a powerful model system to study human retinal development

Studies of humans and model organisms connected thyroid hormone signaling to cone subtype specification during retinal development. Yet, testing human retinal development had been hindered by the lack of an experimentally tractable human model system. Recently, human retinal organoids have emerged as a system to observe development with high temporal resolution and test mechanisms through genetic and pharmacological approaches (Nakano et al., 2012).

Human organoids are three-dimensional tissues, derived from embryonic or induced pluripotent stem cells, that can model developmental events in vitro (Fig. 7). They are small and free-floating in media which enables experimental study and manipulation. A variety of protocols have been developed for generating human retinal organoids. All approaches involve the aggregation of human stem cells (Fig. 7A) to generate a 3D structure. Aggregation is facilitated by the addition of Matrigel, which is composed of extracellular matrix components, during the first few days of differentiation (Fig. 7B). These aggregates are subjected to a series of small molecules that replicate signals that promote retinal development, by suppressing endodermal fate and promoting neural development. In the initial days of differentiation, neuronal vesicles form on the organoid (Fig. 7C), marking the start of eye field specification. Next, the neural tissue is pushed to neural retina fate, and the organoids develop optic vesicles (i.e., early retinal tissue). Successful optic vesicles are excised and cultured individually. Finally, differentiation of retinal cells is promoted, and the organoids are cultured

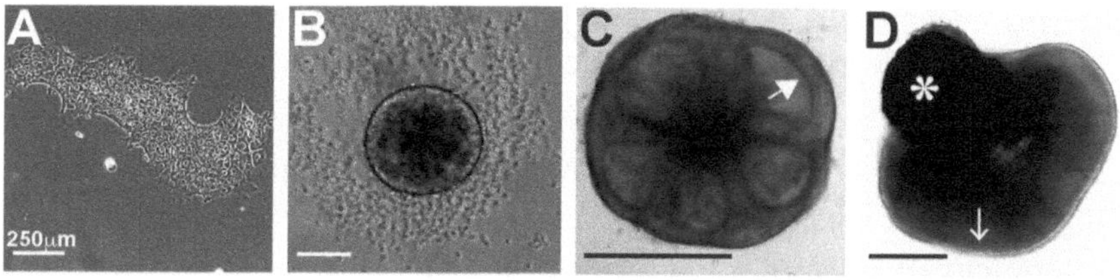

Fig. 7 Differentiation of stem cells to human retinal organoids. (A) Stem cells colonies are cultured (iPSCs or ESCs). (B) Day 3 of differentiation. Aggregation of stem cells is almost complete. (C) Day 7 of differentiation. Neuronal vesicles, indicated with a white arrowhead, begin to differentiate. (D) Day 200 of differentiation. The lighter outer edge of the organoid, indicated by the white arrow, is retinal tissue. Photoreceptors are present on the outer edge of the organoid. RPE is also present, marked by a white asterisk.

long-term (Fig. 7D) (Capowski et al., 2016; Chichagova et al., 2019; Eiraku et al., 2011; Eldred et al., 2018; Nakano et al., 2012; Wahlin et al., 2017; Zhong et al., 2014).

Organoids can be cultured for several hundred days. They appear to develop on approximately the same time scale as human retinas. Importantly, they faithfully recapitulate the generation of the neuronal types of the retina, including photoreceptors with proper gene expression, morphologies, and functions (Chichagova et al., 2019; Eldred et al., 2018; Hoshino et al., 2017; Kaewkhaw et al., 2015; Nakano et al., 2012; Phillips et al., 2018; Sridhar et al., 2020; Wahlin et al., 2017; Zhong et al., 2014). Some protocols promote certain cell fates, yielding photoreceptor-rich organoids (Eldred et al., 2018; Wahlin et al., 2017). Molecular mechanisms have been interrogated through CRISPR-Cas mutagenesis of stem cells and differentiation of mutant organoids (Eldred et al., 2018; Wahlin et al., 2017), addition of pharmacological agonists and antagonists (Eldred et al., 2018), and viral infection for sparse labeling and lineage tracing (Garita-Hernandez et al., 2018, 2020; Quinn et al., 2019). Additionally, patient-derived stem cells can be differentiated into organoids to model diseases (Deng et al., 2018; Foltz & Clegg, 2019; Gao et al., 2020; Guo et al., 2019; Li et al., 2019).

Human retinal organoids have great potential to address questions about human biology. Our work identifying the role of thyroid hormone signaling in human cone subtype specification demonstrated how organoids could be used as a tractable model for studying mechanisms of human development.

Summary: Human retinal organoids are a powerful system to study the molecular mechanisms controlling the development of the human retina.

5.4 Thyroid hormone signaling regulates the choice between S and L/M cone fates in humans

The Johnston lab utilized human retinal organoid technology to investigate the mechanisms controlling S vs L/M cone subtype specification (Eldred et al., 2018). Our studies started by observing the timing of S and L/M cone development in organoids. Tracking the generation of cone subtypes in organoids over time showed that S cones are generated between days 150 and 170 of differentiation. L/M cones are specified later, between days 170 and 300. The temporality of cone specification in organoids mirrors that in the developing human retina, with S cones generated before L/M cones. The approximately 20-day delay between the onset of S and L/M cone specification is similar to the timing observed in fetal development (Hoshino et al., 2017).

The similarities between cones in human retinas and organoids extend beyond timing and opsin expression. Cones, like other neurons, have axons and cell bodies. Additionally, they have functional structures extending opposite the axons. The outer segments are farthest from the cell body and are rich in the opsin proteins that detect light. Between the outer segments and the cell body are inner segments, which are responsible for carrying out the downstream phototransduction cascade to transmit the signal to the interneurons (Richardson, 1969; Wassle, 2004). S and L/M cones are morphologically distinct and can be identified by their unique shapes and sizes of their outer and inner segments. S cones have short outer segments and thin inner segments, while L/M cones have long outer segments and wide inner segments (Curcio et al., 1991). These morphological features are recapitulated in organoids at day 200 of differentiation, showing that organoids grow cones that resemble those in humans (Eldred et al., 2018).

Across the human retina, cone subtypes are found in a ratio of approximately 13% S cones to 87% L/M cones. Human retinal organoids had a similar ratio, with 29% S cones and 71% L/M cones. Therefore, the proportion of cone subtypes is similar in human retinas and organoids, with more L/M cones than S cones (Eldred et al., 2018).

Using human retinal organoids as a model for human retinal development, we next evaluated the role of thyroid hormone signaling in human cone subtype specification. Organoids treated with T3 were comprised of almost exclusively L/M cones and few S cones, suggesting that thyroid hormone is sufficient to promote L/M cone fate and suppress S cone fate.

THRβ mutant retinal organoids displayed a complete loss of L/M cones, generating an S cone-only organoid, showing that THRβ is necessary for the L/M cone fate. Interestingly, knocking out THRβ2 alone showed no changes in cone ratios, suggesting functional differences between human and mouse isoforms of the receptor. Importantly, THRβ knockouts treated with T3 still only generated S cones, showing that T3 signals through THRβ to promote L/M fate and suppress S cone fate in human cone subtype specification (Eldred et al., 2018).

As retinal organoids grow in the absence of other extrinsic signals, regulation of thyroid hormone signaling must be intrinsic to the retina. *DIO3* expression peaks early in development, and then decreases. Inversely, *DIO2* expression is low early in development and peaks later, coinciding with the switch from S to L/M cone subtype specification. This developmental progression resembles that observed in chick and mouse. Other thyroid hormone signaling regulators show varying expression patterns. *SLC16A2* (MCT8), the thyroid hormone membrane transporter, as well as the thyroid hormone receptor cofactors *NCOR2* and *MED1*, are expressed at constant levels throughout development. *THRβ* and some thyroid hormone transport genes are lowly expressed early in development but increase over time (Eldred et al., 2018). While transporters and coregulators may still play an important role in thyroid hormone signaling regulation, the temporality of deiodinase expression with *DIO3* expression peaking early and *DIO2* expression peaking late, correlating with the time that L/M cones begin to be specified, suggests that the deiodinases are the primary regulators of thyroid hormone levels in the developing human retina.

Taken together, these data support a model in which DIO3 degrades T4 and T3 to keep thyroid hormone levels low early in development, creating an environment permissive for specification of S cones (Fig. 8A and B). Later in development, DIO3 expression decreases and DIO2 expression increases, promoting the generation of T3 that signals through THRβ to specify L/M cones (Fig. 8C and D) (Eldred et al., 2018).

Summary: Human retinal organoids recapitulate human cone development temporally and morphologically. In humans, S cones are specified prior to M cones. T3-driven THRβ signaling promotes L/M cone specification and suppression of S cone fate. *DIO3* is highly expressed early to restrict thyroid hormone signaling and allow for specification of S cones, while *DIO2* is highly expressed later in development to promote thyroid hormone signaling and L/M cone specification.

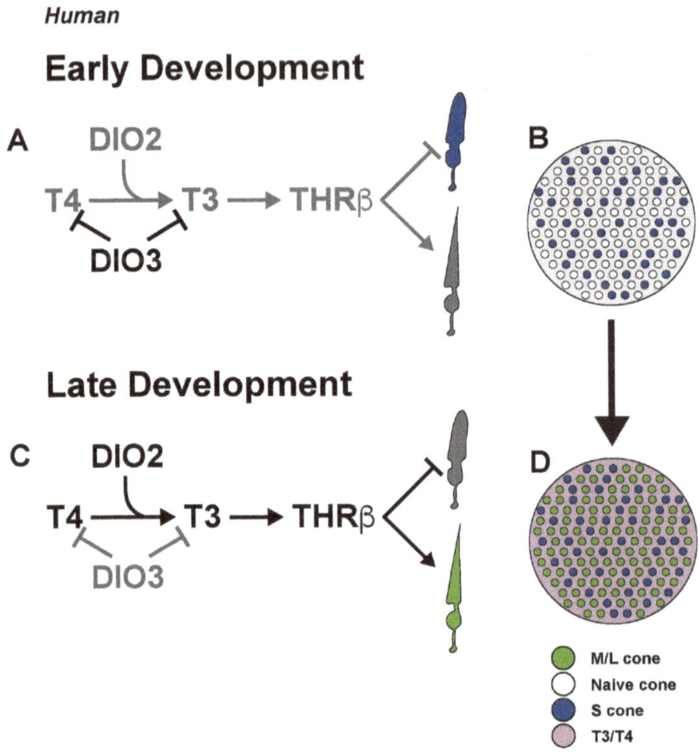

Fig. 8 Human cone subtype specification is controlled by thyroid hormone signaling. (A) Early in retinal development, DIO3 is expressed and thyroid hormones are degraded, allowing for S cone specification. (B) The early retina is populated by S cones and unspecified naïve cones in a low-thyroid hormone environment. (C) Late in retinal development, DIO2 is expressed to promote active thyroid hormone signaling. T3 signals through THRβ to inhibit S cone fate and promote L/M cone specification. (D) The late retina is populated by both S cones and L/M cones in a high-thyroid hormone environment. *Adapted from Eldred, K. C., Hadyniak, S. E., Hussey, K. A., Brenerman, B., Zhang, P. W., Chamling, X., et al. (2018). Thyroid hormone signaling specifies cone subtypes in human retinal organoids.* Science, *362(6411), eaau6348. https://doi.org/10.1126/science.aau6348.*

5.5 Challenges in understanding human cone subtype specification

Human retinal organoids provide a powerful system to study human retinal development. As a genetically and pharmacologically tractable system, human retinal organoids are a model to address new challenges and open questions regarding cone subtype specification. How is the temporal switch between generating S and L/M cone fates regulated? Thyroid hormone signaling through THRβ is required for this process, but how does the retina temporally control thyroid hormone signaling? If the deiodinases are the primary factors controlling the timing, which cell types express the deiodinases and other important regulators? Additionally, is the dynamic expression of the deiodinases during development sufficient to drive this cone fate change? Expression and functional studies in human retinal organoids have the potential to answer these questions.

Experiments on retinal organoids in conjunction with clinical studies would impact our understanding of how differences in thyroid hormone levels affect human vision. If human cones are plastic during development as suggested by observations from fetal retinas (Cornish et al., 2004; Xiao & Hendrickson, 2000), how are S cones maintained after DIO2 begins to be expressed? There may be a distinct mechanism that prevents opsin switching in S cones late in development. Alternatively, plasticity may persist into adulthood, and hyper- or hypothyroidism may cause changes in cone subtype fates, with implications for color vision therapeutics.

Human retinal organoids provide a potential platform to not only study, but also treat, retinal diseases. Current methods involve differentiating photoreceptors from hiPSC or hESCs and transplanting them into the retinas of mouse models of retinal degeneration, which are able to restore some function (Assawachananont et al., 2014; Lamba, Gust, & Reh, 2009; Lamba, Karl, & Reh, 2009; Lamba, Karl, Ware, & Reh, 2006; Lamba et al., 2010; Reh, 2016; Singh et al., 2013); however, this technique requires proper migration and connectivity of the transplanted photoreceptors. Organoids may provide large numbers of photoreceptors and interneurons to allow for higher efficiency of transplantation (Llonch, Carido, & Ader, 2018). Organoid technology provides a tool to address questions in the context of human retinal development, and the possibility of subretinal transplantation of organoids bridges the gap between developmental studies and therapeutic use.

Summary: Human retinal organoids provide a system to address the many remaining questions about the mechanisms of cone subtype specification and the regulation of thyroid hormone signaling during human retinal development.

Acknowledgment

This work was funded by R01EY030872.

References

Alkemade, A., Vuijst, C. L., Unmehopa, U. A., Bakker, O., Vennstrom, B., Wiersinga, W. M., et al. (2005). Thyroid hormone receptor expression in the human hypothalamus and anterior pituitary. *The Journal of Clinical Endocrinology and Metabolism*, *90*(2), 904–912. https://doi.org/10.1210/jc.2004-0474.

Allison, W. T., Dann, S. G., Helvik, J. V., Bradley, C., Moyer, H. D., & Hawryshyn, C. W. (2003). Ontogeny of ultraviolet-sensitive cones in the retina of rainbow trout (Oncorhynchus mykiss). *The Journal of Comparative Neurology*, *461*(3), 294–306. https://doi.org/10.1002/cne.10682.

Applebury, M. L., Antoch, M. P., Baxter, L. C., Chun, L. L., Falk, J. D., Farhangfar, F., et al. (2000). The murine cone photoreceptor: A single cone type expresses both S and M opsins with retinal spatial patterning. *Neuron*, *27*(3), 513–523.

Applebury, M. L., Farhangfar, F., Glosmann, M., Hashimoto, K., Kage, K., Robbins, J. T., et al. (2007). Transient expression of thyroid hormone nuclear receptor TRbeta2 sets S opsin patterning during cone photoreceptor genesis. *Developmental Dynamics*, *236*(5), 1203–1212. https://doi.org/10.1002/dvdy.21155.

Assawachananont, J., Mandai, M., Okamoto, S., Yamada, C., Eiraku, M., Yonemura, S., et al. (2014). Transplantation of embryonic and induced pluripotent stem cell-derived 3D retinal sheets into retinal degenerative mice. *Stem Cell Reports*, *2*(5), 662–674. https://doi.org/10.1016/j.stemcr.2014.03.011.

Astapova, I. (2016). Role of co-regulators in metabolic and transcriptional actions of thyroid hormone. *Journal of Molecular Endocrinology*, *56*(3), 73–97. https://doi.org/10.1530/JME-15-0246.

Baden, T., Schubert, T., Chang, L., Wei, T., Zaichuk, M., Wissinger, B., et al. (2013). A tale of two retinal domains: Near-optimal sampling of achromatic contrasts in natural scenes through asymmetric photoreceptor distribution. *Neuron*, *80*(5), 1206–1217. https://doi.org/10.1016/j.neuron.2013.09.030.

Barron, M. G. (1986). Endocrine control of smoltification in anadromous salmonids. *The Journal of Endocrinology*, *108*(2), 313–319. https://doi.org/10.1677/joe.0.1080313.

Beaudet, L., Browman, H. I., & Hawryshyn, C. W. (1993). Optic nerve response and retinal structure in rainbow trout of different sizes. *Vision Research*, *33*(13), 1739–1746. https://doi.org/10.1016/0042-6989(93)90164-r.

Beaudet, L., Novales Flamarique, I., & Hawryshyn, C. W. (1997). Cone photoreceptor topography in the retina of sexually mature Pacific salmonid fishes. *The Journal of Comparative Neurology*, *383*(1), 49–59.

Benvenga, S., & Robbins, J. (1996). Altered thyroid hormone binding to plasma lipoproteins in hypothyroidism. *Thyroid*, *6*(6), 595–600. https://doi.org/10.1089/thy.1996.6.595.

Bernal, J., Guadano-Ferraz, A., & Morte, B. (2015). Thyroid hormone transporters-functions and clinical implications. *Nature Reviews. Endocrinology*, *11*(12), 690. https://doi.org/10.1038/nrendo.2015.186.

Bianco, A. C., Salvatore, D., Gereben, B., Berry, M. J., & Larsen, P. R. (2002). Biochemistry, cellular and molecular biology, and physiological roles of the iodothyronine selenodeiodinases. *Endocrine Reviews*, *23*(1), 38–89. https://doi.org/10.1210/edrv.23.1.0455.

Billon, N., Tokumoto, Y., Forrest, D., & Raff, M. (2001). Role of thyroid hormone receptors in timing oligodendrocyte differentiation. *Developmental Biology*, *235*(1), 110–120. https://doi.org/10.1006/dbio.2001.0293.

Bowmaker, J. K., & Kunz, Y. W. (1987). Ultraviolet receptors, tetrachromatic colour vision and retinal mosaics in the brown trout (Salmo trutta): Age-dependent changes. *Vision Research*, *27*(12), 2101–2108. https://doi.org/10.1016/0042-6989(87)90124-6.

Branchek, T., & Bremiller, R. (1984). The development of photoreceptors in the zebrafish, Brachydanio rerio. I. Structure. *The Journal of Comparative Neurology*, *224*(1), 107–115. https://doi.org/10.1002/cne.902240109.

Brent, G. A. (2012). Mechanisms of thyroid hormone action. *The Journal of Clinical Investigation, 122*(9), 3035–3043. https://doi.org/10.1172/JCI60047.

Bruhn, S. L., & Cepko, C. L. (1996). Development of the pattern of photoreceptors in the chick retina. *The Journal of Neuroscience, 16*(4), 1430–1439. http://www.jneurosci.org/content/16/4/1430.full.pdf.

Butler, A. J., & Parker, M. G. (1995). COUP-TF II homodimers are formed in preference to heterodimers with RXR alpha or TR beta in intact cells. *Nucleic Acids Research, 23*(20), 4143–4150. https://doi.org/10.1093/nar/23.20.4143.

Cakir, M., Turgut Ozturk, B., Turan, E., Gonulalan, G., Polat, I., & Gunduz, K. (2015). The effect of hypothyroidism on color contrast sensitivity: A prospective study. *European Thyroid Journal, 4*(1), 43–47. https://doi.org/10.1159/000371549.

Calderone, J. B., & Jacobs, G. H. (1995). Regional variations in the relative sensitivity to UV light in the mouse retina. *Visual Neuroscience, 12*(3), 463–468.

Campos-Barros, A., Amma, L. L., Faris, J. S., Shallam, R., Kelley, M. W., & Forrest, D. (1999). Type 2 iodothyronine deiodinase expression in the cochlea before the onset of hearing. *Proceedings of the National Academy of Sciences of the United States of America, 97*, 1287–1292.

Canaris, G. J., Manowitz, N. R., Mayor, G., & Ridgway, E. C. (2000). The Colorado thyroid disease prevalence study. *Archives of Internal Medicine, 160*(4), 526–534. https://doi.org/10.1001/archinte.160.4.526.

Capowski, E. E., Wright, L. S., Liang, K., Phillips, M. J., Wallace, K., Petelinsek, A., et al. (2016). Regulation of WNT signaling by VSX2 during optic vesicle patterning in human induced pluripotent stem cells. *Stem Cells, 34*(11), 2625–2634. https://doi.org/10.1002/stem.2414.

Cepko, C. (2014). Intrinsically different retinal progenitor cells produce specific types of progeny. *Nature Reviews. Neuroscience, 15*(9), 615–627. https://doi.org/10.1038/nrn3767.

Cheng, C. L., Gan, K. J., & Flamarique, I. N. (2009). Thyroid hormone induces a time-dependent opsin switch in the retina of salmonid fishes. *Investigative Ophthalmology & Visual Science, 50*(6), 3024–3032. https://doi.org/10.1167/iovs.08-2713.

Chiamolera, M. I., & Wondisford, F. E. (2009). Minireview: Thyrotropin-releasing hormone and the thyroid hormone feedback mechanism. *Endocrinology, 150*(3), 1091–1096. https://doi.org/10.1210/en.2008-1795.

Chichagova, V., Dorgau, B., Felemban, M., Georgiou, M., Armstrong, L., & Lako, M. (2019). Differentiation of retinal organoids from human pluripotent stem cells. *Current Protocols in Stem Cell Biology, 50*(1), e95. https://doi.org/10.1002/cpsc.95.

Chinen, A., Hamaoka, T., Yamada, Y., & Kawamura, S. (2003). Gene duplication and spectral diversification of cone visual pigments of zebrafish. *Genetics, 163*(2), 663–675. https://www.ncbi.nlm.nih.gov/pubmed/12618404.

Cornish, E. E., Hendrickson, A. E., & Provis, J. M. (2004). Distribution of short-wavelength-sensitive cones in human fetal and postnatal retina: Early development of spatial order and density profiles. *Vision Research, 44*(17), 2019–2026. https://doi.org/10.1016/j.visres.2004.03.030.

Crantz, F. R., Silva, J. E., & Larsen, P. R. (1982). An analysis of the sources and quantity of 3,5,3′-triiodothyronine specifically bound to nuclear receptors in rat cerebral cortex and cerebellum. *Endocrinology, 110*(2), 367–375. https://doi.org/10.1210/endo-110-2-367.

Curcio, C. A., Allen, K. A., Sloan, K. R., Lerea, C. L., Hurley, J. B., Klock, I. B., et al. (1991). Distribution and morphology of human cone photoreceptors stained with anti-blue opsin. *The Journal of Comparative Neurology, 312*(4), 610–624. https://doi.org/10.1002/cne.903120411.

Curcio, C. A., Sloan, K. R., Kalina, R. E., & Hendrickson, A. E. (1990). Human photoreceptor topography. *The Journal of Comparative Neurology, 292*(4), 497–523. https://doi.org/10.1002/cne.902920402.

Deeb, S. S. (2005). The molecular basis of variation in human color vision. *Clinical Genetics*, *67*(5), 369–377. https://doi.org/10.1111/j.1399-0004.2004.00343.x.

Deng, W. L., Gao, M. L., Lei, X. L., Lv, J. N., Zhao, H., He, K. W., et al. (2018). Gene correction reverses ciliopathy and photoreceptor loss in iPSC-derived retinal organoids from retinitis pigmentosa patients. *Stem Cell Reports*, *10*(6), 2005. https://doi.org/10.1016/j.stemcr.2018.05.012.

Easter, S. S., Jr., & Malicki, J. J. (2002). The zebrafish eye: Developmental and genetic analysis. *Results and Problems in Cell Differentiation*, *40*, 346–370. https://doi.org/10.1007/978-3-540-46041-1_17.

Ebrey, T., & Koutalos, Y. (2001). Vertebrate photoreceptors. *Progress in Retinal and Eye Research*, *20*(1), 49–94. https://doi.org/10.1016/s1350-9462(00)00014-8.

Eiraku, M., Takata, N., Ishibashi, H., Kawada, M., Sakakura, E., Okuda, S., et al. (2011). Self-organizing optic-cup morphogenesis in three-dimensional culture. *Nature*, *472*(7341), 51–56. https://doi.org/10.1038/nature09941.

Eldred, K. C., Avelis, C., Johnston, R. J., Jr., & Roberts, E. (2020). Modeling binary and graded cone cell fate patterning in the mouse retina. *PLoS Computational Biology*, *16*(3), e1007691. https://doi.org/10.1371/journal.pcbi.1007691.

Eldred, K. C., Hadyniak, S. E., Hussey, K. A., Brenerman, B., Zhang, P. W., Chamling, X., et al. (2018). Thyroid hormone signaling specifies cone subtypes in human retinal organoids. *Science*, *362*(6411), eaau6348. https://doi.org/10.1126/science.aau6348.

Fei, Y. (2003). Development of the cone photoreceptor mosaic in the mouse retina revealed by fluorescent cones in transgenic mice. *Molecular Vision*, *9*, 31–42. https://www.ncbi.nlm.nih.gov/pubmed/12592228.

Flamarique, I. I., & Hawryshyn, C. (1996). Retinal development and visual sensitivity of young Pacific sockeye salmon (Oncorhynchus nerka). *The Journal of Experimental Biology*, *199*(Pt. 4), 869–882.

Fliers, E., Unmehopa, U. A., & Alkemade, A. (2006). Functional neuroanatomy of thyroid hormone feedback in the human hypothalamus and pituitary gland. *Molecular and Cellular Endocrinology*, *251*(1–2), 1–8. https://doi.org/10.1016/j.mce.2006.03.042.

Foltz, L. P., & Clegg, D. O. (2019). Patient-derived induced pluripotent stem cells for modelling genetic retinal dystrophies. *Progress in Retinal and Eye Research*, *68*, 54–66. https://doi.org/10.1016/j.preteyeres.2018.09.002.

Forrest, D., Sjoberg, M., & Vennstrom, B. (1990). Contrasting developmental and tissue-specific expression of alpha and beta thyroid hormone receptor genes. *The EMBO Journal*, *9*(5), 1519–1528.

Friesema, E. C., Ganguly, S., Abdalla, A., Manning Fox, J. E., Halestrap, A. P., & Visser, T. J. (2003). Identification of monocarboxylate transporter 8 as a specific thyroid hormone transporter. *The Journal of Biological Chemistry*, *278*(41), 40128–40135. https://doi.org/10.1074/jbc.M300909200.

Friesema, E. C., Jansen, J., & Visser, T. J. (2005). Thyroid hormone transporters. *Biochemical Society Transactions*, *33*(Pt. 1), 228–232. https://doi.org/10.1042/BST0330228.

Fujieda, H., Bremner, R., Mears, A. J., & Sasaki, H. (2009). Retinoic acid receptor-related orphan receptor alpha regulates a subset of cone genes during mouse retinal development. *Journal of Neurochemistry*, *108*(1), 91–101. https://doi.org/10.1111/j.1471-4159.2008.05739.x.

Galton, V. A. (2017). The ups and downs of the thyroxine pro-hormone hypothesis. *Molecular and Cellular Endocrinology*, *458*, 105–111. https://doi.org/10.1016/j.mce.2017.01.029.

Gan, K. J., & Novales Flamarique, I. (2010). Thyroid hormone accelerates opsin expression during early photoreceptor differentiation and induces opsin switching in differentiated TRalpha-expressing cones of the salmonid retina. *Developmental Dynamics*, *239*(10), 2700–2713. https://doi.org/10.1002/dvdy.22392.

Gao, M. L., Lei, X. L., Han, F., He, K. W., Jin, S. Q., Zhang, Y. Y., et al. (2020). Patient-specific retinal organoids recapitulate disease features of late-onset retinitis pigmentosa. *Frontiers in Cell and Development Biology*, *8*, 128. https://doi.org/10.3389/fcell.2020.00128.

Garita-Hernandez, M., Guibbal, L., Toualbi, L., Routet, F., Chaffiol, A., Winckler, C., et al. (2018). Optogenetic light sensors in human retinal organoids. *Frontiers in Neuroscience*, *12*, 789. https://doi.org/10.3389/fnins.2018.00789.

Garita-Hernandez, M., Routet, F., Guibbal, L., Khabou, H., Toualbi, L., Riancho, L., et al. (2020). AAV-mediated gene delivery to 3D retinal organoids derived from human induced pluripotent stem cells. *International Journal of Molecular Sciences*, *21*(3), 994. https://doi.org/10.3390/ijms21030994.

Gelfand, R. A., Hutchinson-Williams, K. A., Bonde, A. A., Castellino, P., & Sherwin, R. S. (1987). Catabolic effects of thyroid hormone excess: The contribution of adrenergic activity to hypermetabolism and protein breakdown. *Metabolism*, *36*(6), 562–569. https://doi.org/10.1016/0026-0495(87)90168-5.

Gereben, B., Zavacki, A. M., Ribich, S., Kim, B. W., Huang, S. A., Simonides, W. S., et al. (2008). Cellular and molecular basis of deiodinase-regulated thyroid hormone signaling. *Endocrine Reviews*, *29*(7), 898–938. https://doi.org/10.1210/er.2008-0019.

Gereben, B., Zeold, A., Dentice, M., Salvatore, D., & Bianco, A. C. (2008). Activation and inactivation of thyroid hormone by deiodinases: Local action with general consequences. *Cellular and Molecular Life Sciences*, *65*(4), 570–590. https://doi.org/10.1007/s00018-007-7396-0.

Glaschke, A., Glosmann, M., & Peichl, L. (2010). Developmental changes of cone opsin expression but not retinal morphology in the hypothyroid Pax8 knockout mouse. *Investigative Ophthalmology & Visual Science*, *51*(3), 1719–1727. https://doi.org/10.1167/iovs.09-3592.

Glaschke, A., Weiland, J., Del Turco, D., Steiner, M., Peichl, L., & Glosmann, M. (2011). Thyroid hormone controls cone opsin expression in the retina of adult rodents. *The Journal of Neuroscience*, *31*(13), 4844–4851. https://doi.org/10.1523/JNEUROSCI.6181-10.2011.

Govardovskii, V. I., & Zueva, L. V. (1977). Visual pigments of chicken and pigeon. *Vision Research*, *17*(4), 537–543.

Gowdy, P. D., & Cicerone, C. M. (1998). The spatial arrangement of the L and M cones in the central fovea of the living human eye. *Vision Research*, *38*(17), 2575–2589. https://doi.org/10.1016/s0042-6989(97)00416-1.

Greenblatt, M., Brown, C. L., Lee, M., Dauder, S., & Bern, H. A. (1989). Changes in thyroid hormone levels in eggs and larvae and in iodide uptake by eggs of coho and chinook salmon, Oncorhynchus kisutsch and O. tschawytscha. *Fish Physiology and Biochemistry*, *6*(5), 261–278. https://doi.org/10.1007/BF01881680.

Guadano-Ferraz, A., Obregon, M. J., St. Germain, D. L., & Bernal, J. (1997). The type 2 iodothyronine deiodinase is expressed primarily in glial cells in the neonatal rat brain. *Proceedings of the National Academy of Sciences of the United States of America*, *94*, 10391–10396.

Guo, Y., Wang, P., Ma, J. H., Cui, Z., Yu, Q., Liu, S., et al. (2019). Modeling retinitis pigmentosa: Retinal organoids generated from the iPSCs of a patient with the USH2A mutation show early developmental abnormalities. *Frontiers in Cellular Neuroscience*, *13*, 361. https://doi.org/10.3389/fncel.2019.00361.

Haverkamp, S., Wassle, H., Duebel, J., Kuner, T., Augustine, G. J., Feng, G., et al. (2005). The primordial, blue-cone color system of the mouse retina. *The Journal of Neuroscience*, *25*(22), 5438–5445. https://doi.org/10.1523/jneurosci.1117-05.2005.

Hawryshyn, C. W., Arnold, M. G., Chaisson, D. J., & Martin, P. C. (1989). The ontogeny of ultraviolet photosensitivity in rainbow trout (Salmo gairdneri). *Visual Neuroscience*, *2*(3), 247–254. https://doi.org/10.1017/s0952523800001164.

Hawryshyn, C. W., Martens, G., Allison, W. T., & Anholt, B. R. (2003). Regeneration of ultraviolet-sensitive cones in the retinal cone mosaic of thyroxin-challenged post-juvenile rainbow trout (Oncorhynchus mykiss). *The Journal of Experimental Biology*, *206*(Pt. 15), 2665–2673. https://doi.org/10.1242/jeb.00470.

Hoar, W. S. (1988). The physiology of smolting salmonids. *Fish Physiology*, *11*, 275–343.

Hofer, H., Carroll, J., Neitz, J., Neitz, M., & Williams, D. R. (2005). Organization of the human trichromatic cone mosaic. *The Journal of Neuroscience*, *25*(42), 9669–9679. https://doi.org/10.1523/JNEUROSCI.2414-05.2005.

Hoshino, A., Ratnapriya, R., Brooks, M. J., Chaitankar, V., Wilken, M. S., Zhang, C., et al. (2017). Molecular anatomy of the developing human retina. *Developmental Cell*, *43*(6), 763–779.e764. https://doi.org/10.1016/j.devcel.2017.10.029.

Hu, M., & Easter, S. S. (1999). Retinal neurogenesis: The formation of the initial central patch of postmitotic cells. *Developmental Biology*, *207*(2), 309–321. https://doi.org/10.1006/dbio.1998.9031.

Hunt, D. M. (2001). Seeing rainbows. *Biologist (London, England)*, *48*(2), 67–71.

Jones, I., Rogers, S. A., Kille, P., & Sweeney, G. E. (2002). Molecular cloning and expression of thyroid hormone receptor alpha during salmonid development. *General and Comparative Endocrinology*, *125*(2), 226–235. https://doi.org/10.1006/gcen.2001.7745.

Kaewkhaw, R., Kaya, K. D., Brooks, M., Homma, K., Zou, J., Chaitankar, V., et al. (2015). Transcriptome dynamics of developing photoreceptors in three-dimensional retina cultures recapitulates temporal sequence of human cone and rod differentiation revealing cell surface markers and gene networks. *Stem Cells*, *33*(12), 3504–3518. https://doi.org/10.1002/stem.2122.

Kainz, P. M., Neitz, J., & Neitz, M. (1998). Recent evolution of uniform trichromacy in a New World monkey. *Vision Research*, *38*(21), 3315–3320. https://doi.org/10.1016/s0042-6989(98)00078-9.

Katoh, K., Omori, Y., Onishi, A., Sato, S., Kondo, M., & Furukawa, T. (2010). Blimp1 suppresses Chx10 expression in differentiating retinal photoreceptor precursors to ensure proper photoreceptor development. *The Journal of Neuroscience*, *30*(19), 6515–6526. https://doi.org/10.1523/JNEUROSCI.0771-10.2010.

Kawamura, S., & Tachibanaki, S. (2008). Rod and cone photoreceptors: Molecular basis of the difference in their physiology. *Comparative Biochemistry and Physiology. Part A, Molecular & Integrative Physiology*, *150*(4), 369–377. https://doi.org/10.1016/j.cbpa.2008.04.600.

Koenig, R. J. (1998). Thyroid hormone receptor coactivators and corepressors. *Thyroid*, *8*(8), 703–713. https://doi.org/10.1089/thy.1998.8.703.

Korcek, L., & Tabachnick, M. (1976). Thyroxine-protein interactions. Interaction of thyroxine and triiodothyronine with human thyroxine-binding globulin. *The Journal of Biological Chemistry*, *251*(12), 3558–3562.

Kram, Y. A., Mantey, S., & Corbo, J. C. (2010). Avian cone photoreceptors tile the retina as five independent, self-organizing mosaics. *PLoS One*, *5*(2), e8992. https://doi.org/10.1371/journal.pone.0008992.

Kunz, Y. W. (1987). Tracts of putative ultraviolet receptors in the retina of the two-year-old brown trout (Salmo trutta) and the Atlantic salmon (Salmo salar). *Experientia*, *43*(11 – 12), 1202–1204. https://doi.org/10.1007/BF01945524.

Kunz, Y. W., Wildenburg, G., Goodrich, L., & Callaghan, E. (1994). The fate of ultraviolet receptors in the retina of the Atlantic salmon (Salmo salar). *Vision Research*, *34*(11), 1375–1383. https://doi.org/10.1016/0042-6989(94)90136-8.

Lamba, D. A., Gust, J., & Reh, T. A. (2009). Transplantation of human embryonic stem cell-derived photoreceptors restores some visual function in Crx-deficient mice. *Cell Stem Cell*, *4*(1), 73–79. https://doi.org/10.1016/j.stem.2008.10.015.

Lamba, D. A., Karl, M. O., & Reh, T. A. (2009). Strategies for retinal repair: Cell replacement and regeneration. *Progress in Brain Research, 175*, 23–31. https://doi.org/10.1016/S0079-6123(09)17502-7.

Lamba, D. A., Karl, M. O., Ware, C. B., & Reh, T. A. (2006). Efficient generation of retinal progenitor cells from human embryonic stem cells. *Proceedings of the National Academy of Sciences of the United States of America, 103*(34), 12769–12774. https://doi.org/10.1073/pnas.0601990103.

Lamba, D. A., McUsic, A., Hirata, R. K., Wang, P. R., Russell, D., & Reh, T. A. (2010). Generation, purification and transplantation of photoreceptors derived from human induced pluripotent stem cells. *PLoS One, 5*(1), e8763. https://doi.org/10.1371/journal.pone.0008763.

Langmann, T. (2007). Microglia activation in retinal degeneration. *Journal of Leukocyte Biology, 81*(6), 1345–1351. https://doi.org/10.1189/jlb.0207114.

Larsen, P. R. (1982). Thyroid-pituitary interaction: Feedback regulation of thyrotropin secretion by thyroid hormones. *The New England Journal of Medicine, 306*(1), 23–32. https://doi.org/10.1056/NEJM198201073060107.

Li, G., Gao, G., Wang, P., Song, X., Xu, P., Xie, B., et al. (2019). Generation and characterization of induced pluripotent stem cells and retinal organoids from a Leber's congenital amaurosis patient with novel RPE65 mutations. *Frontiers in Molecular Neuroscience, 12*, 212. https://doi.org/10.3389/fnmol.2019.00212.

Li, K. Y., & Roorda, A. (2007). Automated identification of cone photoreceptors in adaptive optics retinal images. *Journal of the Optical Society of America. A, Optics, Image Science, and Vision, 24*(5), 1358–1363. https://doi.org/10.1364/josaa.24.001358.

Llonch, S., Carido, M., & Ader, M. (2018). Organoid technology for retinal repair. *Developmental Biology, 433*(2), 132–143. https://doi.org/10.1016/j.ydbio.2017.09.028.

Loew, E. R., & Wahl, C. M. (1991). A short-wavelength sensitive cone mechanism in juvenile yellow perch, *Perca flavescens*. *Vision Research, 31*(3), 353–360. https://doi.org/10.1016/0042-6989(91)90088-m.

Lu, A., Ng, L., Ma, M., Kefas, B., Davies, T. F., Hernandez, A., et al. (2009). Retarded developmental expression and patterning of retinal cone opsins in hypothyroid mice. *Endocrinology, 150*(3), 1536–1544. https://doi.org/10.1210/en.2008-1092.

Lyall, A. H. (1957). The growth of the trout retina. *Journal of Cell Science, 3*(98), 101–110.

Ma, H., Thapa, A., Morris, L., Redmond, T. M., Baehr, W., & Ding, X. Q. (2014). Suppressing thyroid hormone signaling preserves cone photoreceptors in mouse models of retinal degeneration. *Proceedings of the National Academy of Sciences of the United States of America, 111*(9), 3602–3607. https://doi.org/10.1073/pnas.1317041111.

Mackin, R. D., Frey, R. A., Gutierrez, C., Farre, A. A., Kawamura, S., Mitchell, D. M., et al. (2019). Endocrine regulation of multichromatic color vision. *Proceedings of the National Academy of Sciences of the United States of America, 116*(34), 16882–16891. https://doi.org/10.1073/pnas.1904783116.

Masland, R. H. (2012). The neuronal organization of the retina. *Neuron, 76*(2), 266–280. https://doi.org/10.1016/j.neuron.2012.10.002.

Morris, V. B. (1982). An afoveate area centralis in the chick retina. *The Journal of Comparative Neurology, 210*(2), 198–203. https://doi.org/10.1002/cne.902100210.

Muller, J., Mayerl, S., Visser, T. J., Darras, V. M., Boelen, A., Frappart, L., et al. (2014). Tissue-specific alterations in thyroid hormone homeostasis in combined Mct10 and Mct8 deficiency. *Endocrinology, 155*(1), 315–325. https://doi.org/10.1210/en.2013-1800.

Nakano, T., Ando, S., Takata, N., Kawada, M., Muguruma, K., Sekiguchi, K., et al. (2012). Self-formation of optic cups and storable stratified neural retina from human ESCs. *Cell Stem Cell, 10*(6), 771–785. https://doi.org/10.1016/j.stem.2012.05.009.

Nathans, J., Davenport, C. M., Maumenee, I. H., Lewis, R. A., Hejtmancik, J. F., Litt, M., et al. (1989). Molecular genetics of human blue cone monochromacy. *Science*, *245*(4920), 831–838. https://doi.org/10.1126/science.2788922.

Nathans, J., & Hogness, D. S. (1984). Isolation and nucleotide sequence of the gene encoding human rhodopsin. *Proceedings of the National Academy of Sciences of the United States of America*, *81*(15), 4851–4855. https://doi.org/10.1073/pnas.81.15.4851.

Nathans, J., Thomas, D., & Hogness, D. S. (1986). Molecular genetics of human color vision: The genes encoding blue, green, and red pigments. *Science*, *232*(4747), 193–202. https://doi.org/10.1126/science.2937147.

Ng, L., Goodyear, R. J., Woods, C. A., Schneider, M. J., Diamond, E., Richardson, G. P., et al. (2004). Hearing loss and retarded cochlear development in mice lacking type 2 iodothyronine deiodinase. *Proceedings of the National Academy of Sciences of the United States of America*, *101*(10), 3474–3479. https://doi.org/10.1073/pnas.0307402101.

Ng, L., Hernandez, A., He, W., Ren, T., Srinivas, M., Ma, M., et al. (2009). A protective role for type 3 deiodinase, a thyroid hormone-inactivating enzyme, in cochlear development and auditory function. *Endocrinology*, *150*(4), 1952–1960. https://doi.org/10.1210/en.2008-1419.

Ng, L., Hurley, J. B., Dierks, B., Srinivas, M., Salto, C., Vennstrom, B., et al. (2001). A thyroid hormone receptor that is required for the development of green cone photoreceptors. *Nature Genetics*, *27*, 94–98.

Ng, L., Liu, H., St Germain, D. L., Hernandez, A., & Forrest, D. (2017). Deletion of the thyroid hormone-activating type 2 deiodinase rescues cone photoreceptor degeneration but not deafness in mice lacking type 3 deiodinase. *Endocrinology*, *158*(6), 1999–2010. https://doi.org/10.1210/en.2017-00055.

Ng, L., Lu, A., Swaroop, A., Sharlin, D. S., Swaroop, A., & Forrest, D. (2011). Two transcription factors can direct three photoreceptor outcomes from rod precursor cells in mouse retinal development. *The Journal of Neuroscience*, *31*(31), 11118–11125. https://doi.org/10.1523/jneurosci.1709-11.2011.

Ng, L., Lyubarsky, A., Nikonov, S. S., Ma, M., Srinivas, M., Kefas, B., et al. (2010). Type 3 deiodinase, a thyroid-hormone-inactivating enzyme, controls survival and maturation of cone photoreceptors. *The Journal of Neuroscience*, *30*(9), 3347–3357. https://doi.org/10.1523/jneurosci.5267-09.2010.

Novales Flamarique, I. (2000). The ontogeny of ultraviolet sensitivity, cone disappearance and regeneration in the sockeye salmon Oncorhynchus nerka. *The Journal of Experimental Biology*, *203*, 1161–1172.

Novales Flamarique, I., Sayed Ahmed, A., Cheng, C. L., Molday, R. S., & Devlin, R. H. (2019). Growth hormone regulates opsin expression in the retina of a salmonid fish. *Journal of Neuroendocrinology*, *31*(11), e12804. https://doi.org/10.1111/jne.12804.

Oppenheimer, J. H., & Schwartz, H. L. (1997). Molecular basis of thyroid hormone-dependent brain development. *Endocrine Reviews*, *18*(4), 462–475. https://doi.org/10.1210/edrv.18.4.0309.

Otake, S., Gowdy, P. D., & Cicerone, C. M. (2000). The spatial arrangement of L and M cones in the peripheral human retina. *Vision Research*, *40*(6), 677–693. https://doi.org/10.1016/s0042-6989(99)00202-3.

Palha, J. A. (2002). Transthyretin as a thyroid hormone carrier: Function revisited. *Clinical Chemistry and Laboratory Medicine*, *40*(12), 1292–1300. https://doi.org/10.1515/CCLM.2002.223.

Pessoa, C. N., Santiago, L. A., Santiago, D. A., Machado, D. S., Rocha, F. A., Ventura, D. F., et al. (2008). Thyroid hormone action is required for normal cone opsin expression during mouse retinal development. Comparative Study Research Support, Non-U.S. Gov't *Investigative Ophthalmology & Visual Science*, *49*(5), 2039–2045. https://doi.org/10.1167/iovs.07-0908.

Phillips, M. J., Jiang, P., Howden, S., Barney, P., Min, J., York, N. W., et al. (2018). A novel approach to single cell RNA-sequence analysis facilitates in silico gene reporting of human pluripotent stem cell-derived retinal cell types. *Stem Cells*, *36*(3), 313–324. https://doi.org/10.1002/stem.2755.

Provis, J. M., Dubis, A. M., Maddess, T., & Carroll, J. (2013). Adaptation of the central retina for high acuity vision: Cones, the fovea and the avascular zone. *Progress in Retinal and Eye Research*, *35*, 63–81. https://doi.org/10.1016/j.preteyeres.2013.01.005.

Quinn, P. M., Buck, T. M., Mulder, A. A., Ohonin, C., Alves, C. H., Vos, R. M., et al. (2019). Human iPSC-derived retinas recapitulate the fetal CRB1 CRB2 complex formation and demonstrate that photoreceptors and Muller glia are targets of AAV5. *Stem Cell Reports*, *12*(5), 906–919. https://doi.org/10.1016/j.stemcr.2019.03.002.

Racheva, K., Totev, T., Natchev, E., Bocheva, N., Beirne, R., & Zlatkova, M. (2020). Color discrimination assessment in patients with hypothyroidism using the Farnsworth-Munsell 100 hue test. *Journal of the Optical Society of America. A, Optics, Image Science, and Vision*, *37*(4), A18–A25. https://doi.org/10.1364/JOSAA.382390.

Raine, J. C., & Hawryshyn, C. W. (2009). Changes in thyroid hormone reception precede SWS1 opsin downregulation in trout retina. *The Journal of Experimental Biology*, *212*(17), 2781–2788. https://doi.org/10.1242/jeb.030866.

Ramírez, J. M., Triviño, A., Ramírez, A. I., & Salazar, J. J. (1998). Organization and function of astrocytes in human retina. In *Understanding glial cells* (pp. 47–62). Springer. https://doi.org/10.1007/978-1-4615-5737-1_3.

Rashid, K., Akhtar-Schaefer, I., & Langmann, T. (2019). Microglia in Retinal Degeneration. *Frontiers in Immunology*, *10*, 1975. https://doi.org/10.3389/fimmu.2019.01975.

Raymond, P. A., Barthel, L. K., & Curran, G. A. (1995). Developmental patterning of rod and cone photoreceptors in embryonic zebrafish. *The Journal of Comparative Neurology*, *359*(4), 537–550. https://doi.org/10.1002/cne.903590403.

Raymond, P. A., Barthel, L. K., Rounsifer, M. E., Sullivan, S. A., & Knight, J. K. (1993). Expression of rod and cone visual pigments in goldfish and zebrafish: A rhodopsin-like gene is expressed in cones. *Neuron*, *10*(6), 1161–1174. https://doi.org/10.1016/0896-6273(93)90064-x.

Reh, T. A. (2016). Photoreceptor transplantation in late stage retinal degeneration. *Investigative Ophthalmology & Visual Science*, *57*(5), ORSFg1-7. https://doi.org/10.1167/iovs.15-17659.

Reichenbach, A., & Bringmann, A. (2013). New functions of Muller cells. *Glia*, *61*(5), 651–678. https://doi.org/10.1002/glia.22477.

Richardson, T. M. (1969). Cytoplasmic and ciliary connections between the inner and outer segments of mammalian visual receptors. *Vision Research*, *9*(7), 727–731. https://doi.org/10.1016/0042-6989(69)90010-8.

Roberts, M. R., Hendrickson, A., McGuire, C. R., & Reh, T. A. (2005). Retinoid X receptor (gamma) is necessary to establish the S-opsin gradient in cone photoreceptors of the developing mouse retina. *Investigative Ophthalmology & Visual Science*, *46*(8), 2897–2904. https://doi.org/10.1167/iovs.05-0093.

Roberts, M. R., Srinivas, M., Forrest, D., Morreale de Escobar, G., & Reh, T. A. (2006). Making the gradient: Thyroid hormone regulates cone opsin expression in the developing mouse retina. *Proceedings of the National Academy of Sciences of the United States of America*, *103*(16), 6218–6223. https://doi.org/10.1073/pnas.0509981103.

Rohlich, P., van Veen, T., & Szel, A. (1994). Two different visual pigments in one retinal cone cell. *Neuron*, *13*(5), 1159–1166. https://www.ncbi.nlm.nih.gov/pubmed/7946352.

Roorda, A. W., & David, R. (1999). The arrangement of the three cone classes in the living human eye. *Nature*, *397*, 520–522.

Roorda, A., Metha, A. B., Lennie, P., & Williams, D. R. (2001). Packing arrangement of the three cone classes in primate retina. *Vision Research*, *41*(10 – 11), 1291–1306. https://doi.org/10.1016/s0042-6989(01)00043-8.

Rossi, E. A., Chung, M., Dubra, A., Hunter, J. J., Merigan, W. H., & Williams, D. R. (2011). Imaging retinal mosaics in the living eye. *Eye (London, England)*, *25*(3), 301–308. https://doi.org/10.1038/eye.2010.221.

Samuels, H. H., Tsai, J. S., Casanova, J., & Stanley, F. (1974). Thyroid hormone action: In vitro characterization of solubilized nuclear receptors from rat liver and cultured GH1 cells. *The Journal of Clinical Investigation*, *54*(4), 853–865. https://doi.org/10.1172/JCI107825.

Sawant, O. B., Horton, A. M., Zucaro, O. F., Chan, R., Bonilha, V. L., Samuels, I. S., et al. (2017). The circadian clock gene Bmal1 controls thyroid hormone-mediated spectral identity and cone photoreceptor function. *Cell Reports*, *21*(3), 692–706. https://doi.org/10.1016/j.celrep.2017.09.069.

Sawin, C. T., Castelli, W. P., Hershman, J. M., McNamara, P., & Bacharach, P. (1985). The aging thyroid. Thyroid deficiency in the Framingham study. *Archives of Internal Medicine*, *145*(8), 1386–1388.

Schmitt, E. A., & Dowling, J. E. (1996). Comparison of topographical patterns of ganglion and photoreceptor cell differentiation in the retina of the zebrafish, Danio rerio. *The Journal of Comparative Neurology*, *371*(2), 222–234. https://doi.org/10.1002/(SICI)1096-9861(19960722)371:2<222::AID-CNE3>3.0.CO;2-4.

Schwartz, C. E., & Stevenson, R. E. (2007). The MCT8 thyroid hormone transporter and Allan-Herndon-Dudley syndrome. *Best Practice & Research. Clinical Endocrinology & Metabolism*, *21*(2), 307–321. https://doi.org/10.1016/j.beem.2007.03.009.

Singh, M. S., Charbel Issa, P., Butler, R., Martin, C., Lipinski, D. M., Sekaran, S., et al. (2013). Reversal of end-stage retinal degeneration and restoration of visual function by photoreceptor transplantation. *Proceedings of the National Academy of Sciences of the United States of America*, *110*(3), 1101–1106. https://doi.org/10.1073/pnas.1119416110.

Sjoberg, M., Vennström, B., & Forrest, D. (1992). Thyroid hormone receptors in chick retinal development: Differential expression of mRNAs for a and N-terminal variant B receptors. *Development*, *114*, 39–47.

Specker, J. L., Eales, J. G., Tagawa, M., & Tyler, W. A. (2000). Parr-smolt transformation in Atlantic salmon: Thyroid hormone deiodination in liver and brain and endocrine correlates of change in rheotactic behavior. *Canadian Journal of Zoology*, *78*, 696–705.

Sridhar, A., Hoshino, A., Finkbeiner, C. R., Chitsazan, A., Dai, L., Haugan, A. K., et al. (2020). Single-cell transcriptomic comparison of human fetal retina, hPSC-derived retinal organoids, and long-term retinal cultures. *Cell Reports*, *30*(5), 1644–1659.e1644. https://doi.org/10.1016/j.celrep.2020.01.007.

Stenkamp, D. L., Barthel, L. K., & Raymond, P. A. (1997). Spatiotemporal coordination of rod and cone photoreceptor differentiation in goldfish retina. *The Journal of Comparative Neurology*, *382*(2), 272–284. https://doi.org/10.1002/(sici)1096-9861(19970602)382:2.

Sun, Z., Feng, D., Fang, B., Mullican, S. E., You, S. H., Lim, H. W., et al. (2013). Deacetylase-independent function of HDAC3 in transcription and metabolism requires nuclear receptor corepressor. *Molecular Cell*, *52*(6), 769–782. https://doi.org/10.1016/j.molcel.2013.10.022.

Suzuki, S. C., Bleckert, A., Williams, P. R., Takechi, M., Kawamura, S., & Wong, R. O. (2013). Cone photoreceptor types in zebrafish are generated by symmetric terminal divisions of dedicated precursors. *Proceedings of the National Academy of Sciences of the United States of America*, *110*(37), 15109–15114. https://doi.org/10.1073/pnas.1303551110.

Szel, A., Csorba, G., Caffe, A. R., Szel, G., Rohlich, P., & van Veen, T. (1994). Different patterns of retinal cone topography in two genera of rodents, Mus and Apodemus. *Cell and Tissue Research*, *276*(1), 143–150. https://doi.org/10.1007/BF00354793.

Szel, A., Rohlich, P., Mieziewska, K., Aguirre, G., & van Veen, T. (1993). Spatial and temporal differences between the expression of short- and middle-wave sensitive cone pigments in the mouse retina: A developmental study. *The Journal of Comparative Neurology*, *331*(4), 564–577. https://doi.org/10.1002/cne.903310411.

Trimarchi, J. M., Harpavat, S., Billings, N. A., & Cepko, C. L. (2008). Thyroid hormone components are expressed in three sequential waves during development of the chick retina. *BMC Developmental Biology*, *8*, 101. https://doi.org/10.1186/1471-213X-8-101.

Tu, H. M., Kim, S. W., Salvatore, D., Bartha, T., Legradi, G., Larsen, P. R., et al. (1997). Regional distribution of type 2 thyroxine deiodinase messenger ribonucleic acid in rat hypothalamus and pituitary and its regulation by thyroid hormone. *Endocrinology*, *138*(8), 3359–3368. https://doi.org/10.1210/endo.138.8.5318.

Tu, H. M., Legradi, G., Bartha, T., Salvatore, D., Lechan, R. M., & Larsen, P. R. (1999). Regional expression of the type 3 iodothyronine deiodinase messenger ribonucleic acid in the rat central nervous system and its regulation by thyroid hormone. *Endocrinology*, *140*(2), 784–790. https://doi.org/10.1210/endo.140.2.6486.

Tunbridge, W. M., Evered, D. C., Hall, R., Appleton, D., Brewis, M., Clark, F., et al. (1977). The spectrum of thyroid disease in a community: The Whickham survey. *Clinical Endocrinology*, 7(6), 481–493. https://doi.org/10.1111/j.1365-2265.1977.tb01340.x.

Vancamp, P., Bourgeois, N. M. A., Houbrechts, A. M., & Darras, V. M. (2019). Knockdown of the thyroid hormone transporter MCT8 in chicken retinal precursor cells hampers early retinal development and results in a shift towards more UV/blue cones at the expense of green/red cones. *Experimental Eye Research*, *178*, 135–147. https://doi.org/10.1016/j.exer.2018.09.018.

Vecino, E., Rodriguez, F. D., Ruzafa, N., Pereiro, X., & Sharma, S. C. (2016). Glia-neuron interactions in the mammalian retina. *Progress in Retinal and Eye Research*, *51*, 1–40. https://doi.org/10.1016/j.preteyeres.2015.06.003.

Viets, K., Eldred, K., & Johnston, R. J., Jr. (2016). Mechanisms of photoreceptor patterning in vertebrates and invertebrates. *Trends in Genetics*, *32*(10), 638–659. https://doi.org/10.1016/j.tig.2016.07.004.

Vihtelic, T. S., Doro, C. J., & Hyde, D. R. (1999). Cloning and characterization of six zebrafish photoreceptor opsin cDNAs and immunolocalization of their corresponding proteins. *Visual Neuroscience*, *16*(3), 571–585. https://doi.org/10.1017/s0952523899163168.

Wahlin, K. J., Maruotti, J. A., Sripathi, S. R., Ball, J., Angueyra, J. M., Kim, C., et al. (2017). Photoreceptor outer segment-like structures in long-term 3D retinas from human pluripotent stem cells. *Scientific Reports*, 7(1), 766. https://doi.org/10.1038/s41598-017-00774-9.

Wassle, H. (2004). Parallel processing in the mammalian retina. *Nature Reviews. Neuroscience*, *5*(10), 747–757. https://doi.org/10.1038/nrn1497.

Weiss, A. H., Kelly, J. P., Bisset, D., & Deeb, S. S. (2012). Reduced L- and M- and increased S-cone functions in an infant with thyroid hormone resistance due to mutations in the THRbeta2 gene. *Ophthalmic Genetics*, *33*(4), 187–195. https://doi.org/10.3109/13816810.2012.681096.

Wikler, K. C., Szel, A., & Jacobsen, A. L. (1996). Positional information and opsin identity in retinal cones. *The Journal of Comparative Neurology*, *374*(1), 96–107. https://doi.org/10.1002/(SICI)1096-9861(19961007)374:1<96::AID-CNE7>3.0.CO;2-I.

Williams, G. R. (2008). Neurodevelopmental and neurophysiological actions of thyroid hormone. *Journal of Neuroendocrinology*, *20*(6), 784–794. https://doi.org/10.1111/j.1365-2826.2008.01733.x.

Xiao, M., & Hendrickson, A. (2000). Spatial and temporal expression of short, long/medium, or both opsins in human fetal cones. *The Journal of Comparative Neurology*, *425*(4), 545–559.

Yokoyama, S. (2000). Molecular evolution of vertebrate visual pigments. *Progress in Retinal and Eye Research*, *19*(4), 385–419. https://doi.org/10.1016/s1350-9462(00)00002-1.

You, S. H., Lim, H. W., Sun, Z., Broache, M., Won, K. J., & Lazar, M. A. (2013). Nuclear receptor co-repressors are required for the histone-deacetylase activity of HDAC3 in vivo. *Nature Structural & Molecular Biology*, *20*(2), 182–187. https://doi.org/10.1038/nsmb.2476.

Zhong, X., Gutierrez, C., Xue, T., Hampton, C., Vergara, M. N., Cao, L. H., et al. (2014). Generation of three-dimensional retinal tissue with functional photoreceptors from human iPSCs. *Nature Communications*, *5*, 4047. https://doi.org/10.1038/ncomms5047.

CHAPTER FOUR

Thyroid hormone and hypothalamic stem cells in seasonal functions

Hugues Dardente* and Martine Migaud
PRC, INRAE, CNRS, IFCE, Université de Tours, Nouzilly, France
*Corresponding author: e-mail address: hugues.dardente@inrae.fr

Contents

Abstract

Seasonal rhythms are a pervasive feature of most living organisms, which underlie yearly timeliness in breeding, migration, hibernation or weight gain and loss. To achieve this, organisms have developed inner timing devices (circannual clocks) that endow them

Vitamins and Hormones, Volume 116
ISSN 0083-6729
https://doi.org/10.1016/bs.vh.2021.02.005

with the ability to predict then anticipate changes to come, usually using daylength as the proximate cue. In Vertebrates, daylength interpretation involves photoperiodic control of TSH production by the *pars tuberalis* (PT) of the pituitary, which governs a seasonal switch in thyroid hormone (TH) availability in the neighboring hypothalamus. Tanycytes, specialized glial cells lining the third ventricle (3V), are responsible for this TH output through the opposite, PT-TSH-driven, seasonal control of deiodinases 2/3 (Dio 2/3). Tanycytes comprise a photoperiod-sensitive stem cell niche and TH is known to play major roles in cell proliferation and differentiation, which suggests that seasonal control of tanycyte proliferation may be involved in the photoperiodic synchronization of seasonal rhythms. Here we review our current knowledge of the molecular and neuroendocrine pathway linking photoperiodic information to seasonal changes in physiological functions and discuss the potential implication of tanycytes, TH and cell proliferation in seasonal timing.

1. Daily and seasonal rhythms

1.1 Overview

Life on Earth is inherently rhythmic, which is unsurprising since life appeared and evolved in a rhythmic environment. Rotation of the Earth around its axis leads to daily cycles of light and darkness. Days are now longer than they used to be 500 million years ago (~21 h) and exposure to UV has considerably diminished in the meantime. However, in these early days, when the DNA-damaging impact of UV radiation during daytime was quite high, night was a more favorable time-window for DNA replication. It is widely accepted that this "escape from (UV) light" has been the driving force behind the appearance and evolution of daily rhythms in unicellular procaryotes (Hut & Beersma, 2011; Pittendrigh, 1993; Rosbash, 2009). Segregating different biological functions at distinct times of the day presents advantages for eucaryotes as well, which explains why daily rhythms are so pervasive. These daily rhythms are generated by inner timing devices known as circadian clocks (Takahashi, 2017). The key physiological roles of these molecular clocks are evidenced by the fact that their environmental or genetic perturbation leads to an array of diseases in humans and other mammals (Bedrosian, Fonken, & Nelson, 2016; Cederroth et al., 2019; Foster & Roenneberg, 2008; Hastings, Reddy, & Maywood, 2003; Levi & Schibler, 2007).

Due to the 23.5° tilt of its axis relative to its orbital plane, the Earth is further exposed to a cyclic annual variation in daily light exposure as it gravitates around the Sun. This is particularly true for places located outside equatorial regions, at temperate and polar latitudes, where marked seasonal variations in food availability have driven the evolution of specific life-history strategies. Much like daily functions, seasonal functions are pervasive across plants and animals, with hibernation, torpor and migration representing the most spectacular and iconic examples. Many mammals also display seasonal variations in breeding and associated functions such as lactation, molt, growth and cast of antlers, but also in their ability to put on and lose weight (Barrett & Bolborea, 2012; Dardente, Hazlerigg, & Ebling, 2014; Dardente, Wood, Ebling, & Saenz de Miera, 2019; Ebling, 2014; Goldman, 2001; Gorman, 2020; Hazlerigg & Simonneaux, 2015; Helfer, Barrett, & Morgan, 2019).

1.2 One hundred years of photoperiodism

Photoperiod, which corresponds to the daily length of the light phase, is the key proximal cue driving seasonal functions at temperate latitudes. For this reason, the term "photoperiodism" was coined just a century ago by Garner and Allard, who studied plants, to describe seasonal variation in one or several phenotypic traits, directly dependent upon changes in the duration of daily light exposure (Garner & Allard, 1920). In the following decades, photoperiodism has been characterized in multiple animal species, including chicken, duck, junco, ferret, vole, goat and sheep, mostly with regards to its potent impact upon the reproductive axis (Baker & Ranson, 1933; Benoit, 1934; Bissonnette, 1936; Hafez, 1952; Karsch et al., 1984; Lincoln & Short, 1980; Marshall, 1936; Rowan, 1938; Whetham, 1933; Yeates, 1949). Of course, these early studies merely described, often in great detail, how photoperiod impacts the phenotype of interest but did not explain how plants and animals actually sense and interpret gradual changes in daylength to time their seasonal functions.

Here we first provide a brief overview of the major features of seasonal functions, then focus on melatonin and thyroid hormone (TH), the two key hormones involved in the photoperiodic response in mammals. We then review our current understanding of the molecular and neuroendocrine underpinnings of photoperiodism, with emphasis on the role played by TH in the medio-basal hypothalamus (MBH). Since TH metabolism within the MBH appears to be chiefly performed by tanycytes, a very peculiar type of glial cells that shows stem-cell properties, we discuss the possibility that autocrine regulation of tanycyte proliferation by TH is an important feature of seasonal programs.

1.3 Organization of seasonal rhythms

For seasonal breeders, the ability to use photoperiod to time seasonal physiology, including reproduction, is key to the survival of the species. As best exemplified in birds and hamsters, these changes are profound and require weeks to months to reach completion: gonads in these species typically exhibit a 10- to 100-fold increase in size between the non-breeding and breeding seasons. Therefore, anticipation rather than passive day-to-day adaptation to the changing photoperiod is necessary. And anticipation requires clocks. Animals and plants have therefore evolved molecular clocks to anticipate changes to come, both at the daily level—circadian clocks—and at the annual level—circannual clocks, by analogy. The molecular make-up of circadian clocks has been elucidated and their function is being systematically assessed and revealed through tissue-specific genetic manipulations (Dardente & Cermakian, 2007; Dibner, Schibler, & Albrecht, 2010; Hogenesch & Ueda, 2011; Mohawk, Green, & Takahashi, 2012; Takahashi, 2017). The molecular underpinnings and role of circannual timing mechanism(s) have also been recently characterized, even though much remains to be done (Dardente, Wood, et al., 2019; Hazlerigg & Lincoln, 2011; Lincoln, 2019; Wood & Loudon, 2018). Indeed, even the relevance of these long-term rhythms has been a matter of debate for quite some time. Half a century ago, Jim Enright wrote: "To suggest that the long-term persistence of a biological rhythm is simply an evolutionary accident makes severe demand on our credulity; however, to propose, in concrete terms, some plausible selective advantage which could account for this persistence is an unmet challenge" (Enright, 1970). There are indeed several such selective advantages including the ability of hibernators to exit the burrow in spring, with gonads already fully grown (Hut, Dardente, & Riede, 2014) or the timing of the annual breeding season for trans-equatorial bird migrants (Rowan, 1938).

Akin to the circadian clock, the circannual clock requires light input to be synchronized. A circadian clock cannot be reset when light is provided during the subjective day (dead-zone). Likewise, not all daylengths are equally efficient at synchronizing the circannual clock; exposure to long days for a few weeks appears to be both necessary and sufficient to prime the circannual program (Dardente, 2012; Woodfill, Wayne, Moenter, & Karsch, 1994). Once synchronized by the appropriate photoperiodic cue, the circannual clock may run autonomously for one or several full cycles, even though animals are maintained in a constant photoperiod thereafter (Dardente, 2012; Woodfill, Robinson, Malpaux, & Karsch, 1991; Woodfill et al., 1994).

This phenomenon has mostly been studied with respect to breeding, using gonad size and LH/FSH or PRL hormonal profiles to track neuroendocrine changes. Under such circumstances, species such as starlings, sheep or chipmunks display recurrent switches between sexually active and inactive states, with a period shorter than a year, typically ~10 months (Dardente, Wood, et al., 2019; Gwinner, 1986; Gwinner & Wozniak, 1982; Kondo et al., 2006; Lincoln, 2019; Lincoln, Clarke, Hut, & Hazlerigg, 2006).

One key underlying feature of these circannual functions is photorefractoriness, which defines the ability to no longer respond to the photoperiod that triggered the initial switch in physiology, and spontaneously revert to the opposite physiological state (Butler et al., 2010; Dardente, Wood, et al., 2019; Goldman, 2001; Gorman, 2020). This has been exemplified in sheep, a species that breeds under short days: ewes kept under inhibitory long days from the summer solstice onwards spontaneously resume breeding activity and ewes kept under stimulatory short days from the winter solstice onwards do not maintain sexual activity beyond the expected date (Robinson & Karsch, 1984; Robinson, Wayne, & Karsch, 1985). Photorefractoriness is observed in other species, such as hamsters, which display spontaneous gonadal reactivation at the end of winter even though maintained under short photoperiods (Butler et al., 2010; Dardente et al., 2014; Reiter, 1993).

2. Integrated mechanisms of photoperiodism

2.1 Implication of melatonin in photoperiodism: Early observations

In birds, photoperiod acts in a direct manner through extra-ocular, deep-brain photoreceptors located in the MBH, without any overt involvement of the eye, the pineal gland and melatonin (Halford et al., 2009; Hanlon, Ramachandran, Zuidhof, & Bedecarrats, 2020; Kosonsiriluk, Mauro, Chaiworakul, Chaiseha, & El Halawani, 2013; Nakane et al., 2010; Yasuo & Yoshimura, 2009). In mammals, photoperiodism requires light detection through the eyes and there is no extra-ocular photoreception. Light activates melanopsin-expressing cells of the retina, which project to the hypothalamic suprachiasmatic nuclei (SCN). In turn, the SCN use a polysynaptic pathway that ultimately projects to the pineal gland to direct melatonin production and secretion exclusively at night; the longer the night, the longer the melatonin signal (Bartness, Powers, Hastings, Bittman, & Goldman, 1993;

Simonneaux & Ribelayga, 2003). Therefore, melatonin is the endocrine transducer of photoperiod, which carries both time-of-day and time-of-year information to melatonin-responsive cells (Reiter, 1993). The localization of these melatonin-responsive cells was made possible by the development of an assay to label melatonin with the ^{125}I isotope (Vakkuri, Leppaluoto, & Vuolteenaho, 1984). This led to a flurry of studies using ^{125}I-Melatonin for binding studies on cell membranes and autoradiography on tissue sections (Morgan, Barrett, Howell, & Helliwell, 1994; Vanecek, 1998).

Then, molecular cloning led to the identification of two high-affinity melatonin receptors, now known as MT1 and MT2 (Dubocovich, Rivera-Bermudez, Gerdin, & Masana, 2003; Reppert et al., 1995; Reppert, Weaver, & Ebisawa, 1994). The MT1 subtype is both necessary and sufficient for photoperiodic responses governed by melatonin (Liu et al., 1997; Reppert et al., 1994; Weaver, Liu, & Reppert, 1996; Yasuo, Yoshimura, Ebihara, & Korf, 2009). So, where are these MT1-expressing, melatonin-responsive cells that trigger the photoperiodic response? The early studies mentioned above had identified melatonin-binding sites in numerous central and peripheral tissues, with variability among species (Malpaux, Migaud, Tricoire, & Chemineau, 2001; Masson-Pevet et al., 1994; Simonneaux & Ribelayga, 2003). However, in the brain, two structures showed consistent labeling across species: the SCN and the *pars tuberalis* (PT) of the pituitary, with a much higher density of binding sites in the PT (Dardente, 2012; Morgan & Hazlerigg, 2008). Furthermore, in the photoperiodic ferret melatonin-binding sites are found in the PT, but are absent in the central nervous system (Weaver & Reppert, 1990). Interestingly, earlier work in hamsters had demonstrated photoperiodic changes in the ultrastructure of PT cells (Wittkowski, Hewing, Hoffmann, Bergmann, & Fechner, 1984) and in their expression of the *Tshb* gene (Wittkowski, Bergmann, Hoffmann, & Pera, 1988). Indeed, the PT comprises a population of specific thyrotropic cells that express the MT1 receptor and these cells are the target of melatonin for the control of seasonal functions (Dardente, Klosen, Pevet, & Masson-Pevet, 2003; Klosen et al., 2002). Compared to the *pars distalis* (PD), the PT presents several remarkable features pertaining to its distinct embryonic origin and the expression of endocrine markers very early during development (Dardente, 2007; Stoeckel, Hindelang, Klein, Poissonnier, & Felix, 1993). To summarize, photoperiodic entrainment of seasonal functions in mammals involves the action of melatonin in the PT.

2.2 Implication of TH in photoperiodism: Early observations

An early study in ducks showed that thyroidectomy (THX), the removal of the thyroid gland, disrupts photorefractoriness (Benoit & Aron, 1934). Similar findings were later reported in starlings (Wieselthier & van Tienhoven, 1972; Woitkewitsch, 1940). Even though results are at variance across species, it appears that TH are somehow required for spontaneous switches of the reproductive axis (Dardente, 2012; Dawson, 2015; Follett & Nicholls, 1984; Nicholls, Follett, Goldsmith, & Pearson, 1988; Yoshimura, 2013).

A potential role for TH in the ovine seasonal cycle has been investigated in great detail throughout the late 80s and 90s. Since this has already been reviewed elsewhere (Dardente, 2012; Karsch, Dahl, Hachigian, & Thrun, 1995), only the most salient points will be mentioned here. When ewes are THX at the end of the anestrus season or during the breeding season, anestrus fails to occur, as assessed by persistently high LH/FSH levels (Lomet et al., 2018; Moenter, Woodfill, & Karsch, 1991; Nicholls et al., 1988; see Fig. 1A). Similar observations have been made in rams (Parkinson & Follett, 1994, 1995). In both sexes, THX does not affect the seasonal profiles of either melatonin or PRL; both are thus independent from TH. Thyroxine (T4) supplementation in THX ewes reinstates the normal transition to anestrus as assessed by a sharp decrease in LH levels (Webster, Moenter, Woodfill, & Karsch, 1991). The root of persistently high LH levels during anestrus seems to be a chronic activation of the Gonadotropin Releasing Hormone (GnRH) pulse generator (Webster, Moenter, Barrell, Lehman, & Karsch, 1991). Finally, an elegant series of studies using THX ewes and T4 replacement demonstrated that T4 does not need to be present all the time, it is only required for a discrete time-window (<90 days) at the end of winter, beginning of spring (Thrun, Dahl, Evans, & Karsch, 1996, 1997). The impact of T4 upon seasonal rhythms is mediated at the central level (Viguie, Battaglia, Krasa, Thrun, & Karsch, 1999), more specifically within the MBH as demonstrated using T4 intracerebral implants (Anderson et al., 2003). Similar studies later undertaken on Siberian hamsters led to a similar conclusion (Barrett et al., 2007; Murphy et al., 2012). It is worth mentioning that the PT, even though it is an endocrine tissue, is part of the MBH, and is located in close apposition to the median eminence (ME), which receives GnRH nerve endings, and to the arcuate nucleus (ARC) that comprises *Kiss1*-expressing neurons (Beltramo, Dardente, Cayla, & Caraty, 2014; Fitzgerald, 1979; Gross, 1984; see Fig. 1B).

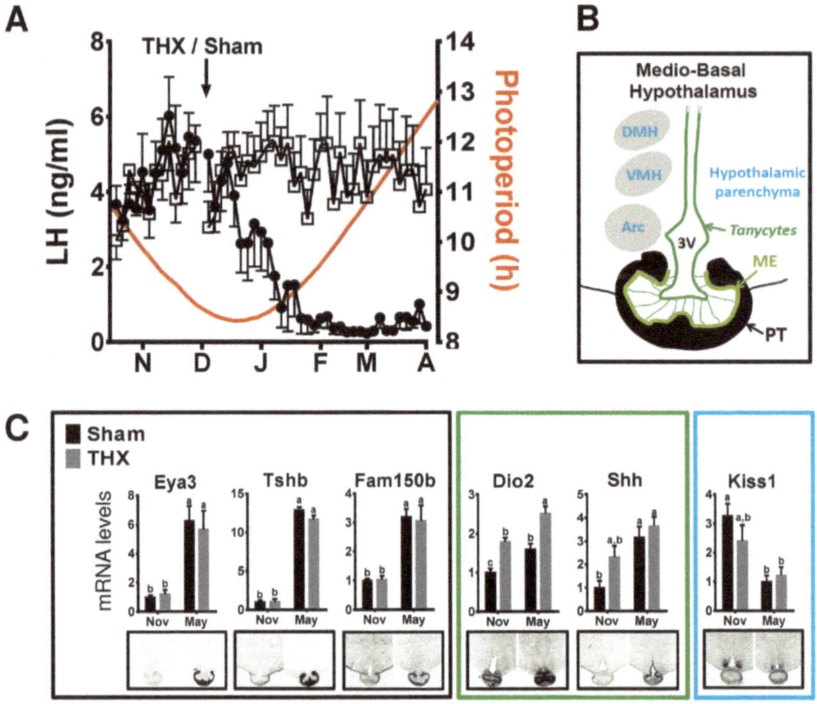

Fig. 1 Endocrine, neuroanatomical and molecular features of seasonal timing in the ewe: a few examples. (A) Thyroidectomy (THX, hollow squares) at the end of the breeding season prevents the normal drop in plasma LH levels (Sham-operated ewes, black dots) in an ovariectomized, estradiol-implanted model of ewe. The orange trace shows duration of photoperiod. (B) Schematics of the medio-basal hypothalamus (MBH), the key neuroendocrine orchestrator of the photoperiodic response. Arc, arcuate nucleus; DMH, dorso-medial hypothalamus; ME, median eminence; PT, pars tuberalis; VMH, ventro-medial hypothalamus; 3V, third ventricle. (C) Examples of "seasonal genes" rhythmically expressed in the PT (Eya3, Tshb and Fam150b; black rectangle), tanycytes (Dio2 and Shh; green rectangle) or the Arc (Kiss1; blue rectangle). Note that THX has no impact upon PT markers, while it differentially affects seasonal expression of markers in tanycytes and hypothalamic parenchyma. Images below graphs show representative results of autoradiographic films. *Adapted from Lomet, D., Cognie, J., Chesneau, D., Dubois, E., Hazlerigg, D., & Dardente, H. (2018). The impact of thyroid hormone in seasonal breeding has a restricted transcriptional signature.* Cellular and Molecular Life Sciences, *75(5), 905–919.*

Overall these studies in birds, sheep and hamsters have demonstrated that T4 is required for photorefractoriness to occur. T4 does not need to be present for long periods of time but its presence is crucial as daylength increases by the end of winter/early spring. To achieve this, T4 needs to be present within the MBH. In many respects the impact of T4 upon the reproductive axis mimics that exerted by exposure to long days (Dardente, 2012). In summary, photoperiodic entrainment of seasonal functions in mammals not only requires melatonin acting at the PT, but also involves the action of T4 within the MBH.

2.3 Basics of TH signaling

The thyroid gland mostly produces T4, while T3 is the active form. T3 binds to nuclear receptors TRα and TRβ, which heterodimerize with the retinoid X receptor (RXR) and recognize specific DNA motifs known as TRE (TH Response Elements). The unliganded TR/RXR heterodimer typically recruits nuclear co-repressors and associated histone deacetylases (NCoR and HDAC), which prevents transcription of target genes. Upon T3 binding, the NcoR/HDAC complex is released and replaced by co-activators, and transcription is initiated (Wu & Koenig, 2000). However, this model does not apply to the *Tshb* and *Trh* genes, which display inhibition rather than activation by T3 and, despite decades of intense investigation, the underlying transcriptional mechanism remains unclear (Weitzel, 2008). Non-genomic effects of TH, some of which depend on TRα/TRβ, have also been described but these are less well characterized (Davis, Goglia, & Leonard, 2016; Gauthier & Flamant, 2014).

Neither T3 nor T4 passively diffuse into the cell cytoplasm: both are actively carried inside and outside cells through dedicated transporters, including OATP1C1, MCT8 and MCT10 (Bianco, Salvatore, Gereben, Berry, & Larsen, 2002; Mullur, Liu, & Brent, 2014; Visser, Friesema, & Visser, 2011; Zhang, Boelen, Kalsbeek, & Fliers, 2018). The local actions of T4/T3 not only depend on the expression of TH transporters and receptors, but also on the presence or absence of specific enzymes, known as TH deiodinases (DIO). *Dio1* expression is restricted to the liver, kidney and thyroid gland. Within the adult brain, both *Dio2* and *Dio3* are expressed according to very specific patterns (Tu et al., 1997, 1999). DIO2 converts T4 into active T3, while DIO3 converts T4 into rT3 and T3 into T2, both "inactive" metabolites. Therefore, provided that TH transporters and receptors are present, local T3 signaling is governed by the DIO2/DIO3 balance.

2.4 Integrating melatonin and TH: Current model for photoperiodic read-out

In their seminal work on quail, Yoshimura et al. (2003) showed that MBH cells lining the third ventricle (3V) display a marked opposite seasonal regulation of *Dio2* and *Dio3* expressions; *Dio2* is highest under long days while *Dio3* is highest under short days (see Fig. 1C). The net result is an increase in local T3 production in the MBH under long days, which drives the gonadal response (Yoshimura et al., 2003). Further work in quail (Nakao et al., 2008) and sheep (Hanon et al., 2008) demonstrated that this conserved response of *Dio2/Dio3* depends upon TSH produced by the PT. In mammals, the current model for photoperiodic entrainment could be summarized as follows: (i) a long day melatonin signal triggers expression of *Tshb* in the PT (see Fig. 1C), (ii) PT-derived TSH binds to its cognate TSHR expressed by neighboring tanycytes of the 3V, (iii) TSHR signaling leads to P-CREB activation of *Dio2* expression and (iv) the local increase in DIO2 leads to heightened T3 signaling under long days; this is possible since tanycytes express the main T4 transporters OATP1C and MCT8. In contrast to *Dio2*, the transcriptional mechanisms underlying the opposite photoperiodic regulation of *Dio3* remain enigmatic (Milesi, Simonneaux, & Klosen, 2017; Petri et al., 2016). The TSH/DIO2/T3 backbone of this cascade is conserved in all species of birds and mammals examined thus far, and it has been detailed in many recent reviews (Barrett & Bolborea, 2012; Dardente et al., 2014; Dardente, Wood, et al., 2019; Ebling, 2014; Gorman, 2020; Hazlerigg & Simonneaux, 2015; Helfer et al., 2019; Korf, 2018; Yoshimura, 2013; see Fig. 2). In contrast, evidence for this cascade in reptiles and amphibians is still fragmentary or missing (Dardente, Wood, et al., 2019). Even though fish do not seem to have a proper PT (Fitzgerald, 1979), roles for TSH, DIO2 and T3 have been suggested, in particular with respect to the smoltification process (Dardente, Wood, et al., 2019; Fleming et al., 2019; Lorgen et al., 2015; Nakane et al., 2013).

To provide a consistent photoperiodic TSH output, PT-specific thyrotropic cells possess several morphological and molecular characteristics, such as their lack of responsiveness to either TRH or T3, as they are void of the cognate receptors (Bockmann et al., 1997; Dardente, 2007; Lomet et al., 2018). Furthermore, this PT-derived TSH has a unique glycosylation signature, which prevents it from acting at the thyroid gland, thereby restraining its action on the neighboring tanycytes (Ikegami et al., 2014). It remains to be investigated whether TSHR expressed by tanycytes differs from that expressed by thyrocytes; this would be expected in order to prevent thyroid-derived TSH action upon tanycytes. Therefore, the PT/tanycytes communication through TH signaling really is unique, disconnected from the classical hypothalamo-pituitary-thyroid axis. It is important to stress that this long day triggered TSH/DIO2/T3 axis is identical in short-day

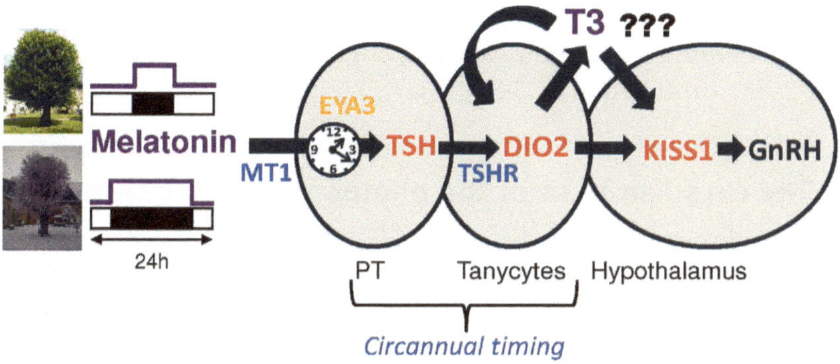

Fig. 2 Schematic of the neuroendocrine response to long days in the MBH. Melatonin is the endocrine messenger of daylength. A short melatonin signal, typical of long days, impacts the PT circadian clock, which triggers *Tshb* expression, induction of *Dio2* in tanycytes and a local increase in T3 levels. The exact role of T3 remains unclear (question marks) but likely involves an impact upon *Kiss1* expressing neurons of the Arc, which eventually impact GnRH/LH secretion. The PT appears to be the site of the circannual clock, even though a participation of tanycytes can not be formally excluded. *Adapted from Lomet, D., Cognie, J., Chesneau, D., Dubois, E., Hazlerigg, D., & Dardente, H. (2018). The impact of thyroid hormone in seasonal breeding has a restricted transcriptional signature.* Cellular and Molecular Life Sciences, *75(5), 905–919; Lomet, D., Druart, X., Hazlerigg, D., Beltramo, M., & Dardente, H. (2020). Circuit-level analysis identifies target genes of sex steroids in ewe seasonal breeding.* Molecular and Cellular Endocrinology, 512, *110825.*

(e.g., sheep and goats) and long-day breeders (e.g., hamsters and quail), and is also conserved in non-photoperiodic (e.g., mouse) or weakly photoperiodic (e.g., F344 rats) species (Ono et al., 2008; Ross, Helfer, Russell, Darras, & Morgan, 2011). This implies that divergent phenologies arise downstream of this TSH/DIO2/T3 axis (Dardente, Wood, et al., 2019; Helfer et al., 2019).

Finally, besides the TSH output, it seems likely that photoperiod activates several other autocrine/paracrine pathways within the PT, in a species-specific manner (Helfer et al., 2019; Korf, 2018). We and others have recently characterized the seasonal transcriptome of the ewe MBH and found that ~10% of all expressed genes display statistically significant seasonal expression (Lomet et al., 2018; Wood et al., 2015). Many of these genes encode secreted factors and are specifically expressed in the PT (Dardente, Lomet, Chesneau, Pellicer-Rubio, & Hazlerigg, 2019; Lomet et al., 2018). The expression of these PT-specific seasonal genes appears unaffected by TH (Lomet et al., 2018) or by sex steroids (Lomet, Druart, Hazlerigg, Beltramo, & Dardente, 2020), which strengthens the notion that the PT acts as a circannual clock, exclusively responsive to photoperiod through melatonin (Dardente, Wood, et al., 2019; Lincoln et al., 2006; Lincoln, Johnston, Andersson, Wagner, & Hazlerigg, 2005; Saenz de Miera et al., 2013, 2014; Wood et al., 2015).

2.5 The circadian basis of the photoperiodic response

The crux of the photoperiodic axis is PT-derived TSH. But how does the long day melatonin signal lead to increased expression of *Tshb* in the PT? Photoperiodism is based on daylength comparison throughout the course of the year. This requires a daily timer, which can then acts as a comparator: is today's daylength longer or shorter than yesterday's? The net physiological output of this measure will not only depend on the absolute daylength, but also on the direction of the photoperiodic change; this corresponds to the concepts of critical photoperiod and photoperiodic history. These will not be developed here as reviews and papers on that topic abound (Appenroth et al., 2020; Dardente, Lomet, et al., 2019; Dardente, Wood, et al., 2019; Goldman, 2001; Gorman, 2020; Hazlerigg, Lomet, Lincoln, & Dardente, 2018; Saenz de Miera, 2019). However, one can deduce that any tissue involved in photoperiodism is expected to harbor a circadian clock. Melatonin being the endocrine messenger of photoperiod, this circadian clock is also expected to be sensitive to melatonin. Indeed, the PT rhythmically expresses a complete set of clock genes and melatonin acutely induces the expression of the circadian clock gene *Cry1* in this tissue (Dardente, 2007; Dardente, Menet, et al., 2003; Fustin et al., 2009; Hazlerigg, Andersson, Johnston, & Lincoln, 2004; Lincoln, Messager, Andersson, & Hazlerigg, 2002; Messager, Ross, Barrett, & Morgan, 1999). Therefore, melatonin acts as a potent synchronizer of the PT circadian clock (see Fig. 3). If *Cry1* induction marks "night onset," how is "light onset" signaled? Actually, is it necessary to have both to measure duration? Probably not. The best model to account for the photoperiodic response is the classical "external coincidence model" developed by Büning (1936) and further refined in the early 70s (see Appenroth et al., 2020; Hazlerigg & Wagner, 2006). In this model, an output of the circadian clock may be produced within or outside a temporal window of sensitivity, such that a physiological response is triggered or is not. In this context, the *Eya3* gene is the prime candidate: it is controlled by the circadian clock (i.e., a clock-controlled gene), its expression is blunted by melatonin and EYA3 is a strong co-activator of *Tshb* expression (Dardente et al., 2010; Masumoto et al., 2010; Wood et al., 2020). All these features allow an increase in *Eya3* expression, hence PT-TSH output, when

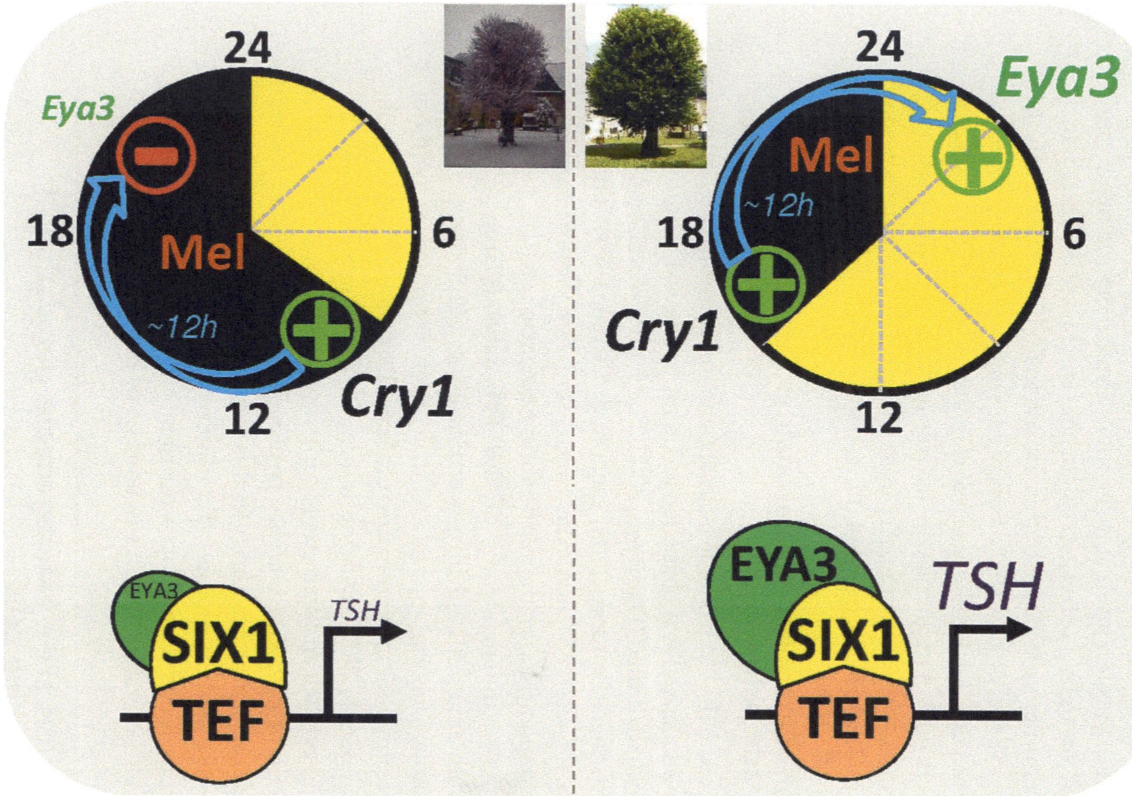

Fig. 3 Schematic illustrating how melatonin impacts the PT circadian clock to govern differential TSH seasonal output: the external coincidence model. Under both short (left panels) and long (right panels) daylengths, melatonin triggers a rapid increase in *Cry1* expression, which resets the PT clock. The expression of *Eya3* is phase-locked to the clock, as it is a clock-controlled gene. Hence, the peak of *Eya3* expression always occurs ~12 h after the *Cry1* peak. As *Eya3* expression is also repressed by melatonin (through unclear mechanisms) it follows that *Eya3* levels do not rise under short days. EYA3 is a co-activator of the TEF/SIX1 heterodimer, which transactivates the *Tshb* gene at a specific D-Box motif (bottom panels). This accounts for the increased *Tshb* expression under long daylengths. *Adapted from Dardente, H., Hazlerigg, D. G., & Ebling, F. J. (2014). Thyroid hormone and seasonal rhythmicity.* Frontiers in Endocrinology, *5, 19.*

daylengths get longer than 12–14h (Dardente et al., 2014; Dardente, Wood, et al., 2019; Hazlerigg et al., 2018; Wood & Loudon, 2018; see Fig. 3), making it possible to translate coincidence into amplitude of the photoperiodic response (Hazlerigg & Wagner, 2006). Interestingly, the profound photoperiodic changes in expression levels at loci coding *Eya3*, *Tshb* and many other "seasonal genes" are tightly correlated to changes in chromatin state, which demonstrates genome-wide orchestration of seasonal changes down to the epigenetic level (Wood et al., 2020). Finally, definitive functional evidence for the implication of EYA3 in the seasonal response in mammals is still missing but we note that EYA might play such a role in drosophila (Abrieux et al., 2020), which points to some degree of conservation in the molecular bases for seasonality throughout the animal kingdom.

2.6 Photoperiodic control of seasonality: The classical anterograde pathway

Two decades before this "retrograde" signaling pathway (from the pituitary back to the hypothalamus) was characterized, a more classical "anterograde" pathway had been identified (Dardente, 2007; Dardente, Wood, et al., 2019; Lincoln, Andersson, & Loudon, 2003; Lincoln & Clarke, 1994). This pathway involves a PT-derived endocrine factor, tentatively called tuberalin, which appears responsible for the photoperiodic control of PRL secretion by PD lactotrophs, independently of the TH status (Dardente, 2007; Dardente, Wood, et al., 2019; Morgan & Hazlerigg, 2008; Morgan et al., 1996). Even though several candidates have been proposed throughout the years, tuberalin's identity remains unclear (Dardente, Wood, et al., 2019). To summarize, PT thyrotrophs govern two signaling pathways which operate in parallel: an anterograde pathway, which targets PRL secretion and a retrograde pathway, which is responsible for the seasonal control of LH/FSH release, hence breeding. In this retrograde pathway, tanycytes play a pivotal role as they produce T3 under the control of PT-derived TSH. In Sections 3 and 4, we focus on these peculiar type of glial cells and explore their potential roles in the control of breeding and feeding.

3. Tanycytes are a heterogeneous population of multi-tasking cells

3.1 Morphology

Tanycytes, initially described by Ramon y Cajal in the adult mammalian brain as cells highly reminiscent of embryonic radial glia cells, were later named from the Greek "tanus" for elongated (Prevot et al., 2018; Rodriguez,

Guerra, Peruzzo, & Blazquez, 2019). These specialized ependymoglial cells differ ontogenetically, morphologically and functionally from the neighboring cuboidal ependymal cells, which line the dorsal part of the third ventricular wall and whose main role appears to be the control of the flow of cerebro-spinal fluid (CSF). At first glance, the whole tanycytic population share common morphological features: a cell body forming the ventral part of the third ventricular wall and a unique elongated basal process extending into the hypothalamic parenchyma or the ME. However, they were first divided into two classes according to observations made by photonic microscopy:

i. Dorsal tanycytes that exhibit a long process forming a bow through the ARC. In this region, tanycytes lining the third ventricular wall have recently been shown to be biciliated ependymal cells (Mirzadeh et al., 2017). This tanycyte subtype makes contacts with blood-brain barrier (BBB) vessels in the ARC, the ventro-median and dorso-median hypothalamus (VMH, DMH).

ii. Ventral tanycytes located on the lower third and floor of the 3V, in close proximity to the ARC. Their apical pole contacts the ventricular lumen and they display a rather short process that either forms contact with the ME capillary network, linking the floor of the 3V to the portal system, or extend toward the pial surface of the brain. In contrast to the dorsal tanycytes, ventral ones only have a primary cilium (uniciliated cells).

Later on, these two groups of tanycytes were classified as α and β tanycytes, according to their dorsal or ventral position along the border of the 3V, respectively. These two classes were further split into two subtypes (Fig. 4; Akmayev, Fidelina, Kabolova, Popov, & Schitkova, 1973) based on their topology along the 3V, the location of their unique extension, their ultrastructure, their histochemical features and their functions (Rodriguez et al., 2019). α1 tanycytes send their long processes (300 μm in the rat brain) to the DMH and VMH, which also establish contact with blood capillary basement membrane as well as *en passant* contact. α2 tanycytes have basal processes that spread into the ARC where they contact blood capillaries. β1 tanycytes line the infundibular recess, or pituitary stalk, and their processes mainly reach the pial surface of the brain, more specifically the perivascular space of the fenestrated vessels located laterally to the ME. Ultrastructural analysis revealed that their basal process exhibit endocytosis/transcytosis and transport. They could further be subdivided into dorsal β1 and ventral β1 tanycytes (Rodriguez et al., 2019). Finally, β2 tanycytes line the floor of the 3V and extend processes through the ME toward the pial surface. Closer to the external border of the ME, their processes branch into multiple

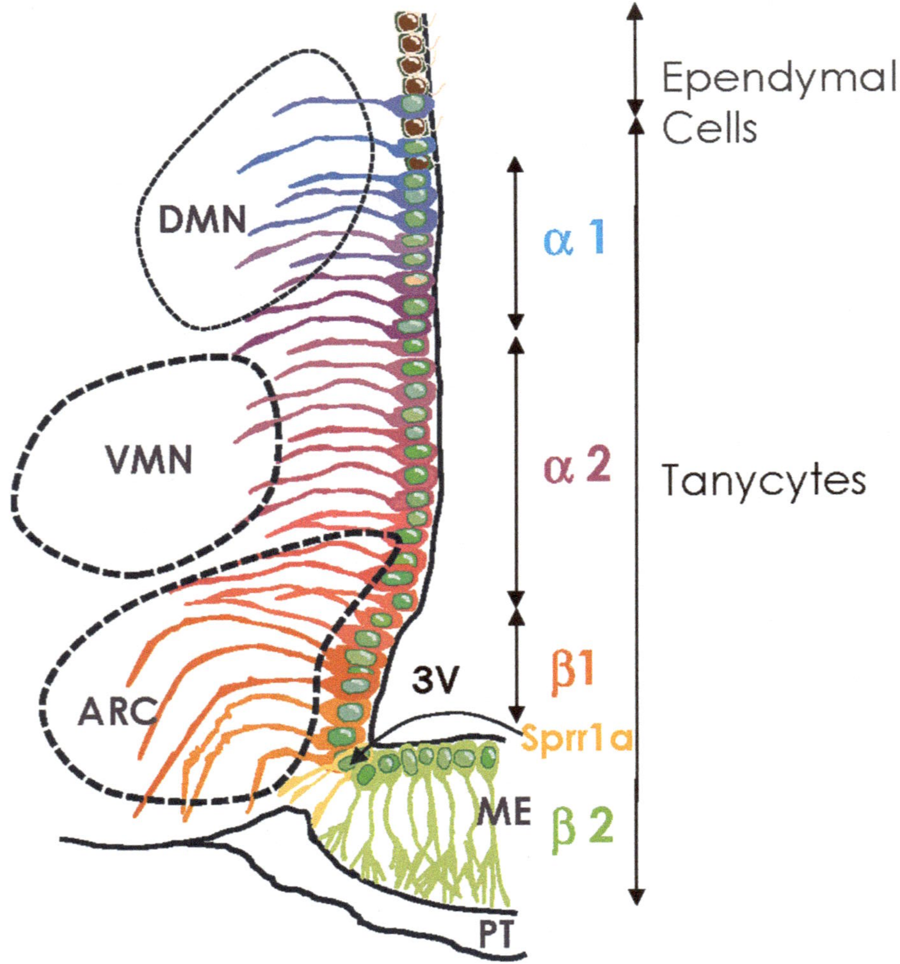

Fig. 4 See figure legend on opposite page.

parallel ramifications forming large pyramidal "endfeet" that connect pituitary portal vessels (Rodriguez, Gonzalez, & Delannoy, 1979).

Although widely used because of its practicality, this subdivision of tanycyte subtypes based on histological criteria is somewhat incomplete and recent thorough single-cell RNAseq analyses demonstrate a much larger diversity within these cells (Campbell et al., 2017; Chen, Wu, Jiang, & Zhang, 2017; Romanov et al., 2020, 2017). For instance, a new small subgroup of tanycytes located at the limit of the arcuate nucleus and the ME was recently reported (Campbell et al., 2017). This peculiar subtype may display unexpected characteristics in its ability to form a barrier to diffusion. Initially thought to lack tight junctions, these tanycytes specifically express *Sprr1a*, a small prolin-rich protein, which participates in the formation of an insoluble protective cornified cell envelope and provides optimal barrier function to epithelia (Fig. 4). These data clearly illustrate that there are more than four subtypes of tanycytes. This complexity has been nicely summarized by Chen et al. (2017): "Notably, although specific marker genes (or combinations of marker genes) can be used to roughly separate tanycyte subtypes, many genes exhibited a gradient, rather than a clear-cut distribution across tanycyte subpopulations consistent with the notion that tanycytes may be composed of continuous cell trajectory with transition zones between different subtypes." Finally, what is being described in detail using mouse as an animal model may well apply to other mammals, so that this cellular complexity must be kept in mind when analyzing tanycytes. This complexity might hamper the identification of the exact tanycyte subtype, if any, that corresponds to neural stem cells (NSCs).

3.2 Tanycytes as key actors of the interaction between the brain and the periphery

Due to their position at the interface between the CSF, the hypothalamic parenchyma and fenestrated blood vessels of the pituitary plexus, tanycytes

Fig. 4 Schematic representation of the different populations of tanycytes in the tuberal region of the hypothalamus. Ventrally to the cuboidal ependymal cells, four subtypes of tanycytes (α1, α2, β1 and β2) can be distinguished, mainly based on their location along the dorso-ventral axis. An additional fifth subgroup was recently reported (Sprr1a), which is specifically located in the zone limiting the ME and the ARC (Campbell et al., 2017). The color grades reflect the notion that most of the genes present in tanycytes show a gradient of expression, rather than a well-defined distribution between the tanycyte subpopulations (see text). ARC, arcuate nucleus; DMN, dorsomedial nucleus; ME, median eminence; VMN, ventromedial nucleus; 3V, third ventricle.

are ideally located to relay information from the periphery to the brain, and vice versa. Indeed, over the last decade, tanycytes located in the infundibular recess and at the ventral surface of the 3V have been under close scrutiny after their roles in the regulation of reproduction and energy expenditure were disclosed. Tanycytes seem to play an integrative and bi-directional function: they relay peripheral information to the neural brain circuits and also impinge on peripheral signaling through an impact upon central processes. The BBB tightly controls the entry of blood-borne molecules into the brain. However, seven structures called circumventricular organs (CVO) make exceptions. In these CVO, the BBB is interrupted (Miyata, 2015) since the endothelia of their capillaries are fenestrated (Duvernoy & Risold, 2007). Therefore, molecules can be transferred from the periphery to the brain and vice versa. In addition, within the secretory CVOs, among which the ME, tanycytes are acting as ventricular barriers since they selectively express tight junction proteins such as occludin, zonula occludens-1 (ZO-1) and Claudin 1 and 5 (Langlet, Mullier, Bouret, Prevot, & Dehouck, 2013; Mullier, Bouret, Prevot, & Dehouck, 2010). This tanycytic physical belt precludes circulating molecules from diffusing to the brain via the CSF.

3.3 Tanycytes and metabolic regulation

The potential role of tanycytes in the regulation of food intake and energy balance in general has been discussed thoroughly in many recent reviews (Bolborea & Dale, 2013; Dardente, Wood, et al., 2019; Ebling, 2015; Ebling & Lewis, 2018; Goodman & Hajihosseini, 2015; Helfer et al., 2019; Prevot et al., 2018). Here, we will only briefly discuss the most salient findings. Until recently, it was assumed that the impact of T3 upon metabolism was dependent upon its action in adipose tissue (Mullur et al., 2014). However, recent findings indicate that T3 might act within the MBH to trigger long-lasting changes in energy metabolism (Kong et al., 2004; Lopez, Alvarez, Nogueiras, & Dieguez, 2013; Lopez et al., 2010; Zhang et al., 2018). Therefore, since tanycytes display photoperiodic control of T3 production, they are well positioned to govern seasonal switches in energy storage and expenditure. Considering the pivotal role of the neighboring ARC, which comprises both orexigenic NPY/AgRP and anorexigenic POMC/CART neurons, a straightforward model would be that T3 organizes swings in expression of these antagonistic systems. There is evidence that expression of POMC is higher under long days compared to short days in the jerboa (Talbi, Laran-Chich, Magoul, El Ouezzani, & Simonneaux, 2016) and the Siberian hamster (Mercer et al., 2001; Reddy, Cronin, Ford, & Ebling, 1999; Rousseau et al., 2002), and it has indeed been proposed that this long-day increase in POMC is governed by T3 (Bao et al., 2019). However, this finding is paradoxical since siberian hamsters actually gain weight upon transfer from short days to long days. Overall, the current view is that this ARC neuropeptidergic circuit, which accounts for the homeostatic control of feeding, is not primarily involved in driving the seasonal body weight response (Dardente, Wood, et al., 2019; Ebling, 2015; Morgan, Ross, Mercer, & Barrett, 2006). In this respect, a novel mechanism involving the transfer of leptin from the CSF to the ARC through tanycytes had been proposed (Balland et al., 2014). However, further extensive experiments using intersectional genetics with the CRE-LoxP system and single-cell RNAseq failed to detect leptin receptor in tanycytes and showed a tanycyte-independent regulation of leptin signaling (Yoo, Cha, Kim, Hoang, & Blackshaw, 2019). Furthermore, ablation of tanycytes themselves has a surprisingly mild impact upon energy balance (Yoo et al., 2020). Overall, even though tanycytes have the potential to impact the ARC neuronal network (Bolborea, Pollatzek, Benford, Sotelo-Hitschfeld, & Dale, 2020), their exact role in the control of metabolism remains ill-defined. Is the impact of tanycytes achieved through T3 or does it (also) depend upon other hormones or peptides? Much remains to be learnt.

3.4 Tanycytic plasticity, Kiss1 neurons and GnRH secretion

Ultrastructural studies have demonstrated that β1 tanycytes play a key role in the neuroendocrine control of GnRH secretion in vertebrates, by ensheathing and thus preventing the GnRH endfeet from contacting the basal lamina of the fenestrated capillaries of the portal vessels (Prevot et al., 1999; Prevot, Dutoit, Croix, Tramu, & Beauvillain, 1998). This tanycytic plasticity, which is key to the neuroendocrine control of reproduction involves semaphorins, a family of soluble and membrane-bound proteins constitutively expressed postnatally. For instance, Semaphorin7A expression triggers retraction of GnRH terminals (Parkash et al., 2015). The above-mentioned studies pertained to the GnRH surge that triggers ovulation. However, a similar plasticity mechanism has been found for the seasonal regulation of GnRH output at the level of the ME, both in quail (Yamamura, Hirunagi, Ebihara, & Yoshimura, 2004) and sheep (Wood et al., 2015). At least in quail, this plasticity is T3-dependent (Yamamura, Yasuo, Hirunagi, Ebihara, & Yoshimura, 2006). Besides GnRH, plasticity at the endfeet of tanycytes appears to regulate the release into the pituitary portal blood system of other hypothalamic secretagogues such as corticotropin-releasing hormone and thyrotropin-releasing hormone (Muller-Fielitz et al., 2017; Prevot et al., 2018; Rodriguez-Rodriguez et al., 2019).

To govern seasonal GnRH release in mammals, it is still unclear whether tanycyte-derived T3 acts directly at the level of neuronal endfeet within the ME and/or upstream, at the level of the cell bodies of neurons expressing *Kiss1* and *Npvf*, which respectively encode the neuropeptides KISS1 and RFRP3. KISS1 is the main GnRH secretagogue, expressed by a subpopulation of neurons in the ARC. In contrast, the role of RFRP3 in the control of GnRH has not been forthcoming, perhaps because it seems to vary according to multiple parameters including species, sex and season. Nevertheless, T3 infusion in the MBH of short-day inhibited hamsters reinstates a long-day expression pattern of *Kiss1* and *Npvf* and reactivates the gonadotropic axis (Angelopoulou, Quignon, Kriegsfeld, & Simonneaux, 2019; Dardente, Wood, et al., 2019; Henson, Carter, & Freeman, 2013; Klosen, Sebert, Rasri, Laran-Chich, & Simonneaux, 2013; Simonneaux, 2020). These two neuronal populations are therefore prime candidates in the T3-dependent control of seasonal breeding. Indeed, recent data point to a TRα-mediated role for T3 in the transcriptional control of the murine *Npvf* gene (Quignon, Beymer, Gauthier, Gauer, & Simonneaux, 2020). However, we wish to stress that increased expression of these neuropeptides without concomitant cytoskeletal rearrangements at the ME might not be efficacious; both mechanisms might work together. Furthermore, our data in sheep demonstrate that when high LH levels typical of the breeding season are blunted by exposure to long days, this occurs without any concomitant change in expression of either *Kiss1* or *Npvf* (Lomet et al., 2018). Since the impact of long days is very rapid, these findings seem to indicate that rapid, non-transcriptional mechanisms may be at play. In short, transcriptional changes at *Kiss1* and *Npvf* loci might not be primary events in the response to daylength.

4. Tanycytes as NSCs: Implication for long-term circannual rhythms?

4.1 Overview

Adult neurogenesis consists in the generation of new neurons from NSCs located in restricted regions of the brain within a specialized microenvironment, called a neurogenic niche. The two most scrutinized NSCs niches are the subgranular zone of the hippocampal dentate gyrus (SGZ) and the subventricular zone (SVZ) of the lateral ventricles (Fuentealba, Obernier, & Alvarez-Buylla, 2012; Migaud et al., 2010; Obernier & Alvarez-Buylla, 2019). More recently, the existence of neurogenic niches was demonstrated in other brain regions including the hypothalamus. Within the hypothalamus, there is mounting evidence that tanycytes are the source of these NSCs (Sousa-Ferreira, de Almeida, & Cavadas, 2014; Yoo & Blackshaw, 2018).

First, numerous NSCs markers are expressed by tanycytes such as glial fibrillary acidic protein (GFAP), Sox2, nestin, vimentin, and the glutamate/aspartate transporter (GLAST) (Batailler et al., 2014; Lee et al., 2012; Rodriguez et al., 2005). This feature is conserved among a variety of species including mouse (Haan et al., 2013; Robins et al., 2013), rat (Perez-Martin et al., 2010; Xu et al., 2005), sheep (Batailler et al., 2014), Siberian hamster (Barrett et al., 2006; Bolborea et al., 2011), gray mouse lemur (Pellegrino et al., 2018) and humans (Koopman, Taziaux, & Bakker, 2017; Pellegrino et al., 2018). Second, *in vitro*, tanycytes are capable of generating neurospheres and can differentiate into neurons and glial cells (Bennett, Yang, Enikolopov, & Iacovitti, 2009; Li, Tang, & Cai, 2012; Markakis, Palmer, Randolph-Moore, Rakic, & Gage, 2004; Robins et al., 2013; Xu et al., 2005). Using isolated cells from transgenic animals expressing reporter genes specifically in tanycytes has shown that they recapitulate cardinal properties of the NSCs. In the presence of epidermal growth factor (EGF) and fibroblast growth factor (FGF), isolated tanycytes can proliferate and form neurospheres that can be passaged, indicating their self-renewal potential. When cultured in media without growth factors, tanycytic neurospheres differentiate into neurons, astrocytes and oligodendrocytes (Bennett et al., 2009).

In vivo, tanycytes can proliferate at a slow rate (Chaker et al., 2016; Haan et al., 2013; Robins et al., 2013; Xu et al., 2005), similar to the proliferation rate reported in canonical neurogenic niches of the DG and SVZ (Alvarez-Buylla & Lim, 2004). *In vivo* lineage tracing experiments have clearly confirmed the stemness potential of tanycytes (Chaker et al., 2016; Haan et al., 2013; Lee et al., 2012; Li et al., 2012; Robins et al., 2013). However, not all tanycyte subtypes retain the capacity to generate new neurons and glial cells, and the incriminated subtype is still a matter of debate. Furthermore, considering the staggering diversity of tanycytes (*vide supra*), the identification of the "NSC-subtype," if any, promises to be challenging.

There is ultrastructural evidence that some tanycytes exhibit two (9 + 2) cilia, whereas others are endowed with either a single cilium (9 + 0) or none

(Del Brio, Riera, Garcia, & Alvarez-Uria, 1991; Flament-Durand & Brion, 1985). The biciliated and monocliated tanycytes were found to belong to the α and β tanycytes, respectively (Mirzadeh et al., 2017). Since the presence of a single cilium (9+0) is a conserved feature of the NSCs located in the SVZ, β tanycytes were then suggested to be NSCs (Mirzadeh et al., 2017). Beyond ultrastructural studies indicating that tanycytes are proliferating cells (Rodriguez et al., 2005), the use of transgenic mouse models expressing reporter genes under the control of tanycyte-specific promoters has provided insights regarding the stemness properties and the cell fate of the various tanycyte subtypes *in vivo*. Lineage tracing of β2 tanycytes using the Nestin-CreERT2 mouse line leads to the gradual appearance of neurons in the ME (Lee et al., 2012), suggesting that β2 tanycytes are the neurogenic progenitors of the ME. However, an experiment using GLAST-CreERT2 lineage tracing has evidenced that, in addition to be the only tanycyte subtype to express GLAST and to form neurospheres, the α2 tanycyte subtype also appears to be the source of β1 tanycytes subtype regeneration (Robins et al., 2013). Interestingly, during adulthood, the α tanycyte subtypes keep on expressing SHH, a morphogen involved in the control of cell proliferation and fate of neural progenitors, while β1-β2 subtypes of the ME do not (Traiffort et al., 1999). Moreover, EGF and FGF2, growth factors known to tightly control cell proliferation and neurogenesis in the SVZ and SGZ (Nieto-Estevez, Defterali, & Vicario-Abejon, 2016), increase adult α tanycytes proliferation level (Oya et al., 2008; Robins et al., 2013). Finally, Prss56, a serine protease specifically expressed in the SVZ and SGZ neurogenic niches was also detected in the α2-tanycytes subpopulation (Jourdon et al., 2016). In summary, the identity of the source of NSCs among tanycytes remains unclear.

4.2 Functional relevance for energy balance

The MBH is involved in the control of energy balance through modulation of appetite, food intake, regulation of fat storage and energy expenditure. A potential role for the newly born hypothalamic cells in the regulation of energy metabolism has been suggested (Haan et al., 2013; Kokoeva, Yin, & Flier, 2005; Lee et al., 2012; Li et al., 2012). Recent studies have pointed the interrelationship between the homeostatic role of MBH and adult neurogenesis. In human, obese individuals treated with Ciliary neurotrophic factor (CNTF) keep on losing weight beyond the end of the treatment. In mice, there is evidence that adult neurogenesis in the hypothalamus mediates the impact of CNTF on weight (Kokoeva et al., 2005). In addition, mice fed a high-fat diet display reduced proliferation levels of hypothalamic NSCs associated with a lower survival rate and decreased differentiation of newly generated cells into neurons, oligodendrocytes and astrocytes (Li et al., 2012). This effect appears to be mediated by an inflammatory process triggered by microglia-derived TNFα and IL-1β. This inflammatory state activates pro-apoptotic pathways such as the IKK-β/NF-κB pathway in NSCs and also inhibits neuronal differentiation through the activation of the Notch pathway. The long-lasting activation of the IKK-β/NF-κB pathway in NSCs causes obesity and diabetic symptoms in normally fed mice (Li et al., 2012). Moreover, the decreased neurogenesis associated with the micro-inflammation triggered by the IKK-β/NF-κB pathway also activates aging mechanisms (Zhang et al., 2013). In conclusion, there is now substantial evidence that hypothalamic NSCs play diverse roles in the regulation of energy balance.

4.3 Functional relevance for seasonal breeding

A striking and unique feature of the hypothalamic neurogenic niche resides in the fact that it is impacted by photoperiod in seasonal mammals such as sheep. Under short photoperiod, when ewes are sexually active, the cell layer that lines the 3V and comprises tanycyte cell bodies is broader that during long photoperiod. Furthermore, numerous protrusions or blebs budding of the tanycytic cytoplasm are observed (Butruille, Batailler, Mazur, Prevot, & Migaud, 2018). These cytoarchitectural rearrangements are thought to reflect changes in the activity of tanycytic NSCs (Butruille et al., 2018). Indeed, proliferation of these NSCs depends upon photoperiod. It is consistently higher during short photoperiod compared to long photoperiod as assessed by short-term BrdU incorporation (Fig. 5A; Batailler et al., 2018; Hazlerigg, Wyse, Dardente, Hanon, & Lincoln, 2013; Migaud, Batailler, Pillon, Franceschini, & Malpaux, 2011). Survival studies show that 4 weeks after BrdU injection, 17% of the BrdU-positive cells also express the mature neuronal marker NeuN while 70% express either GFAP or S100b, both markers of astrocytes. Most importantly, as assessed through the expression of doublecortin (DCX), the number of young neurons or neuroblasts in the ARC and VMH is significantly higher under short photoperiod than long photoperiod (Batailler, Derouet, Butruille, & Migaud, 2016; Migaud, Butruille, & Batailler, 2015). Therefore, the period of sexual activity is associated with increased hypothalamic neurogenesis (Fig. 5B). Finally, removing

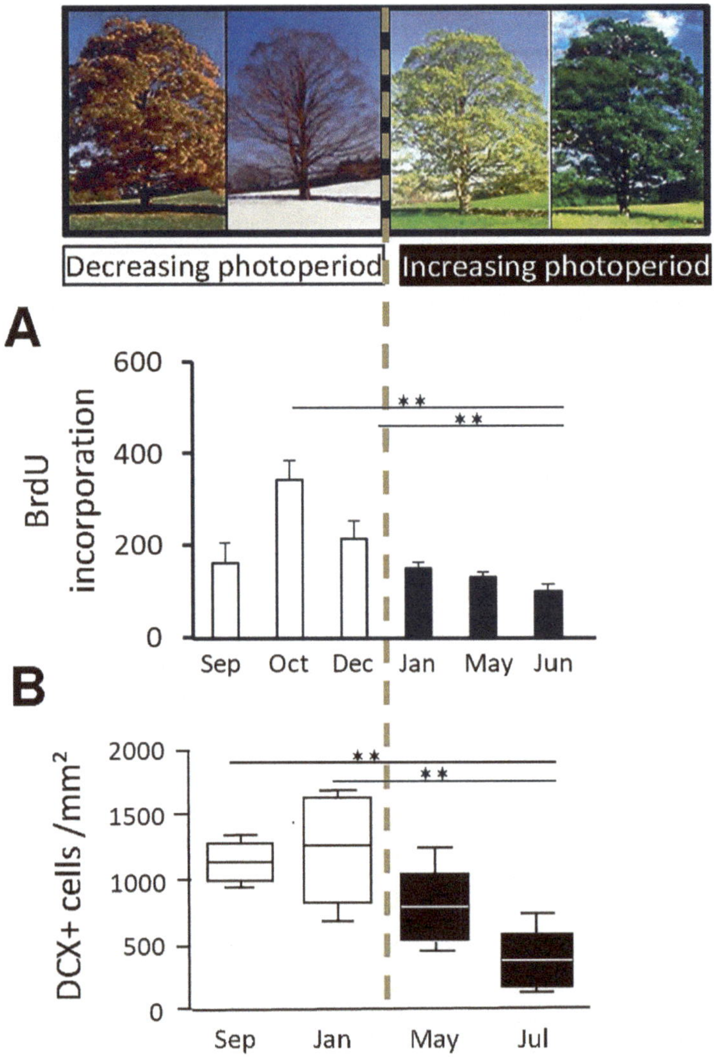

Fig. 5 Seasonal variation of neuronal proliferation in the hypothalamus of the ewe. (A) Longitudinal cell proliferation in sheep hypothalamus at six time points across the year, as assessed through BrdU incorporation in cell nuclei. A peak of cell proliferation consistently occurs under decreasing daylengths; (B) The density of DCX-positive cells in the ARC is higher under short than long photoperiods. Results are median and interquartiles ranges of the number of DCX-positive cells per mm^2. *Adapted from Batailler, M., Derouet, L., Butruille, L., & Migaud, M. (2016). Sensitivity to the photoperiod and potential migratory features of neuroblasts in the adult sheep hypothalamus.* Brain Structure & Function, 221*(6), 3301–3314; Batailler, M., Chesneau, D., Derouet, L., Butruille, L., Segura, S., Cognie, J., et al. (2018). Pineal-dependent increase of hypothalamic neurogenesis contributes to the timing of seasonal reproduction in sheep.* Scientific Reports, 8*(1), 1–13. 6188-018-24381-4.*

melatonin through pinealectomy abolishes this seasonal neurogenic activity in the hypothalamus, which demonstrates that this rhythm is indeed photoperiod-dependent (Batailler et al., 2018).

The functional relevance of this acute neurogenic activity was tested through intracerebroventricular infusion of the antimitotic drug Ara-C, which prevents NSCs proliferation. The Ara-C treatment decreased the number of DCX-immunopositive neurons and altered the timing of reproduction (Fig. 6A–C; Batailler et al., 2018). Ewes treated with Ara-C stopped their breeding season earlier than untreated ewes and also resumed their next breeding season earlier, as if the inner circannual rhythm had been shifted by ~1 month (Fig. 6B). These data suggest that the hypothalamic newborn neurons are involved in the physiology of seasonal reproduction as previously proposed (Hazlerigg & Lincoln, 2011).

How hypothalamic NSCs impact the timing of seasonal breeding remains an enigma. One hypothesis could be that newborn cells integrate specific circuits, which control GnRH neurons, either directly in the preoptic area and/or the ME, or indirectly. The indirect control of GnRH neurons could involve the RF-amide peptide circuits, namely KISS1 and/or RFRP3 (*vide supra*). Interestingly, lineage tracing studies in mouse have shown that mouse β tanycyte-derived cells can migrate long distances across the hypothalamus, more specifically toward anterior areas (Haan et al., 2013). Remarkably, in sheep, the DCX-positive cells located nearby the 3V are small, round shaped cells, reminiscent of immature neuronal precursors, whereas the neuroblasts located in the parenchymal regions are larger and show a fusiform bipolar morphology. The hypothalamic neuroblasts co-express Cdk5 and p35, two proteins involved in DCX phosphorylation, which governs the migration process (Batailler et al., 2016). In addition, DCX-labeled cells are also detected in rostral parts of the hypothalamus such as the preoptic area, which contains GnRH neurons (Batailler et al., 2016). These data indicate that cells derived from hypothalamic NSCs may migrate rather long distances within the hypothalamus. Therefore, at this stage both direct and indirect regulation of GnRH are plausible hypotheses.

4.4 Tanycytes, NSCs and TH in circannual timing

Could T3 produced by tanycytes be responsible for the aforementioned seasonal changes in the proliferation of hypothalamic NSCs? In the SVZ of adult mice, T3 controls cell proliferation and the commitment of progenitors into neurons or glial cells. Indeed, T3 is known to promote neuronal

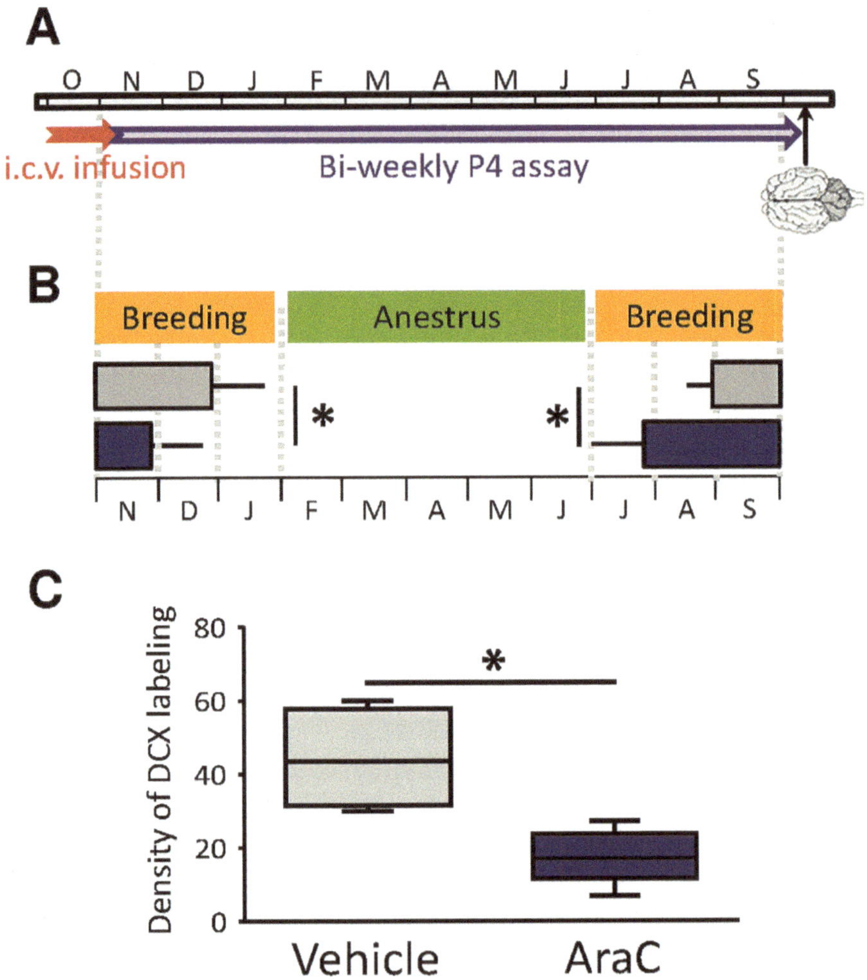

Fig. 6 Involvement of newly generated hypothalamic neurons in the seasonal schedule of reproduction. (A) Schedule of the experiment on reproductive programming: 10 ewes were treated for a month (i.c.v) with either vehicle ($n=5$) or the Ara-C (cytosine-b-D-arabinofuranoside, $n=5$) antimitotic drug at the beginning of October. Ovarian activity was assessed through plasma progesterone (P4) assay twice weekly. At the end of the treatments, brains were collected and assessed for neurogenesis. (B) Dates of the end of ovulatory activity followed by the dates of resuming ovulatory activity in vehicle (gray bars) and Ara-C (blue bars) treated animals. Ara-C treatment leads to a significant phase advance in the seasonal programing of reproduction. (C) Densities of DCX-positive cells in vehicle (gray bars) and Ara-C (blue bars) treated animals. Ara-C treatment diminishes the number of young neurons. *Adapted from Batailler, M., Chesneau, D., Derouet, L., Butruille, L., Segura, S., Cognie, J., et al. (2018). Pineal-dependent increase of hypothalamic neurogenesis contributes to the timing of seasonal reproduction in sheep.* Scientific Reports, 8*(1), 1–13. 6188-018-24381-4.*

differentiation of progenitor cells (Fanibunda, Desouza, Kapoor, Vaidya, & Vaidya, 2018; Gothie et al., 2017; Lopez-Juarez et al., 2012; Remaud, Gothie, Morvan-Dubois, & Demeneix, 2014; Remaud et al., 2017). The impact of T3 in this process involves transcripitonal repression of *Sox2*, Ccnd1a and c-Myc in progenitors (Lemkine et al., 2005; Lopez-Juarez et al., 2012). In sheep, *Dio2* expression increases dramatically during long days, leading to a greater local availability of T3 in the MBH. Our recent RNAseq analysis of the ewe MBH transcriptome revealed that the morphogen SHH is a β tanycyte marker whose expression is robustly regulated by season and T3 (Dardente, Lomet, et al., 2019; Lomet et al., 2018), consistent with prior findings in the adult rodent brain (Desouza et al., 2011; Gil-Ibanez, Bernal, & Morte, 2014). In the canonical neurogenic niches, SHH signaling is implicated in self-renewal processes and specification of the NSCs (Alvarez-Buylla & Ihrie, 2014). In the sheep hypothalamus, *Shh* expression is increased following THX, suggesting that T3 inhibits its expression (Lomet et al., 2018). We further showed that *Dct* (Dopachrome tautomerase), a major regulator of neural progenitor cell proliferation (Jiao et al., 2006), also displays strong photoperiodic, TH-dependent, regulation of expression in tanycytes (Dardente & Lomet, 2018). Finally, the gene most impacted by THX is *Tmem252* (a transmembrane protein of unknown function), which completely fails to show LP induction in THX animals. Interestingly, *Tmem 252* is exclusively expressed by β2 tanycytes. Considering that THX only has a very restricted impact in terms of transcriptional regulation within the MBH (<50 genes affected; Lomet et al., 2018, 2020), it seems that tanycytes constitute a hotspot, and cell proliferation of these peculiar glial cells might be an important feature of the photoperiodic control of seasonality. In conclusion, there are converging lines of evidence in favor of a role for T3 in controlling the expression of a few key genes involved in cell proliferation. Interestingly, these genes are expressed by tanycytes themselves, which suggests that an autocrine T3 feedback loop may be part of the seasonal timing mechanism. Exploring hypothalamic neurogenesis in THX ewes and in THX-T3 replaced ewes appears to be an interesting avenue for future research.

5. Conclusion

Seasonal rhythms are present in most living organisms inhabiting temperate regions of the Earth. At these latitudes, photoperiod is the main environmental cue responsible for the entrainment of seasonal rhythms, with other cues such as food availability, stress and social interactions acting as modulators (Dardente et al., 2016). Seasonal rhythms are not passively driven by changes in photoperiod, they rely on built-in molecular and neuroendocrine circuits, which keep a form of time-memory. Photoperiod synchronizes these inner timing devices known as circannual clocks. Our current understanding of photoperiodic entrainment involves a circadian clock located in melatonin-responsive cells of the PT, which produce more TSH under long days than under short days. This specific PT-TSH acts on tanycytes of the neighboring hypothalamus to turn on and off expression of Dio2/3, leading to heightened T3 signaling under long days. These swings in local T3 production are required for the progression of the annual breeding cycle. How T3 ultimately impinges on GnRH secretion remains unclear but this may involve neurons expressing *Kiss1* and *Npvf*, which yields the RFRP3 precursor (Dardente, Wood, et al., 2019; Simonneaux, 2020). Tanycytes constitute a very peculiar and heterogeneous population of glial cells, which are responsive to photoperiod through TSH and perhaps other PT-derived signaling factors. Some of these tanycytes are NSCs, sensitive to an array of growth factors. The exact identity of the tanycyte subtype corresponding to NSC, if any, remains unclear. Considering the well-established role of T3 in cell proliferation and differentiation, the notion that photoperiod governs seasonal functions through T3-driven changes in NSCs proliferation has some appeal. While definitive functional evidence for this is still scarce, our recent findings appear consistent with this hypothesis (Batailler et al., 2018). Interestingly, transcriptomic analyses of the sheep MBH revealed that genes displaying robust photoperiodic switches in expression keep on doing so under constant light conditions (Dardente & Lomet, 2018; Dardente, Wood, et al., 2019; Lomet et al., 2018; Wood et al., 2015). This seems to indicate that the EYA3-TSH-DIO2 backbone (and many other genes) involved in photoperiodic read-out is also part of the circannual clock. There is accumulating evidence that the PT harbors this circannual clock, with tanycytes acting as relays toward select hypothalamic targets (Dardente, Wood, et al., 2019; Lincoln et al., 2005, 2006; Saenz de Miera et al., 2014; Wood et al., 2015). Much has been learnt over a "century of photoperiodism," yet there still is plenty of work ahead.

Acknowledgments

H.D. wishes to thank the members of the lab Molecular Neuroendocrinology of Reproduction and dedicates this review to the memory of Gerald Lincoln, gentleman farmer, enthusiast biologist and inspiring collaborator.

References

Abrieux, A., Xue, Y., Cai, Y., Lewald, K. M., Nguyen, H. N., Zhang, Y., et al. (2020). Eyes absent and timeless integrate photoperiodic and temperature cues to regulate seasonal physiology in drosophila. *Proceedings of the National Academy of Sciences of the United States of America, 117*(26), 15293–15304.

Akmayev, I. G., Fidelina, O. V., Kabolova, Z. A., Popov, A. P., & Schitkova, T. A. (1973). Morphological aspects of the hypothalamic-hypophyseal system. IV. Medial basal hypothalamus. An experimental morphological study. *Zeitschrift für Zellforschung und Mikroskopische Anatomie, 137*(4), 493–512.

Alvarez-Buylla, A., & Ihrie, R. A. (2014). Sonic hedgehog signaling in the postnatal brain. *Seminars in Cell & Developmental Biology, 33*, 105–111.

Alvarez-Buylla, A., & Lim, D. A. (2004). For the long run: Maintaining germinal niches in the adult brain. *Neuron, 41*(5), 683–686.

Anderson, G. M., Hardy, S. L., Valent, M., Billings, H. J., Connors, J. M., & Goodman, R. L. (2003). Evidence that thyroid hormones act in the ventromedial preoptic area and the premammillary region of the brain to allow the termination of the breeding season in the ewe. *Endocrinology, 144*(7), 2892–2901.

Angelopoulou, E., Quignon, C., Kriegsfeld, L. J., & Simonneaux, V. (2019). Functional implications of RFRP-3 in the central control of daily and seasonal rhythms in reproduction. *Frontiers in Endocrinology, 10*, 183.

Appenroth, D., Melum, V. J., West, A. C., Dardente, H., Hazlerigg, D. G., & Wagner, G. C. (2020). Photoperiodic induction without light-mediated circadian entrainment in a high arctic resident bird. *The Journal of Experimental Biology, 223*(Pt. 16), 1–8. https://doi.org/10.1242/jeb.220699.

Baker, J. R., & Ranson, R. M. (1933). Factors affecting the breeding of the field mouse (*Microtus agrestis*). Part III.—Locality. *Proceedings of the Royal Society of London. Series B, 113*(784), 486–495.

Balland, E., Dam, J., Langlet, F., Caron, E., Steculorum, S., Messina, A., et al. (2014). Hypothalamic tanycytes are an ERK-gated conduit for leptin into the brain. *Cell Metabolism, 19*(2), 293–301.

Bao, R., Onishi, K. G., Tolla, E., Ebling, F. J. P., Lewis, J. E., Anderson, R. L., et al. (2019). Genome sequencing and transcriptome analyses of the siberian hamster hypothalamus identify mechanisms for seasonal energy balance. *Proceedings of the National Academy of Sciences of the United States of America, 116*(26), 13116–13121.

Barrett, P., & Bolborea, M. (2012). Molecular pathways involved in seasonal body weight and reproductive responses governed by melatonin. *Journal of Pineal Research, 52*(4), 376–388.

Barrett, P., Ebling, F. J., Schuhler, S., Wilson, D., Ross, A. W., Warner, A., et al. (2007). Hypothalamic thyroid hormone catabolism acts as a gatekeeper for the seasonal control of body weight and reproduction. *Endocrinology, 148*(8), 3608–3617.

Barrett, P., Ivanova, E., Graham, E. S., Ross, A. W., Wilson, D., Ple, H., et al. (2006). Photoperiodic regulation of cellular retinol binding protein, CRBP1 [corrected] and nestin in tanycytes of the third ventricle ependymal layer of the siberian hamster. *The Journal of Endocrinology, 191*(3), 687–698.

Bartness, T. J., Powers, J. B., Hastings, M. H., Bittman, E. L., & Goldman, B. D. (1993). The timed infusion paradigm for melatonin delivery: What has it taught us about the melatonin signal, its reception, and the photoperiodic control of seasonal responses? *Journal of Pineal Research, 15*(4), 161–190.

Batailler, M., Chesneau, D., Derouet, L., Butruille, L., Segura, S., Cognie, J., et al. (2018). Pineal-dependent increase of hypothalamic neurogenesis contributes to the timing of seasonal reproduction in sheep. *Scientific Reports, 8*(1), 1–13. 6188-018-24381-4.

Batailler, M., Derouet, L., Butruille, L., & Migaud, M. (2016). Sensitivity to the photoperiod and potential migratory features of neuroblasts in the adult sheep hypothalamus. *Brain Structure & Function*, *221*(6), 3301–3314.

Batailler, M., Droguerre, M., Baroncini, M., Fontaine, C., Prevot, V., & Migaud, M. (2014). DCX-expressing cells in the vicinity of the hypothalamic neurogenic niche: A comparative study between mouse, sheep, and human tissues. *The Journal of Comparative Neurology*, *522*(8), 1966–1985.

Bedrosian, T. A., Fonken, L. K., & Nelson, R. J. (2016). Endocrine effects of circadian disruption. *Annual Review of Physiology*, *78*, 109–131.

Beltramo, M., Dardente, H., Cayla, X., & Caraty, A. (2014). Cellular mechanisms and integrative timing of neuroendocrine control of GnRH secretion by kisspeptin. *Molecular and Cellular Endocrinology*, *382*(1), 387–399.

Bennett, L., Yang, M., Enikolopov, G., & Iacovitti, L. (2009). Circumventricular organs: A novel site of neural stem cells in the adult brain. *Molecular and Cellular Neurosciences*, *41*(3), 337–347.

Benoit, J. (1934). Activation sexuelle obtenue chez le canard par l'éclairement artificiel pendant la période de repos génital. *Comptes Rendus de l'Académie des Sciences (Paris)*, *199*, 1671–1674.

Benoit, J., & Aron, M. (1934). Sur le conditionnement hormonique du développement testiculaire, chez les oiseaux. Résultats de la thyroïdectomie chez le coq et le canard. *Comptes Rendus des séances de la Société de Biologie*, *116*, 218–220.

Bianco, A. C., Salvatore, D., Gereben, B., Berry, M. J., & Larsen, P. R. (2002). Biochemistry, cellular and molecular biology, and physiological roles of the iodothyronine selenodeiodinases. *Endocrine Reviews*, *23*(1), 38–89.

Bissonnette, T. H. (1936). Sexual photoperiodicity. *The Quarterly Review of Biology*, *11*(4), 371–386.

Bockmann, J., Bo Ckers, T. M., Winter, C., Wittkowski, W., Winterhoff, H., Deufel, T., et al. (1997). Thyrotropin expression in hypophyseal pars tuberalis-specific cells is 3,5,3′-triiodothyronine, thyrotropin-releasing hormone, and pit-1 independent. *Endocrinology*, *138*(3), 1019–1028.

Bolborea, M., & Dale, N. (2013). Hypothalamic tanycytes: Potential roles in the control of feeding and energy balance. *Trends in Neurosciences*, *36*(2), 91–100.

Bolborea, M., Laran-Chich, M. P., Rasri, K., Hildebrandt, H., Govitrapong, P., Simonneaux, V., et al. (2011). Melatonin controls photoperiodic changes in tanycyte vimentin and neural cell adhesion molecule expression in the djungarian hamster (*Phodopus sungorus*). *Endocrinology*, *152*(10), 3871–3883.

Bolborea, M., Pollatzek, E., Benford, H., Sotelo-Hitschfeld, T., & Dale, N. (2020). Hypothalamic tanycytes generate acute hyperphagia through activation of the arcuate neuronal network. *Proceedings of the National Academy of Sciences of the United States of America*, *117*(25), 14473–14481.

Büning, E. (1936). Die endogene tagesrhythmik als grundlage der photoperiodischen reaktion. *Berichte der Deutschen Botanischen Gesellschaft*, *54*, 590–607.

Butler, M. P., Turner, K. W., Park, J. H., Schoomer, E. E., Zucker, I., & Gorman, M. R. (2010). Seasonal regulation of reproduction: Altered role of melatonin under naturalistic conditions in hamsters. *Proceedings of the Biological Sciences*, *277*(1695), 2867–2874.

Butruille, L., Batailler, M., Mazur, D., Prevot, V., & Migaud, M. (2018). Seasonal reorganization of hypothalamic neurogenic niche in adult sheep. *Brain Structure & Function*, *223*(1), 91–109.

Campbell, J. N., Macosko, E. Z., Fenselau, H., Pers, T. H., Lyubetskaya, A., Tenen, D., et al. (2017). A molecular census of arcuate hypothalamus and median eminence cell types. *Nature Neuroscience*, *20*(3), 484–496.

Cederroth, C. R., Albrecht, U., Bass, J., Brown, S. A., Dyhrfjeld-Johnsen, J., Gachon, F., et al. (2019). Medicine in the fourth dimension. *Cell Metabolism*, *30*(2), 238–250.

Chaker, Z., George, C., Petrovska, M., Caron, J. B., Lacube, P., Caille, I., et al. (2016). Hypothalamic neurogenesis persists in the aging brain and is controlled by energy-sensing IGF-I pathway. *Neurobiology of Aging*, *41*, 64–72.

Chen, R., Wu, X., Jiang, L., & Zhang, Y. (2017). Single-cell RNA-seq reveals hypothalamic cell diversity. *Cell Reports*, *18*(13), 3227–3241.

Dardente, H. (2007). Does a melatonin-dependent circadian oscillator in the pars tuberalis drive prolactin seasonal rhythmicity? *Journal of Neuroendocrinology*, *19*(8), 657–666.

Dardente, H. (2012). Melatonin-dependent timing of seasonal reproduction by the pars tuberalis: Pivotal roles for long daylengths and thyroid hormones. *Journal of Neuroendocrinology*, *24*(2), 249–266.

Dardente, H., & Cermakian, N. (2007). Molecular circadian rhythms in central and peripheral clocks in mammals. *Chronobiology International*, *24*(2), 195–213.

Dardente, H., Hazlerigg, D. G., & Ebling, F. J. (2014). Thyroid hormone and seasonal rhythmicity. *Frontiers in Endocrinology*, *5*, 19.

Dardente, H., Klosen, P., Pevet, P., & Masson-Pevet, M. (2003). MT1 melatonin receptor mRNA expressing cells in the pars tuberalis of the european hamster: Effect of photoperiod. *Journal of Neuroendocrinology*, *15*(8), 778–786.

Dardente, H., & Lomet, D. (2018). Photoperiod and thyroid hormone regulate expression of l-dopachrome tautomerase (dct), a melanocyte stem-cell marker, in tanycytes of the ovine hypothalamus. *Journal of Neuroendocrinology*, *30*(9), e12640.

Dardente, H., Lomet, D., Chesneau, D., Pellicer-Rubio, M. T., & Hazlerigg, D. (2019). Discontinuity in the molecular neuroendocrine response to increasing daylengths in Ile-de-France ewes: Is transient Dio2 induction a key feature of circannual timing? *Journal of Neuroendocrinology*, *31*(8), e12775.

Dardente, H., Lomet, D., Robert, V., Decourt, C., Beltramo, M., & Pellicer-Rubio, M. T. (2016). Seasonal breeding in mammals: From basic science to applications and back. *Theriogenology*, *86*(1), 324–332.

Dardente, H., Menet, J. S., Poirel, V. J., Streicher, D., Gauer, F., Vivien-Roels, B., et al. (2003). Melatonin induces Cry1 expression in the pars tuberalis of the rat. *Molecular Brain Research*, *114*(2), 101–106.

Dardente, H., Wood, S., Ebling, F., & Saenz de Miera, C. (2019). An integrative view of mammalian seasonal neuroendocrinology. *Journal of Neuroendocrinology*, *31*(5), e12729.

Dardente, H., Wyse, C. A., Birnie, M. J., Dupre, S. M., Loudon, A. S., Lincoln, G. A., et al. (2010). A molecular switch for photoperiod responsiveness in mammals. *Current Biology*, *20*(24), 2193–2198.

Davis, P. J., Goglia, F., & Leonard, J. L. (2016). Nongenomic actions of thyroid hormone. *Nature Reviews. Endocrinology*, *12*(2), 111–121.

Dawson, A. (2015). Annual gonadal cycles in birds: Modeling the effects of photoperiod on seasonal changes in GnRH-1 secretion. *Frontiers in Neuroendocrinology*, *37*, 52–64.

Del Brio, M. A., Riera, P., Garcia, J. M., & Alvarez-Uria, M. (1991). Cell types of the third ventricle wall of the rabbit (*Oryctolagus cuniculus*). *Journal of Submicroscopic Cytology and Pathology*, *23*(1), 147–157.

Desouza, L. A., Sathanoori, M., Kapoor, R., Rajadhyaksha, N., Gonzalez, L. E., Kottmann, A. H., et al. (2011). Thyroid hormone regulates the expression of the sonic hedgehog signaling pathway in the embryonic and adult mammalian brain. *Endocrinology*, *152*(5), 1989–2000.

Dibner, C., Schibler, U., & Albrecht, U. (2010). The mammalian circadian timing system: Organization and coordination of central and peripheral clocks. *Annual Review of Physiology*, *72*, 517–549.

Dubocovich, M. L., Rivera-Bermudez, M. A., Gerdin, M. J., & Masana, M. I. (2003). Molecular pharmacology, regulation and function of mammalian melatonin receptors. *Frontiers in Bioscience*, *8*, S1093–S1108.

Duvernoy, H. M., & Risold, P. Y. (2007). The circumventricular organs: An atlas of comparative anatomy and vascularization. *Brain Research Reviews*, *56*(1), 119–147.

Ebling, F. J. (2014). On the value of seasonal mammals for identifying mechanisms underlying the control of food intake and body weight. *Hormones and Behavior*, *66*(1), 56–65.

Ebling, F. J. (2015). Hypothalamic control of seasonal changes in food intake and body weight. *Frontiers in Neuroendocrinology*, *37*, 97–107.

Ebling, F. J. P., & Lewis, J. E. (2018). Tanycytes and hypothalamic control of energy metabolism. *Glia*, *66*(6), 1176–1184.

Enright, J. T. (1970). Ecological aspects of endogenous rhythmicity. *Annual Review of Ecology, Evolution, and Systematics*, *1*, 221–238.

Fanibunda, S. E., Desouza, L. A., Kapoor, R., Vaidya, R. A., & Vaidya, V. A. (2018). Thyroid hormone regulation of adult neurogenesis. *Vitamins and Hormones*, *106*, 211–251.

Fitzgerald, K. T. (1979). The structure and function of the pars tuberalis of the vertebrate adenohypophysis. *General and Comparative Endocrinology*, *37*(3), 383–399.

Flament-Durand, J., & Brion, J. P. (1985). Tanycytes: Morphology and functions: A review. *International Review of Cytology*, *96*, 121–155.

Fleming, M. S., Maugars, G., Lafont, A. G., Rancon, J., Fontaine, R., Nourizadeh-Lillabadi, R., et al. (2019). Functional divergence of thyrotropin beta-subunit paralogs gives new insights into salmon smoltification metamorphosis. *Scientific Reports*, *9*(1), 1–15. 4561-019-40019-5.

Follett, B. K., & Nicholls, T. J. (1984). Photorefractoriness in japanese quail: Possible involvement of the thyroid gland. *The Journal of Experimental Zoology*, *232*(3), 573–580.

Foster, R. G., & Roenneberg, T. (2008). Human responses to the geophysical daily, annual and lunar cycles. *Current Biology*, *18*(17), R784–R794.

Fuentealba, L. C., Obernier, K., & Alvarez-Buylla, A. (2012). Adult neural stem cells bridge their niche. *Cell Stem Cell*, *10*(6), 698–708.

Fustin, J. M., Dardente, H., Wagner, G. C., Carter, D. A., Johnston, J. D., Lincoln, G. A., et al. (2009). Egr1 involvement in evening gene regulation by melatonin. *FASEB Journal*, *23*(3), 764–773.

Garner, W. W., & Allard, H. A. (1920). Effect of the relative length of day and night and other factors of the environment on growth and reproduction in plants. *Journal of Agricultural Research*, *18*, 553–606.

Gauthier, K., & Flamant, F. (2014). Nongenomic, TRbeta-dependent, thyroid hormone response gets genetic support. *Endocrinology*, *155*(9), 3206–3209.

Gil-Ibanez, P., Bernal, J., & Morte, B. (2014). Thyroid hormone regulation of gene expression in primary cerebrocortical cells: Role of thyroid hormone receptor subtypes and interactions with retinoic acid and glucocorticoids. *PLoS One*, *9*(3), e91692.

Goldman, B. D. (2001). Mammalian photoperiodic system: Formal properties and neuroendocrine mechanisms of photoperiodic time measurement. *Journal of Biological Rhythms*, *16*(4), 283–301.

Goodman, T., & Hajihosseini, M. K. (2015). Hypothalamic tanycytes-masters and servants of metabolic, neuroendocrine, and neurogenic functions. *Frontiers in Neuroscience*, *9*, 387.

Gorman, M. R. (2020). Temporal organization of pineal melatonin signaling in mammals. *Molecular and Cellular Endocrinology*, *503*, 110687.

Gothie, J. D., Sebillot, A., Luongo, C., Legendre, M., Nguyen Van, C., Le Blay, K., et al. (2017). Adult neural stem cell fate is determined by thyroid hormone activation of mitochondrial metabolism. *Molecular Metabolism*, *6*(11), 1551–1561.

Gross, D. S. (1984). The mammalian hypophysial pars tuberalis: A comparative immunocytochemical study. *General and Comparative Endocrinology*, *56*(2), 283–298.
Gwinner, E. (1986). *Circannual rhythms*. Berlin: Springer.
Gwinner, E., & Wozniak, J. (1982). Circannual rhythms in european starlings: Why do they stop under long photoperiods? *Journal of Comparative Physiology A*, *146*, 419–421.
Haan, N., Goodman, T., Najdi-Samiei, A., Stratford, C. M., Rice, R., El Agha, E., et al. (2013). Fgf10-expressing tanycytes add new neurons to the appetite/energy-balance regulating centers of the postnatal and adult hypothalamus. *The Journal of Neuroscience*, *33*(14), 6170–6180.
Hafez, E. S. E. (1952). Studies on the breeding season and reproduction of the ewe. Part I: The breeding season in different environments. Part II: The breeding season in one locality. Part III: The breeding season and artificial light. Part IV: Studies on the reproduction of the ewe. Part V: Mating behaviour and pregnancy diagnosis. *Journal of Agricultural Science*, *42*(3), 189–265.
Halford, S., Pires, S. S., Turton, M., Zheng, L., Gonzalez-Menendez, I., Davies, W. L., et al. (2009). VA opsin-based photoreceptors in the hypothalamus of birds. *Current Biology*, *19*(16), 1396–1402.
Hanlon, C., Ramachandran, R., Zuidhof, M. J., & Bedecarrats, G. Y. (2020). Should I lay or should I grow: Photoperiodic versus metabolic cues in chickens. *Frontiers in Physiology*, *11*, 707.
Hanon, E. A., Lincoln, G. A., Fustin, J. M., Dardente, H., Masson-Pevet, M., Morgan, P. J., et al. (2008). Ancestral TSH mechanism signals summer in a photoperiodic mammal. *Current Biology*, *18*(15), 1147–1152.
Hastings, M. H., Reddy, A. B., & Maywood, E. S. (2003). A clockwork web: Circadian timing in brain and periphery, in health and disease. *Nature Reviews. Neuroscience*, *4*(8), 649–661.
Hazlerigg, D. G., Andersson, H., Johnston, J. D., & Lincoln, G. (2004). Molecular characterization of the long-day response in the soay sheep, a seasonal mammal. *Current Biology*, *14*(4), 334–339.
Hazlerigg, D. G., & Lincoln, G. A. (2011). Hypothesis: Cyclical histogenesis is the basis of circannual timing. *Journal of Biological Rhythms*, *26*(6), 471–485.
Hazlerigg, D., Lomet, D., Lincoln, G., & Dardente, H. (2018). Neuroendocrine correlates of the critical day length response in the soay sheep. *Journal of Neuroendocrinology*, *30*(9), e12631.
Hazlerigg, D. G., & Simonneaux, V. (2015). Seasonal regulation of reproduction in mammals. In *Vol. 2. Knobil and Neill's physiology of reproduction* (4th ed., pp. 1575–1604). Elsevier.
Hazlerigg, D. G., & Wagner, G. C. (2006). Seasonal photoperiodism in vertebrates: From coincidence to amplitude. *Trends in Endocrinology and Metabolism*, *17*(3), 83–91.
Hazlerigg, D. G., Wyse, C. A., Dardente, H., Hanon, E. A., & Lincoln, G. A. (2013). Photoperiodic variation in CD45-positive cells and cell proliferation in the mediobasal hypothalamus of the soay sheep. *Chronobiology International*, *30*(4), 548–558.
Helfer, G., Barrett, P., & Morgan, P. J. (2019). A unifying hypothesis for control of body weight and reproduction in seasonally breeding mammals. *Journal of Neuroendocrinology*, *31*(3), e12680.
Henson, J. R., Carter, S. N., & Freeman, D. A. (2013). Exogenous T(3) elicits long day-like alterations in testis size and the RFamides kisspeptin and gonadotropin-inhibitory hormone in short-day siberian hamsters. *Journal of Biological Rhythms*, *28*(3), 193–200.
Hogenesch, J. B., & Ueda, H. R. (2011). Understanding systems-level properties: Timely stories from the study of clocks. *Nature Reviews. Genetics*, *12*(6), 407–416.
Hut, R. A., & Beersma, D. G. (2011). Evolution of time-keeping mechanisms: Early emergence and adaptation to photoperiod. *Philosophical Transactions of the Royal Society of London. Series B, Biological Sciences*, *366*(1574), 2141–2154.

Hut, R. A., Dardente, H., & Riede, S. J. (2014). Seasonal timing: How does a hibernator know when to stop hibernating? *Current Biology*, *24*(13), R602–R605.

Ikegami, K., Liao, X. H., Hoshino, Y., Ono, H., Ota, W., Ito, Y., et al. (2014). Tissue-specific posttranslational modification allows functional targeting of thyrotropin. *Cell Reports*, *9*(3), 801–810.

Jiao, Z., Zhang, Z. G., Hornyak, T. J., Hozeska, A., Zhang, R. L., Wang, Y., et al. (2006). Dopachrome tautomerase (dct) regulates neural progenitor cell proliferation. *Developmental Biology*, *296*(2), 396–408.

Jourdon, A., Gresset, A., Spassky, N., Charnay, P., Topilko, P., & Santos, R. (2016). Prss56, a novel marker of adult neurogenesis in the mouse brain. *Brain Structure & Function*, *221*(9), 4411–4427.

Karsch, F. J., Bittman, E. L., Foster, D. L., Goodman, R. L., Legan, S. J., & Robinson, J. E. (1984). Neuroendocrine basis of seasonal reproduction. *Recent Progress in Hormone Research*, *40*, 185–232.

Karsch, F. J., Dahl, G. E., Hachigian, T. M., & Thrun, L. A. (1995). Involvement of thyroid hormones in seasonal reproduction. *Journal of Reproduction and Fertility. Supplement*, *49*, 409–422.

Klosen, P., Bienvenu, C., Demarteau, O., Dardente, H., Guerrero, H., Pevet, P., et al. (2002). The mt1 melatonin receptor and RORbeta receptor are co-localized in specific TSH-immunoreactive cells in the pars tuberalis of the rat pituitary. *The Journal of Histochemistry and Cytochemistry*, *50*(12), 1647–1657.

Klosen, P., Sebert, M. E., Rasri, K., Laran-Chich, M. P., & Simonneaux, V. (2013). TSH restores a summer phenotype in photoinhibited mammals via the RF-amides RFRP3 and kisspeptin. *FASEB Journal*, *27*(7), 2677–2686.

Kokoeva, M. V., Yin, H., & Flier, J. S. (2005). Neurogenesis in the hypothalamus of adult mice: Potential role in energy balance. *Science*, *310*(5748), 679–683.

Kondo, N., Sekijima, T., Kondo, J., Takamatsu, N., Tohya, K., & Ohtsu, T. (2006). Circannual control of hibernation by HP complex in the brain. *Cell*, *125*(1), 161–172.

Kong, W. M., Martin, N. M., Smith, K. L., Gardiner, J. V., Connoley, I. P., Stephens, D. A., et al. (2004). Triiodothyronine stimulates food intake via the hypothalamic ventromedial nucleus independent of changes in energy expenditure. *Endocrinology*, *145*(11), 5252–5258.

Koopman, A. C. M., Taziaux, M., & Bakker, J. (2017). Age-related changes in the morphology of tanycytes in the human female infundibular nucleus/median eminence. *Journal of Neuroendocrinology*, *29*(5), 1–9. https://doi.org/10.1111/jne.12467.

Korf, H. W. (2018). Signaling pathways to and from the hypophysial pars tuberalis, an important center for the control of seasonal rhythms. *General and Comparative Endocrinology*, *258*, 236–243.

Kosonsiriluk, S., Mauro, L. J., Chaiworakul, V., Chaiseha, Y., & El Halawani, M. E. (2013). Photoreceptive oscillators within neurons of the premammillary nucleus (PMM) and seasonal reproduction in temperate zone birds. *General and Comparative Endocrinology*, *190*, 149–155.

Langlet, F., Mullier, A., Bouret, S. G., Prevot, V., & Dehouck, B. (2013). Tanycyte-like cells form a blood-cerebrospinal fluid barrier in the circumventricular organs of the mouse brain. *The Journal of Comparative Neurology*, *521*(15), 3389–3405.

Lee, D. A., Bedont, J. L., Pak, T., Wang, H., Song, J., Miranda-Angulo, A., et al. (2012). Tanycytes of the hypothalamic median eminence form a diet-responsive neurogenic niche. *Nature Neuroscience*, *15*(5), 700–702.

Lemkine, G. F., Raj, A., Alfama, G., Turque, N., Hassani, Z., Alegria-Prevot, O., et al. (2005). Adult neural stem cell cycling in vivo requires thyroid hormone and its alpha receptor. *FASEB Journal*, *19*(7), 863–865.

Levi, F., & Schibler, U. (2007). Circadian rhythms: Mechanisms and therapeutic implications. *Annual Review of Pharmacology and Toxicology*, *47*, 593–628.

Li, J., Tang, Y., & Cai, D. (2012). IKKbeta/NF-kappaB disrupts adult hypothalamic neural stem cells to mediate a neurodegenerative mechanism of dietary obesity and pre-diabetes. *Nature Cell Biology*, *14*(10), 999–1012.

Lincoln, G. (2019). A brief history of circannual time. *Journal of Neuroendocrinology*, *31*(3), e12694.

Lincoln, G. A., Andersson, H., & Loudon, A. (2003). Clock genes in calendar cells as the basis of annual timekeeping in mammals—A unifying hypothesis. *The Journal of Endocrinology*, *179*(1), 1–13.

Lincoln, G. A., & Clarke, I. J. (1994). Photoperiodically-induced cycles in the secretion of prolactin in hypothalamo-pituitary disconnected rams: Evidence for translation of the melatonin signal in the pituitary gland. *Journal of Neuroendocrinology*, *6*(3), 251–260.

Lincoln, G. A., Clarke, I. J., Hut, R. A., & Hazlerigg, D. G. (2006). Characterizing a mammalian circannual pacemaker. *Science*, *314*(5807), 1941–1944.

Lincoln, G. A., Johnston, J. D., Andersson, H., Wagner, G., & Hazlerigg, D. G. (2005). Photorefractoriness in mammals: Dissociating a seasonal timer from the circadian-based photoperiod response. *Endocrinology*, *146*(9), 3782–3790.

Lincoln, G., Messager, S., Andersson, H., & Hazlerigg, D. (2002). Temporal expression of seven clock genes in the suprachiasmatic nucleus and the pars tuberalis of the sheep: Evidence for an internal coincidence timer. *Proceedings of the National Academy of Sciences of the United States of America*, *99*(21), 13890–13895.

Lincoln, G. A., & Short, R. V. (1980). Seasonal breeding: Nature's contraceptive. *Recent Progress in Hormone Research*, *36*, 1–52.

Liu, C., Weaver, D. R., Jin, X., Shearman, L. P., Pieschl, R. L., Gribkoff, V. K., et al. (1997). Molecular dissection of two distinct actions of melatonin on the suprachiasmatic circadian clock. *Neuron*, *19*(1), 91–102.

Lomet, D., Cognie, J., Chesneau, D., Dubois, E., Hazlerigg, D., & Dardente, H. (2018). The impact of thyroid hormone in seasonal breeding has a restricted transcriptional signature. *Cellular and Molecular Life Sciences*, *75*(5), 905–919.

Lomet, D., Druart, X., Hazlerigg, D., Beltramo, M., & Dardente, H. (2020). Circuit-level analysis identifies target genes of sex steroids in ewe seasonal breeding. *Molecular and Cellular Endocrinology*, *512*, 110825.

Lopez, M., Alvarez, C. V., Nogueiras, R., & Dieguez, C. (2013). Energy balance regulation by thyroid hormones at central level. *Trends in Molecular Medicine*, *19*(7), 418–427.

Lopez, M., Varela, L., Vazquez, M. J., Rodriguez-Cuenca, S., Gonzalez, C. R., Velagapudi, V. R., et al. (2010). Hypothalamic AMPK and fatty acid metabolism mediate thyroid regulation of energy balance. *Nature Medicine*, *16*(9), 1001–1008.

Lopez-Juarez, A., Remaud, S., Hassani, Z., Jolivet, P., Pierre Simons, J., Sontag, T., et al. (2012). Thyroid hormone signaling acts as a neurogenic switch by repressing Sox2 in the adult neural stem cell niche. *Cell Stem Cell*, *10*(5), 531–543.

Lorgen, M., Casadei, E., Krol, E., Douglas, A., Birnie, M. J., Ebbesson, L. O., et al. (2015). Functional divergence of type 2 deiodinase paralogs in the Atlantic salmon. *Current Biology*, *25*(7), 936–941.

Malpaux, B., Migaud, M., Tricoire, H., & Chemineau, P. (2001). Biology of mammalian photoperiodism and the critical role of the pineal gland and melatonin. *Journal of Biological Rhythms*, *16*(4), 336–347.

Markakis, E. A., Palmer, T. D., Randolph-Moore, L., Rakic, P., & Gage, F. H. (2004). Novel neuronal phenotypes from neural progenitor cells. *The Journal of Neuroscience*, *24*(12), 2886–2897.

Marshall, F. H. A. (1936). The croonian lecture. Sexual periodicity and the causes which determine it. *Philosophical Transactions of the Royal Society of London. Series B, Biological Sciences, 226*, 423–456.
Masson-Pevet, M., George, D., Kalsbeek, A., Saboureau, M., Lakhdar-Ghazal, N., & Pevet, P. (1994). An attempt to correlate brain areas containing melatonin-binding sites with rhythmic functions: A study in five hibernator species. *Cell and Tissue Research, 278*(1), 97–106.
Masumoto, K. H., Ukai-Tadenuma, M., Kasukawa, T., Nagano, M., Uno, K. D., Tsujino, K., et al. (2010). Acute induction of Eya3 by late-night light stimulation triggers TSHbeta expression in photoperiodism. *Current Biology, 20*(24), 2199–2206.
Mercer, J. G., Moar, K. M., Logie, T. J., Findlay, P. A., Adam, C. L., & Morgan, P. J. (2001). Seasonally inappropriate body weight induced by food restriction: Effect on hypothalamic gene expression in male siberian hamsters. *Endocrinology, 142*(10), 4173–4181.
Messager, S., Ross, A. W., Barrett, P., & Morgan, P. J. (1999). Decoding photoperiodic time through Per1 and ICER gene amplitude. *Proceedings of the National Academy of Sciences of the United States of America, 96*(17), 9938–9943.
Migaud, M., Batailler, M., Pillon, D., Franceschini, I., & Malpaux, B. (2011). Seasonal changes in cell proliferation in the adult sheep brain and pars tuberalis. *Journal of Biological Rhythms, 26*(6), 486–496.
Migaud, M., Batailler, M., Segura, S., Duittoz, A., Franceschini, I., & Pillon, D. (2010). Emerging new sites for adult neurogenesis in the mammalian brain: A comparative study between the hypothalamus and the classical neurogenic zones. *The European Journal of Neuroscience, 32*(12), 2042–2052.
Migaud, M., Butruille, L., & Batailler, M. (2015). Seasonal regulation of structural plasticity and neurogenesis in the adult mammalian brain: Focus on the sheep hypothalamus. *Frontiers in Neuroendocrinology, 37*, 146–157.
Milesi, S., Simonneaux, V., & Klosen, P. (2017). Downregulation of deiodinase 3 is the earliest event in photoperiodic and photorefractory activation of the gonadotropic axis in seasonal hamsters. *Scientific Reports*, 7, 17739.
Mirzadeh, Z., Kusne, Y., Duran-Moreno, M., Cabrales, E., Gil-Perotin, S., Ortiz, C., et al. (2017). Bi- and uniciliated ependymal cells define continuous floor-plate-derived tanycytic territories. *Nature Communications, 8*, 13759.
Miyata, S. (2015). New aspects in fenestrated capillary and tissue dynamics in the sensory circumventricular organs of adult brains. *Frontiers in Neuroscience, 9*, 390.
Moenter, S. M., Woodfill, C. J., & Karsch, F. J. (1991). Role of the thyroid gland in seasonal reproduction: Thyroidectomy blocks seasonal suppression of reproductive neuroendocrine activity in ewes. *Endocrinology, 128*(3), 1337–1344.
Mohawk, J. A., Green, C. B., & Takahashi, J. S. (2012). Central and peripheral circadian clocks in mammals. *Annual Review of Neuroscience, 35*, 445–462.
Morgan, P. J., Barrett, P., Howell, H. E., & Helliwell, R. (1994). Melatonin receptors: Localization, molecular pharmacology and physiological significance. *Neurochemistry International, 24*(2), 101–146.
Morgan, P. J., & Hazlerigg, D. G. (2008). Photoperiodic signalling through the melatonin receptor turns full circle. *Journal of Neuroendocrinology, 20*(6), 820–826.
Morgan, P. J., Ross, A. W., Mercer, J. G., & Barrett, P. (2006). Chapter 19: What can we learn from seasonal animals about the regulation of energy balance? In A. Kalsbeek, E. Fliers, M. A. Hofman, D. F. Swaab, E. J. W. van Someren, & R. M. Buijs (Eds.), *Vol. 153. Progress in brain research* (pp. 325–337). Elsevier.
Morgan, P. J., Webster, C. A., Mercer, J. G., Ross, A. W., Hazlerigg, D. G., MacLean, A., et al. (1996). The ovine pars tuberalis secretes a factor(s) that regulates gene expression in both lactotropic and nonlactotropic pituitary cells. *Endocrinology, 137*(9), 4018–4026.

Muller-Fielitz, H., Stahr, M., Bernau, M., Richter, M., Abele, S., Krajka, V., et al. (2017). Tanycytes control the hormonal output of the hypothalamic-pituitary-thyroid axis. *Nature Communications*, *8*(1), 1–13. 484-017-00604-6.

Mullier, A., Bouret, S. G., Prevot, V., & Dehouck, B. (2010). Differential distribution of tight junction proteins suggests a role for tanycytes in blood-hypothalamus barrier regulation in the adult mouse brain. *The Journal of Comparative Neurology*, *518*(7), 943–962.

Mullur, R., Liu, Y. Y., & Brent, G. A. (2014). Thyroid hormone regulation of metabolism. *Physiological Reviews*, *94*(2), 355–382.

Murphy, M., Jethwa, P. H., Warner, A., Barrett, P., Nilaweera, K. N., Brameld, J. M., et al. (2012). Effects of manipulating hypothalamic triiodothyronine concentrations on seasonal body weight and torpor cycles in siberian hamsters. *Endocrinology*, *153*(1), 101–112.

Nakane, Y., Ikegami, K., Iigo, M., Ono, H., Takeda, K., Takahashi, D., et al. (2013). The saccus vasculosus of fish is a sensor of seasonal changes in day length. *Nature Communications*, *4*, 2108.

Nakane, Y., Ikegami, K., Ono, H., Yamamoto, N., Yoshida, S., Hirunagi, K., et al. (2010). A mammalian neural tissue opsin (opsin 5) is a deep brain photoreceptor in birds. *Proceedings of the National Academy of Sciences of the United States of America*, *107*(34), 15264–15268.

Nakao, N., Ono, H., Yamamura, T., Anraku, T., Takagi, T., Higashi, K., et al. (2008). Thyrotrophin in the pars tuberalis triggers photoperiodic response. *Nature*, *452*(7185), 317–322.

Nicholls, T. J., Follett, B. K., Goldsmith, A. R., & Pearson, H. (1988). Possible homologies between photorefractoriness in sheep and birds: The effect of thyroidectomy on the length of the ewe's breeding season. *Reproduction, Nutrition, Development*, *28*, 375–385.

Nieto-Estevez, V., Defterali, C., & Vicario-Abejon, C. (2016). IGF-I: A key growth factor that regulates neurogenesis and synaptogenesis from embryonic to adult stages of the brain. *Frontiers in Neuroscience*, *10*, 52.

Obernier, K., & Alvarez-Buylla, A. (2019). Neural stem cells: Origin, heterogeneity and regulation in the adult mammalian brain. *Development*, *146*(4), 1–15. https://doi.org/10.1242/dev.156059.

Ono, H., Hoshino, Y., Yasuo, S., Watanabe, M., Nakane, Y., Murai, A., et al. (2008). Involvement of thyrotropin in photoperiodic signal transduction in mice. *Proceedings of the National Academy of Sciences of the United States of America*, *105*(47), 18238–18242.

Oya, S., Yoshikawa, G., Takai, K., Tanaka, J., Higashiyama, S., Saito, N., et al. (2008). Region-specific proliferative response of neural progenitors to exogenous stimulation by growth factors following ischemia. *Neuroreport*, *19*(8), 805–809.

Parkash, J., Messina, A., Langlet, F., Cimino, I., Loyens, A., Mazur, D., et al. (2015). Semaphorin7A regulates neuroglial plasticity in the adult hypothalamic median eminence. *Nature Communications*, *6*, 6385.

Parkinson, T. J., & Follett, B. K. (1994). Effect of thyroidectomy upon seasonality in rams. *Journal of Reproduction and Fertility*, *101*(1), 51–58.

Parkinson, T. J., & Follett, B. K. (1995). Thyroidectomy abolishes seasonal testicular cycles of soay rams. *Proceedings of the Royal Society of London. Series B*, *259*(1354), 1–6.

Pellegrino, G., Trubert, C., Terrien, J., Pifferi, F., Leroy, D., Loyens, A., et al. (2018). A comparative study of the neural stem cell niche in the adult hypothalamus of human, mouse, rat and gray mouse lemur (*Microcebus murinus*). *The Journal of Comparative Neurology*, *526*(9), 1419–1443.

Perez-Martin, M., Cifuentes, M., Grondona, J. M., Lopez-Avalos, M. D., Gomez-Pinedo, U., Garcia-Verdugo, J. M., et al. (2010). IGF-I stimulates neurogenesis in the hypothalamus of adult rats. *The European Journal of Neuroscience*, *31*(9), 1533–1548.

Petri, I., Diedrich, V., Wilson, D., Fernandez-Calleja, J., Herwig, A., Steinlechner, S., et al. (2016). Orchestration of gene expression across the seasons: Hypothalamic gene expression in natural photoperiod throughout the year in the siberian hamster. *Scientific Reports*, *6*, 29689.

Pittendrigh, C. S. (1993). Temporal organization: Reflections of a darwinian clock-watcher. *Annual Review of Physiology*, *55*, 16–54.

Prevot, V., Croix, D., Bouret, S., Dutoit, S., Tramu, G., Stefano, G. B., et al. (1999). Definitive evidence for the existence of morphological plasticity in the external zone of the median eminence during the rat estrous cycle: Implication of neuro-glio-endothelial interactions in gonadotropin-releasing hormone release. *Neuroscience*, *94*(3), 809–819.

Prevot, V., Dehouck, B., Sharif, A., Ciofi, P., Giacobini, P., & Clasadonte, J. (2018). The versatile tanycyte: A hypothalamic integrator of reproduction and energy metabolism. *Endocrine Reviews*, *39*(3), 333–368.

Prevot, V., Dutoit, S., Croix, D., Tramu, G., & Beauvillain, J. C. (1998). Semi-quantitative ultrastructural analysis of the localization and neuropeptide content of gonadotropin releasing hormone nerve terminals in the median eminence throughout the estrous cycle of the rat. *Neuroscience*, *84*(1), 177–191.

Quignon, C., Beymer, M., Gauthier, K., Gauer, F., & Simonneaux, V. (2020). Thyroid hormone receptors are required for the melatonin-dependent control of *Rfrp* gene expression in mice. *FASEB Journal*, *34*(9), 12072–12082.

Reddy, A. B., Cronin, A. S., Ford, H., & Ebling, F. J. (1999). Seasonal regulation of food intake and body weight in the male siberian hamster: Studies of hypothalamic orexin (hypocretin), neuropeptide Y (NPY) and pro-opiomelanocortin (POMC). *The European Journal of Neuroscience*, *11*(9), 3255–3264.

Reiter, R. J. (1993). The melatonin rhythm: Both a clock and a calendar. *Experientia*, *49*(8), 654–664.

Remaud, S., Gothie, J. D., Morvan-Dubois, G., & Demeneix, B. A. (2014). Thyroid hormone signaling and adult neurogenesis in mammals. *Frontiers in Endocrinology*, *5*, 62.

Remaud, S., Ortiz, F. C., Perret-Jeanneret, M., Aigrot, M. S., Gothie, J. D., Fekete, C., et al. (2017). Transient hypothyroidism favors oligodendrocyte generation providing functional remyelination in the adult mouse brain. *eLife*, *6*, e29996.

Reppert, S. M., Godson, C., Mahle, C. D., Weaver, D. R., Slaugenhaupt, S. A., & Gusella, J. F. (1995). Molecular characterization of a second melatonin receptor expressed in human retina and brain: The Mel1b melatonin receptor. *Proceedings of the National Academy of Sciences of the United States of America*, *92*(19), 8734–8738.

Reppert, S. M., Weaver, D. R., & Ebisawa, T. (1994). Cloning and characterization of a mammalian melatonin receptor that mediates reproductive and circadian responses. *Neuron*, *13*(5), 1177–1185.

Robins, S. C., Stewart, I., McNay, D. E., Taylor, V., Giachino, C., Goetz, M., et al. (2013). Alpha-tanycytes of the adult hypothalamic third ventricle include distinct populations of FGF-responsive neural progenitors. *Nature Communications*, *4*, 2049.

Robinson, J. E., & Karsch, F. J. (1984). Refractoriness to inductive day lengths terminates the breeding season of the Suffolk ewe. *Biology of Reproduction*, *31*(4), 656–663.

Robinson, J. E., Wayne, N. L., & Karsch, F. J. (1985). Refractoriness to inhibitory day lengths initiates the breeding season of the suffolk ewe. *Biology of Reproduction*, *32*(5), 1024–1030.

Rodriguez, E. M., Blazquez, J. L., Pastor, F. E., Pelaez, B., Pena, P., Peruzzo, B., et al. (2005). Hypothalamic tanycytes: A key component of brain-endocrine interaction. *International Review of Cytology*, *247*, 89–164.

Rodriguez, E. M., Gonzalez, C. B., & Delannoy, L. (1979). Cellular organization of the lateral and postinfundibular regions of the median eminence in the rat. *Cell and Tissue Research*, *201*(3), 377–408.

Rodriguez, E., Guerra, M., Peruzzo, B., & Blazquez, J. L. (2019). Tanycytes: A rich morphological history to underpin future molecular and physiological investigations. *Journal of Neuroendocrinology*, *31*(3), e12690.

Rodriguez-Rodriguez, A., Lazcano, I., Sanchez-Jaramillo, E., Uribe, R. M., Jaimes-Hoy, L., Joseph-Bravo, P., et al. (2019). Tanycytes and the control of thyrotropin-releasing hormone flux into portal capillaries. *Frontiers in Endocrinology*, *10*, 401.

Romanov, R. A., Tretiakov, E. O., Kastriti, M. E., Zupancic, M., Haring, M., Korchynska, S., et al. (2020). Molecular design of hypothalamus development. *Nature*, *582*(7811), 246–252.

Romanov, R. A., Zeisel, A., Bakker, J., Girach, F., Hellysaz, A., Tomer, R., et al. (2017). Molecular interrogation of hypothalamic organization reveals distinct dopamine neuronal subtypes. *Nature Neuroscience*, *20*(2), 176–188.

Rosbash, M. (2009). The implications of multiple circadian clock origins. *PLoS Biology*, 7(3), e62.

Ross, A. W., Helfer, G., Russell, L., Darras, V. M., & Morgan, P. J. (2011). Thyroid hormone signalling genes are regulated by photoperiod in the hypothalamus of F344 rats. *PLoS One*, *6*(6), e21351.

Rousseau, K., Atcha, Z., Cagampang, F. R., Le Rouzic, P., Stirland, J. A., Ivanov, T. R., et al. (2002). Photoperiodic regulation of leptin resistance in the seasonally breeding siberian hamster (*Phodopus sungorus*). *Endocrinology*, *143*(8), 3083–3095.

Rowan, W. (1938). Light and seasonal reproduction in animals. *Biological Reviews of the Cambridge Philosophical Society*, *13*(4), 374–401.

Saenz de Miera, C. (2019). Maternal photoperiodic programming enlightens the internal regulation of thyroid-hormone deiodinases in tanycytes. *Journal of Neuroendocrinology*, *31*(1), e12679.

Saenz de Miera, C., Hanon, E. A., Dardente, H., Birnie, M., Simonneaux, V., Lincoln, G. A., et al. (2013). Circannual variation in thyroid hormone deiodinases in a short-day breeder. *Journal of Neuroendocrinology*, *25*(4), 412–421.

Saenz de Miera, C., Monecke, S., Bartzen-Sprauer, J., Laran-Chich, M. P., Pevet, P., Hazlerigg, D. G., et al. (2014). A circannual clock drives expression of genes central for seasonal reproduction. *Current Biology*, *24*(13), 1500–1506.

Simonneaux, V. (2020). A kiss to drive rhythms in reproduction. *The European Journal of Neuroscience*, *51*(1), 509–530.

Simonneaux, V., & Ribelayga, C. (2003). Generation of the melatonin endocrine message in mammals: A review of the complex regulation of melatonin synthesis by norepinephrine, peptides, and other pineal transmitters. *Pharmacological Reviews*, *55*(2), 325–395.

Sousa-Ferreira, L., de Almeida, L. P., & Cavadas, C. (2014). Role of hypothalamic neurogenesis in feeding regulation. *Trends in Endocrinology and Metabolism*, *25*(2), 80–88.

Stoeckel, M. E., Hindelang, C., Klein, M. J., Poissonnier, M., & Felix, J. M. (1993). Early expression of the glycoprotein hormone alpha-subunit in the pars tuberalis of the rat pituitary gland during ontogenesis. *Neuroendocrinology*, *58*(6), 616–624.

Takahashi, J. S. (2017). Transcriptional architecture of the mammalian circadian clock. *Nature Reviews. Genetics*, *18*(3), 164–179.

Talbi, R., Laran-Chich, M. P., Magoul, R., El Ouezzani, S., & Simonneaux, V. (2016). Kisspeptin and RFRP-3 differentially regulate food intake and metabolic neuropeptides in the female desert jerboa. *Scientific Reports*, *6*, 36057.

Thrun, L. A., Dahl, G. E., Evans, N. P., & Karsch, F. J. (1996). Time-course of thyroid hormone involvement in the development of anestrus in the ewe. *Biology of Reproduction*, *55*(4), 833–837.

Thrun, L. A., Dahl, G. E., Evans, N. P., & Karsch, F. J. (1997). A critical period for thyroid hormone action on seasonal changes in reproductive neuroendocrine function in the ewe. *Endocrinology*, *138*(8), 3402–3409.

Traiffort, E., Charytoniuk, D., Watroba, L., Faure, H., Sales, N., & Ruat, M. (1999). Discrete localizations of hedgehog signalling components in the developing and adult rat nervous system. *The European Journal of Neuroscience*, *11*(9), 3199–3214.

Tu, H. M., Kim, S. W., Salvatore, D., Bartha, T., Legradi, G., Larsen, P. R., et al. (1997). Regional distribution of type 2 thyroxine deiodinase messenger ribonucleic acid in rat hypothalamus and pituitary and its regulation by thyroid hormone. *Endocrinology*, *138*(8), 3359–3368.

Tu, H. M., Legradi, G., Bartha, T., Salvatore, D., Lechan, R. M., & Larsen, P. R. (1999). Regional expression of the type 3 iodothyronine deiodinase messenger ribonucleic acid in the rat central nervous system and its regulation by thyroid hormone. *Endocrinology*, *140*(2), 784–790.

Vakkuri, O., Leppaluoto, J., & Vuolteenaho, O. (1984). Development and validation of a melatonin radioimmunoassay using radioiodinated melatonin as tracer. *Acta Endocrinologica*, *2*, 152–157.

Vanecek, J. (1998). Cellular mechanisms of melatonin action. *Physiological Reviews*, *78*(3), 687–721.

Viguie, C., Battaglia, D. F., Krasa, H. B., Thrun, L. A., & Karsch, F. J. (1999). Thyroid hormones act primarily within the brain to promote the seasonal inhibition of luteinizing hormone secretion in the ewe. *Endocrinology*, *140*(3), 1111–1117.

Visser, W. E., Friesema, E. C., & Visser, T. J. (2011). Minireview: Thyroid hormone transporters: The knowns and the unknowns. *Molecular Endocrinology*, *25*(1), 1–14.

Weaver, D. R., Liu, C., & Reppert, S. M. (1996). Nature's knockout: The Mel1b receptor is not necessary for reproductive and circadian responses to melatonin in siberian hamsters. *Molecular Endocrinology*, *10*(11), 1478–1487.

Weaver, D. R., & Reppert, S. M. (1990). Melatonin receptors are present in the ferret pars tuberalis and pars distalis, but not in brain. *Endocrinology*, *127*(5), 2607–2609.

Webster, J. R., Moenter, S. M., Barrell, G. K., Lehman, M. N., & Karsch, F. J. (1991). Role of the thyroid gland in seasonal reproduction. III. Thyroidectomy blocks seasonal suppression of gonadotropin-releasing hormone secretion in sheep. *Endocrinology*, *129*(3), 1635–1643.

Webster, J. R., Moenter, S. M., Woodfill, C. J., & Karsch, F. J. (1991). Role of the thyroid gland in seasonal reproduction. II. Thyroxine allows a season-specific suppression of gonadotropin secretion in sheep. *Endocrinology*, *129*(1), 176–183.

Weitzel, J. M. (2008). To bind or not to bind—How to down-regulate target genes by liganded thyroid hormone receptor? *Thyroid Research*, *1*(1), 1–5. 4-6614-1-4.

Whetham, E. O. (1933). Factors modifying egg production with special reference to seasonal changes. *The Journal of Agricultural Science*, *23*(3), 383–418.

Wieselthier, A. S., & van Tienhoven, A. (1972). The effect of thyroidectomy on testicular size and on the photorefractory period in the starling (*Sturnus vulgaris* L.). *The Journal of Experimental Zoology*, *179*(3), 331–338.

Wittkowski, W., Bergmann, M., Hoffmann, K., & Pera, F. (1988). Photoperiod-dependent changes in TSH-like immunoreactivity of cells in the hypophysial pars tuberalis of the djungarian hamster, *Phodopus sungorus*. *Cell and Tissue Research*, *251*(1), 183–187.

Wittkowski, W., Hewing, M., Hoffmann, K., Bergmann, M., & Fechner, J. (1984). Influence of photoperiod on the ultrastructure of the hypophysial pars tuberalis of the djungarian hamster, *Phodopus sungorus*. *Cell and Tissue Research*, *238*(1), 213–216.

Woitkewitsch, A. A. (1940). Dependence of seasonal periodicity in gonadal changes on the thyroid gland in *Sturnus vulgaris*. *Comptes Rendus de l'Académie des Sciences de l'URSS*, *27*, 741–745.

Wood, S. H., Christian, H. C., Miedzinska, K., Saer, B. R., Johnson, M., Paton, B., et al. (2015). Binary switching of calendar cells in the pituitary defines the phase of the circannual cycle in mammals. *Current Biology*, *25*(20), 2651–2662.

Wood, S. H., Hindle, M. M., Mizoro, Y., Cheng, Y., Saer, B. R. C., Miedzinska, K., et al. (2020). Circadian clock mechanism driving mammalian photoperiodism. *Nature Communications*, *11*, 4291.

Wood, S., & Loudon, A. (2018). The pars tuberalis: The site of the circannual clock in mammals? *General and Comparative Endocrinology*, *258*, 222–235.

Woodfill, C. J., Robinson, J. E., Malpaux, B., & Karsch, F. J. (1991). Synchronization of the circannual reproductive rhythm of the ewe by discrete photoperiodic signals. *Biology of Reproduction*, *45*(1), 110–121.

Woodfill, C. J., Wayne, N. L., Moenter, S. M., & Karsch, F. J. (1994). Photoperiodic synchronization of a circannual reproductive rhythm in sheep: Identification of season-specific time cues. *Biology of Reproduction*, *50*(4), 965–976.

Wu, Y., & Koenig, R. J. (2000). Gene regulation by thyroid hormone. *Trends in Endocrinology and Metabolism*, *11*(6), 207–211.

Xu, Y., Tamamaki, N., Noda, T., Kimura, K., Itokazu, Y., Matsumoto, N., et al. (2005). Neurogenesis in the ependymal layer of the adult rat 3rd ventricle. *Experimental Neurology*, *192*(2), 251–264.

Yamamura, T., Hirunagi, K., Ebihara, S., & Yoshimura, T. (2004). Seasonal morphological changes in the neuro-glial interaction between gonadotropin-releasing hormone nerve terminals and glial endfeet in Japanese quail. *Endocrinology*, *145*(9), 4264–4267.

Yamamura, T., Yasuo, S., Hirunagi, K., Ebihara, S., & Yoshimura, T. (2006). T(3) implantation mimics photoperiodically reduced encasement of nerve terminals by glial processes in the median eminence of Japanese quail. *Cell and Tissue Research*, *324*(1), 175–179.

Yasuo, S., & Yoshimura, T. (2009). Comparative analysis of the molecular basis of photoperiodic signal transduction in vertebrates. *Integrative and Comparative Biology*, *49*(5), 507–518.

Yasuo, S., Yoshimura, T., Ebihara, S., & Korf, H. W. (2009). Melatonin transmits photoperiodic signals through the MT1 melatonin receptor. *The Journal of Neuroscience*, *29*(9), 2885–2889.

Yeates, N. T. M. (1949). The breeding season of the sheep with particular reference to its modification by artificial mean using light. *Journal of Agricultural Science*, *39*(1), 1–42.

Yoo, S., & Blackshaw, S. (2018). Regulation and function of neurogenesis in the adult mammalian hypothalamus. *Progress in Neurobiology*, *170*, 53–66.

Yoo, S., Cha, D., Kim, D. W., Hoang, T. V., & Blackshaw, S. (2019). Tanycyte-independent control of hypothalamic leptin signaling. *Frontiers in Neuroscience*, *13*, 240.

Yoo, S., Cha, D., Kim, S., Jiang, L., Cooke, P., Adebesin, M., et al. (2020). Tanycyte ablation in the arcuate nucleus and median eminence increases obesity susceptibility by increasing body fat content in male mice. *Glia*, *68*(10), 1987–2000.

Yoshimura, T. (2013). Thyroid hormone and seasonal regulation of reproduction. *Frontiers in Neuroendocrinology*, *34*(3), 157–166.

Yoshimura, T., Yasuo, S., Watanabe, M., Iigo, M., Yamamura, T., Hirunagi, K., et al. (2003). Light-induced hormone conversion of T4 to T3 regulates photoperiodic response of gonads in birds. *Nature*, *426*(6963), 178–181.

Zhang, Z., Boelen, A., Kalsbeek, A., & Fliers, E. (2018). TRH neurons and thyroid hormone coordinate the hypothalamic response to cold. *European Thyroid Journal*, 7(6), 279–288.

Zhang, G., Li, J., Purkayastha, S., Tang, Y., Zhang, H., Yin, Y., et al. (2013). Hypothalamic programming of systemic ageing involving IKK-beta, NF-kappaB and GnRH. *Nature*, *497*(7448), 211–216.

CHAPTER FIVE

Thyroid hormone regulation of adult neural stem cell fate: A comparative analysis between rodents and primates

Lucile Butruille, Pieter Vancamp, Barbara A. Demeneix, and Sylvie Remaud*
UMR 7221 Phyma, CNRS/Muséum National d'Histoire Naturelle, Paris, France
*Corresponding author: e-mail address: sremaud@mnhn.fr

Contents

Abstract

Thyroid hormone (TH) signaling, a highly conserved pathway across vertebrates, is crucial for brain development and function throughout life. In the adult mammalian brain, including that of humans, multipotent neural stem cells (NSCs) proliferate and generate neuronal and glial progenitors. The role of TH has been intensively investigated in the two main neurogenic niches of the adult mouse brain, the subventricular and the subgranular zone. A key finding is that T3, the biologically active form of THs,

Vitamins and Hormones, Volume 116
ISSN 0083-6729
https://doi.org/10.1016/bs.vh.2021.02.009

promotes NSC commitment toward a neuronal fate. In this review, we first discuss the roles of THs in the regulation of adult rodent neurogenesis, as well as how it relates to functional behavior, notably olfaction and cognition. Most research uncovering these roles of TH in adult neurogenesis was conducted in rodents, whose genetic background, brain structure and rate of neurogenesis are considerably different from that of humans. To bridge the phylogenetic gap, we also explore the similarities and divergences of TH-dependent adult neurogenesis in non-human primate models. Lastly, we examine how photoperiodic length changes TH homeostasis, and how that might affect adult neurogenesis in seasonal species to increase fitness. Several aspects by which TH acts on adult NSCs seem to be conserved among mammals, while we only start to uncover the molecular pathways, as well as how other in- and extrinsic factors are intertwined. A multispecies approach delivering more insights in the matter will pave the way for novel NSC-based therapies to combat neurological disorders.

1. Introduction

In the adult vertebrate, newly formed neural cells contribute to central nervous system (CNS) plasticity and repair throughout life. New neurons and glial cells are generated from neural stem cells (NSCs) from embryonic stages to aging. Neurogenesis enables the generation of mature neurons through various processes including NSC proliferation, cell fate specification, survival, migration, differentiation and maturation of newborn neurons, as well as their functional integration into preexisting neuronal networks (Barnea & Pravosudov, 2011; Tanti, Rainer, Minier, Surget, & Belzung, 2012). In addition, NSCs are able to generate glial progenitors giving rise to both astrocytes and oligodendrocyte precursor cells (OPCs) (Laywell, Rakic, Kukekov, Holland, & Steindler, 2000; Taupin & Gage, 2002). Astrocytes modulate NSC proliferation, neuronal differentiation and maturation (Gengatharan, Bammann, & Saghatelyan, 2016; Lim & Alvarez-Buylla, 1999; Song, Stevens, & Gage, 2002) and regulate the efficiency of synaptic transmission and integrity (Parpura et al., 2012; Sultan et al., 2015; Volterra & Meldolesi, 2005). Oligodendrocytes produce a myelin sheath surrounding neuronal axons, facilitating nerve impulse transduction and supporting neuronal survival (Baumann & Pham-Dinh, 2001; Hughes & Appel, 2016).

In adult mammals, neurogenesis only occurs in specific brain regions, where the balance between NSC self-renewal and fate commitment is dynamically regulated (Alvarez-Buylla, Kirn, & Nottebohm, 1990; Ma, Ming, & Song, 2005; Morrison & Spradling, 2008). Two neurogenic niches have been described in detail, the subventricular zone (SVZ) of the lateral ventricles (Doetsch & Alvarez-Buylla, 1996; Lois & Alvarez-Buylla, 1993, 1994) and the subgranular zone (SGZ) of the hippocampal dentate gyrus (DG) (Cameron, Woolley, McEwen, & Gould, 1993; Eriksson et al., 1998; Kuhn, Dickinson-Anson, & Gage, 1996). In most mammals studied so far, newly generated SVZ-neuronal progenitors migrate tangentially toward the olfactory bulbs (OBs) where they differentiate into glomerular and periglomerular interneurons (Alvarez-Buylla, García-Verdugo, & Tramontin, 2001; Lois & Alvarez-Buylla, 1993). Rodent SVZ neurogenesis regulates various olfactory functions including odor discrimination, and olfactory learning and memory (Bath et al., 2008; Breton-Provencher, Lemasson, Peralta, & Saghatelyan, 2009; Enwere et al., 2004; Gheusi et al., 2000; Rochefort, Gheusi, Vincent, & Lledo, 2002; Zou, Wang, Pan, Lu, & Xia, 2015).

In the SGZ, progenitors migrate a short distance toward the granule cell layer of the hippocampus, and differentiate into granular neurons and interneurons (reviewed in Kempermann, Song, & Gage, 2015). The hippocampus is involved in spatial learning and memory in all mammals, including humans (Imayoshi et al., 2008; Kempermann, Brandon, & Gage, 1998; Kempermann, Kuhn, & Gage, 1997; Snyder, Hong, McDonald, & Wojtowicz, 2005). While oligodendrogenesis in human neurogenic regions seems to be maintained throughout life (Nait-Oumesmar et al., 2007; Weissleder et al., 2016), adult human neurogenesis is still under debate. This issue will be addressed in detail in Section 2. More recently, a third neurogenic niche has been identified within the arcuate nucleus of the hypothalamus and the median eminence (Lee et al., 2012; Li, Tang, & Cai, 2012; Robins et al., 2013).

Cellular and molecular mechanisms underlying adult mammalian neurogenesis were mainly uncovered in rodents. NSC proliferation and differentiation implicate numerous factors such as growth factors, sex hormones and other endocrine factors (Galea et al., 2013; Kapoor, Ghosh, Nordstrom, Vennstrom, & Vaidya, 2011; Rai, Hattiangady, & Shetty, 2007; Remaud, Gothié, Morvan-Dubois, & Demeneix, 2014; Werner et al., 2012). Several studies have shown that a fine-tuning of thyroid hormone (TH) action regulates NSC behavior (Gothié, Demeneix, & Remaud, 2017; López-Juárez et al., 2012; Mayerl, Heuer, & Ffrench-Constant, 2020; Remaud et al., 2017). In particular, T3, the biologically active form of THs, is a key regulator of NSC proliferation and fate decision in the SVZ and SGZ by determining (i) cell-cycle exit and survival of neural progenitor cells, (ii) NSC commitment toward a neuronal or a glial fate, and (iii) differentiation and maturation of both neuronal and glial cells (Gothié, Demeneix, & Remaud, 2017; Kapoor, Desouza, Nanavaty, Kernie, & Vaidya, 2012; Lemkine et al., 2005; Remaud et al., 2017).

In addition to regulating adult neurogenesis, THs control various physiological functions including energy metabolism, growth and reproduction. In seasonal species, whose energy metabolism and reproductive function are linked to variations in day length, TH metabolism is modulated by the photoperiod, with increased serum TH availability during long days in spring/summer (Nakao, Ono, & Yoshimura, 2008; Yoshimura et al., 2003). This seasonal variation in TH levels helps to control seasonal responses, notably reproductive activity (Hanon et al., 2008). Accordingly, thyroidectomy in seasonal mammals prevents the transition from breeding to non-breeding season both in males and females, leading to sustained activation of the gonadotropic axis (Moenter, Woodfill, & Karsch, 1991; Nicholls, Follett, Goldsmith, & Pearson, 1988; Parkinson & Follett, 1995). TH administration successfully restores this transition (Follett & Nicholls, 1985; Goldsmith & Nicholls, 1992), confirming the implication of THs on seasonal regulation of reproduction.

In seasonal mammals, photoperiod also modulates adult neurogenesis and associated behavior (Batailler et al., 2018; Galea & McEwen, 1999; Huang, DeVries, & Bittman, 1998; Migaud et al., 2010; Perrot-Sinal, Kavaliers, & Ossenkopp, 1998), which could possibly be linked to the seasonal variations in melatonin and TH levels. However, if so, then it remains unknown how TH signaling modulates cell fate choice and *a fortiori* behavioral responses to adapt to changing environments. Some hypotheses have been proposed, including the concept that in seasonally breeding mammals, the hormones controlling NSC generation, are also subject to seasonal variation (Hazlerigg & Lincoln, 2011).

Our review first focuses on the molecular and cellular processes underlying adult neurogenesis in the SVZ, the SGZ and the hypothalamus. Second, we discuss the TH-dependent mechanisms that control NSC physiology throughout adult life. Lastly, we address the importance of THs in the regulation of NSC fate in seasonal rodents. Most of what we know about TH-dependent adult neurogenesis so far, has been studied in rodents that have genetic, physiological and structural differences with humans, not the least being brain size. To understand similarities and evolutionary divergences between rodents and humans better, we compare cellular, molecular and environmental mechanisms that finely regulate the NSC proliferation and fate choice in rodent and primate models.

2. Cellular processes underlying adult neurogenesis: Rodents *vs.* primates

2.1 NSC fate determination in the adult SVZ

2.1.1 The adult rodent SVZ

In rodents, pluripotent SVZ-NSCs (B cells) divide slowly (Lim & Alvarez-Buylla, 2016; Reynolds & Weiss, 1992) giving rise to highly proliferative transient amplifying cells (TACs, C cells) (Fig. 1). TACs generate young neuroblasts (A cells) that migrate tangentially into the OBs *via* the rostro-caudal migratory stream (RMS), closely associated with astrocytes and blood

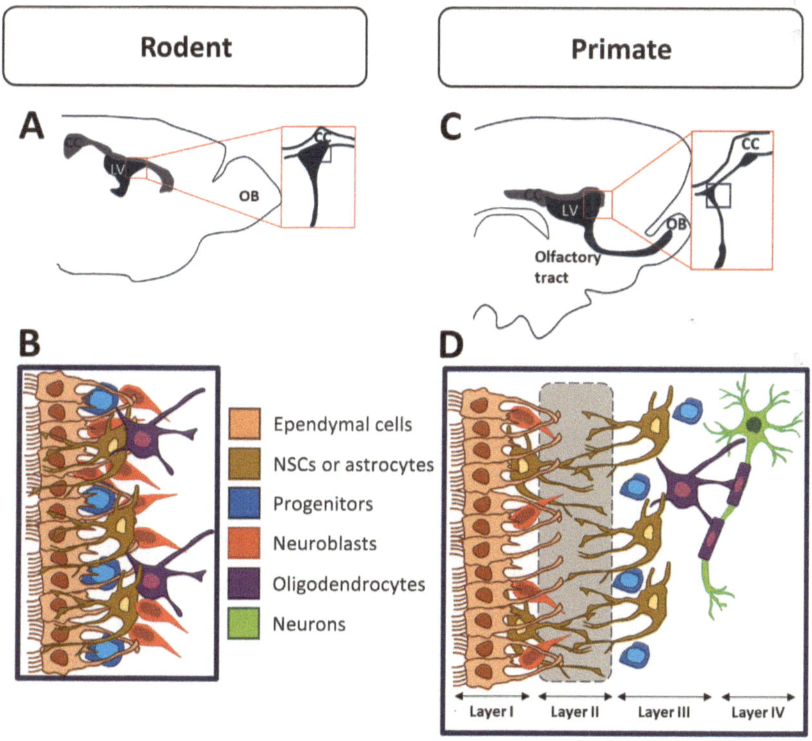

Fig. 1 Cyto-architectural organization of the adult rodent and primate SVZ. (A and C) Sagittal representation of the adult rodent (A) and primate (C) frontal brain, and location of the lateral ventricle (LV), corpus callosum (CC) and olfactory bulb (OB). The red inserts show the coronal view of the lateral ventricle morphology (in black) in rodents and primates. (B and D) Schematic drawing of the different cell populations in the rodent (B) and primate (D) SVZ. *NSCs*, neural stem cells.

vessels (Jankovski & Sotelo, 1996; Lois, García-Verdugo, & Alvarez-Buylla, 1996; Snapyan et al., 2009). In the OBs, migrating neuroblasts differentiate mainly into granular interneurons (~90%), while approximately 10% mature into GABAergic and dopaminergic periglomerular interneurons (Hack et al., 2005; Merkle, Mirzadeh, & Alvarez-Buylla, 2007; Petreanu & Alvarez-Buylla, 2002; Winner, Cooper-Kuhn, Aigner, Winkler, & Kuhn, 2002).

SVZ-derived NSCs also generate oligodendrocytes (Menn et al., 2006). However, SVZ-NSCs can give rise to either neuronal precursors (PAX6+) or OPCs (OLIG2+) in the dorsal part of the SVZ, but not a mix of both (Ortega et al., 2013). This suggests that neuronal or oligodendroglial identity is genetically programmed at the NSC (or TAC) stage (Jones & Connor, 2011). Likewise, Calzolari et al. showed that most SVZ-TACs are committed uniquely to the neuronal differentiation pathway (Calzolari et al., 2015). These results clearly question NSC multipotency, widely demonstrated *in vitro*, where NSCs are able to generate all cell types. However, a recent lineage tracing experiment demonstrated *in vivo* that about 40% of SVZ-TACs were bipotent early after birth, thus generating glial and neuronal cells (Figueres-Oñate, Sánchez-Villalón, Sánchez-González, & López-Mascaraque, 2019). It would be interesting to investigate whether this heterogeneity in the progenitor fate potential reduces as a function of aging.

Furthermore, dorsally situated SVZ-OPCs, constituting a subpopulation of TACs (Menn et al., 2006), migrate to the corpus callosum to differentiate first into the NG2 + non-myelinating, and subsequently into mature myelinating oligodendrocytes (Hack et al., 2005; Menn et al., 2006). They can contribute to myelin turnover and/or myelin repair after a myelin lesion (Remaud et al., 2017; Xing et al., 2014). The oligodendroglial identity of NSCs is promoted by overexpression of *Olig2* (Hack et al., 2005) and *Egfr* (Aguirre & Gallo, 2007; Aguirre, Dupree, Mangin, & Gallo, 2007; Remaud et al., 2017), by a demyelinating lesion (Keirstead & Blakemore, 1999; Nait-Oumesmar et al., 1999), increased expression of *Wnt* (Ortega, Gascón, et al., 2013) or Sonic hedgehog signaling (Ortega, Radonjic, & Zecevic, 2013). Most of these signaling pathways are also controlled by TH (see Section 3). Together, these data demonstrate that SVZ-derived OPCs constitute an endogenous source of remyelinating oligodendrocytes in the adult rodent brain (McKenzie et al., 2014).

2.1.2 The adult primate SVZ

The ultrastructural organization of the non-human primate SVZ (Fig. 1) differs from that of rodents. Notably, the astrocyte ribbon is mainly composed of proliferative astrocytes (Gil-Perotin, Duran-Moreno, Belzunegui, Luquin, & Garcia-Verdugo, 2009) constituting a potential source of quiescent NSCs (Lim & Alvarez-Buylla, 2016). The astrocyte ribbon (layer III) is separated from the ependymal layer (layer I) by a hypocellular gap (layer II), composed of astrocyte processes and some neuronal somata with a so far unknown function (Akter et al., 2020; Gil-Perotin et al., 2009; Quiñones-Hinojosa et al., 2006; Sanai et al., 2004; Sawamoto et al., 2011). A fourth layer (layer IV), described in the macaque monkey (Gil-Perotin et al., 2009) and common marmoset SVZ (Azim et al., 2013; Sawamoto et al., 2006), adjacent to the astrocyte ribbon, is mainly composed of myelinated axons (Quiñones-Hinojosa et al., 2006). In primate species, mitotically active neuroblasts are abundant in the SVZ during the perinatal period, but numbers decrease strongly during the first months of life (Gil-Perotin et al., 2009; Pencea, Bingaman, Freedman, & Luskin, 2001; Sawamoto et al., 2011).

Furthermore, the neuroblast migratory stream is organized around a lateral ventricular extension that reaches as far as the OBs (Fig. 1C). This continuous ventricle has been identified in macaques (Gil-Perotin et al., 2009), rhesus monkeys (Wang et al., 2011) and common marmosets (Sawamoto et al., 2011). As in rodents, migrating neuroblasts are observed in the RMS-like pathway and the OB (Bédard, Lévesque, Bernier, & Parent, 2002; Fasemore et al., 2018; Pencea et al., 2001; Sawamoto et al., 2011; Wang et al., 2011). In marmosets, the incorporation of new olfactory neurons decreased by more than 50% during the first months of life (Akter et al., 2020), suggesting apoptosis occurs during neuronal maturation and not during neuroblast migration, as in rodents (Platel et al., 2019). Moreover, and in contrast to rodents, the production of periglomerular interneurons is limited to the first months of life (Akter et al., 2020), suggesting that newly generated neurons from then on give rise to granule cells in the OB.

The four-layer organization of the primate SVZ has also been observed in humans (Quiñones-Hinojosa et al., 2006; Sanai et al., 2004). However, the existence of adult human SVZ neurogenesis is still being questioned, largely because of the rapid decline in neuroblast production after birth (Bergmann et al., 2012; Sanai et al., 2011; Wang et al., 2011). Dennis, Suh, Rodriguez, Kril, and Sutherland (2016) reported approximately 234 DCX+ neuroblasts/mm^2 in a 2-month-old individual, whereas numbers dropped to 4.6 DCX+ neuroblasts/mm^2 in young adults.

Similarly, although SVZ-derived cells with a migratory phenotype were found in adult humans, the existence of a RMS as it exists in rodents is still controversial (Curtis et al., 2007; Sanai, Berger, Garcia-Verdugo, & Alvarez-Buylla, 2007; Wang et al., 2011). Adult human studies using electron microscopy (Kam et al., 2009), immunohistochemistry (Sawamoto et al., 2006) and magnetic resonance imaging (Curtis et al., 2007) suggested the existence of a continuous ventricle along a migratory stream filled with cerebrospinal fluid, as observed in human fetuses (Guerrero-Cázares et al., 2011) and non-human primates. Unlike rodents, few human neuroblasts migrate through the RMS-like ventricular extension to the OBs (Dennis et al., 2016; Sanai et al., 2011; Wang et al., 2011) and do not form a migration chain (Quiñones-Hinojosa et al., 2006; Sanai et al., 2004).

Furthermore, the integration of new mature neurons into the preexisting neuronal networks of the human OB seems negligible (Bergmann et al., 2012; Wang et al., 2011). C^{14} dating estimated that the annual turnover of olfactory interneurons is less than 1% per year in humans, whereas it reaches 50% in mice (Bergmann et al., 2012). Surprisingly, some adult-born interneurons generated from non-human and human primate SVZ-NSCs migrate toward the striatum (Bédard, Gravel, & Parent, 2006; Ernst et al., 2014), neocortex (Bernier, Bédard, Vinet, Lévesque, & Parent, 2002; Gould et al., 1999), ventral medial prefrontal cortex (Paredes et al., 2016; Sanai et al., 2004), amygdala and piriform cortex (Bernier et al., 2002). However, their function is still unknown.

SVZ-oligodendrogenesis seems to be maintained throughout human life (Ernst et al., 2014; Weissleder et al., 2016). Additionally, myelination of the corpus callosum occurs gradually from birth in marmosets (Akter et al., 2020), as it does in humans (Yeung et al., 2014).

Finally, given that the ventricular walls of the mammalian CNS consist largely of NSC-like cells, the presence of a continuous ventricle in some primates suggests that neurogenic processes continue in these species. Indeed, Kam et al. have shown proliferating cells along the continuous ventricle and hypothesized they could as such provide new neurons to the human forebrain (Kam et al., 2009). Future investigations are necessary to understand whether regions like the RMS have neurogenic potential as well, in addition to the SVZ/SGZ/hypothalamus, the main NSC-harboring the main regions, and which mechanisms regulate proliferation in such other neurogenic niches observed in several species including humans.

2.1.3 Functional relevance of SVZ neurogenesis in rodents and primates

Olfactory inhibitory interneurons are essential for distinguishing and memorizing odors underlying rodent social interactions, foraging, and sexual partner selection (Lepousez, Nissant, & Lledo, 2015; Lledo, Alonso, & Grubb, 2006; Lledo, Merkle, & Alvarez-Buylla, 2008; Ming & Song, 2011). In rodents, selective ablation of SVZ neurogenesis reduced short-term olfactory memory (Breton-Provencher et al., 2009), while increased SVZ-NSC proliferation by olfactory stimulation enhanced memorization (Rochefort et al., 2002). Additionally, mice null for the NSC cycle inhibitor *p21* exhibited improved olfactory abilities and memory during physical activity (Nicolis di Robilant et al., 2019). These results showed that olfactory memory depends on SVZ neurogenesis in rodents.

In contrast, the ablation of SVZ neurogenesis by the antimitotic drug AraC (28-day infusion period) in adult mice did not alter their ability to discriminate two distinct odors, but affected the discrimination of similar odors (Breton-Provencher et al., 2009; Enwere et al., 2004; Gheusi et al., 2000). Likewise, increased SVZ neurogenesis following overexpression of the cell-cycle regulators *Cdk4/cyclinD1* improved discrimination of similar odors (Bragado Alonso et al., 2019). Together, these results indicate that bulbar preexisting interneurons are involved in simple odor discrimination, while newborn olfactory interneurons could be required exclusively for more challenging olfactory tasks in which animals need to distinguish very similar odors (Bragado Alonso et al., 2019). Finally, increased complexity of olfactory perceptual learning, an improvement in odor discrimination induced by previous experiences, required the recruitment of neurons generated throughout life (Forest et al., 2019).

Hence, olfactory neuroplasticity could be an evolutionary conserved mechanism in nocturnal mammals to increase fitness and survival chances when confronted with continuously changing diets or sexual partners. Indeed, the continuous loss of newly-formed neurons in genetically engineered mice reduced maternal care and nursing in females, and led to less aggressive and altered sexual behaviors in males (Sakamoto et al., 2011).

In mice, high prolactin concentrations during gestation increased both SVZ proliferation and the generation of olfactory interneurons (Larsen & Grattan, 2010; Shingo et al., 2003). Absence of prolactin increased maternal anxiety and altered behavior (Larsen & Grattan, 2010). Together, this suggests prolactin-dependent regulation of SVZ neurogenesis is vital for normal maternal behavior. Future studies could determine whether maternal olfactory performance and SVZ neurogenesis are altered during pregnancy in humans, in a way that is similar to mice and possibly other mammals.

Human OBs are relatively smaller in size compared to other mammals: 0.01% of the whole brain volume *vs.* 2% in mice (Kavoi & Jameela, 2011).

Comparative analyses of olfactory performance have nevertheless demonstrated that primates, including humans, are able to detect and distinguish a wide variety of odors with larger or smaller detection thresholds compared to other mammalian species, depending on the presented odor (Güven & Laska, 2012; McGann, 2017; Sarrafchi, Odhammer, Hernandez Salazar, & Laska, 2013).

Furthermore, primate SVZ-neuroblasts could be involved in another CNS functionality as they migrate to other structures such as the striatum (Ernst et al., 2014), a structure notably involved in motor and executive functions. It seems necessary to determine why primate SVZ-neuroblasts have not the same destination and functionality as in rodents, which would help to understand mechanisms which govern adult human neurogenesis, and its implication in the development of neurological disorders.

2.2 NSC fate determination in the adult SGZ

2.2.1 The adult rodent SGZ

Adult hippocampal neurogenesis has been extensively studied in rodents (reviewed in Toda & Gage, 2018). SGZ-derived NSCs, also called radial astrocytes or type I cells (Fuentealba, Obernier, & Alvarez-Buylla, 2012; Seri, García-Verdugo, Collado-Morente, McEwen, & Alvarez-Buylla, 2004; Seri, Garcı´a-Verdugo, McEwen, & Alvarez-Buylla, 2001), give rise to two populations of TACs (Kempermann, Gast, Kronenberg, Yamaguchi, & Gage, 2003; Seri et al., 2001), type IIa and type IIb (Kronenberg et al., 2003). Under physiological conditions, type IIa cells divide more actively than type I cells, which are mostly quiescent. Cell commitment toward a neuronal fate is determined by the expression of several genes including *Prox1*, *NeuroD1* and *Dcx* in type III post-mitotic progenitors. Newborn neurons derived from SGZ-NSCs migrate a short distance to the granular layer of the DG, where they differentiate into granular interneurons (Cameron & Mckay, 2001; Kempermann, 2012; Lledo et al., 2006). These immature early post-mitotic cells transiently express Calretinin and then mature into Calbindin+ granular neurons.

2.2.2 The adult primate SGZ

Adult hippocampal neurogenesis persists in several non-human primates studied so far, although the proliferation rate is higher in newborns and gradually declines to negligible levels from sexual maturity (Aizawa, Ageyama, Yokoyama, & Hisatsune, 2009; Amrein et al., 2015; Eckenhoff & Rakic, 1988; Gould et al., 1999; Kornack & Rakic, 1999; Ngwenya, Rosene, & Peters, 2008). Contrary to rodents that generate hippocampal neurons throughout life (Amrein, Isler, & Lipp, 2011), most granular interneurons in rhesus monkeys are mainly produced during the postnatal period (Jabès, Lavenex, Amaral, & Lavenex, 2010). Cameron and Mckay (2001) estimated that rats (~lifespan 2–3 years) produce about 2.250 new granule neurons per day, corresponding to 0.2% of the total number of neurons. In comparison, primates (~lifespan 10 times greater than rodents) produce 10 times less new granule cells than rodents (0.02% of the total number of neurons per day). Additionally, the number of hippocampal proliferative cells, neuroblasts and granular cells remained similar in adult mice and adult marmosets (Amrein et al., 2015).

Adult neurogenesis also persists in the human hippocampus (Boldrini et al., 2018; Moreno-Jiménez et al., 2019; Tobin et al., 2019), although the proliferation rate appears extremely low in adults and older individuals (Dennis et al., 2016). Immnohistochemical analysis demonstrated the presence of type IIb and type III cells up to 79 years of age, and new granule cells up to 100 years of age (Boldrini et al., 2018; Knoth et al., 2010). Despite this, Sorrells and collaborators reported that human hippocampal neurogenesis was restricted to the first years of life, and that proliferating progenitors and young immature neurons were absent in adults (Sorrells et al., 2018). However, several methodological factors could have contributed to the discrepancies (Flor-García et al., 2020) between the studies of Boldrini and Sorrells. In the paper from Sorrells et al. (2018), the patients had chronic epilepsy that could have impacted hippocampal neurogenesis, whereas Boldrini et al. used healthy patients. Second, there was a larger post-mortem interval in Sorrells' study that could have altered tissue integrity and degraded certain antigens. Further, Sorrells et al. used a limited number of hippocampal sections (approximatively 3 sections per subject) whereas Boldrini and colleagues analyzed the whole DG. Finally, Boldrini et al. used an unbiased stereological approach to estimate the total number of immunoreactive cells, an approach not used in the Sorrells study. Therefore, the results obtained by Boldrini's team, have been recently confirmed by Moreno-Jiménez et al. (2019) and Tobin et al. (2019).

2.2.3 Functional relevance of SGZ neurogenesis in rodents and primates

In rodents, hippocampal neurogenesis actively participates in hippocampal-dependent learning and memory (Aimone et al., 2014; Gros, Veyrac, & Laroche, 2015; Sahay et al., 2011), and is likely required to establish and maintain long-term spatial memory (Imayoshi et al., 2008; Pan et al., 2019; Snyder et al., 2005). Consequently, overexpression of the cell-cycle regulators *Cdk4/cyclinD1* enhanced adult hippocampal neurogenesis in old mice, favoring rejuvenation of hippocampal functions, especially spatial memory and navigational strategies (Berdugo-Vega et al., 2020).

In macaque monkeys, the hippocampal proliferation rate and cognitive functions are also linked. Macaques with a higher level of SGZ neurogenesis had better learning skills (Aizawa et al., 2009). The age-related decline in adult hippocampal neurogenesis correlated with impaired cognitive performance, suggesting that it participates in learning and spatial memory in this species too. Interestingly, a study on human epileptic patients showed that those with low SGZ-NSC proliferation rates *in vitro* exhibited learning and memory deficits (Coras et al., 2010), as compared to patients with a normal proliferation capacity. Human SGZ neurogenesis probably associates with many other cognitive functions as well, since decreased hippocampal neurogenesis has been observed in cognitive deficits, neuropsychiatric disorders and neurodegenerative diseases (Apple, Solano-fonseca, & Kokovay, 2017; Kaneko & Sawamoto, 2009).

2.3 NSC fate determination in the adult hypothalamus

The mediobasal hypothalamus is the most recently identified neurogenic niche, more specifically the adult rodent arcuate nucleus and the median eminence (Lee et al., 2012; Li et al., 2012; Nogueira et al., 2014; Robins et al., 2013). Both of these structures are involved in the control of basic physiological functions such as reproduction, energy balance and food intake. Cellular characterization of the adult rodent mediobasal hypothalamus showed that tanycytes lining the third ventricle form a progenitor cell population (Xu et al., 2005). They are able to generate multipotent neurospheres and differentiate into neurons and glial cells *in vitro*, indicating multipotency (Li et al., 2012; Robins et al., 2013; Zilkha-Falb, Kaushansky, & Ben-Nun, 2020). Tanycytes generate the different neuronal populations regulating metabolism, i.e., orexigenic neurons which express neuropeptide-Y (NPY) and agouti-related protein (AgRP) neurotransmitters to stimulate food intake, and anorexigenic neurons expressing proopiomelanocortin (POMC), all sensitive to leptin, ghrelin and fasting (Haan et al., 2013; Kokoeva, Yin, & Flier, 2005; Li et al., 2012; Werner et al., 2012).

Furthermore, in mice, the proliferative cells of the median eminence gave rise to OPCs and pre-myelinating oligodendrocytes (Zilkha-Falb et al., 2020), showing that the hypothalamus is also a source of oligodendrocytes. However, their destination and function has yet to be elucidated.

Interestingly, the proportion of OPCs generated from median eminence-NSCs was twice as high as compared to those derived from SVZ-NSCs both *in vivo* and *in vitro* (Zilkha-Falb et al., 2020), indicating that the adult hypothalamus favors OPC generation more than the SVZ.

An immunohistochemical study revealed a four-layer organization of the human hypothalamus (Pellegrino et al., 2018), highly reminiscent of the human SVZ (Sanai et al., 2004) and the sheep hypothalamus (Butruille, Batailler, Mazur, Prévot, & Migaud, 2018). This cyto-architectural organization was not observed in the gray mouse lemur, a basal primate whose hypothalamus is more identical to that of rodents (Pellegrino et al., 2018). In addition, the human and non-human primate hypothalamus is composed of progenitor cells expressing similar neural stem/progenitor and neuronal markers as do rodents (Batailler et al., 2014; Nogueira et al., 2014; Pellegrino et al., 2018).

The role of hypothalamic neurogenesis as implicated in metabolic regulation, food intake, obesity and the development of metabolic disorders has been the subject of intense research (Kokoeva et al., 2005; Pierce & Xu, 2010; Li et al., 2012; Sousa-Ferreira, de Almeida, & Cavadas, 2014; for review, see Rojczyk-Gołębiewska, Pałasz, & Wiaderkiewicz, 2014). Tanycytes play a key role in regulating fat metabolism (Yoo et al., 2020), while different diets such as those enriched in fats modify hypothalamic tanycyte proliferation and cell fate determination (Lee et al., 2012; Li et al., 2012; McNay, Briançon, Kokoeva, Maratos-Flier, & Flier, 2012). For example, proliferation of median eminence-tanycytes as well as the production of new neurons are increased following a high-fat diet (HFD) over 2 months (Lee et al., 2012). In contrast, a 4-month HFD altered arcuate nucleus-neurogenesis decreasing proliferation, survival and cell differentiation (Li et al., 2012; McNay et al., 2012). These data showed that HFD effects were dependent on the hypothalamic region and the diet length, and that a chronic HFD adversely affected hypothalamic neurogenesis.

The role of hypothalamic neurogenesis in reproduction and sexual behavior is still unclear. A subpopulation of mouse hypothalamic cells expresses the estrogen receptor α (ERα) which regulates reproductive function at multiple levels (Bless et al., 2016). To date, the regulation of reproductive function by hypothalamic neurogenesis has only been demonstrated in sheep, whose reproductive function is also controlled by day length (Batailler et al., 2018; see Section 4.3.3). Effectively, the ablation of hypothalamic neurogenesis by an antimitotic treatment decreased neuroblast production and altered the timing of seasonal reproduction in this species.

Future studies might analyze the existence, regulation and rate of human hypothalamic neurogenesis, and how it could be implicated food intake control and energy metabolism. This might provide new targets and strategies dealing with the alarming increase of diseases linked to metabolic syndrome or even infertility.

3. How does thyroid hormone contribute to adult neurogenesis in rodents and primates throughout life?

3.1 TH, a key signal involved in brain evolution and cognition

The TH signaling pathway is well conserved over mammalian evolution, regulating brain development and its functions throughout life. An emerging hypothesis suggests that TH was a key neuro-hormonal signal driving evolutionary expansion of the primate brain, in particular the neocortex. That could have paved the way for improved cognitive abilities and the emergence of human intelligence (Saniotis, Grantham, Kumaratilake, & Henneberg, 2020; Stenzel & Huttner, 2013).

Interestingly, higher T3 and T4 serum levels were found in humans and their closest living relatives, the chimpanzees and bonobos (Aliesky, Courtney, Rapoport, & McLachlan, 2013). In parallel, T3 and T4 levels were lower in orangutans and gorillas that are phylogenetically more divergent. Accordingly, the inverse pattern was observed for TSH levels. This suggests either that the hypothalamic-pituitary-thyroid (HPT) axis set point might have changed among primate species, or that liver DIO1-dependent T4 to T3 conversion increased in some, with higher serum TH concentrations as a result. The latter possibility could for instance explain why a lower T4/T3 ratio was measured in chimpanzees (Gagneux et al., 2001; see comment Previc, 2002). They also had higher TTR levels in the CSF as compared to more basal primates (Gagneux et al., 2001), suggesting local T4 supply to the brain via this route might have changed as well. The ventricular system is particularly large in embryonic stages, offering an alternative TH supply route via the blood-cerebrospinal fluid barrier, in times the blood-brain barrier (BBB) has not matured yet. Hence, both altered HPT axis homeostasis and systemic TH levels, as well as varying expression of local regulators defining TH availability, could have been the perfect cocktail for enhanced neurogenic processes underlying brain development. In such a way, brain size gradually increased resulting in the appearance of more human-like primates. Detailed data on the local expression patterns of TH regulators across primate species could back up such hypotheses, but are hard to obtain. We do however know that the human fetal brain expresses all TH transporters and TRs from early stages of brain development, and that TR-dependent actions occur from the first gestational trimester (López-Espíndola et al., 2019; Moog et al., 2017).

Another important aspect in this proposed evolutionary role of TH in brain expansion is timing. Human brain development is prolonged as compared to non-human primates. Millions of neurons are added each second from gestational week 5 up to adulthood. Our extended development left a greater window for THs to act on neuroprogenitors (Moog et al., 2017; Zoeller & Rovet, 2004). For instance, the extremely large size of the primate, and especially human neocortex is assumed to originate from a much greater NSC pool that had the chance to expand over a longer period of time as compared to other mammalian species (Charvet & Finlay, 2014). Furthermore, this allometric isocortical expansion is accelerated relative to other brain regions. The strong expression of *MCT8* and other TH regulators in neural progenitors from early fetal development (Chan et al., 2014; López-Espíndola et al., 2019) suggests early TH-dependent stimulation of neurogenesis could underlie this disproportionate growth. Besides, an intermediate type of NSCs that only exists in primates, the outer radial glia, allow for an extra round of proliferation, increasing the cell pool size even more (Pollen et al., 2015). Whether they are responsive to TH is still unknown.

Several other cell types found in highly developed regions in the human brain are also responsive to TH in developmental and adult stages. For instance, the development of greater mental capacities could be due to TH-dependent expansion of dopaminergic systems. TH converts tyrosine to L-DOPA, the precursor of dopamine in the CNS (Claustre, Balende, & Pujol, 1996). Increased dopaminergic activity is key for higher cognitive skills such as working memory, abstract reasoning and language (Previc, 2002, 2009; Saniotis et al., 2020).

Another important factor are the neurotrophins. They are essential for various stages of adult neurogenesis (Vilar & Mira, 2016). In particular brain-derived neurotrophic factor (BDNF) is key in regulating plastic processes in the hippocampus, improving learning and spatial memory (Miranda, Morici, Zanoni, & Bekinschtein, 2019). The same neurotrophins could have enhanced fetal and postnatal neurogenesis in primates, and broadened associated cognitive functions, during hominid brain evolution (Hill & Polk, 2019; Raichlen & Polk, 2013). *BDNF* mRNA expression is higher throughout human life as compared to other primates (Liu et al., 2012). *Bdnf* knockout mice have smaller brains, indicating BDNF indeed favors brain growth (Fox, Biddinger, Jones, McAdams, & Worman, 2013; Vigers et al., 2012). Interestingly, TH positively regulates *Bdnf* expression in the developing rodent brain (Camboni, Roskoden, & Schwegler, 2003; Lüesse et al., 1998), and this TH-dependent action was neuroprotective after traumatic brain injury by promoting neuronal survival in postnatal mice (Shulga et al., 2009). Natural selection of a better cross talk between TH and BDNF in humans could have promoted these molecular actions underlying brain size expansion and structural complexity, laying the foundation for novel functions.

Lastly, vascular endothelial growth factor (VEGF) and insulin-like growth factor 1 (IGF1)—two other evolutionary conserved growth factors that control brain growth and function (Kappeler et al., 2008; Rosenstein, Krum, & Ruhrberg, 2010)—may have been vital for TH-dependent primate/hominid brain evolution. TH regulates the expression of both of them, as observed in several mammals (Zhang, Cooper-Kuhn, Nannmark, Blomgren, & Kuhn, 2010). It is interesting to note that exercise upregulates hippocampal expression of *BDNF*, *VEGF* and *IFG1*, promoting neurogenesis and hippocampal-dependent learning (Fabel et al., 2003; Liu & Nusslock, 2018; LLorens-Martín, Torres-Alemán, & Trejo, 2010). That suggests that adult NSCs still respond to these factors in a way that embryonic NSCs might do too, and that their expression can be altered by TH in adults. Such molecular interactions could underlie cognitive improvements in patients suffering from major depression disorder that received TH treatment (Biondi & Wartofsky, 2012).

It is tempting to speculate that the acquisition of bipedalism in the earliest hominids, that facilitated running and access to iodine-rich food (e.g. consumption of iodine-containing aquatic herbs by Bonobos in a poor iodine environment, Hohmann, Ortmann, Remer, & Fruth, 2019), provided the essential amount of iodine that allowed such changes in TH homeostasis to facilitate relatively rapid hominid evolution. Over millions of years, that could have paved the way for the appearance of higher cognitive skills that led the first ancestors of the *Homo* genus to develop and utilize tools, and invent culture and art, characteristics we now consider as unique and typical to humans.

Data on TH status and metabolism in great apes is incomplete, stressing the need of further research that can explain if, and how, an evolutionary-driven change in TH homeostasis and the set point of the HPT axis, as well as mechanisms regulating local TH availability, could have engaged structural and functional differences in the CNS of human and non-human primates.

3.2 From NSCs to olfactory behavior: Roles of TH in the adult SVZ

In the adult mouse SVZ, TH deficiency blocked NSC and progenitor cell-cycle progression, an effect rescued by a short exogenous administration of T3 (Lemkine et al., 2005). The same phenotype was observed in adult $TR\alpha^{0/0}$ mice, suggesting that T3 through its receptor TRα can directly regulate the proliferative potential of SVZ-NSCs and progenitors. The accumulation of NSCs/progenitors blocked during the S-phase after adult-onset hypothyroidism can also be caused by decreased apoptosis. Indeed, the brain's lower TH availability due to a decrease in T4 uptake into the brain of $TTR^{-/-}$ mice, reduced apoptosis of NSCs and progenitors compared to wild-type mice, but was similar to hypothyroid mice (Richardson, Lemkine, Alfama, Hassani, & Demeneix, 2007).

Furthermore, T3 can act as a neurogenic factor by promoting NSC and progenitor commitment preferentially toward a neuronal fate in the adult mouse SVZ (López-Juárez et al., 2012; Remaud et al., 2017). In presence of T3, TRα1 downregulated several genes involved in the pluripotency of NSCs and progenitors such as *c-Myc, Ccnd1, Sox2, Egfr* and *Musashi1* (Hassani et al., 2007; Lemkine et al., 2005; López-Juárez et al., 2012; Remaud et al., 2017), thus promoting the generation of new migrating neuroblasts. Accordingly, SVZ-neuroblasts expressed high levels of TRα1 and low levels of DIO3, enabling them to respond to T3 (López-Juárez et al., 2012; Remaud et al., 2017).

Mice null for the efficient TH distributor in the CSF, TTR, had lower T3 and T4 levels in the CSF (Almeida Palha et al., 2000), decreasing the generation of new SVZ-neuroblasts but promoting SVZ-oligodendrogenesis (Vancamp et al., 2019). Similar results have been reported in double knock-out mice for MCT8 and OATP1C1, two key TH transporters, leading to impairments in olfactory function (Luongo et al., 2021). This bolsters previous results showing that brain hypothyroidism regulates the neuron/glia ratio in favor of newly generated SVZ-OPCs. Moreover, MCT8 has been shown to regulate function of hippocampus (Mayerl et al., 2020) and skeletal muscle stem cells (Mayerl et al., 2018) and is therefore a promising candidate to control cellular entry and efflux of TH in SVZ cells too. Similarly, while previous data showed that *Dio3* is expressed in both Nestin+ NSCs and proliferating progenitors expressing either SOX2 or EGFR, nothing is known about the expression pattern of *Dio2* in the SVZ.

A changing TH status modulates olfactory performance, as experimental induction of hypothyroidism induced a loss of the sense of smell (i.e. anosmia) in mice (Beard & Mackay-Sim, 1987) and was also reported in some human cases of hypothyroidism (Deniz et al., 2016; McConnell, Menendez, Smith, Henkin, & Rivlin, 1975). Accordingly, the smell sense in hypothyroid and anosmic adult mice was recovered by daily injections of T4 (Beard & Mackay-Sim, 1987). Mechanisms governing TH-dependent olfaction are not established so far, but in salmons, elevated T3 or T4 levels increased the number of BrdU+ proliferating cells in the olfactory epithelium that could allow generating more olfactory receptor neurons (Lema & Nevitt, 2004). Likewise, the number of mature olfactory interneurons was reduced in hypothyroid newborn (Paternostro & Meisami, 1991) and adult rats (Dhong, Kim, & Ha, 2003), suggesting a key role of TH in the maturation of olfactory interneurons.

Additionally, THs are well-known to promote the generation of tyrosine hydroxylase interneurons in the rodent midbrain and hypothalamus (Ayala et al., 2019; Chen et al., 2015; Kincaid, 2001) through TRα1 (Chen et al., 2015). Since *Thra* is expressed in the olfactory epithelium of salmons (Kudo, Eto, Abe, & Mochida, 2018), one could analyze olfactory memory and odor discrimination in $TR\alpha^{0/0}$ adult mice that are devoid of the two isoforms, TRα1 and TRα2 (Flamant, Gauthier, & Richard, 2017). Furthermore, as SVZ neurogenesis tightly regulates olfaction in rodents, we could hypothesize that T3 action on SVZ neurogenesis alters olfactory function. In that line, adult-onset hypothyroidism reduced the number of migrating neuroblasts along the RMS of adult mice (Lemkine et al., 2005b) whereas the overexpression of TRα1 in SVZ cells increased the generation of clusters of DCX+ neuroblasts at the beginning of the RMS, 4 days after transfection of a DNA-plasmid encoding *Thra1* (López-Juárez et al., 2012). This strongly suggests that any change in (local) TH availability and action could modulate the production of neuroblasts. However, it remains to be investigated how TH functionally contributes to producing the different subpopulations of mature interneurons in the OB of adult mice.

Although the role of TH on the adult primate SVZ has not been studied yet, some evidence suggests a potential relevant function that deserves detailed analysis. First, SVZ-proliferating neural progenitors expressing SOX2 in postnatal marmosets (Homman-Ludiye, Merson, & Bourne, 2012) or Musashi in adult macaque monkeys (Tonchev, Yamashima,

Sawamoto, & Okano, 2005) could be targeted by TH to drive NSC fate choice, as in adult mice (López-Juárez et al., 2012). Furthermore, expression of some TH regulators such as MCT8 was detected in brain microvessels of the adult Cynomolgus monkey whereas OATP1C1 was largely absent (Ito et al., 2011). This suggests that TH supply across the primate BBB is mainly mediated by MCT8. In humans, mutations in the *SLC16A2/MCT8* gene result in the Allan-Herndon-Dudley syndrome (AHDS), a rare X-linked psychomotor retardation syndrome due to severe brain hypothyroidism (Friesema et al., 2004).

In contrast, TH entry into the rodent brain via the BBB is mediated by MCT8 and OATP1C1 (Mayerl et al., 2014), resulting in only mild symptoms in *Mct8* KO mice due to residual TH uptake *via* OATP1C1 (Trajkovic et al., 2007). This underlines key species differences in the regulation of TH availability at the BBB, and possibly also the brain cellular level (Vancamp & Darras, 2018). A better idea of the distinct profiles of TH transporters, deiodinases and the TH distributor protein TTR in adult rodents and primates are necessary to advance our understanding of how local TH action differentially regulates SVZ neurogenesis in both animal groups. A last important issue is whether SVZ-neuroblasts contribute to olfactory behavior in primates as they do in rodents and in particular, whether optimal olfaction is dependent on TH. The existence of a homolog of the rodent SVZ/RMS/OB axis was observed in the adult macaque monkey (Kornack & Rakic, 2001), but how TH influences the production of newly generated neuroblasts and *a fortiori* olfactory cues is an open question.

Oligodendrogenesis is a well-established TH-dependent process (Baas et al., 1997) and SVZ-OPCs are a key source of remyelinating cells to promote endogenous brain repair (Vancamp, Butruille, Demeneix, & Remaud, 2020). Promoting endogenous myelin repair by targeting SVZ-OPCs could provide new therapeutic strategies to regenerate myelin loss that occurs in several neurodegenerative and neurological disorders as multiple sclerosis (MS), Alzheimer's disease and schizophrenia (Cole, Early, & Lyons, 2017; Vancamp et al., 2020). We demonstrated that a lack of T_3 favored the generation of new SVZ-derived OPCs in WT adult mice (Gothié et al., 2017; Remaud et al., 2017). Adult SVZ-OPCs are protected from the neutralizing effects of T3 by the absence of TRα1 and strong expression of *Dio3* (Remaud et al., 2017). After cuprizone-induced demyelination, resulting in a complete loss of mature oligodendrocytes in the corpus callosum, newly generated SVZ-OPCs migrated toward the lesion site and participated to myelin repair (Brousse, Magalon, Durbec, & Cayre, 2015;

Remaud et al., 2017; Xing et al., 2014). Fate-mapping experiments showed that a TH-free window applied during the demyelination phase, when new SVZ-OPCs are generated (Picard-Riera, Nait-Oumesmar, & Baron-Van Evercooren, 2004), followed by three T3/T4 pulses during the recovery phase, enhanced the generation of mature oligodendrocytes derived from SVZ-OPCs as assessed 1 week after cuprizone withdrawal (Remaud et al., 2017). Intriguingly, myelin sheath thickness of remyelinated axons in the corpus callosum was normal with a complete rescue of electrophysiological function (Remaud et al., 2017). Our hypothesis is that hypothyroidism promotes the generation of new SVZ-OPCs both under pathological and physiological conditions. Accordingly, *Dio3* knock-down using shRNA directed against *Dio3* transcripts decreased the number of OPCs in the SVZ (Remaud et al., 2017). In contrast, T3/T4 pulses accelerated oligodendrocyte differentiation similar as to developmental myelination (Barres, Lazar, & Raff, 1994).

Other studies in different demyelinating rodent models also identified a key role of TH on SVZ-OPCs and myelination repair. In a rat model of cuprizone-induced demyelination, expression of mature oligodendrocyte markers was restored in the corpus callosum when treated with T3 1 week after cuprizone diet, whereas the number of Nestin+ uncommitted progenitors was reduced in the SVZ (Franco, Silvestroff, Soto, & Pasquini, 2008). This suggests that T3 promotes the emergence of SVZ-derived remyelinating cells from the pool of proliferating TACs. Similarly, exogenous T4 administration during the acute phase of demyelination in a rat experimental autoimmune encephalomyelitis (or EAE) model, the most frequently used *in vivo* model for human MS, decreased SVZ-cell proliferation, and enhanced mRNA expression of platelet-derived growth factor α receptor (*Pdgfrα*) and myelin basic protein (*Mbp*), promoting oligodendroglial differentiation and improving remyelination (Calza, Fernandez, Giuliani, Aloe, & Giardino, 2002; Fernandez et al., 2004).

This phenotype was mimicked in the common marmoset EAE model, a relevant non-human primate model as it reliably shares clinical symptoms with human MS patients, thus constituting a "bridge between mouse and man" (t Hart, van Kooyk, Geurts, & Gran, 2015). Exogenous T3 administration restored expression of the oligodendrocyte markers *Pdgfrα* and *Mbp* in the cerebral cortex (D'Intino et al., 2011), suggesting that mechanisms underlying T3-dependent remyelination are preserved in non-human primates. However, the relative contribution of SVZ-OPCs in primate brain repair is unknown, demanding more research to determine how T3 modulates SVZ-derived NSC fate and especially how TH targets SVZ-OPCs to improve the demyelination/remyelination ratio toward remyelination in non-human primate models of MS. Furthermore, it is unsure whether TH also facilitates migration of new OPCs toward lesion sites in primate brains, in which they have to span enormous distances (Dulamea, 2017). The additional use of non-human primate models could therefore deliver new insights and approaches that translate easier into clinically testable approaches that exploit this role of TH in SVZ-derived oligodendrocyte regeneration. EAE can be induced in macaques (Peng et al., 2018) and rhesus monkeys as well (Haanstra et al., 2013), making this experimental paradigm replicable to analyze the functional response of primate SVZ-OPCs to a demyelinating insult with or without THs.

3.3 From NSCs to cognition: Roles of TH in the adult SGZ

TH signaling influences the generation of neurons from hippocampal NSCs. Thyroidectomy in rodents reduced proliferative cells in the SGZ and impaired the generation and maturation of neuroblasts into granular neurons. This phenotype was rescued by adding physiological levels of T3 and T4 in the drinking water (Montero-Pedrazuela et al., 2006). Similarly, the number of BrdU+ cells in the adult hippocampus locally increased after T3 treatment of another murine model of hypothyroidism (the hypothyroidal growth-retarded mouse) (Uchida, Yonezawa, Nakamura, Kobayashi, & Machida, 2005). In contrast, hypothyroidism induced by the goitrogen methimazole did not alter cell proliferation as assessed by BrdU incorporation (Ambrogini et al., 2005; Desouza et al., 2005). Likewise, progenitor proliferation was not affected in the adult SGZ of mice harboring a TRαl null allele ($TR\alpha1^{-/-}$) as well as in $TR\alpha2^{-/-}$ mice that overexpress TRα1 (Kapoor et al., 2010). These data suggest that TH mainly impacts the post-mitotic stages of hippocampal progenitors toward the neuronal lineage (Kapoor et al., 2012; Kapoor, Fanibunda, Desouza, Guha, & Vaidya, 2015). Accordingly, TRα1 is expressed in DCX+ progenitors within the DG (Mayerl et al., 2020), confirming previous results in TRα1-GFP-knock-in mice (Kapoor et al., 2010; Wallis et al., 2010). Moreover, *Mct8* mRNA and protein were detected in DCX+ cells. Deletion of *Mct8* specifically in the adult SGZ-NSCs reduced the number of granular cells, highlighting a cell-autonomous action of TH promoting cell transition from neuroblast to granular neuron (Mayerl et al., 2020). As in hypothyroid rats (Montero-Pedrazuela et al., 2006), overexpression of TRα1 in $TR\alpha2^{-/-}$ mutant mice, which exhibit low levels of FT3 and FT4 (Saltó et al., 2001), decreased the number of DCX+ immature neurons in the adult DG (Kapoor et al., 2010). This phenotype, linked to a limited T3 availability because of the overexpressed TRα1, was rescued by exogenous TH administration, thus showing the reversible effect of the unliganded TRα1 aporeceptor on SGZ neurogenesis. In contrast to TRα2$^{-/-}$ mice, in TRα1$^{-/-}$ mice lacking TRα1 (Kapoor et al., 2010), as well as in T3-treated hyperthyroid mice (Kapoor et al., 2011), more DCX+ and PSA-NCAM+ cells were counted.

Altogether, these data show that TH directly acts on neuronal progenitor differentiation and promotes the generation of new neurons in the adult rodent hippocampus. Moreover, it has been postulated that TH uptake by MCT8 within SGZ-neuroblasts positively regulates the expression of the key cell-cycle regulator *P27kip1*, that inhibits the cell cycle thereby promoting the generation of newborn granular cells (Mayerl et al., 2020). This molecular mechanism coincides with data showing T3-dependent upregulation of several genes involved in neuronal differentiation such as *Dlx2* and *Tis21*, and this in parallel of increased *Thra* expression (Kapoor et al., 2012).

Mechanisms controlling hippocampal progenitor cell and neuroblast proliferation have been poorly investigated in primates. A special focus on a potential hormonal control of neuronal processes should shed new light on how the evolutionary conserved TH signaling pathway could regulate SGZ neurogenesis in these mammalian species. Single-cell RNA-seq analysis enables analyzing how the transcriptional machinery is conserved among the different SGZ-cell types. A previous genome-wide transcriptional analysis of the entire SGZ of the DG indicated the preservation of multiple molecular signatures in mice and rhesus monkeys during postnatal development (Miller et al., 2013).

SGZ neurogenesis is involved in some hippocampus-dependent cognitive abilities as learning and memory in rodents and primates. As in rodents (Kuhn et al., 1996), an age-related decline of adult neurogenesis in the DG of several non-human primate species associated with cognitive impairments (Aizawa et al., 2009; Ngwenya, Heyworth, Shwe, Moore, & Rosene, 2015). SGZ neurogenesis has been recently described in the adult DG of the mouse lemur as well (Royo et al., 2018), but further analysis is needed to determine whether it can be linked to decreased cognitive abilities in old mouse lemurs (Picq, 2007). Moreover, while some hormonal secretions decline with aging in mouse lemurs (Perret & Aujard, 2005), nothing is known about the levels of THs in old *vs.* young lemurs, and more generally in non-human primates. Filling this gap could connect evidence acquired in both rodents and humans where a link between TH and cognition is suspected.

In humans, a dysregulated thyroid axis can cause numerous mood disorders characterized by altered anxiety, memory and depression (Chakrabarti, 2011; Lass, Slawek, Derejko, & Rubello, 2008). In particular, hypothyroidism is often associated with these neuropsychiatric disabilities while TH therapy bolsters the antidepressant strategy (Bauer, 2002; Joffe, 2002). While causal links between alterations in TH levels, cognitive deficits and hippocampal neurogenesis remain elusive in humans, a reduced hippocampal volume in adult-onset hypothyroid patients (Cooke, Mullally, Correia, O'Mara, & Gibney, 2013) strongly suggests that TH could influence neurogenic processes and *a fortiori* hippocampal-dependent affective behavior.

4. How TH signaling and photoperiodism could interact to regulate adult neurogenesis

At temperate latitudes, animals must adapt their physiology, such as reproduction, to seasonally-driven environmental variations to ensure survival. This reproductive seasonality leads to a peak of births and weaning during spring (Bronson, 1985; Bronson & Heideman, 1994). For short-day (SD) breeders, species with a long gestational period such as small ruminants and non-human primates (Ortavant, Pelletier, Ravault, Thimonier, & Volland-Nail, 1985; Wickings & Nieschlag, 1980), the breeding season occurs in fall and winter when the day length decreases. In contrast, long-day (LD) breeders are species with a short gestational period, like rodents, for which breeding takes place during spring, when the day length increases (Ortavant et al., 1985). Evolutionary selection led to adapted hormonal mechanisms that control reproduction in accordance with these circannual rhythms.

Seasonal mammals use annual variations in day length (or photoperiod) as a temporal landmark to adjust breeding behavior. Light is transmitted to the suprachiasmatic nuclei (SCN) of the hypothalamus and transformed into an endocrine signal in the form of melatonin, that is nocturnally released by the pineal gland (Malpaux, Daveau, Maurice, Gayrard, & Thiery, 1993) (Fig. 2A). The duration of melatonin secretion is directly proportional to the night length with a longer release during SD, and a shorter one during LD (Karsch et al., 1986). Pinealectomy in LD-exposed rodents, such as Indian palm squirrels or hamsters, maintained animals in a sexually active state, while melatonin injection downregulated the hypothalamic-pituitary-gonadal axis (Carter & Goldman, 1983a, 1983b; Haldar, Shavali, & Singh, 1992;

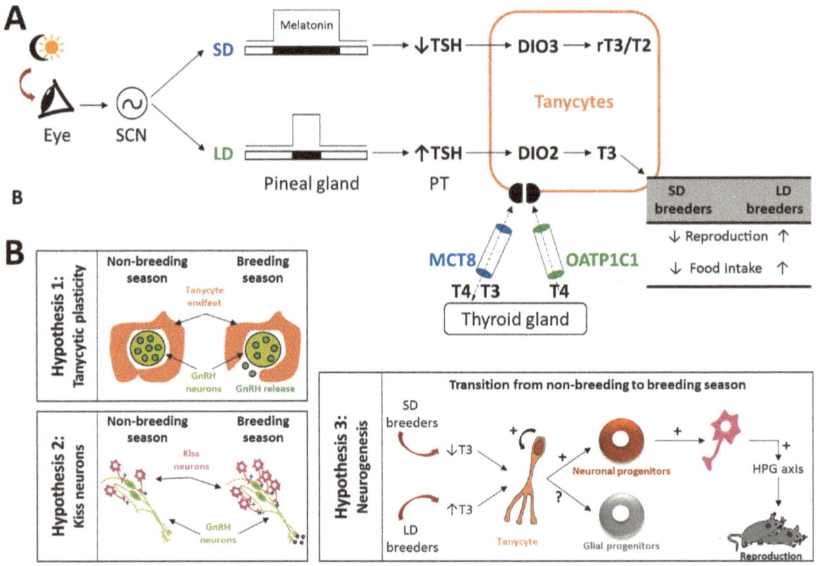

Fig. 2 Photoperiodic regulation of reproductive function. (A) Light information, perceived by the retinal photoreceptors, is transmitted to the suprachiasmatic nucleus (SCN) of the hypothalamus and transformed into an endocrine signal by nocturnal secretion of melatonin by the pineal gland. Short secretion of melatonin during long days (LD) stimulates the release of TSH by the pars tuberalis (PT), which leads to activation of *DIO2* and inhibition of *DIO3* expression in tanycytes. DIO2 converts T4 produced by the thyroid gland into T3, while DIO3 degrades T4 and T3 into inactive metabolites. Upregulation of TSH during LD induces local T3 production, which stimulates reproduction and food intake in LD breeders. (B) Three modes of action were proposed to explain the photoperiodic control of reproduction in seasonal mammals. Hypothesis 1: T3 indirectly modulates the retraction of tanycyte endfeet that allows GnRH release during breeding season. Hypothesis 2: T3 acts on the neuronal networks, which control GnRH synthesis, consisting in part of kisspeptin (kiss) neurons, whose number is increased during the breeding season. Hypothesis 3: increased T3 availability during LD stimulates hypothalamic neurogenesis, while its effect on oligodendrogenesis is not yet known. Neural progenitors derived from tanycytes could thus integrate the neuronal networks involved in the stimulation and/or inhibition of the gonadotropic axis (HPG axis), leading to changes in the reproductive status of animals.

Haldar-Misra & Srivastava, 1987; Revel, Saboureau, Pévet, Mikkelsen, & Simonneaux, 2006). Hence, melatonin functionally regulates circannual rhythms of reproduction.

4.1 Photoperiodic modulation of TH metabolism

In all seasonal mammals, short secretion of melatonin during LD increased (i) *TSHβ* expression in the pars tuberalis which binds to its TSH receptors expressed by tanycytes (Bolborea, Helfer, Ebling, & Barrett, 2015; Shinomiya, Shimmura, Nishiwaki-Ohkawa, & Yoshimura, 2014), (ii) *Dio2* expression in tanycytes and (iii) T3 availability in the hypothalamus (Helfer et al., 2012; Klosen, Sébert, Rasri, Laran-Chich, & Simonneaux, 2013; Ono, Nakao, & Yoshimura, 2009). In contrast, *Dio3* was highly expressed in the hypothalamus during SD, thus decreasing intracellular T3 availability (Ebling, 2015). The photoperiodic modulation in the expression of hypothalamic deiodinases implies a tight regulation of intracellular T3 levels that vary over seasons. Therefore, THs have opposing effects in seasonal animals according to their breeding season: high T3 levels stimulate the gonadotropic axis and favor reproductive behavior in LD breeders, but inhibit reproductive function in SD breeders (Dahl, Evans, Thrun, & Karsch, 1995; Parkinson & Follett, 1995; Thrun, Dahl, Evans, & Karsch, 1997; Viguié, Battaglia, Krasa, Thrun, & Karsch, 1999; Webster, Suttie, & Corson, 1991). This opposite effect can be explained by the dynamic activity of deiodinases in tanycytes (Sáenz de Miera et al., 2013).

Tanycytes express the two main TH transporters MCT8 and OATP1C1 (Ceballos et al., 2009; Ross, Helfer, Russell, Darras, & Morgan, 2011). SD-exposure increased *Mct8* expression in hamsters and photoperiodically sensitive Fischer 344 (F344) rats (Milesi, Simonneaux, & Klosen, 2017; Ross et al., 2011). During SD in F344 rats, *Ttr* is also highly expressed in the arcuate nucleus (Ross et al., 2011), where tanycytes take up the TTR-T4 complex (Rodríguez, Blázquez, & Guerra, 2010). Conversely, *Oatp1c1* is upregulated during LD (Ross et al., 2011), suggesting that this transporter, with a higher affinity for T4 than T3 (Sugiyama et al., 2003), could transport T4 into the tanycytes during LD (Fig. 2A). Whether, and how TH levels locally interact with expression levels of TH transporters in tanycytes remains unknown.

Although we know that melatonin controls circannual production of T3, the actual cellular and molecular targets that govern hypothalamic-pituitary-gonadal axis activity are not fully elucidated (Fig. 2B). A first hypothesis is that T3 indirectly modulates structural plasticity of tanycytic processes (Bolborea et al., 2011; Kameda, Arai, & Nishimaki, 2003), influencing release of the GnRH (or gonadotropin-releasing hormone), as observed in hamsters (Jansen, Cutter, Hardy, Lehman, & Goodman, 2003). Second, T3 could modify wiring of neuronal networks that control GnRH synthesis, for instance through RFamide-related peptides (RFRP) and kisspeptin neurons (Caraty et al., 2007; Revel et al., 2006; Revel, Saboureau, Pévet, Simonneaux, & Mikkelsen, 2008; Smith, Clay, Caraty, & Clarke, 2007), important secretagogues of GnRH (Beltramo, Dardente, Cayla, & Caraty, 2014; Messager et al., 2005). Lastly, increased T3 availability could stimulate tanycyte proliferation and neuronal differentiation within the mediobasal hypothalamus. These tanycyte-derived progenitors could then integrate neuronal networks that modulate the hypothalamic-pituitary-gonadal axis.

To conclude, although gradual fluctuations in T3 are required to adapt animal reproductive behavior to changing environmental conditions (Dardente, Hazlerigg, & Ebling, 2014), future studies are needed to determine whether T3 alone underlies seasonal rhythmicity, as well as to understand the T3-dependent cellular and molecular mechanisms that regulate reproductive function in seasonal mammals. Furthermore, it would be interesting to extend such results to humans, who are prone to disrupted natural day lengths by exposure to artificial light.

4.2 Photoperiodic regulation of reproductive function in mammals

THs control reproductive function in mammals, including humans. Exogenous administration of T3 stimulates spermatogenesis (Turner & Bagnara, 1971) by enhancing Sertoli cell maturation through TRα (Van Haaster, De Jong, Docter, & De Rooij, 1992). In females, THs regulate the development of ovarian and uterine tissues, ovulation, sex hormone secretion and sexual behavior (for a review, see Silva, Ocarino, & Serakides, 2018), especially in seasonal mammals. In photoperiodic rodents, the breeding season comes with a simultaneous increase in gonadal volume (Ben Saad & Maurel, 2001; Chemineau et al., 1992; Gorman, Goldman, & Zucker, 2001), sperm activity (Thimonier, Mauléon, Bézard, DE Reviers, & Cornu, 1969), testosterone and estrogen concentrations, the establishment of estrus cycles and sexual behavior (Gorman et al., 2001), and the metabolic alterations (Bartness & Wade, 1984; Mercer, Moar, Ross, Hoggard, & Morgan, 2000; Ross et al., 2009).

The existence of circannual rhythms in humans and non-human primates is discussed in Wehr's review (Wehr, 2001). In squirrel monkeys, rhesus monkeys and baboons, the anatomical structures involved in transmission of light information as well as melatonin binding sites in the pars tuberalis are present (Albers, Lydic, Gander, & Moore-Ede, 1984; Reppert et al., 1981; Wehr, 2001). Accordingly, selective ablation of the SCN altered circadian rhythms in both squirrels and rhesus monkeys (Albers et al., 1984; Reppert et al., 1981). These results strongly suggest that anatomical and molecular elements required for circadian and circannual rhythms are evolutionary conserved in primates. Moreover, several hormones are also seasonally regulated in primates, including TH (Kaack, Walker, & Walker, 1980), prolactin and testosterone (Chik et al., 1992; Perret & Aujard, 2005; Rostal, Glick, Eaton, & Resko, 1986), which in turn can stimulate reproductive function. However, experimental pinealectomy or thyroidectomy are required to determine whether these hormones are implicated in the various photoperiodic changes observed in primates.

Humans, like other mammals, secrete melatonin overnight (Wehr, 1997). Several reproductive parameters including conception and birth rates, are also seasonally regulated (Mathers & Harris, 1983; ØDegård, 1977). In men, FT3, testosterone and luteinizing hormone (LH) levels are higher in summer (Leppäluoto, Sikkilä, & Hassi, 1998; Meriggiola, Noonan, Paulsen, & Bremner, 1996), even though high testicular temperatures can negatively affect sperm quality and quantity, and thus decrease fertility (Bronson, 1995; Lam & Miron, 1994; Roenneberg & Aschoff, 1990; Rojansky, Brzezinski, & Schenker, 1992).

Moreover, photoperiodic transitions altered the reproductive axis of sheep only after a latency of about 1 month (Malpaux, Daveau, Maurice-Mandon, Duarte, & Chemineau, 1998), what might suggest slow mechanisms of cell plasticity such as neuro- or gliogenesis. Therefore, in the next part of this section, we will discuss the photoperiodic regulation of adult neurogenesis in seasonal mammals.

4.3 Can seasonal variations in TH availability modulate adult neurogenesis?

Melatonin regulates various aspects of adult neurogenesis, including proliferation, survival and neuronal determination of SVZ- (Sotthibundhu, Phansuwan-Pujito, & Govitrapong, 2010) and SGZ-derived NSCs (Ramírez-Rodríguez, Klempin, Babu, Benítez-King, & Kempermann, 2009; Tocharus et al., 2014), as well as hippocampal dendritic arborization (Ramírez-Rodríguez et al., 2009; Ramírez-Rodríguez, Olvera-Hernández, Vega-Rivera, & Ortiz-López, 2019; Ramírez-Rodríguez, Palacios-Cabriales, Ortiz-López, Estrada-Camarena, & Vega-Rivera, 2020). Given the regulation of NSC proliferation and neuronal fate commitment by melatonin and THs, we hypothesize that photoperiod-dependent T3 levels could influence adult neurogenesis.

4.3.1 Photoperiodic regulation of SVZ neurogenesis

A few studies have investigated the photoperiodic-dependent regulation of adult neurogenesis in seasonal rodents (Table 1). One pioneering study in

Table 1 Photoperiodic regulation of adult neurogenesis and associated behaviors.

Species	Breeding	SVZ/OB	SGZ	Hypothalamus	Behavior	References
Syrian hamster	LD	SD ↑ proliferation	SD ↑ proliferation	SD ↑ proliferation		Huang et al. (1998)
		SD ↑ neuronal survival	SD ↑ neuronal survival			
Siberian hamster	LD			SD ↓ *Nestin* expression		Kameda et al. (2003), Barrett et al. (2006), Bolborea et al. (2011), and Herwig, Petri, and Barrett (2012)
				SD ↓ Vimentin		
F344 rat	LD		SD = LD proliferation	SD ↑ proliferation		Shearer et al. (2012)
				SD ↓ *RALDH1*		
White-footed mouse	LD	SD = LD proliferation in the OB	Survival ↓ until 8 weeks after SD		SD ↑ olfactory discrimination and ↓ spatial learning and memory	Perrot-Sinal et al. (1998), Pyter, Reader, and Nelson (2005), Pyter, Trainor, and Nelson (2006), Workman, Bowers, and Nelson (2009), Walton et al. (2011), Walton, Pyter, Weil, and Nelson (2012), Walton, Chen, Travers, and Nelson (2013), and Walton, Aubrecht, Weil, Leuner, and Nelson (2014)
		SD ↑ survival in the OB	SD ↑ survival at 10 weeks			
		Neuronal morphology =	SD ↓ synaptic transmission and spine density			
			SD ↓ LTP			
Deer mice	LD				SD ↓ spatial learning and memory	Perrot-Sinal et al. (1998)

Meadow vole male	LD		SD ↑ proliferation and apoptosis		SD ↓ odor preference for female and spatial learning	Ferkin and Gorman (1992), Galea, Kavaliers, Ossenkopp, and Hampson (1995), Jacobs (1996), Galea and McEwen (1999), Shors et al. (2001), Ormerod and Galea (2001, 2003), and Spritzer et al. (2017)
			SD ↓ neuronal survival			
Mongolian gerbil	LD	SD ↓ proliferation	SD ↓ proliferation	SD ↓ proliferation	SD ↑ object recognition memory	Yao et al. (2019)
Shrew	LD	SD ↓ proliferation	SD ↓ proliferation			Bartkowska, Djavadian, Taylor, and Turlejski (2008)
		SD ↓ neuronal survival	SD ↓ neuronal survival			
Gray squirrel	LD		SD = LD proliferation		SD ↑ spatial learning and memory	Lavenex, Steele, and Jacobs (2000)
			SD = LD neuron number			
Sheep	SD	SD = LD proliferation	SD ↓ proliferation	SD ↑ proliferation		Batailler et al. (2014, 2018), Batailler, Derouet, Butruille, and Migaud (2016), Butruille et al. (2018), Hazlerigg, Wyse, Dardente, Hanon, and Lincoln (2013), and Migaud, Batailler, Pillon, Franceschini, and Malpaux (2011)
				SD ↑ NSC marker expression		
				SD ↑ neuroblast density		

Abbreviations: SVZ, subventricular zone; OB, olfactory bulb; SGZ, subgranular zone; LD, long days; SD, short days; LTP, long-term potentiation.

Syrian hamsters showed increasing numbers of both dividing cells and newly-formed neurons in the SVZ, SGZ and hypothalamus during SD as compared to LD exposure (Huang et al., 1998). In seasonal white-footed mice, SD-exposure promoted survival of newborn neurons, especially in the internal part of the caudal OB, even though the proliferation rate did not correlate with photoperiod. Morphological analysis of the OB showed that seasonal changes did not impact dendritic length and density, nor dendritic intersections and branching (Walton et al., 2012). In contrast to hamsters and white-footed mice, proliferation and neuronal survival in the SVZ/OB of Sorex shrews and Mongolian gerbils were reduced following SD exposure (Bartkowska et al., 2008; Yao et al., 2019).

In hibernating species, for example Syrian hamsters, the torpor phase (i.e. a phase of controlled reduced physiological activity) inhibited progenitor proliferation and strongly decreased numbers of migrating neurons, impairing olfactory function (León-Espinosa et al., 2016). These data suggest a shutdown of the neuroendocrine system during torpor. The transition from torpor to breeding season occurs in concert with increased TH levels. Unlike hibernating ground squirrels and marmots, in which circulating TH concentrations are higher in autumn and winter than in spring and summer (Demeneix & Henderson, 1978; Henderson & Demeneix, 1981; Young, Rajatanavin, Braverman, & Tennant, 1986), the torpor phase in Djungarian hamsters is characterized by low T3 levels and decreased hypothalamic *Dio2* expression (Bank, Kemmling, Rijntjes, Wirth, & Herwig, 2015). This is consistent with studies showing strongly reduced TH metabolism in hibernating bats, and its reactivation induces the transition to breeding season (Damassa, Gustafson, Kwiecinski, & Gagin, 1995; Krishna & Singh, 1998). It is possible that strongly reduced TH metabolism during hibernation decreases TH-dependent SVZ neurogenesis, while its reactivation allows the recrudescence of cell proliferation (Hazlerigg & Lincoln, 2011; León-Espinosa et al., 2016).

Furthermore, photoperiodic changes also induce a seasonal modulation of olfactory discrimination capabilities (Walton et al., 2012) and olfactory preference for female *vs.* male odors (Ferkin & Gorman, 1992) in white-footed mice and meadow voles, respectively. Olfactory behavior is influenced by TH signaling as inactivating mutations in the *Dio3* gene induced odor discrimination deficits in mice (Stohn et al., 2018). This suggests that photoperiodic regulation of olfactory behavior in seasonal mammals could depend on intracellular TH availability.

During the non-breeding season, photoperiodic rodents live in groups and display prosocial behaviors useful for winter nesting. Transition to the breeding season involves changes in the social group structure: males establish a territory and behave aggressively toward intruders (Wolff & Durr, 1986), both enforced by increased testosterone and TH levels (Popova & Vasyl'ieva, 2010). *Dio3* is essential for normal social behaviors, as *Dio3*$^{-/-}$ mice display increased aggressive behavior (Stohn et al., 2018). In seasonal rodents, *Dio3* is reduced in LD, a period when animals are aggressive and defend their breeding territory. Furthermore, improved olfactory function in spring and higher prolactin levels (Johnston, 2004; Johnston et al., 2003; Wood & Loudon, 2014) trigger maternal behavior (e.g., maternal attraction and maternal care) during parturition. Therefore, functional changes in SVZ/OB neurogenesis, influenced by altered TH availability, could enable integrating environmental odors and establishing social ties in rodents (Walton et al., 2012).

In non-human primates, seasonal regulation of adult SVZ neurogenesis has not been studied yet. However, olfaction is crucial for human and non-human primates such as gray mouse lemurs, which have highly developed OBs (Séguy & Perret, 2005). In this species, TH levels wane during winter, an effect revised by bulbectomy (Séguy & Perret, 2005). Likewise, bulbectomy performed during LD decreased high testosterone levels observed during the breeding season (Perret & Schilling, 1993; Schilling & Perret, 1993). These results indicate that olfaction participates in the seasonal control of reproductive function in primates, as it does in rodents (Smith, Pencea, Wang, Luskin, & Insel, 2001). Knowing that gray mouse lemurs seasonally regulate physiological functions like rodents (Génin & Perret, 2000; Hozer, Pifferi, Aujard, & Perret, 2019; Perret, Aujard, & Vannier, 1998; Séguy & Perret, 2005), it is tempting to speculate that adult neurogenesis in non-human primates could be also influenced by the photoperiod.

A remaining question is how adult SVZ neurogenesis is modulated by photoperiod in SD breeders. The SVZ proliferation rate was not affected by photoperiod in sheep, the only SD breeder studied so far (Batailler et al., 2018). That might suggest that photoperiodic-dependent regulation of SVZ neurogenesis differs among mammalian species and breeder types.

4.3.2 Photoperiodic regulation of SGZ neurogenesis

Seasonal modulations of adult hippocampal neurogenesis are species-specific (Table 1). SGZ-NSC proliferation in the DG is enhanced in hamsters (Huang et al., 1998) and meadow voles (Galea & McEwen, 1999; Spritzer et al., 2017), but decreased in gerbils and shrews (Bartkowska et al., 2008; Yao et al., 2019), while not affected by SD in gray squirrels and F344 rats (Lavenex et al., 2000; Shearer et al., 2012). In sheep, proliferative activity reached a maximum in the middle of the non-breeding season, and then decreased at the end of anestrus, reaching lower levels than during breeding (Batailler et al., 2018). In addition, levels of stress-related steroids are higher during the breeding season of female meadow voles (Galea & McEwen, 1999; Spritzer et al., 2017). As in other species, high levels of glucocorticoids inhibit hippocampal neurogenesis and induce dendritic atrophy (Mirescu & Gould, 2006; Woon, Sood, & Hedges, 2010).

In addition, neuronal survival also differs depending on the type of seasonal species (Table 1). In white-footed mice, the recrudescence of hippocampal progenitor proliferation occurs before the transition to the breeding season, which might improve spatial orientation, learning and memory, required in territory maintenance and mating (Walton et al., 2014). It would be interesting to determine whether this upregulation is associated with increased TH levels before the transition to estrus. In hibernating species, hippocampal mitotic activity and the generation of new neuronal progenitors are inhibited during the torpor phase (León-Espinosa et al., 2016; Popov, Kraev, Ignat'ev, & Stewart, 2011) could be due to low TH availability (Bank et al., 2015).

Hippocampal electrophysiological functions are also altered by the photoperiod (Cao et al., 2009; Collins & Davies, 1997; Feng, Zhang, & Chao, 2002). First, the *in vitro* efficiency of synaptic transmission increased long-term potentiation (LTP) amplitudes of CA1 neurons during high TH levels in LD (Pyter et al., 2005). LTP is a synaptic plasticity process based on an increased rate of signal transmission between neurons. As memory is encoded by the modification of synaptic strength, LTP is a cellular mechanism underlying learning and memory (Pastalkova et al., 2006). In hypothyroid rats, the ability to evoke LTP was impaired and therefore spatial learning and memory capabilities as well (Artis et al., 2012; Fernández-Lamo, Montero-Pedrazuela, Delgado-García, Guadaño-Ferraz, & Gruart, 2009; Vara, Muñoz-Cuevas, & Colino, 2003), demonstrating that THs modulate hippocampal synaptic plasticity (Taşkın et al., 2011). Second, the increase in dendritic spines observed in LD (Pyter et al., 2005; Workman et al., 2009) improved spatial performances of seasonal rodents (Ormerod, Lee, & Galea, 2004; Pyter et al., 2005; Walton et al., 2011; Workman et al., 2009). Lower circulating THs (hypothyroidism, Artis et al., 2012; Chaalal, Poirier, Blum, Laroche, & Enderlin, 2019), and a lack of DIO2 (Bárez-López, Montero-Pedrazuela, Bosch-García, Venero, & Guadaño-Ferraz, 2017) or TRα1 (Wilcoxon, Nadolski, Samarut, Chassande, & Redei, 2007) impaired learning and memorizing, while increasing anxious behavior. Altogether, the data demonstrate a crucial role for TH signaling in spatial learning and memory, and also behavior related to hippocampal neurogenesis.

During the breeding season in photoperiodic rodents, high T3 levels increased spatial memory (Table 1) allowing males to establish their breeding territory (Jacobs, 1996) and actively search for food and sexual partners (Gaulin & Fitzgerald, 1989; Jacobs, 1996). In contrast, autumn simultaneously reduced TH levels and hippocampal neurogenesis and induced (i) a decrease of male territoriality and home range size, (ii) memory for recognition and familiarization with the group's congeners (Yao et al., 2019), and (iii) the elaboration of social groups necessary for nesting during winter. Unlike the SVZ, in which the seasonal regulation of olfactory function has not been clearly investigated, the involvement of the photoperiodic regulation of TH availability has been widely demonstrated in the hippocampus of photoperiodic rodents.

4.3.3 Photoperiodic regulation of hypothalamic neurogenesis

A recently performed thyroidectomy in sheep upregulated several T3-dependent genes involved in cell plasticity and proliferation of tanycytes (Dardente, Wood, Ebling, & Sáenz de Miera, 2019; Lomet et al., 2018), suggesting the involvement of THs in hypothalamic neurogenesis. In hamsters, continual darkness or SD-exposure reduced the density and the size of tanycyte processes (Barrett et al., 2007; Bolborea et al., 2011; Ebling et al., 2008; Ebling & Barrett, 2008; Herwig et al., 2012; Kameda et al., 2003) (Table 1), reflecting a strong sensitivity of hypothalamic neurogenesis to photoperiodic variations. Moreover, expression of *Raldh1,* encoding the proliferation inhibitor retinaldehyde dehydrogenase 1 (Bonnet et al., 2008; Crandall et al., 2004; Jacobs et al., 2006), decreased in SD and increased hypothalamic cell proliferation in F344 rats (Shearer et al., 2012) and Syrian hamsters (Huang et al., 1998). Interestingly, hamster and rat hypothalamic *Raldh1* was induced by both melatonin (Helfer et al., 2012; Ross et al., 2004) and T4 treatment (Stoney, Helfer, Rodrigues, Morgan, & Mccaffery, 2016). In parallel, oligodendrogenesis has been only investigated in sheep and seemed not to be sensitive to seasonal variation in day length (Butruille et al., 2018). Thus, photoperiod-dependent alterations in TH availability could finely regulate hypothalamic neurogenesis and induce a rise of new neurons involved in reproduction or metabolism.

GnRH synthesis and secretion is controlled by hypothalamic T3 availability through kisspeptin and RFRP neurons, which are more abundant during the breeding season, thus stimulating reproduction (Ansel et al., 2010; Clarke, Smith, Caraty, Goodman, & Lehman, 2009; Dardente et al., 2019; Greives et al., 2008; Henningsen, Ancel, Mikkelsen, Gauer, & Simonneaux, 2016; Klosen et al., 2013; Revel et al., 2006; Smith et al., 2007, 2008). However, Batailler et al. showed that the ablation of hypothalamic neurogenesis by an antimitotic agent injected in the third ventricle of sheep decreases the number of neuroblasts and alters seasonal reproduction timing (Batailler et al., 2018). Treated sheep stopped breeding earlier than untreated sheep, and also started their next breeding season earlier. Therefore, seasonal regulation of hypothalamic neurogenesis might allow the synchronization of reproduction in photoperiodic mammals. This suggests that T3 indirectly acts on kisspeptin or RFRP neurons, and that TH-dependent plasticity processes such as the production of new hypothalamic progenitors might expand neuronal networks involved in GnRH stimulation (Fig. 2B). Whether these changes are dependent on seasonal variations of hypothalamic neurogenesis needs to be explored in future studies.

Although *POMC*, *AgRP* and *NPY* expression are photoperiodically regulated in F344 rats (Ross et al., 2009) and Siberian hamsters (Mercer et al., 2000), there is no evidence for a direct involvement of adult hypothalamic neurogenesis in the seasonal control of metabolism. Recently, Helfer et al. suggested that the lower body weight of SD-exposed hamsters could be the consequence of progressive neuronal degeneration due to low T3 levels (Helfer, Barrett, & Morgan, 2019). However, T3 modulates NSC commitment and neuronal survival, and data obtained from muscle tissue suggests that THs also regulate muscle stem cell quiescence (Mayerl et al., 2018). Thus, high T3 levels in LD breeders, could activate quiescent NSCs to increase the progenitor pool, thereby expanding neuronal networks involved in the various behaviors controlled by the three neurogenic niches. This requires further investigation.

4.3.4 TH-dependent photoperiodic regulation of neurogenesis in primates and humans

About 5% of the adult population in the United States and 10% in Alaska (Nolen-Hoeksema, 2014) develop seasonal affective disorder (SAD) in fall and winter, due to the lack of sufficient sunlight (Lukmanji, Williams, Bulloch, Bhattarai, & Patten, 2019; Wirz-Justice et al., 1996) or a dysregulation of the melatoninergic system (Comai, Ochoa-Sanchez, Dominguez-Lopez, Bambico, & Gobbi, 2015; Weil, Hotchkiss, Gatien, Pieke-Dahl, & Nelson, 2006; Wu et al., 2013). Similarly, exposure to a short light cycle or a melatonin-based treatment caused a depressive state in diurnal Fat sand rats. In addition to inducing a depressive state, light deprivation altered cell proliferation in the SVZ and SGZ of Mongolian gerbils and human depressed patients (Boldrini et al., 2009; Lau et al., 2011; Lucassen, Stumpel, Wang, & Aronica, 2010), decreasing neuronal differentiation and dendritic arborization in the DG. However, light therapy has been able to counteract depressive symptoms in Fat sand rats (Ashkenazy, Einat, & Kronfeld-Schor, 2009; Einat, Kronfeld-Schor, & Eilam, 2006) and humans (Lukmanji et al., 2019; for review, see Kurlansik & Ibay, 2012) and thus could be an appropriate treatment to prevent depressive illness, or seasonal-dysregulation of adult neurogenesis.

In addition, SAD patients have slightly lower FT4 (Sher, Rosenthal, & Wehr, 1999) and higher TSH levels (Raitiere, 1992). This suggests that SAD patients exhibit a mild form of hypothyroidism, potentially contributing to seasonal variations of SAD prevalence. SAD is also associated with metabolic disorders in humans (Rosenthal et al., 1984). However, SD-exposure or melatonin injections reduced body weight in Fat sand rats (Ashkenazy et al., 2009) but increased that of humans (Rosenthal et al., 1984). Since the hypothalamus controls energy metabolism, hypothalamic neurogenesis might also be impaired in SAD patients. The prevention of metabolic disorders could also help to limit the occurrence of other metabolic syndrome-related diseases, such as diabetes. Thus, there is clinical potential to fine-tune TH-dependent NSC activity pharmacologically and counteract the seasonal development of depression or obesity-related diseases, a major issue for human health care. Clinical trials with THs in patients suffering from bipolar depression improved mental state (Stamm et al., 2014), although large cohort studies should still evaluate long-term efficacy and safety of such a treatment.

Photoperiodic regulation of adult neurogenesis in non-human primates has only been investigated in a few studies. Like rodents, primates establish territories and live in well-defined social groups (Deputte, 1987, 1992), but whether these behaviors are photoperiodically regulated has not been demonstrated yet.

In view of the data above, photoperiodic-dependent TH levels play a crucial role in the seasonal regulation of reproduction, probably by influencing adult neurogenesis and behavior of most seasonal mammals.

Overall, elevated TH levels promote proliferation and survival of neuronal progenitors in the three neurogenic niches, promoting associated physiological functions. Nevertheless, there remains an incomplete understanding why and when gender affects these parameters differentially, since in certain species opposite effects were observed. In addition, transition from breeding to non-breeding season induces a controlled peripheral and central hypothyroidism, associated with a dormant state of TH-dependent peripheral tissues such as the reproductive organs, but also the brain as observed by reduced adult neurogenesis. This hypothyroid state is followed by a gradual rise of T3 levels to reinstall a euthyroid state and restore T3-dependent physiological processes, such as adult neurogenesis. Hence, data so far indicate that photoperiodic regulation of adult neurogenesis is a well-conserved mechanism in all seasonal mammals at the origin of multiple adaptive endocrine, metabolic and behavioral responses. However, the functional relevance of seasonal neurogenesis requires further investigation to determine its function as well as a T3-dependent regulation. A better understanding of the species-specific differences is also necessary, including in humans.

5. Conclusion

Besides controlling many other physiological functions, TH is a key regulator of NSC proliferation and fate determination, allowing for a certain degree of neuroplasticity in mammals, including humans. While we start to decipher the molecular pathways by which TH acts, thorough investigation of the contribution of components regulating TH availability in the neurogenic niches will be necessary to understand better how TH modulates NSC physiology throughout life. A special interest lies in how the balance between neuro- and oligodendrogenesis is determined, as both processes contribute to behavioral functions and hold repair potential in case of disease. As discussed, the role of this signaling pathway in so far less explored models that are phylogenetically closer to humans, such as non-human primates, will shed more light on how THs might regulate human neurogenesis, and whether that could affect similar behaviors. Lastly, we tried to grasp how an environmental factor such as the photoperiod, which directly affects TH homeostasis, might alter neurogenesis as well. Such knowledge allows for a better understanding of the etiology of many diseases related to depression, infertility or metabolism, and could one day provide a better basis for developing more adequate and sophisticated treatments.

Acknowledgments

Lucile Butruille was supported by the Centre National de la Recherche Scientifique (CNRS) and the EU H2020 contract Thyrage (Grant n°666869). Pieter Vancamp was supported by the European Thyroid Association (basic science grant) and the Fondation pour la Recherche Médicale (Grant: SPF201909009111).

References

Aguirre, A., Dupree, J. L., Mangin, J. M., & Gallo, V. (2007). A functional role for EGFR signaling in myelination and remyelination. *Nature Neuroscience*, *10*, 990–1002.

Aguirre, A., & Gallo, V. (2007). Reduced EGFR signaling in progenitor cells of the adult subventricular zone attenuates oligodendrogenesis after demyelination. *Neuron Glia Biology*, *3*, 209–220.

Aimone, J. B., Li, Y., Lee, S. W., Clemenson, G. D., Deng, W., & Gage, F. H. (2014). Regulation and function of adult neurogenesis: From genes to cognition. *Physiological Reviews*, *94*, 991–1026.

Aizawa, K., Ageyama, N., Yokoyama, C., & Hisatsune, T. (2009). Age-dependent alteration in hippocampal neurogenesis correlates with learning performance of macaque monkeys. *Experimental Animals*, *58*, 403–407.

Akter, M., Kaneko, N., Herranz-Pérez, V., Nakamura, S., Oishi, H., García-Verdugo, J. M., et al. (2020). Dynamic changes in the neurogenic potential in the ventricular–subventricular zone of common marmoset during postnatal brain development. *Cerebral Cortex*, *30*, 4092–4109.

Albers, H. E., Lydic, R., Gander, P. H., & Moore-Ede, M. C. (1984). Role of the suprachiasmatic nuclei in the circadian timing system of the squirrel monkey. I. The generation of rhythmicity. *Brain Research*, *300*, 275–284.

Aliesky, H., Courtney, C. L., Rapoport, B., & McLachlan, S. M. (2013). Thyroid autoantibodies are rare in nonhuman great apes and hypothyroidism cannot be attributed to thyroid autoimmunity. *Endocrinology*, *154*, 4896–4907.

Almeida Palha, J., Fernandes, R., Morreale de Escobar, G., Episkopou, V., Gottesman, M., & Saraiva, M. J. (2000). Transthyretin regulates thyroid hormone levels in the choroid plexus, but not in the brain parenchyma: Study in a transthyretin-null mouse model. *Endocrinology*, *141*, 3267–3272.

Alvarez-Buylla, A., García-Verdugo, J. M., & Tramontin, A. D. (2001). A unified hypothesis on the lineage of neural stem cells. *Nature Reviews. Neuroscience*, *2*(4), 287–293.

Alvarez-Buylla, A., Kirn, J. R., & Nottebohm, F. (1990). Birth of projection neurons in adult avian brain may be related to perceptual or motor learning. *Science*, *249*, 1444–1446.

Ambrogini, P., Cuppini, R., Ferri, P., Mancini, C., Ciaroni, S., Voci, A., et al. (2005). Thyroid hormones affect neurogenesis in the dentate gyrus of adult rat. *Neuroendocrinology*, *81*, 244–253.

Amrein, I., Isler, K., & Lipp, H.-P. (2011). Comparing adult hippocampal neurogenesis in mammalian species and orders: Influence of chronological age and life history stage. *The European Journal of Neuroscience*, *34*, 978–987.

Amrein, I., Nosswitz, M., Slomianka, L., van Dijk, R. M., Engler, S., Klaus, F., et al. (2015). Septo-temporal distribution and lineage progression of hippocampal neurogenesis in a primate (Callithrix jacchus) in comparison to mice. *Frontiers in Neuroanatomy*, *9*, 85.

Ansel, L., Bolborea, M., Bentsen, A. H., Klosen, P., Mikkelsen, J. D., & Simonneaux, V. (2010). Differential regulation of Kiss1 expression by melatonin and gonadal hormones in male and female Syrian hamsters. *Journal of Biological Rhythms*, *25*, 81–91.

Apple, D. M., Solano-fonseca, R., & Kokovay, E. (2017). Neurogenesis in the aging brain. *Biochemical Pharmacology*, *141*, 77–85.

Artis, A. S., Bitiktas, S., Taşkın, E., Dolu, N., Liman, N., & Suer, C. (2012). Experimental hypothyroidism delays field excitatory post-synaptic potentials and disrupts hippocampal long-term potentiation in the dentate gyrus of hippocampal formation and Y-maze performance in adult rats. *Journal of Neuroendocrinology*, *24*, 422–433.

Ashkenazy, T., Einat, H., & Kronfeld-Schor, N. (2009). We are in the dark here: Induction of depression- and anxiety-like behaviours in the diurnal fat sand rat, by short daylight or melatonin injections. *The International Journal of Neuropsychopharmacology*, *12*, 83–93.

Ayala, C., Pennacchio, G. E., Soaje, M., Bittencourt, J. C., Celis, M. E., Jahn, G. A., et al. (2019). Differential effects of hypo- and hyperthyroidism on remodeling of contacts between neurons expressing the neuropeptide EI and tyrosine hydroxylase in hypothalamic areas of the male rat. *Peptides*, *113*, 1–10.

Azim, K., Zweifel, S., Klaus, F., Yoshikawa, K., Amrein, I., & Raineteau, O. (2013). Early decline in progenitor diversity in the marmoset lateral ventricle. *Cerebral Cortex*, *23*, 922–931.

Baas, D., Bourbeau, D., Sarlième, L. L., Ittel, M. E., Dussault, J. H., & Puymirat, J. (1997). Oligodendrocyte maturation and progenitor cell proliferation are independently regulated by thyroid hormone. *Glia*, *19*, 324–332.

Bank, J. H. H., Kemmling, J., Rijntjes, E., Wirth, E. K., & Herwig, A. (2015). Thyroid hormone status affects expression of daily torpor and gene transcription in Djungarian hamsters (Phodopus sungorus). *Hormones and Behavior*, *75*, 120–129.

Bárez-López, S., Montero-Pedrazuela, A., Bosch-García, D., Venero, C., & Guadaño-Ferraz, A. (2017). Increased anxiety and fear memory in adult mice lacking type 2 deiodinase. *Psychoneuroendocrinology*, *84*, 51–60.

Barnea, A., & Pravosudov, V. (2011). Birds as a model to study adult neurogenesis: Bridging evolutionary, comparative and neuroethological approaches. *The European Journal of Neuroscience*, *34*, 884–907.

Barres, B. A., Lazar, M. A., & Raff, M. C. (1994). A novel role for thyroid hormone, glucocorticoids and retinoic acid in timing oligodendrocyte development. *Development*, *120*, 1097–1108.

Barrett, P., Ebling, F. J. P., Schuhler, S., Wilson, D., Ross, A. W., Warner, A., et al. (2007). Hypothalamic thyroid hormone catabolism acts as a gatekeeper for the seasonal control of body weight and reproduction. *Endocrinology*, *148*, 3608–3617.

Barrett, P., Ivanova, E., Graham, E. S., Ross, A. W., Wilson, D., Plé, H., et al. (2006). Photoperiodic regulation of cellular retinoic acid-binding protein 1, GPR50 and nestin in tanycytes of the third ventricle ependymal layer of the Siberian hamster. *The Journal of Endocrinology*, *191*, 687–698.

Bartkowska, K., Djavadian, R. L., Taylor, J. R. E., & Turlejski, K. (2008). Generation recruitment and death of brain cells throughout the life cycle of Sorex shrews (Lipotyphla). *The European Journal of Neuroscience*, *27*, 1710–1721.

Bartness, T. J., & Wade, G. N. (1984). Photoperiodic control of body weight and energy metabolism in Syrian hamsters (Mesocricetus auratus): Role of pineal gland, melatonin, gonads, and diet. *Endocrinology*, *114*, 492–498.

Batailler, M., Chesneau, D., Derouet, L., Butruille, L., Segura, S., Cognié, J., et al. (2018). Pineal-dependent increase of hypothalamic neurogenesis contributes to the timing of seasonal reproduction in sheep. *Scientific Reports*, *8*, 6188.

Batailler, M., Derouet, L., Butruille, L., & Migaud, M. (2016). Sensitivity to the photoperiod and potential migratory features of neuroblasts in the adult sheep hypothalamus. *Brain Structure & Function*, *221*, 3301–3314.

Batailler, M., Droguerre, M., Baroncini, M., Fontaine, C., Prevot, V., & Migaud, M. (2014). DCX-expressing cells in the vicinity of the hypothalamic neurogenic niche: A comparative study between mouse, sheep, and human tissues. *The Journal of Comparative Neurology*, *522*, 1966–1985.

Bath, K. G., Mandairon, N., Jing, D., Rajagopal, R., Kapoor, R., Chen, Z. Y., et al. (2008). Variant brain-derived neurotrophic factor (Val66Met) alters adult olfactory bulb neurogenesis and spontaneous olfactory discrimination. *The Journal of Neuroscience*, *28*(10), 2383–2393.

Bauer, M. (2002). Thyroid hormone augmentation with levothyroxine in bipolar depression. *Bipolar Disorders*, *4*, 109–110.

Baumann, N., & Pham-Dinh, D. (2001). Biology of oligodendrocyte and myelin in the mammalian central nervous system. *Physiological Reviews*, *81*(2), 871–927.

Beard, M. D., & Mackay-Sim, A. (1987). Loss of sense of smell in adult, hypothyroid mice. *Developmental Brain Research*, *433*, 181–189.

Bédard, A., Gravel, C., & Parent, A. (2006). Chemical characterization of newly generated neurons in the striatum of adult primates. *Experimental Brain Research*, *170*, 501–512.

Bédard, A., Lévesque, M., Bernier, P. J., & Parent, A. (2002). The rostral migratory stream in adult squirrel monkeys: Contribution of new neurons to the olfactory tubercle and involvement of the antiapoptotic protein Bcl-2. *The European Journal of Neuroscience*, *16*, 1917–1924.

Beltramo, M., Dardente, H., Cayla, X., & Caraty, A. (2014). Cellular mechanisms and integrative timing of neuroendocrine control of GnRH secretion by kisspeptin. *Molecular and Cellular Endocrinology*, *382*, 387–399.

Ben Saad, M. M., & Maurel, D. L. (2001). Effects of bilateral ganglionectomy and melatonin replacement on seasonal rhythm of testicular activity in Zembra Island wild rabbits (Oryctolagus cuniculus). *Reproduction*, *121*, 323–329.

Berdugo-Vega, G., Arias-Gil, G., López-Fernández, A., Artegiani, B., Wasielewska, J. M., Lee, C. C., et al. (2020). Increasing neurogenesis refines hippocampal activity rejuvenating navigational learning strategies and contextual memory throughout life. *Nature Communications*, *11*, 135.

Bergmann, O., Liebl, J., Bernard, S., Alkass, K., Yeung, M. S. Y., Steier, P., et al. (2012). The age of olfactory bulb neurons in humans. *Neuron*, *74*, 634–639.

Bernier, P. J., Bédard, A., Vinet, J., Lévesque, M., & Parent, A. (2002). Newly generated neurons in the amygdala and adjoining cortex of adult primates. *Proceedings of the National Academy of Sciences of the United States of America*, *99*, 11464–11469.

Biondi, B., & Wartofsky, L. (2012). Combination treatment with T4 and T3: Toward personalized replacement therapy in hypothyroidism? *The Journal of Clinical Endocrinology and Metabolism*, *97*, 2256–2271.

Bless, E. P., Yang, J., Acharya, K. D., Nettles, S. A., Vassoler, F. M., Byrnes, E. M., et al. (2016). Adult neurogenesis in the female mouse hypothalamus: Estradiol and high-fat diet Alter the generation of newborn neurons expressing estrogen receptor α. *3*(4), ENEURO.0027-16.2016.

Bolborea, M., Helfer, G., Ebling, F. J. P., & Barrett, P. (2015). Dual signal transduction pathways activated by TSH receptors in rat primary tanycyte cultures. *Journal of Molecular Endocrinology*, *54*, 241–250.

Bolborea, M., Laran-Chich, M.-P., Rasri, K., Hildebrandt, H., Govitrapong, P., Simonneaux, V., et al. (2011). Melatonin controls photoperiodic changes in tanycyte vimentin and neural cell adhesion molecule expression in the Djungarian hamster (Phodopus sungorus). *Endocrinology*, *152*, 3871–3883.

Boldrini, M., Fulmore, C. A., Tartt, A. N., Simeon, L. R., Pavlova, I., Poposka, V., et al. (2018). Human hippocampal neurogenesis persists throughout aging. *Cell Stem Cell*, *22*, 589–599.e5.

Boldrini, M., Underwood, M. D., Hen, R., Rosoklija, G. B., Dwork, A. J., John Mann, J., et al. (2009). Antidepressants increase neural progenitor cells in the human hippocampus. *Neuropsychopharmacology*, *34*, 2376–2389.

Bonnet, E., Touyarot, K., Alfos, S., Pallet, V., Higueret, P., & Abrous, D. N. (2008). Retinoic acid restores adult hippocampal neurogenesis and reverses spatial memory deficit in vitamin A deprived rats. *PLoS One*, *3*, e3487.

Bragado Alonso, S., Reinert, J. K., Marichal, N., Massalini, S., Berninger, B., Kuner, T., et al. (2019). An increase in neural stem cells and olfactory bulb adult neurogenesis improves discrimination of highly similar odorants. *The EMBO Journal*, *38*(6), e98791.

Breton-Provencher, V., Lemasson, M., Peralta, M. R., & Saghatelyan, A. (2009). Interneurons produced in adulthood are required for the normal functioning of the olfactory bulb network and for the execution of selected olfactory behaviors. *The Journal of Neuroscience*, *9*(48), 15245–15257.

Bronson, F. H. (1985). Mammalian reproduction: An ecological Perspective1. *Biology of Reproduction*, *32*, 1–26.

Bronson, F. H. (1995). Seasonal variation in human reproduction: Environmental factors. *The Quarterly Review of Biology*, *70*(2), 141–164.

Bronson, F. H., & Heideman, P. (1994). Seasonal regulation of reproduction in mammals. *Physiological Reports*, 541–584.

Brousse, B., Magalon, K., Durbec, P., & Cayre, M. (2015). Region and dynamic specificities of adult neural stem cells and oligodendrocyte precursors in myelin regeneration in the mouse brain. *Biology Open*, *4*, 980–992.

Butruille, L., Batailler, M., Mazur, D., Prévot, V., & Migaud, M. (2018). Seasonal reorganization of hypothalamic neurogenic niche in adult sheep. *Brain Structure & Function*, *223*, 91–109.

Calza, L., Fernandez, M., Giuliani, A., Aloe, L., & Giardino, L. (2002). Thyroid hormone activates oligodendrocyte precursors and increases a myelin-forming protein and NGF content in the spinal cord during experimental allergic encephalomyelitis. *Proceedings of the National Academy of Sciences of the United States of America*, *99*, 3258–3263.

Calzolari, F., Michel, J., Baumgart, E. V., Theis, F., Götz, M., & Ninkovic, J. (2015). Fast clonal expansion and limited neural stem cell self-renewal in the adult subependymal zone. *Nature Neuroscience*, *18*, 490–492.

Camboni, D., Roskoden, T., & Schwegler, H. (2003). Effect of early thyroxine treatment on brain-derived neurotrophic factor mRNA expression and protein amount in the rat medial septum/diagonal band of broca. *Neuroscience Letters*, *350*, 141–144.

Cameron, H. A., & Mckay, R. D. G. (2001). Adult neurogenesis produces a large pool of new granule cells in the dentate gyrus. *The Journal of Comparative Neurology*, *435*, 406–417.

Cameron, H. A., Woolley, C. S., McEwen, B. S., & Gould, E. (1993). Differentiation of newly born neurons and glia in the dentate gyrus of the adult rat. *Neuroscience*, *56*(2), 337–344.

Cao, X. J., Ming, W., Chen, W. H., Zhu, D. M., She, J. Q., & Ruan, D. Y. (2009). Effects of chronic administration of melatonin on spatial learning ability and long-term potentiation in lead-exposed and control rats. *Biomedical and Environmental Sciences*, *22*, 70–75.

Caraty, A., Smith, J. T., Lomet, D., Ben Saïd, S., Morrissey, A., Cognie, J., et al. (2007). Kisspeptin synchronizes preovulatory surges in cyclical ewes and causes ovulation in seasonally acyclic ewes. *Endocrinology*, *148*, 5258–5267.

Carter, D. S., & Goldman, B. D. (1983a). Antigonadal effects of timed melatonin infusion in pinealectomized male Djungarian hamsters (Phodopus sungorus sungorus): Duration is the critical parameter. *Endocrinology*, *113*, 1261–1267.

Carter, D. S., & Goldman, B. D. (1983b). Progonadal role of the pineal in the Djungarian hamster (Phodopus sungorus sungorus): Mediation by melatonin. *Endocrinology*, *113*, 1268–1273.

Ceballos, A., Belinchon, M. M., Sanchez-Mendoza, E., Grijota-Martinez, C., Dumitrescu, A. M., Refetoff, S., et al. (2009). Importance of monocarboxylate transporter 8 for the blood-brain barrier-dependent availability of 3,5,3′-triiodo-L-thyronine. *Endocrinology*, *150*, 2491–2496.

Chaalal, A., Poirier, R., Blum, D., Laroche, S., & Enderlin, V. (2019). Thyroid hormone supplementation restores spatial memory, hippocampal markers of neuroinflammation, plasticity-related signaling molecules, and β-amyloid peptide load in hypothyroid rats. *Molecular Neurobiology*, *56*, 722–735.

Chakrabarti, S. (2011). Thyroid functions and bipolar affective disorder. *Journal of Thyroid Research*, *2011*, 306367.

Chan, S. Y., Hancox, L. A., Martín-Santos, A., Loubière, L. S., Walter, M. N. M., González, A.-M., et al. (2014). MCT8 expression in human fetal cerebral cortex is reduced in severe intrauterine growth restriction. *The Journal of Endocrinology*, *220*, 85–95.

Charvet, C. J., & Finlay, B. L. (2014). Evo-devo and the primate isocortex: The central organizing role of intrinsic gradients of neurogenesis. *Brain, Behavior and Evolution*, *84*, 81–92.

Chemineau, P., Malpaux, B., Guérin, Y., Maurice, F., Daveau, A., & Pelletier, J. (1992). Lumière et mélatonine pour la maîtrise de la reproduction des ovins et des caprins. *Annales de Zootechnie*, *41*, 247–261.

Chen, C., Ma, Q., Chen, X., Zhong, M., Deng, P., Zhu, G., et al. (2015). Thyroid hormone-Otx2 signaling is required for embryonic ventral midbrain neural stem cells differentiated into dopamine neurons. *Stem Cells and Development*, *24*, 1751–1765.

Chik, C. L., Almeida, O. F., Libré, E. A., Booth, J. D., Renquist, D., & Merriam, G. R. (1992). Photoperiod-driven changes in reproductive function in male rhesus monkeys. *The Journal of Clinical Endocrinology and Metabolism*, *74*, 1068–1074.

Clarke, I. J., Smith, J. T., Caraty, A., Goodman, R. L., & Lehman, M. N. (2009). Kisspeptin and seasonality in sheep. *Peptides*, *30*, 154–163.

Claustre, J., Balende, C., & Pujol, J. F. (1996). Influence of the thyroid hormone status on tyrosie hydroxylase in central and peripheral catecholaminergic structures. *Neurochemistry International*, *28*, 277–281.

Cole, K. L. H., Early, J. J., & Lyons, D. A. (2017). Drug discovery for remyelination and treatment of MS. *Glia*, *65*, 1565–1589.

Collins, D. R., & Davies, S. N. (1997). Melatonin blocks the induction of long-term potentiation in an N-methyl-d-aspartate independent manner. *Brain Research*, *767*, 162–165.

Comai, S., Ochoa-Sanchez, R., Dominguez-Lopez, S., Bambico, F. R., & Gobbi, G. (2015). Melancholic-like behaviors and circadian neurobiological abnormalities in melatonin MT1 receptor knockout mice. *The International Journal of Neuropsychopharmacology*, *18*(3), 1–10.

Cooke, G. E., Mullally, S., Correia, N., O'Mara, S. M., & Gibney, J. (2013). Hippocampal volume is decreased in adults with hypothyroidism. *Thyroid*, *24*, 433–440.

Coras, R., Siebzehnrubl, F. A., Pauli, E., Huttner, H. B., Njunting, M., Kobow, K., et al. (2010). Low proliferation and differentiation capacities of adult hippocampal stem cells correlate with memory dysfunction in humans. *Brain*, *133*, 3359–3372.

Crandall, J., Sakai, Y., Zhang, J., Koul, O., Mineur, Y., Crusio, W. E., et al. (2004). 13-cis-retinoic acid suppresses hippocampal cell division and hippocampal-dependent learning in mice. *Proceedings of the National Academy of Sciences of the United States of America*, *101*, 5111–5116.

Curtis, M. A., Kam, M., Nannmark, U., Anderson, M. F., Axell, M. Z., Wikkelso, C., et al. (2007). Human neuroblasts migrate to the olfactory bulb via a lateral ventricular extension. *Science*, *315*, 1243–1249.

Dahl, G. E., Evans, N. P., Thrun, L. A., & Karsch, F. J. (1995). Thyroxine is permissive to seasonal transitions in reproductive neuroendocrine activity in the Ewe1. *Biology of Reproduction*, *52*, 690–696.

Damassa, D. A., Gustafson, A. W., Kwiecinski, G. G., & Gagin, G. A. (1995). Seasonal influences on the control of plasma sex hormone-binding globulin by T4 in male little brown bats. *American Journal of Physiology Integrative and Comparative Physiology*, *268*, R1303–R1309.

Dardente, H., Hazlerigg, D., & Ebling, F. (2014). Thyroid hormone and seasonal rhythmicity. *Frontiers in Endocrinology*, *5*, 19.
Dardente, H., Wood, S., Ebling, F., & Sáenz de Miera, C. (2019). An integrative view of mammalian seasonal neuroendocrinology. *Journal of Neuroendocrinology*, *31*, e12729.
Demeneix, B. A., & Henderson, N. E. (1978). Serum T4 and T3 in active and torpid ground squirrels, Spermophilus richardsoni. *General and Comparative Endocrinology*, *35*, 77–85.
Deniz, F., Ay, S. A., Salihoglu, M., Kurt, O., Baskoy, K., Altundag, A., et al. (2016). Thyroid hormone replacement therapy improves olfaction and taste sensitivity in primary hypothyroid patients: A prospective randomised clinical trial. *Experimental and Clinical Endocrinology & Diabetes*, *124*, 562–567.
Dennis, C. V., Suh, L. S., Rodriguez, M. L., Kril, J. J., & Sutherland, G. T. (2016). Human adult neurogenesis across the ages: An immunohistochemical study. *Neuropathology and Applied Neurobiology*, *42*, 621–638.
Deputte, B. (1987). L'évitement de l'inceste chez les primates. *Recherche*, *193*, 1332–1342.
Deputte, B. (1992). La cognition sociale chez les primates. *Psychologie Française*, *37*, 91–97.
Desouza, L. A., Ladiwala, U., Daniel, S. M., Agashe, S., Vaidya, R. A., & Vaidya, V. A. (2005). Thyroid hormone regulates hippocampal neurogenesis in the adult rat brain. *Molecular and Cellular Neurosciences*, *29*, 414–426.
Dhong, H. J., Kim, H. Y., & Ha, B. S. (2003). Histologic changes to olfactory epithelium in hypothyroid rats. *Otolaryngology – Head and Neck Surgery*, *129*, 24–32.
D'Intino, G., Lorenzini, L., Fernandez, M., Taglioni, A., Perretta, G., Del Vecchio, G., et al. (2011). Triiodothyronine administration ameliorates the demyelination/remyelination ratio in a non-human primate model of multiple sclerosis by correcting tissue hypothyroidism. *Journal of Neuroendocrinology*, *23*, 778–790.
Doetsch, F., & Alvarez-Buylla, A. (1996). Network of tangential pathways for neuronal migration in adult mammalian brain. *Proceedings of the National Academy of Sciences of the United States of America*, *93*(25), 14895–14900.
Dulamea, A. (2017). The contribution of oligodendrocytes and oligodendrocyte progenitor cells to central nervous system repair in multiple sclerosis: Perspectives for remyelination therapeutic strategies PT—INVI. *Neural Regeneration Research*, *12*, 1939–1944.
Ebling, F. J. P. (2015). Hypothalamic control of seasonal changes in food intake and body weight. *Frontiers in Neuroendocrinology*, *37*, 97–107.
Ebling, F. J. P., & Barrett, P. (2008). The regulation of seasonal changes in food intake and body weight. *Journal of Neuroendocrinology*, *20*, 827–833.
Ebling, F. J. P., Wilson, D., Wood, J., Hughes, D., Mercer, J. G., Morgan, P. J., et al. (2008). The thyrotropin-releasing hormone secretory system in the hypothalamus of the Siberian hamster in long and short photoperiods. *Journal of Neuroendocrinology*, *20*, 576–586.
Eckenhoff, M. F., & Rakic, P. (1988). Nature and fate of proliferative cells in the hippocampal dentate gyrus during the life span of the rhesus monkey. *The Journal of Neuroscience*, *8*, 2729–2747.
Einat, H., Kronfeld-Schor, N., & Eilam, D. (2006). Sand rats see the light: Short photoperiod induces a depression-like response in a diurnal rodent. *Behavioural Brain Research*, *173*, 153–157.
Enwere, E., Shingo, T., Gregg, C., Fujikawa, H., Ohta, S., & Weiss, S. (2004). Aging results in reduced epidermal growth factor receptor signaling, diminished olfactory neurogenesis, and deficits in fine olfactory discrimination. *The Journal of Neuroscience*, *24*, 8354–8365.
Eriksson, P. S., Perfilieva, E., Björk-Eriksson, T., Alborn, A.-M., Nordborg, C., Peterson, D. A., et al. (1998). Neurogenesis in the adult human hippocampus. *Nature Medicine*, *4*, 1313–1317.
Ernst, A., Alkass, K., Bernard, S., Salehpour, M., Perl, S., Tisdale, J., et al. (2014). Neurogenesis in the striatum of the adult human brain. *Cell*, *156*, 1072–1083.

Fabel, K., Fabel, K., Tam, B., Kaufer, D., Baiker, A., Simmons, N., et al. (2003). VEGF is necessary for exercise-induced adult hippocampal neurogenesis. *The European Journal of Neuroscience, 18*, 2803–2812.

Fasemore, T. M., Patzke, N., Kaswera-Kyamakya, C., Gilissen, E., Manger, P. R., & Ihunwo, A. O. (2018). The distribution of Ki-67 and Doublecortin-immunopositive cells in the brains of three Strepsirrhine primates: Galago demidoff, Perodicticus potto, and Lemur catta. *Neuroscience, 372*, 46–57.

Feng, Y., Zhang, L. X., & Chao, D. M. (2002). Role of melatonin in spatial learning and memory in rats and its mechanism. *Acta Physiology Sinica, 54*, 65–70.

Ferkin, M. H., & Gorman, M. R. (1992). Photoperiod and gonadal hormones influence odor preferences of the male meadow vole, Microtus pennsylvanicus. *Physiology & Behavior, 51*, 1087–1091.

Fernandez, M., Giuliani, A., Pirondi, S., D'Intino, G., Giardino, L., Aloe, L., et al. (2004). Thyroid hormone administration enhances remyelination in chronic demyelinating inflammatory disease. *Proceedings of the National Academy of Sciences of the United States of America, 101*, 16363–16368.

Fernández-Lamo, I., Montero-Pedrazuela, A., Delgado-García, J. M., Guadaño-Ferraz, A., & Gruart, A. (2009). Effects of thyroid hormone replacement on associative learning and hippocampal synaptic plasticity in adult hypothyroid rats. *The European Journal of Neuroscience, 30*, 679–692.

Figueres-Oñate, M., Sánchez-Villalón, M., Sánchez-González, R., & López-Mascaraque, L. (2019). Lineage tracing and cell potential of postnatal single progenitor cells. *Stem Cell Reports, 13*, 700–712.

Flamant, F., Gauthier, K., & Richard, S. (2017). Chapter eleven—genetic investigation of thyroid hormone receptor function in the developing and adult brain. In D. Forrest, & S. Tsai (Eds.), *Nuclear receptors in development and disease* (pp. 303–335). Academic Press.

Flor-García, M., Terreros-Roncal, J., Moreno-Jiménez, E. P., Ávila, J., Rábano, A., & Llorens-Martín, M. (2020). Unraveling human adult hippocampal neurogenesis. *Nature Protocols, 15*, 668–693.

Follett, B. K., & Nicholls, T. J. (1985). Influences of thyroidectomy and thyroxine replacement on photoperiodically controlled reproduction in quail. *The Journal of Endocrinology, 107*, 211–221.

Forest, J., Chalençon, L., Midroit, M., Terrier, C., Caillé, I., Sacquet, J., et al. (2019). Role of adult-born versus preexisting neurons born at P0 in olfactory perception in a complex olfactory environment in mice. *Cerebral Cortex, 30*, 534–549.

Fox, E. A., Biddinger, J. E., Jones, K. R., McAdams, J., & Worman, A. (2013). Mechanism of hyperphagia contributing to obesity in brain-derived neurotrophic factor knockout mice. *Neuroscience, 229*, 176–199.

Franco, P. G., Silvestroff, L., Soto, E. F., & Pasquini, J. M. (2008). Thyroid hormones promote differentiation of oligodendrocyte progenitor cells and improve remyelination after cuprizone-induced demyelination. *Experimental Neurology, 212*, 458–467.

Friesema, E. C. H., Grueters, P. A., Biebermann, H., Krude, H., Von Moers, A., Reeser, M., et al. (2004). Association between mutations in a thyroid hormone transporter and severe X-linked psychomotor retardation. *Lancet, 364*, 1435–1437.

Fuentealba, L. C., Obernier, K., & Alvarez-Buylla, A. (2012). Adult neural stem cells bridge their niche. *Cell Stem Cell, 10*, 698–708.

Gagneux, P., Amess, B., Diaz, S., Moore, S., Patel, T., Dillmann, W., et al. (2001). Proteomic comparison of human and great ape blood plasma reveals conserved glycosylation and differences in thyroid hormone metabolism. *American Journal of Physical Anthropology, 115*, 99–109.

Galea, L. A. M., Kavaliers, M., Ossenkopp, K.-P., & Hampson, E. (1995). Gonadal hormone levels and spatial learning performance in the morris water maze in male and female meadow voles, Microtus pennsylvanicus. *Hormones and Behavior, 29*, 106–125.

Galea, L. A. M., & McEwen, B. S. (1999). Sex and seasonal changes in the rate of cell proliferation in the dentate gyrus of adult wild meadow voles. *Neuroscience*, *89*, 955–964.

Galea, L. A. M., Wainwright, S. R., Roes, M. M., Duarte-Guterman, P., Chow, C., & Hamson, D. K. (2013). Sex, hormones and neurogenesis in the hippocampus: Hormonal modulation of neurogenesis and potential functional implications. *Journal of Neuroendocrinology*, *25*, 1039–1061.

Gaulin, S. J. C., & Fitzgerald, R. W. (1989). Sexual selection for spatial-learning ability. *Animal Behaviour*, *32*, 322–331.

Gengatharan, A., Bammann, R. R., & Saghatelyan, A. (2016). The role of astrocytes in the generation, migration, and integration of new neurons in the adult olfactory bulb. *Frontiers in Neuroscience*, *10*, 149.

Génin, F., & Perret, M. (2000). Photoperiod-induced changes in energy balance in gray mouse lemurs. *Physiology & Behavior*, *71*, 315–321.

Gheusi, G., Cremer, H., McLean, H., Chazal, G., Vincent, J. D., & Lledo, P. M. (2000). Importance of newly generated neurons in the adult olfactory bulb for odor discrimination. *Proceedings of the National Academy of Sciences of the United States of America*, *97*, 1823–1828.

Gil-Perotin, S., Duran-Moreno, M., Belzunegui, S., Luquin, M. R., & Garcia-Verdugo, J. M. (2009). Ultrastructure of the subventricular zone in Macaca fascicularis and evidence of a mouse-like migratory stream. *The Journal of Comparative Neurology*, *514*, 533–554.

Goldsmith, A. R., & Nicholls, T. J. (1992). Thyroxine effects upon reproduction, prolactin secretion and plumage moult in intact and in thyroidectomised European starlings Sturnus vulgaris. *Ornis Scandinavica*, *23*, 398–404.

Gorman, M. R., Goldman, B. D., & Zucker, I. (2001). Mammalian photoperiodism. In R. Y. Moore, J. S. Akahashi, & F. W. Turek (Eds.), *Vol. 12. Circadian clocks, handbook of behavioral neurobiology* (pp. 481–508). New York, NY: Kluwer Academic/Plenum Publishers.

Gothié, J. D., Demeneix, B., & Remaud, S. (2017). Comparative approaches to understanding thyroid hormone regulation of neurogenesis. *Molecular and Cellular Endocrinology*, *459*, 104–115.

Gothié, J. D., Sébillot, A., Luongo, C., Legendre, M., Nguyen Van, C., Le Blay, K., et al. (2017). Adult neural stem cell fate is determined by thyroid hormone activation of mitochondrial metabolism. *Molecular Metabolism*, *6*, 1551–1561.

Gould, E., Reeves, A. J., Fallah, M., Tanapat, P., Gross, C. G., & Fuchs, E. (1999). Hippocampal neurogenesis in adult Old World primates. *Proceedings of the National Academy of Sciences*, *96*, 5263–5267.

Greives, T. J., Humber, S. A., Goldstein, A. N., Scotti, M.-A. L., Demas, G. E., & Kriegsfeld, L. J. (2008). Photoperiod and testosterone interact to drive seasonal changes in kisspeptin expression in Siberian hamsters (Phodopus sungorus). *Journal of Neuroendocrinology*, *20*, 1339–1347.

Gros, A., Veyrac, A., & Laroche, S. (2015). Brain and memory: New neurons to remember. *Biologie Aujourd'hui*, *209*, 229–248.

Guerrero-Cázares, H., Gonzalez-Perez, O., Soriano-Navarro, M., Zamora-Berridi, G., García-Verdugo, J. M., & Quinoñes-Hinojosa, A. (2011). Cytoarchitecture of the lateral ganglionic eminence and rostral extension of the lateral ventricle in the human fetal brain. *The Journal of Comparative Neurology*, *519*, 1165–1180.

Güven, S., & Laska, M. (2012). Olfactory sensitivity and odor structure-activity relationships for aliphatic carboxylic acids in CD-1 mice. *PLoS One*, 7(3), e34301. https://doi.org/10.1371/journal.pone.0034301.

Haan, N., Goodman, T., Najdi-Samiei, A., Stratford, C. M., Rice, R., El Agha, E., et al. (2013). Fgf10-expressing tanycytes add new neurons to the appetite/energy-balance regulating centers of the postnatal and adult hypothalamus. *The Journal of Neuroscience*, *33*, 6170–6180.

Haanstra, K. G., Jagessar, S. A., Bauchet, A.-L., Doussau, M., Fovet, C.-M., Heijmans, N., et al. (2013). Induction of experimental autoimmune encephalomyelitis with recombinant human myelin oligodendrocyte glycoprotein in incomplete Freund's adjuvant in three non-human primate species. *Journal of Neuroimmune Pharmacology, 8,* 1251–1264.

Hack, M. A., Saghatelyan, A., de Chevigny, A., Pfeifer, A., Ashery-Padan, R., Lledo, P.-M., et al. (2005). Neuronal fate determinants of adult olfactory bulb neurogenesis. *Nature Neuroscience, 8,* 865–872.

Haldar, C., Shavali, S. S., & Singh, S. (1992). Photoperiodic response of pineal-thyroid axis of the female Indian palm squirrel, Funambulus pennanti. *Journal of Neural Transmission, 90,* 45–52.

Haldar-Misra, C., & Srivastava, M. (1987). Effect of pinealectomy on photoperiodic gonadal response of Indian palm squirrel, Funambulus pennanti. *Canadian Journal of Zoology, 65,* 833–836.

Hanon, E. A., Lincoln, G. A., Fustin, J.-M., Dardente, H., Masson-Pévet, M., Morgan, P. J., et al. (2008). Ancestral TSH mechanism signals summer in a photoperiodic mammal. *Current Biology, 18,* 1147–1152.

Hassani, Z., François, J.-C., Alfama, G., Dubois, G. M., Paris, M., Giovannangeli, C., et al. (2007). A hybrid CMV-H1 construct improves efficiency of PEI-delivered shRNA in the mouse brain. *Nucleic Acids Research, 35,* e65.

Hazlerigg, D. G., & Lincoln, G. A. (2011). Hypothesis: Cyclical histogenesis is the basis of circannual timing. *Journal of Biological Rhythms, 26,* 471–485.

Hazlerigg, D. G., Wyse, C. A., Dardente, H., Hanon, E. A., & Lincoln, G. A. (2013). Photoperiodic variation in CD45-positive cells and cell proliferation in the mediobasal hypothalamus of the Soay sheep. *Chronobiology International, 30,* 548–558.

Helfer, G., Barrett, P., & Morgan, P. J. (2019). A unifying hypothesis for control of body weight and reproduction in seasonally breeding mammals. *Journal of Neuroendocrinology, 31,* e12680.

Helfer, G., Ross, A. W., Russell, L., Thomson, L. M., Shearer, K. D., Goodman, T. H., et al. (2012). Photoperiod regulates vitamin A and Wnt/β-catenin signaling in F344 rats. *Endocrinology, 153*(2), 815–824.

Henderson, N. E., & Demeneix, B. A. (1981). Hibernation in thyroidectomized ground squirrels, Spermophilus richardsoni. *General and Comparative Endocrinology, 43,* 543–548.

Henningsen, J. B., Ancel, C., Mikkelsen, J. D., Gauer, F., & Simonneaux, V. (2016). Roles of RFRP-3 in the daily and seasonal regulation of reproductive activity in female Syrian hamsters. *Endocrinology, 158,* 652–663.

Herwig, A., Petri, I., & Barrett, P. (2012). Hypothalamic gene expression rapidly changes in response to photoperiod in juvenile Siberian hamsters (Phodopus sungorus). *Journal of Neuroendocrinology, 24,* 991–998.

Hill, T., & Polk, J. D. (2019). BDNF, endurance activity, and mechanisms underlying the evolution of hominin brains. *American Journal of Physical Anthropology, 168,* 47–62.

Hohmann, G., Ortmann, S., Remer, T., & Fruth, B. (2019). Fishing for iodine: What aquatic foraging by bonobos tells us about human evolution. *BMC Zoology, 4,* 5.

Homman-Ludiye, J., Merson, T. D., & Bourne, J. A. (2012). The early postnatal nonhuman primate neocortex contains self-renewing multipotent neural progenitor cells. *PLoS One, 7,* e34383.

Hozer, C., Pifferi, F., Aujard, F., & Perret, M. (2019). The biological clock in gray mouse lemur: Adaptive, evolutionary and aging considerations in an emerging non-human primate model. *Frontiers in Physiology, 10,* 1033.

Huang, L., DeVries, G. J., & Bittman, E. L. (1998). Photoperiod regulates neuronal bromodeoxyuridine labeling in the brain of a seasonally breeding mammal. *Journal of Neurobiology, 36,* 410–420.

Hughes, E. G., & Appel, B. (2016). The cell biology of CNS myelination. *Current Opinion in Neurobiology*, *39*, 93–100.

Imayoshi, I., Sakamoto, M., Ohtsuka, T., Takao, K., Miyakawa, T., Yamaguchi, M., et al. (2008). Roles of continuous neurogenesis in the structural and functional integrity of the adult forebrain. *Nature Neuroscience*, *11*(10), 1153–1161.

Ito, K., Uchida, Y., Ohtsuki, S., Aizawa, S., Kawakami, H., Katsukura, Y., et al. (2011). Quantitative membrane protein expression at the blood-brain barrier of adult and younger cynomolgus monkeys. *Journal of Pharmaceutical Sciences*, *100*, 3939–3950.

Jabès, A., Lavenex, P. B., Amaral, D. G., & Lavenex, P. (2010). Quantitative analysis of postnatal neurogenesis and neuron number in the macaque monkey dentate gyrus. *The European Journal of Neuroscience*, *31*, 273–285.

Jacobs, L. F. (1996). The economy of winter: Phenotypic plasticity in behavior and brain structure. *The Biological Bulletin*, *191*, 92–100.

Jacobs, S., Lie, D. C., DeCicco, K. L., Shi, Y., DeLuca, L. M., Gage, F. H., et al. (2006). Retinoic acid is required early during adult neurogenesis in the dentate gyrus. *Proceedings of the National Academy of Sciences of the United States of America*, *103*, 3902–3907.

Jankovski, A., & Sotelo, C. (1996). Subventricular zone-olfactory bulb migratory pathway in the adult mouse: Cellular composition and specificity as determined by heterochronic and heterotopic transplantation. *Journal of Comparative Neurology*, *371*(3), 376–696. https://doi.org/10.1002/(SICI)1096-9861(19960729)371:3<376::AID-CNE3>3.0.CO;2-#.

Jansen, H. T., Cutter, C., Hardy, S., Lehman, M. N., & Goodman, R. L. (2003). Seasonal plasticity within the gonadotropin-releasing hormone (GnRH) system of the ewe: Changes in identified GnRH inputs and glial association. *Endocrinology*, *144*, 3663–3676.

Joffe, R. T. (2002). Should thyroid replacement therapy be considered for patients with treatment-refractory depression? *Journal of Psychiatry & Neuroscience*, *27*, 80.

Johnston, J. D. (2004). Photoperiodic regulation of prolactin secretion: Changes in intra-pituitary signalling and lactotroph heterogeneity. *The Journal of Endocrinology*, *180*, 351–356.

Johnston, J. D., Cagampang, F. R. A., Stirland, J. A., Carr, A.-J. F., White, M. R. H., Davis, J. R. E., et al. (2003). Evidence for an endogenous per1- and ICER-independent seasonal timer in the hamster pituitary gland. *The FASEB Journal*, *17*, 810–815.

Jones, K. S., & Connor, B. (2011). Intrinsic regulation of adult subventricular zone neural progenitor cells and the effect of brain injury. *American Journal of Stem Cells*, *1*, 48–58.

Kaack, B., Walker, M., & Walker, L. (1980). Seasonal changes in the thyroid hormones of the male squirrel monkey. *Systems Biology in Reproductive Medicine*, *4*, 133–136.

Kam, M., Curtis, M. A., McGlashan, S. R., Connor, B., Nannmark, U., & Faull, R. L. M. (2009). The cellular composition and morphological organization of the rostral migratory stream in the adult human brain. *Journal of Chemical Neuroanatomy*, *37*, 196–205.

Kameda, Y., Arai, Y., & Nishimaki, T. (2003). Ultrastructural localization of vimentin immunoreactivity and gene expression in tanycytes and their alterations in hamsters kept under different photoperiods. *Cell and Tissue Research*, *314*, 251–262.

Kaneko, N., & Sawamoto, K. (2009). Adult neurogenesis and its alteration under pathological conditions. *Neuroscience Research*, *63*, 155–164.

Kapoor, R., Desouza, L. A., Nanavaty, I. N., Kernie, S. G., & Vaidya, V. A. (2012). Thyroid hormone accelerates the differentiation of adult hippocampal progenitors. *Journal of Neuroendocrinology*, *24*, 1259–1271.

Kapoor, R., Fanibunda, S. E., Desouza, L. A., Guha, S. K., & Vaidya, V. A. (2015). Perspectives on thyroid hormone action in adult neurogenesis. *Journal of Neurochemistry*, *133*, 599–616.

Kapoor, R., Ghosh, H., Nordstrom, K., Vennstrom, B., & Vaidya, V. A. (2011). Loss of thyroid hormone receptor beta is associated with increased progenitor proliferation and NeuroD positive cell number in the adult hippocampus. *Neuroscience Letters*, *487*, 199–203.

Kapoor, R., van Hogerlinden, M., Wallis, K., Ghosh, H., Nordstrom, K., Vennstrom, B., et al. (2010). Unliganded thyroid hormone receptor α1 impairs adult hippocampal neurogenesis. *The FASEB Journal*, *24*, 4793–4805.

Kappeler, L., De Magalhaes Filho, C., Dupont, J., Leneuve, P., Cervera, P., Périn, L., et al. (2008). Brain IGF-1 receptors control mammalian growth and lifespan through a neuroendocrine mechanism. *PLoS Biology*, *6*, e254.

Karsch, F. J., Bittman, E. L., Robinson, J. E., Yellon, S. M., Wayne, N. L., Olster, D. H., et al. (1986). Melatonin and photorefractoriness: Loss of response to the melatonin signal leads to seasonal reproductive transitions in the Ewe1. *Biology of Reproduction*, *34*, 265–274.

Kavoi, B. M., & Jameela, H. (2011). Comparative morphometry of the olfactory bulb, tract and stria in the human, dog and goat. *International Journal of Morphology*, *29*, 939–946.

Keirstead, H. S., & Blakemore, W. F. (1999). The role of oligodendrocytes and oligodendrocyte progenitors in CNS remyelination. *Adv. Exp. Med. Biol.*, *468*, 183–197.

Kempermann, G. (2012). New neurons for "survival of the fittest". *Nature Reviews. Neuroscience*, *13*, 727–736.

Kempermann, G., Brandon, E. P., & Gage, F. H. (1998). Environmental stimulation of 129/SvJ mice causes increased cell proliferation and neurogenesis in the adult dentate gyrus. *Current Biology*, *8*, 939–944.

Kempermann, G., Gast, D., Kronenberg, G., Yamaguchi, M., & Gage, F. H. (2003). Early determination and long-term persistence of adult-generated new neurons in the hippocampus of mice. *Development*, *130*, 391–399.

Kempermann, G., Kuhn, H. G., & Gage, F. H. (1997). More hippocampal neurons in adult mice living in an enriched environment. *Nature*, *386*(6624), 493–495.

Kempermann, G., Song, H., & Gage, F. H. (2015). Neurogenesis in the adult hippocampus. *Cold Spring Harbor Perspectives in Biology*, 7(9), a018812.

Kincaid, A. E. (2001). Spontaneous circling behavior and dopamine neuron loss in a genetically hypothyroid mouse. *Neuroscience*, *105*, 891–898.

Klosen, P., Sébert, M.-E., Rasri, K., Laran-Chich, M.-P., & Simonneaux, V. (2013). TSH restores a summer phenotype in photoinhibited mammals via the RF-amides RFRP3 and kisspeptin. *The FASEB Journal*, *27*, 2677–2686.

Knoth, R., Singec, I., Ditter, M., Pantazis, G., Capetian, P., Meyer, R. P., et al. (2010). Murine features of neurogenesis in the human hippocampus across the lifespan from 0 to 100 years. *PLoS One*, *5*, e8809.

Kokoeva, M. V., Yin, H., & Flier, J. S. (2005). Neurogenesis in the hypothalamus of adult mice: Potential role in energy balance. *Science*, *310*, 679–683.

Kornack, D. R., & Rakic, P. (1999). Continuation of neurogenesis in the hippocampus of the adult macaque monkey. *Proceedings of the National Academy of Sciences*, *96*, 5768–5773.

Kornack, D. R., & Rakic, P. (2001). Cell proliferation without neurogenesis in adult primate neocortex. *Science*, *294*, 2127–2130.

Krishna, A., & Singh, K. (1998). Changes in the thyroid gland during the reproductive cycle of the male vespertilionid bat, Scotophilus heathi. *Revista Brasileira de Biologia*, *58*, 707–716.

Kronenberg, G., Reuter, K., Steiner, B., Brandt, M. D., Jessberger, S., Yamaguchi, M., et al. (2003). Subpopulations of proliferating cells of the adult hippocampus respond differently to physiologic neurogenic stimuli. *The Journal of Comparative Neurology*, *467*, 455–463.

Kudo, H., Eto, A., Abe, T., & Mochida, K. (2018). Detection and localization of the thyroid hormone receptor beta mRNA in the immature olfactory receptor neurons of chum salmon. *Heliyon*, *4*, e00744.

Kuhn, H. G., Dickinson-Anson, H., & Gage, F. H. (1996). Neurogenesis in the dentate gyrus of the adult rat: Age-related decrease of neuronal progenitor proliferation. *The Journal of Neuroscience*, *16*, 2027–2033.

Kurlansik, S. L., & Ibay, A. D. (2012). Seasonal affective disorder. *American Family Physician*, *86*(11), 1037–1041.

Lam, D. A., & Miron, J. A. (1994). Global patterns of seasonal variation in human Fertilitya. *Annals of the New York Academy of Sciences, 709*, 9–28.

Larsen, C. M., & Grattan, D. R. (2010). Prolactin-induced mitogenesis in the subventricular zone of the maternal brain during early pregnancy is essential for normal postpartum behavioral responses in the mother. *Endocrinology, 151*, 3805–3814.

Lass, P., Slawek, J., Derejko, M., & Rubello, D. (2008). Neurological and psychiatric disorders in thyroid dysfunctions. The role of nuclear medicine: SPECT and PET imaging. *Minerva Endocrinologica, 33*, 75–84.

Lau, B. W.-M., Ren, C., Yang, J., Yan, S. W. L., Chang, R. C.-C., Pu, M., et al. (2011). Light deprivation induces depression-like behavior and suppresses neurogenesis in diurnal Mongolian gerbil (Meriones unguiculatus). *Cell Transplantation, 20*, 871–882.

Lavenex, P., Steele, M. A., & Jacobs, L. F. (2000). The seasonal pattern of cell proliferation and neuron number in the dentate gyrus of wild adult eastern grey squirrels. *The European Journal of Neuroscience, 12*, 643–648.

Laywell, E. D., Rakic, P., Kukekov, V. G., Holland, E. C., & Steindler, D. A. (2000). Identification of a multipotent astrocytic stem cell in the immature and adult mouse brain. *Proceedings of the National Academy of Sciences of the United States of America, 97*(25), 13883–13888.

Lee, D. A., Bedont, J. L., Pak, T., Wang, H., Song, J., Miranda-Angulo, A., et al. (2012). Tanycytes of the hypothalamic median eminence form a diet-responsive neurogenic niche. *Nature Neuroscience, 15*, 700–702.

Lema, S. C., & Nevitt, G. A. (2004). Evidence that thyroid hormone induces olfactory cellular proliferation in salmon during a sensitive period for imprinting. *The Journal of Experimental Biology, 207*, 3317–3327.

Lemkine, G. F., Raji, A., Alfama, G., Turque, N., Hassani, Z., Alegria-Prévot, O., et al. (2005). Adult neural stem cell cycling in vivo requires thyroid hormone and its alpha receptor. *The FASEB Journal, 19*(7), 863–865.

León-Espinosa, G., García, E., Gómez-Pinedo, U., Hernández, F., DeFelipe, J., & Ávila, J. (2016). Decreased adult neurogenesis in hibernating Syrian hamster. *Neuroscience, 333*, 181–192.

Lepousez, G., Nissant, A., & Lledo, P.-M. (2015). Adult neurogenesis and the future of the rejuvenating brain circuits. *Neuron, 86*, 387–401.

Leppäluoto, J., Sikkilä, K., & Hassi, J. (1998). Seasonal variation of serum TSH and thyroid hormones in males living in subarctic environmental conditions. *International Journal of Circumpolar Health, 57*, 383–385.

Li, J., Tang, Y., & Cai, D. (2012). IKKβ/NF-κB disrupts adult hypothalamic neural stem cells to mediate a neurodegenerative mechanism of dietary obesity and pre-diabetes. *Nature Cell Biology, 14*, 999–1012.

Lim, D. A., & Alvarez-Buylla, A. (1999). Interaction between astrocytes and adult subventricular zone precursors stimulates neurogenesis. *Proceedings of the National Academy of Sciences of the United States of America, 96*(13), 7526–7531.

Lim, D. A., & Alvarez-Buylla, A. (2016). The adult ventricular–subventricular zone (V-SVZ) and olfactory bulb (OB) neurogenesis. *Cold Spring Harbor Perspectives in Biology, 8*(5), a018820.

Liu, P. Z., & Nusslock, R. (2018). Exercise-mediated neurogenesis in the hippocampus via BDNF. *Frontiers in Neuroscience, 12*, 52.

Liu, J., Zhou, Y., Wang, C., Wang, T., Zheng, Z., & Chan, P. (2012). Brain-derived neurotrophic factor (BDNF) genetic polymorphism greatly increases risk of leucine-rich repeat kinase 2 (LRRK2) for Parkinson's disease. *Parkinsonism & Related Disorders, 18*, 140–143.

Lledo, P.-M., Alonso, M., & Grubb, M. S. (2006). Adult neurogenesis and functional plasticity in neuronal circuits. *Nature Reviews. Neuroscience, 7*, 179–193.

Lledo, P.-M., Merkle, F. T., & Alvarez-Buylla, A. (2008). Origin and function of olfactory bulb interneuron diversity. *Trends in Neurosciences*, *31*, 392–400.
LLorens-Martín, M., Torres-Alemán, I., & Trejo, J. L. (2010). Exercise modulates insulin-like growth factor 1-dependent and -independent effects on adult hippocampal neurogenesis and behaviour. *Molecular and Cellular Neurosciences*, *44*, 109–117.
Lois, C., & Alvarez-Buylla, A. (1993). Proliferating subventricular zone cells in the adult mammalian forebrain can differentiate into neurons and glia. *Proceedings of the National Academy of Sciences*, *90*, 2074–2077.
Lois, C., & Alvarez-Buylla, A. (1994). Long-distance neuronal migration in the adult mammalian brain. *Science*, *264*(5162), 1145–1148.
Lois, C., García-Verdugo, J.-M., & Alvarez-Buylla, A. (1996). Chain migration of neuronal precursors. *Science*, *271*, 978–981.
Lomet, D., Cognié, J., Chesneau, D., Dubois, E., Hazlerigg, D., & Dardente, H. (2018). The impact of thyroid hormone in seasonal breeding has a restricted transcriptional signature. *Cellular and Molecular Life Sciences*, *75*, 905–919.
López-Espíndola, D., García-Aldea, Á., Gómez de la Riva, I., Rodríguez-García, A. M., Salvatore, D., Visser, T. J., et al. (2019). Thyroid hormone availability in the human fetal brain: Novel entry pathways and role of radial glia. *Brain Structure & Function*, *224*, 2103–2119.
López-Juárez, A., Remaud, S., Hassani, Z., Jolivet, P., Pierre Simons, J., Sontag, T., et al. (2012). Thyroid hormone signaling acts as a neurogenic switch by repressing Sox2 in the adult neural stem cell niche. *Cell Stem Cell*, *10*, 531–543.
Lucassen, P. J., Stumpel, M. W., Wang, Q., & Aronica, E. (2010). Decreased numbers of progenitor cells but no response to antidepressant drugs in the hippocampus of elderly depressed patients. *Neuropharmacology*, *58*, 940–949.
Lüesse, H.-G., Roskoden, T., Linke, R., Otten, U., Heese, K., & Schwegler, H. (1998). Modulation of mRNA expression of the neurotrophins of the nerve growth factor family and their receptors in the septum and hippocampus of rats after transient postnatal thyroxine treatmentI. Expression of nerve growth factor, brain-derived neurotrophic. *Experimental Brain Research*, *119*, 1–8.
Lukmanji, A., Williams, J. V. A., Bulloch, A. G. M., Bhattarai, A., & Patten, S. B. (2019). Seasonal variation in symptoms of depression: A Canadian population based study. *Journal of Affective Disorders*, *255*, 142–149.
Luongo, C., Butruille, L., Sébillot, A., Le Blay, K., Schwaninger, M., Heuer, H., et al. (2021). Absence of both thyroid hormone transporters MCT8 and OATP1C1 impairs neural stem cell fate in the adult mouse subventricular zone. *Stem Cell Reports*, *16*(2), 337–353. https://doi.org/10.1016/j.stemcr.2020.12.009. PMID: 33450189; PMCID: PMC7878696.
Ma, D. K., Ming, G. L., & Song, H. (2005). Glial influences on neural stem cell development: Cellular niches for adult neurogenesis. *Current Opinion in Neurobiology*, *15*(5), 514–520.
Malpaux, B., Daveau, A., Maurice, F., Gayrard, V., & Thiery, J.-C. (1993). Short-day effects of melatonin on luteinizing hormone secretion in the ewe: Evidence for central sites of action in the mediobasal hypothalamus. *Biology of Reproduction*, *48*, 752–760.
Malpaux, B., Daveau, A., Maurice-Mandon, F., Duarte, G., & Chemineau, P. (1998). Evidence that melatonin acts in the premammillary hypothalamic area to control reproduction in the ewe: Presence of binding sites and stimulation of luteinizing hormone secretion by in situ microimplant delivery. *Endocrinology*, *139*, 1508–1516.
Mathers, C. D., & Harris, R. S. (1983). Seasonal distribution of births in Australia. *International Journal of Epidemiology*, *12*, 326–331.
Mayerl, S., Heuer, H., & Ffrench-Constant, C. (2020). Hippocampal neurogenesis requires cell-autonomous thyroid hormone signaling. *Stem Cell Reports*, *14*, 845–860.

Mayerl, S., Müller, J., Bauer, R., Richert, S., Kassmann, C. M., Darras, V. M., et al. (2014). Transporters MCT8 and OATP1C1 maintain murine brain thyroid hormone homeostasis. *The Journal of Clinical Investigation*, *124*, 1987–1999.

Mayerl, S., Schmidt, M., Doycheva, D., Darras, V. M., Hüttner, S. S., Boelen, A., et al. (2018). Thyroid hormone transporters MCT8 and OATP1C1 control skeletal muscle regeneration. *Stem Cell Reports*, *10*, 1959–1974.

McConnell, R. J., Menendez, C. E., Smith, F. R., Henkin, R. I., & Rivlin, R. S. (1975). Defects of taste and smell in patients with hypothyroidism. *The American Journal of Medicine*, *59*, 354–364.

McGann, J. P. (2017). Poor human olfaction is a 19th-century myth. *Science*, *356*(6338), eaam7263.

McKenzie, I. A., Ohayon, D., Li, H., De Faria, J. P., Emery, B., Tohyama, K., et al. (2014). Motor skill learning requires active central myelination. *Science*, *346*, 318–322.

McNay, D. E. G., Briançon, N., Kokoeva, M. V., Maratos-Flier, E., & Flier, J. S. (2012). Remodeling of the arcuate nucleus energy-balance circuit is inhibited in obese mice. *The Journal of Clinical Investigation*, *122*, 142–152.

Menn, B., Garcia-Verdugo, J. M., Yaschine, C., Gonzalez-Perez, O., Rowitch, D., & Alvarez-Buylla, A. (2006). Origin of oligodendrocytes in the subventricular zone of the adult brain. *The Journal of Neuroscience*, *26*, 7907–7918.

Mercer, J. G., Moar, K. M., Ross, A. W., Hoggard, N., & Morgan, P. J. (2000). Photoperiod regulates arcuate nucleus POMC, AGRP, and leptin receptor mRNA in Siberian hamster hypothalamus. *American Journal of Physiology. Regulatory, Integrative and Comparative Physiology*, *278*, R271–R281.

Meriggiola, M. C., Noonan, E. A., Paulsen, C. A., & Bremner, W. (1996). Annual patterns of luteinizing hormone, follicle stimulating hormone, testosterone and inhibin in normal men. *Human Reproduction*, *11*, 248–252.

Merkle, F. T., Mirzadeh, Z., & Alvarez-Buylla, A. (2007). Mosaic organization of neural stem cells in the adult brain. *Science*, *317*, 381–384.

Messager, S., Chatzidaki, E. E., Ma, D., Hendrick, A. G., Zahn, D., Dixon, J., et al. (2005). Kisspeptin directly stimulates gonadotropin-releasing hormone release via G protein-coupled receptor 54. *Proceedings of the National Academy of Sciences of the United States of America*, *102*, 1761–1766.

Migaud, M., Batailler, M., Pillon, D., Franceschini, I., & Malpaux, B. (2011). Seasonal changes in cell proliferation in the adult sheep brain and pars tuberalis. *Journal of Biological Rhythms*, *26*, 486–496.

Migaud, M., Batailler, M., Segura, S., Duittoz, A., Franceschini, I., & Pillon, D. (2010). Emerging new sites for adult neurogenesis in the mammalian brain: A comparative study between the hypothalamus and the classical neurogenic zones. *The European Journal of Neuroscience*, *32*, 2042–2052.

Milesi, S., Simonneaux, V., & Klosen, P. (2017). Downregulation of deiodinase 3 is the earliest event in photoperiodic and photorefractory activation of the gonadotropic axis in seasonal hamsters. *Scientific Reports*, *7*, 17739.

Miller, J. A., Nathanson, J., Franjic, D., Shim, S., Dalley, R. A., Shapouri, S., et al. (2013). Conserved molecular signatures of neurogenesis in the hippocampal subgranular zone of rodents and primates. *Development*, *140*, 4633–4644.

Ming, G., & Song, H. (2011). Adult neurogenesis in the mammalian brain: significant answers and significant questions. *Neuron*, *70*, 687–702.

Miranda, M., Morici, J. F., Zanoni, M. B., & Bekinschtein, P. (2019). Brain-derived neurotrophic factor: A key molecule for memory in the healthy and the pathological brain. *Frontiers in Cellular Neuroscience*, *13*, 363.

Mirescu, C., & Gould, E. (2006). Stress and adult neurogenesis. *Hippocampus*, *16*, 233–238.

Moenter, S. M., Woodfill, C. J. I., & Karsch, F. J. (1991). Role of the thyroid gland in seasonal reproduction: Thyroidectomy blocks seasonal suppression of reproductive neuroendocrine activity in ewes. *Endocrinology*, *128*, 1337–1344.

Montero-Pedrazuela, A., Venero, C., Lavado-Autric, R., Fernández-Lamo, I., García-Verdugo, J. M., Bernal, J., et al. (2006). Modulation of adult hippocampal neurogenesis by thyroid hormones: Implications in depressive-like behavior. *Molecular Psychiatry*, *11*, 361–371.

Moog, N. K., Entringer, S., Heim, C., Wadhwa, P. D., Kathmann, N., & Buss, C. (2017). Influence of maternal thyroid hormones during gestation on fetal brain development. *Neuroscience*, *342*, 68–100.

Moreno-Jiménez, E. P., Flor-García, M., Terreros-Roncal, J., Rábano, A., Cafini, F., Pallas-Bazarra, N., et al. (2019). Adult hippocampal neurogenesis is abundant in neurologically healthy subjects and drops sharply in patients with Alzheimer's disease. *Nature Medicine*, *25*, 554–560.

Morrison, S. J., & Spradling, A. C. (2008). Stem cells and niches: Mechanisms that promote stem cell maintenance throughout life. *Cell*, *132*, 598–611.

Nait-Oumesmar, B., Decker, L., Lachapelle, F., Avellana-Adalid, V., Bachelin, C., & Van Evercooren, A. B. (1999). Progenitor cells of the adult mouse subventricular zone proliferate, migrate and differentiate into oligodendrocytes after demyelination. *The European Journal of Neuroscience*, *11*, 4357–4366.

Nait-Oumesmar, B., Picard-Riera, N., Kerninon, C., Decker, L., Seilhean, D., Höglinger, G. U., et al. (2007). Activation of the subventricular zone in multiple sclerosis: Evidence for early glial progenitors. *Proceedings of the National Academy of Sciences of the United States of America*, *104*, 4694–4699.

Nakao, N., Ono, H., & Yoshimura, T. (2008). Thyroid hormones and seasonal reproductive neuroendocrine interactions. *Reproduction*, *136*, 1–8.

Ngwenya, L. B., Heyworth, N. C., Shwe, Y., Moore, T. L., & Rosene, D. L. (2015). Age-related changes in dentate gyrus cell numbers, neurogenesis, and associations with cognitive impairments in the rhesus monkey. *Frontiers in Systems Neuroscience*, *9*, 102.

Ngwenya, L. B., Rosene, D. L., & Peters, A. (2008). An ultrastructural characterization of the newly generated cells in the adult monkey dentate gyrus. *Hippocampus*, *18*, 210–220.

Nicholls, T. J., Follett, B. K., Goldsmith, A. R., & Pearson, H. (1988). Possible homologies between photorefractoriness in sheep and birds: The effect of thyroidectomy on the length of the ewe's breeding season. *Reproduction, Nutrition, Development*, *28*, 375–385.

Nicolis di Robilant, V., Scardigli, R., Strimpakos, G., Tirone, F., Middei, S., Scopa, C., et al. (2019). Running-activated neural stem cells enhance subventricular neurogenesis and improve olfactory behavior in p21 knockout mice. *Molecular Neurobiology*, *56*, 7534–7556.

Nogueira, A. B., Sogayar, M. C., Colquhoun, A., Siqueira, S. A., Nogueira, A. B., Marchiori, P. E., et al. (2014). Existence of a potential neurogenic system in the adult human brain. *Journal of Translational Medicine*, *12*, 75.

Nolen-Hoeksema, S. (2014). *Abnormal psychology*. New York: McGraw-Hill Education.

ØDegård, Ø. (1977). Season of birth in the population of Norway, with particular reference to the September birth maximum. *The British Journal of Psychiatry*, *131*, 339–344.

Ono, H., Nakao, N., & Yoshimura, T. (2009). Identification of the photoperiodic signaling pathway regulating seasonal reproduction using the functional genomics approach. *General and Comparative Endocrinology*, *163*, 2–6.

Ormerod, B. K., & Galea, L. A. M. (2001). Reproductive status influences cell proliferation and cell survival in the dentate gyrus of adult female meadow voles: A possible regulatory role for estradiol. *Neuroscience*, *102*, 369–379.

Ormerod, B. K., & Galea, L. A. M. (2003). Reproductive status influences the survival of new cells in the dentate gyrus of adult male meadow voles. *Neuroscience Letters*, *346*, 25–28.

Ormerod, B. K., Lee, T. T.-Y., & Galea, L. A. M. (2004). Estradiol enhances neurogenesis in the dentate gyri of adult male meadow voles by increasing the survival of young granule neurons. *Neuroscience*, *128*, 645–654.

Ortavant, R., Pelletier, J., Ravault, J. P., Thimonier, J., & Volland-Nail, P. (1985). Photoperiod: Main proximal and distal factor of the circannual cycle of reproduction in farm mammals. *Oxford Reviews of Reproductive Biology*, 7, 305–345.

Ortega, F., Gascón, S., Masserdotti, G., Deshpande, A., Simon, C., Fischer, J., et al. (2013). Oligodendrogliogenic and neurogenic adult subependymal zone neural stem cells constitute distinct lineages and exhibit differential responsiveness to Wnt signalling. *Nature Cell Biology*, *15*, 602–613.

Ortega, J. A., Radonjic, N., & Zecevic, N. (2013). Sonic hedgehog promotes generation and maintenance of human forebrain Olig2 progenitors. *Frontiers in Cellular Neuroscience*, 7, 254.

Pan, W. L., Chopp, M., Fan, B., Zhang, R., Wang, X., Hu, J., et al. (2019). Ablation of the microRNA-17-92 cluster in neural stem cells diminishes adult hippocampal neurogenesis and cognitive function. *The FASEB Journal*, *33*, 5257–5267.

Paredes, M. F., James, D., Gil-Perotin, S., Kim, H., Cotter, J. A., Ng, C., et al. (2016). Extensive migration of young neurons into the infant human frontal lobe. *Science*, *354*(6308), aaf7073.

Parkinson, T. J., & Follett, B. K. (1995). Thyroidectomy abolishes seasonal testicular cycles of Soay rams. *Proceedings of the Royal Society of London. Series B: Biological Sciences*, *259*, 1–6.

Parpura, V., Heneka, M. T., Montana, V., Oliet, S. H. R., Schousboe, A., Haydon, P. G., et al. (2012). Glial cells in (patho)physiology. *Journal of Neurochemistry*, *121*, 4–27.

Pastalkova, E., Serrano, P., Pinkhasova, D., Wallace, E., Fenton, A. A., & Sacktor, T. C. (2006). Storage of spatial information by the maintenance mechanism of LTP. *Science*, *313*, 1141–1144.

Paternostro, M. A., & Meisami, E. (1991). Lack of thyroid hormones but not their excess affects the maturation of olfactory receptor neurons: A quantitative morphologic study in the postnatal rat. *International Journal of Developmental Neuroscience*, *9*, 439–452.

Pellegrino, G., Trubert, C., Terrien, J., Pifferi, F., Leroy, D., Loyens, A., et al. (2018). A comparative study of the neural stem cell niche in the adult hypothalamus of human, mouse, rat and gray mouse lemur (Microcebus murinus). *The Journal of Comparative Neurology*, *526*, 1419–1443.

Pencea, V., Bingaman, K. D., Freedman, L. J., & Luskin, M. B. (2001). Neurogenesis in the subventricular zone and rostral migratory stream of the neonatal and adult primate forebrain. *Experimental Neurology*, *172*, 1–16.

Peng, Z., Zhang, L., Wang, H., He, X., Peng, X., Zhang, Q., et al. (2018). Experimental autoimmune encephalomyelitis (EAE) model of cynomolgus macaques induced by recombinant human MOG1-125 (rhMOG1-125) protein and MOG34-56 peptide. *Protein and Peptide Letters*, *24*, 1166–1178.

Perret, M., & Aujard, F. (2005). Aging and season affect plasma dehydroepiandrosterone sulfate (DHEA-S) levels in a primate. *Experimental Gerontology*, *40*, 582–587.

Perret, M., Aujard, F., & Vannier, G. (1998). Influence of daylength on metabolic rate and daily water loss in the male prosimian primate Microcebus murinus. *Comparative Biochemistry and Physiology, Part A: Molecular & Integrative Physiology*, *119*, 981–989.

Perret, M., & Schilling, A. (1993). Response to short photoperiod and spontaneous sexual recrudescence in the lesser mouse lemur: Role of olfactory bulb removal. *The Journal of Endocrinology*, *137*, 511–518.

Perrot-Sinal, T. S., Kavaliers, M., & Ossenkopp, K.-P. (1998). Spatial learning and hippocampal volume in male deer mice: Relations to age, testosterone and adrenal gland weight. *Neuroscience*, *86*, 1089–1099.
Petreanu, L., & Alvarez-Buylla, A. (2002). Maturation and death of adult-born olfactory bulb granule neurons: Role of olfaction. *The Journal of Neuroscience*, *22*, 6106–6113.
Picard-Riera, N., Nait-Oumesmar, B., & Baron-Van Evercooren, A. (2004). Endogenous adult neural stem cells: Limits and potential to repair the injured central nervous system. *Journal of Neuroscience Research*, *76*, 223–231.
Picq, J.-L. (2007). Aging affects executive functions and memory in mouse lemur primates. *Experimental Gerontology*, *42*, 223–232.
Pierce, A. A., & Xu, A. W. (2010). De novo neurogenesis in adult hypothalamus as a compensatory mechanism to regulate energy balance. *The Journal of Neuroscience*, *30*, 723–730.
Platel, J.-C., Angelova, A., Bugeon, S., Wallace, J., Ganay, T., Chudotvorova, I., et al. (2019). Neuronal integration in the adult mouse olfactory bulb is a non-selective addition process. *eLife*, *8*, e44830.
Pollen, A. A., Nowakowski, T. J., Chen, J., Retallack, H., Sandoval-Espinosa, C., Nicholas, C. R., et al. (2015). Molecular identity of human outer radial glia during cortical development. *Cell*, *163*, 55–67.
Popov, V. I., Kraev, I. V., Ignat'ev, D. A., & Stewart, M. G. (2011). Suspension of mitotic activity in dentate gyrus of the hibernating ground squirrel. *Neural Plasticity*, *2011*, 867525.
Popova, L. D., & Vasyl'ieva, I. M. (2010). Testosterone and thyroid hormones in blood plasma of rats with submissive and aggressive type of behavior. *Fiziologicheskiĭ Zhurnal*, *56*, 49–52.
Previc, F. H. (2002). Thyroid hormone production in chimpanzees and humans: Implications for the origins of human intelligence. *American Journal of Physical Anthropology*, *118*, 402–403.
Previc, F. H. (2009). The dopaminergic mind in history. In F. H. Previc (Ed.), *The dopaminergic mind in human evolution and history* (pp. 123–154). Cambridge University Press: Cambridge.
Pyter, L. M., Reader, B. F., & Nelson, R. J. (2005). Short photoperiods impair spatial learning and alter hippocampal dendritic morphology in adult male white-footed mice (Peromyscus leucopus). *The Journal of Neuroscience*, *25*, 4521–4526.
Pyter, L. M., Trainor, B. C., & Nelson, R. J. (2006). Testosterone and photoperiod interact to affect spatial learning and memory in adult male white-footed mice (Peromyscus leucopus). *The European Journal of Neuroscience*, *23*, 3056–3062.
Quiñones-Hinojosa, A., Sanai, N., Soriano-Navarro, M., Gonzalez-Perez, O., Mirzadeh, Z., Gil-Perotin, S., et al. (2006). Cellular composition and cytoarchitecture of the adult human subventricular zone: A niche of neural stem cells. *The Journal of Comparative Neurology*, *494*, 415–434.
Rai, K. S., Hattiangady, B., & Shetty, A. K. (2007). Enhanced production and dendritic growth of new dentate granule cells in the middle-aged hippocampus following intracerebroventricular FGF-2 infusions. *The European Journal of Neuroscience*, *26*, 1765–1779.
Raichlen, D. A., & Polk, J. D. (2013). Linking brains and brawn: Exercise and the evolution of human neurobiology. *Proceedings of the Biological Sciences*, *280*, 20122250.
Raitiere, M. N. (1992). Clinical evidence for thyroid dysfunction in patients with seasonal affective disorder. *Psychoneuroendocrinology*, *17*, 231–241.
Ramírez-Rodríguez, G., Klempin, F., Babu, H., Benítez-King, G., & Kempermann, G. (2009). Melatonin modulates cell survival of new neurons in the Hippocampus of adult mice. *Neuropsychopharmacology*, *34*, 2180–2191.

Ramírez-Rodríguez, G. B., Olvera-Hernández, S., Vega-Rivera, N. M., & Ortiz-López, L. (2019). Melatonin influences structural plasticity in the axons of granule cells in the dentate gyrus of Balb/C mice. *International Journal of Molecular Sciences*, *20*, 73.

Ramírez-Rodríguez, G. B., Palacios-Cabriales, D. M., Ortiz-López, L., Estrada-Camarena, E. M., & Vega-Rivera, N. M. (2020). Melatonin mocdulates dendrite maturation and complexity in the dorsal- and ventral- dentate gyrus concomitantly with its antidepressant-like effect in male Balb/C mice. *International Journal of Molecular Sciences*, *21*(5), 1724.

Remaud, S., Gothié, J. D., Morvan-Dubois, G., & Demeneix, B. A. (2014). Thyroid hormone signaling and adult neurogenesis in mammals. *Frontiers in Endocrinology (Lausanne)*, *5*, 62.

Remaud, S., Ortiz, F. C., Perret-Jeanneret, M., Aigrot, M. S., Gothié, J. D., Fekete, C., et al. (2017). Transient hypothyroidism favors oligodendrocyte generation providing functional remyelination in the adult mouse brain. *eLife*, *6*, e29996.

Reppert, S. M., Perlow, M. J., Ungerleider, L. G., Mishkin, M., Tamarkin, L., Orloff, D. G., et al. (1981). Effects of damage to the suprachiasmatic area of the anterior hypothalamus on the daily melatonin and cortisol rhythms in the rhesus monkey. *The Journal of Neuroscience*, *1*, 1414–1425.

Revel, F. G., Saboureau, M., Pévet, P., Mikkelsen, J. D., & Simonneaux, V. (2006). Melatonin regulates type 2 deiodinase gene expression in the Syrian hamster. *Endocrinology*, *147*, 4680–4687.

Revel, F. G., Saboureau, M., Pévet, P., Simonneaux, V., & Mikkelsen, J. D. (2008). RFamide-related peptide gene is a melatonin-driven photoperiodic gene. *Endocrinology*, *149*, 902–912.

Reynolds, B. A., & Weiss, S. (1992). Generation of neurons and astrocytes from isolated cells of the adult mammalian central nervous system. *Science*, *255*, 1707–1710.

Richardson, S. J., Lemkine, G. F., Alfama, G., Hassani, Z., & Demeneix, B. A. (2007). Cell division and apoptosis in the adult neural stem cell niche are differentially affected in transthyretin null mice. *Neuroscience Letters*, *421*, 234–238.

Robins, S. C., Stewart, I., McNay, D. E., Taylor, V., Giachino, C., Goetz, M., et al. (2013). α-Tanycytes of the adult hypothalamic third ventricle include distinct populations of FGF-responsive neural progenitors. *Nature Communications*, *4*, 2049.

Rochefort, C., Gheusi, G., Vincent, J. D., & Lledo, P. M. (2002). Enriched odor exposure increases the number of newborn neurons in the adult olfactory bulb and improves odor memory. *The Journal of Neuroscience*, *22*, 2679–2689.

Rodríguez, E. M., Blázquez, J. L., & Guerra, M. (2010). The design of barriers in the hypothalamus allows the median eminence and the arcuate nucleus to enjoy private milieus: The former opens to the portal blood and the latter to the cerebrospinal fluid. *Peptides*, *31*, 757–776.

Roenneberg, T., & Aschoff, J. (1990). Annual rhythm of human reproduction: II. Environmental correlations. *Journal of Biological Rhythms*, *5*, 217–239.

Rojansky, N., Brzezinski, A., & Schenker, J. G. (1992). Seasonality in human reproduction: An update. *Human Reproduction*, *7*, 735–745.

Rojczyk-Gołębiewska, E., Pałasz, A., & Wiaderkiewicz, R. (2014). Hypothalamic subependymal niche: A novel site of the adult neurogenesis. *Cellular and Molecular Neurobiology*, *34*, 631–642.

Rosenstein, J. M., Krum, J. M., & Ruhrberg, C. (2010). VEGF in the nervous system. *Organogenesis*, *6*, 107–114.

Rosenthal, N. E., Sack, D. A., Gillin, J. C., Lewy, A. J., Goodwin, F. K., Davenport, Y., et al. (1984). Seasonal affective disorder: A description of the syndrome and preliminary findings with light therapy. *Archives of General Psychiatry*, *41*(1), 72–80.

Ross, A. W., Helfer, G., Russell, L., Darras, V. M., & Morgan, P. J. (2011). Thyroid hormone signalling genes are regulated by photoperiod in the hypothalamus of F344 rats. *PLoS One*, *6*, e21351.

Ross, A. W., Johnson, C. E., Bell, L. M., Reilly, L., Duncan, J. S., Barrett, P., et al. (2009). Divergent regulation of hypothalamic neuropeptide Y and agouti-related protein by photoperiod in F344 rats with differential food intake and growth. *Journal of Neuroendocrinology*, *21*, 610–619.

Ross, A. W., Webster, C. A., Mercer, J. G., Moar, K. M., Ebling, F. J., Schuhler, S., et al. (2004). Photoperiodic regulation of hypothalamic retinoid signaling: association of retinoid X receptor γ with body weight. *Endocrinology*, *145*(1), 13–20.

Rostal, D. C., Glick, B. B., Eaton, G. G., & Resko, J. A. (1986). Seasonally of adult male japanese macaques (Macaca fuscata): Androgens and behavior in a confined troop. *Hormones and Behavior*, *20*, 452–462.

Royo, J., Villain, N., Champeval, D., Del Gallo, F., Bertini, G., Aujard, F., et al. (2018). Effects of n-3 polyunsaturated fatty acid supplementation on cognitive functions, electrocortical activity and neurogenesis in a non-human primate, the grey mouse lemur (Microcebus murinus). *Behavioural Brain Research*, *347*, 394–407.

Sáenz de Miera, C., Hanon, E. A., Dardente, H., Birnie, M., Simonneaux, V., Lincoln, G. A., et al. (2013). Circannual variation in thyroid hormone deiodinases in a short-day breeder. *Journal of Neuroendocrinology*, *25*, 412–421.

Sahay, A., Scobie, K. N., Hill, A. S., O'Carroll, C. M., Kheirbek, M. A., Burghardt, N. S., et al. (2011). Increasing adult hippocampal neurogenesis is sufficient to improve pattern separation. *Nature*, *472*, 466–470.

Sakamoto, M., Imayoshi, I., Ohtsuka, T., Yamaguchi, M., Mori, K., & Kageyama, R. (2011). Continuous neurogenesis in the adult forebrain is required for innate olfactory responses. *Proceedings of the National Academy of Sciences*, *108*, 8479–8484.

Saltó, C., Kindblom, J. M., Johansson, C., Wang, Z., Gullberg, H., Nordström, K., et al. (2001). Ablation of TRα2 and a concomitant overexpression of α1 yields a mixed hypo- and hyperthyroid phenotype in mice. *Molecular Endocrinology*, *15*, 2115–2128.

Sanai, N., Berger, M. S., Garcia-Verdugo, J. M., & Alvarez-Buylla, A. (2007). Comment on "Human neuroblasts migrate to the olfactory bulb via a lateral ventricular extension". *Science*, *318*, 393.

Sanai, N., Nguyen, T., Ihrie, R. A., Mirzadeh, Z., Tsai, H.-H., Wong, M., et al. (2011). Corridors of migrating neurons in the human brain and their decline during infancy. *Nature*, *478*, 382–386.

Sanai, H., Tramontin, A. D., Quiñones-Hinojosa, A., Barbaro, N. M., Gupta, H., Kunwar, S., et al. (2004). Unique astrocyte ribbon in adult human brain contains neural stem cells but lacks chain migration. *Nature*, *427*(6976), 740–744.

Saniotis, A., Grantham, J. P., Kumaratilake, J., & Henneberg, M. (2020). Neuro-hormonal regulation is a better Indicator of human cognitive abilities than brain anatomy: The need for a new paradigm. *Frontiers in Neuroanatomy*, *13*, 101.

Sarrafchi, A., Odhammer, A. M. E., Hernandez Salazar, L. T., & Laska, M. (2013). Olfactory sensitivity for six predator odorants in CD-1 mice, human subjects, and spider monkeys. *PLoS One*, *8*, e80621.

Sawamoto, K., Hirota, Y., Alfaro-Cervello, C., Soriano-Navarro, M., He, X., Hayakawa-Yano, Y., et al. (2011). Cellular composition and organization of the subventricular zone and rostral migratory stream in the adult and neonatal common marmoset brain. *The Journal of Comparative Neurology*, *519*, 690–713.

Sawamoto, K., Wichterle, H., Gonzalez-Perez, O., Cholfin, J. A., Yamada, M., Spassky, N., et al. (2006). New neurons follow the flow of cerebrospinal fluid in the adult brain. *Science*, *311*, 629–632.

Schilling, A., & Perret, M. (1993). Removal of the olfactory bulbs modifies the gonadal responses to photoperiod in the lesser mouse Lemur (Microcebus Murinus). *Biology of Reproduction*, *49*, 58–65.

Séguy, M., & Perret, M. (2005). Changes in olfactory inputs modify the energy balance response to short days in male gray mouse lemurs. *Physiology & Behavior, 84*, 23–31.
Seri, B., García-Verdugo, J. M., Collado-Morente, L., McEwen, B. S., & Alvarez-Buylla, A. (2004). Cell types, lineage, and architecture of the germinal zone in the adult dentate gyrus. *The Journal of Comparative Neurology, 478*, 359–378.
Seri, B., Garcı´a-Verdugo, J. M., McEwen, B. S., & Alvarez-Buylla, A. (2001). Astrocytes give rise to new neurons in the adult mammalian hippocampus. *The Journal of Neuroscience, 21*, 7153–7160.
Shearer, K. D., Stoney, P. N., Nanescu, S. E., Helfer, G., Barrett, P., Ross, A. W., et al. (2012). Photoperiodic expression of two RALDH enzymes and the regulation of cell proliferation by retinoic acid in the rat hypothalamus. *Journal of Neurochemistry, 122*, 789–799.
Sher, L., Rosenthal, N. E., & Wehr, T. A. (1999). Free thyroxine and thyroid-stimulating hormone levels in patients with seasonal affective disorder and matched controls. *Journal of Affective Disorders, 56*, 195–199.
Shingo, T., Gregg, C., Enwere, E., Fujikawa, H., Hassam, R., Geary, C., et al. (2003). Pregnancy-stimulated neurogenesis in the adult female forebrain mediated by prolactin. *Science, 299*, 117–120.
Shinomiya, A., Shimmura, T., Nishiwaki-Ohkawa, T., & Yoshimura, T. (2014). Regulation of seasonal reproduction by hypothalamic activation of thyroid hormone. *Frontiers in Endocrinology, 5*, 12.
Shors, T. J., Miesegaes, G., Beylin, A., Zhao, M., Rydel, T., & Gould, E. (2001). Neurogenesis in the adult is involved in the formation of trace memories. *Nature, 410*, 372–376.
Shulga, A., Blaesse, A., Kysenius, K., Huttunen, H. J., Tanhuanpää, K., Saarma, M., et al. (2009). Thyroxin regulates BDNF expression to promote survival of injured neurons. *Molecular and Cellular Neurosciences, 42*, 408–418.
Silva, J. F., Ocarino, N. M., & Serakides, R. (2018). Thyroid hormones and female reproduction. *Biology of Reproduction, 99*, 907–921.
Smith, J. T., Clay, C. M., Caraty, A., & Clarke, I. J. (2007). KiSS-1 messenger ribonucleic acid expression in the hypothalamus of the ewe is regulated by sex steroids and season. *Endocrinology, 148*, 1150–1157.
Smith, J. T., Coolen, L. M., Kriegsfeld, L. J., Sari, I. P., Jaafarzadehshirazi, M. R., Maltby, M., et al. (2008). Variation in kisspeptin and RFamide-related peptide (RFRP) expression and terminal connections to gonadotropin-releasing hormone neurons in the brain: A novel medium for seasonal breeding in the sheep. *Endocrinology, 149*, 5770–5782.
Smith, M. T., Pencea, V., Wang, Z., Luskin, M. B., & Insel, T. R. (2001). Increased number of BrdU-labeled neurons in the rostral migratory stream of the estrous prairie vole. *Hormones and Behavior, 39*, 11–21.
Snapyan, M., Lemasson, M., Brill, M. S., Blais, M., Massouh, M., Ninkovic, J., et al. (2009). Vasculature guides migrating neuronal precursors in the adult mammalian forebrain via brain-derived neurotrophic factor signaling. *The Journal of Neuroscience, 29*, 4172–4188.
Snyder, J. S., Hong, N. S., McDonald, R. J., & Wojtowicz, J. M. (2005). A role for adult neurogenesis in spatial long-term memory. *Neuroscience, 130*(4), 843–852.
Song, H., Stevens, C. F., & Gage, F. H. (2002). Astroglia induce neurogenesis from adult neural stem cells. *Nature, 417*(6884), 39–44.
Sorrells, S. F., Paredes, M. F., Cebrian-Silla, A., Sandoval, K., Qi, D., Kelley, K. W., et al. (2018). Human hippocampal neurogenesis drops sharply in children to undetectable levels in adults. *Nature, 555*, 377–381.

Sotthibundhu, A., Phansuwan-Pujito, P., & Govitrapong, P. (2010). Melatonin increases proliferation of cultured neural stem cells obtained from adult mouse subventricular zone. *Journal of Pineal Research*, *49*, 291–300.
Sousa-Ferreira, L., de Almeida, L. P., & Cavadas, C. (2014). Role of hypothalamic neurogenesis in feeding regulation. *Trends in Endocrinology and Metabolism*, *25*, 80–88.
Spritzer, M. D., Panning, A. W., Engelman, S. M., Prince, W. T., Casler, A. E., Georgakas, J. E., et al. (2017). Seasonal and sex differences in cell proliferation, neurogenesis, and cell death within the dentate gyrus of adult wild-caught meadow voles. *Neuroscience*, *360*, 155–165.
Stamm, T. J., Lewitzka, U., Sauer, C., Pilhatsch, M., Smolka, M. N., Koeberle, U., et al. (2014). Supraphysiologic doses of levothyroxine as adjunctive therapy in bipolar depression: A randomized, double-blind, placebo-controlled study. *The Journal of Clinical Psychiatry*, *75*, 162–168.
Stenzel, D., & Huttner, W. B. (2013). Role of maternal thyroid hormones in the developing neocortex and during human evolution. *Frontiers in Neuroanatomy*, 7, 19.
Stohn, J. P., Martinez, M. E., Zafer, M., López-Espíndola, D., Keyes, L. M., & Hernandez, A. (2018). Increased aggression and lack of maternal behavior in Dio3-deficient mice are associated with abnormalities in oxytocin and vasopressin systems. *Genes, Brain, and Behavior*, *17*, 23–35.
Stoney, P. N., Helfer, G., Rodrigues, D., Morgan, P. J., & Mccaffery, P. (2016). Thyroid hormone activation of retinoic acid synthesis in hypothalamic tanycytes. *Glia*, *64*(3), 425–439.
Sugiyama, D., Kusuhara, H., Taniguchi, H., Ishikawa, S., Nozaki, Y., Aburatani, H., et al. (2003). Functional characterization of rat brain-specific organic anion transporter (Oatp14) at the blood-brain barrier: High affinity transporter for thyroxine. *The Journal of Biological Chemistry*, *278*, 43489–43495.
Sultan, S., Li, L., Moss, J., Petrelli, F., Cassé, F., Gebara, E., et al. (2015). Synaptic integration of adult-born hippocampal neurons is locally controlled by astrocytes. *Neuron*, *88*(5), 957–972.
t Hart, B. A., van Kooyk, Y., Geurts, J. J. G., & Gran, B. (2015). The primate autoimmune encephalomyelitis model; A bridge between mouse and man. *Annals of Clinical Translational Neurology*, *2*, 581–593.
Tanti, A., Rainer, Q., Minier, F., Surget, A., & Belzung, C. (2012). Differential environmental regulation of neurogenesis along the septo-temporal axis of the hippocampus. *Neuropharmacology*, *63*, 374–384.
Taşkın, E., Artis, A. S., Bitiktas, S., Dolu, N., Liman, N., & Süer, C. (2011). Experimentally induced hyperthyroidism disrupts hippocampal long-term potentiation in adult rats. *Neuroendocrinology*, *94*, 218–227.
Taupin, P., & Gage, F. H. (2002). Adult neurogenesis and neural stem cells of the central nervous system in mammals. *Journal of Neuroscience Research*, *69*(6), 745–749.
Thimonier, J., Mauléon, P., Bézard, J., DE Reviers, M.-M., & Cornu, C. (1969). Variations saisonnières du comportement d'oestrus et des activités ovariennes et hypophysaires chez les ovins. *Annales de Biologie Animale, Biochimie, Biophysique*, *9*, 233–250.
Thrun, L. A., Dahl, G. E., Evans, N. P., & Karsch, F. J. (1997). A critical period for thyroid hormone action on seasonal changes in reproductive neuroendocrine function in the ewe. *Endocrinology*, *138*, 3402–3409.
Tobin, M. K., Musaraca, K., Disouky, A., Shetti, A., Bheri, A., Honer, W. G., et al. (2019). Human hippocampal neurogenesis persists in aged adults and Alzheimer's disease patients. *Cell Stem Cell*, *24*, 974–982. e3.
Tocharus, C., Puriboriboon, Y., Junmanee, T., Tocharus, J., Ekthuwapranee, K., & Govitrapong, P. (2014). Melatonin enhances adult rat hippocampal progenitor cell proliferation via ERK signaling pathway through melatonin receptor. *Neuroscience*, *275*, 314–321.

Toda, T., & Gage, F. H. (2018). Review: Adult neurogenesis contributes to hippocampal plasticity. *Cell and Tissue Research*, *373*, 693–709.

Tonchev, A. B., Yamashima, T., Sawamoto, K., & Okano, H. (2005). Enhanced proliferation of progenitor cells in the subventricular zone and limited neuronal production in the striatum and neocortex of adult macaque monkeys after global cerebral ischemia. *Journal of Neuroscience Research*, *81*, 776–788.

Trajkovic, M., Visser, T. J., Mittag, J., Horn, S., Lukas, J., Darras, V. M., et al. (2007). Abnormal thyroid hormone metabolism in mice lacking the monocarboxylate transporter 8. *The Journal of Clinical Investigation*, *117*, 627–635.

Turner, C., & Bagnara, J. (1971). *General endocrinology*. Philadelphia and London: WB Saunders Co.

Uchida, K., Yonezawa, M., Nakamura, S., Kobayashi, T., & Machida, T. (2005). Impaired neurogenesis in the growth-retarded mouse is reversed by T3 treatment. *Neuroreport*, *16*(2), 103–106. https://doi.org/10.1097/00001756-200502080-00005. PMID: 15671855.

Van Haaster, L. H., De Jong, F. H., Docter, R., & De Rooij, D. G. (1992). The effect of hypothyroidism on Sertoli cell proliferation and differentiation and hormone levels during testicular development in the rat. *Endocrinology*, *131*, 1574–1576.

Vancamp, P., Butruille, L., Demeneix, B. A., & Remaud, S. (2020). Thyroid hormone and neural stem cells: Repair potential following brain and spinal cord injury. *Frontiers in Neuroscience*, *14*, 875.

Vancamp, P., & Darras, V. M. (2018). From zebrafish to human: A comparative approach to elucidate the role of the thyroid hormone transporter MCT8 during brain development. *General and Comparative Endocrinology*, *265*, 219–229. https://doi.org/10.1016/j.ygcen.2017.11.023.

Vancamp, P., Gothié, J.-D., Luongo, C., Sébillot, A., Le Blay, K., Butruille, L., et al. (2019). Gender-specific effects of transthyretin on neural stem cell fate in the subventricular zone of the adult mouse. *Scientific Reports*, *9*, 19689.

Vara, H., Muñoz-Cuevas, J., & Colino, A. (2003). Age-dependent alterations of long-term synaptic plasticity in thyroid-deficient rats. *Hippocampus*, *13*, 816–825.

Vigers, A. J., Amin, D. S., Talley-Farnham, T., Gorski, J. A., Xu, B., & Jones, K. R. (2012). Sustained expression of brain-derived neurotrophic factor is required for maintenance of dendritic spines and normal behavior. *Neuroscience*, *212*, 1–18.

Viguié, C., Battaglia, D. F., Krasa, H. B., Thrun, L. A., & Karsch, F. J. (1999). Thyroid hormones act primarily within the brain to promote the seasonal inhibition of luteinizing hormone secretion in the ewe. *Endocrinology*, *140*, 1111–1117.

Vilar, M., & Mira, H. (2016). Regulation of neurogenesis by neurotrophins during adulthood: Expected and unexpected roles. *Frontiers in Neuroscience*, *10*, 26.

Volterra, A., & Meldolesi, J. (2005). Astrocytes, from brain glue to communication elements: The revolution continues. *Nature Reviews. Neuroscience*, *6*(8), 626–640.

Wallis, K., Dudazy, S., van Hogerlinden, M., Nordström, K., Mittag, J., & Vennström, B. (2010). The thyroid hormone receptor α1 protein is expressed in embryonic postmitotic neurons and persists in most adult neurons. *Molecular Endocrinology*, *24*, 1904–1916.

Walton, J. C., Aubrecht, T. G., Weil, Z. M., Leuner, B., & Nelson, R. J. (2014). Photoperiodic regulation of hippocampal neurogenesis in adult male white-footed mice (Peromyscus leucopus). *The European Journal of Neuroscience*, *40*, 2674–2679.

Walton, J. C., Chen, Z., Travers, J. B., & Nelson, R. J. (2013). Exogenous melatonin reproduces the effects of short day lengths on hippocampal function in male white-footed mice, Peromyscus leucopus. *Neuroscience*, *248*, 403–413.

Walton, J. C., Chen, Z., Weil, Z. M., Pyter, L. M., Travers, J. B., & Nelson, R. J. (2011). Photoperiod-mediated impairment of long-term potention and learning and memory in male white-footed mice. *Neuroscience*, *175*, 127–132.

Walton, J. C., Pyter, L. M., Weil, Z. M., & Nelson, R. J. (2012). Photoperiod mediated changes in olfactory bulb neurogenesis and olfactory behavior in male white-footed mice (Peromyscus leucopus). *PLoS One*, 7, e42743.

Wang, C., Liu, F., Liu, Y.-Y., Zhao, C.-H., You, Y., Wang, L., et al. (2011). Identification and characterization of neuroblasts in the subventricular zone and rostral migratory stream of the adult human brain. *Cell Research*, *21*, 1534–1550.

Webster, J. R., Suttie, J. M., & Corson, I. D. (1991). Effects of melatonin implants on reproductive seasonality of male red deer (Cervus elaphus). *Journal of Reproduction and Fertility*, *92*(1), 1–11.

Wehr, T. A. (1997). Melatonin and seasonal rhythms. *Journal of Biological Rhythms*, *12*, 518–527.

Wehr, T. A. (2001). Photoperiodism in humans and other primates: Evidence and implications. *Journal of Biological Rhythms*, *16*, 348–364.

Weil, Z. M., Hotchkiss, A. K., Gatien, M. L., Pieke-Dahl, S., & Nelson, R. J. (2006). Melatonin receptor (MT1) knockout mice display depression-like behaviors and deficits in sensorimotor gating. *Brain Research Bulletin*, *68*, 425–429.

Weissleder, C., Fung, S. J., Wong, M. W., Barry, G., Double, K. L., Halliday, G. M., et al. (2016). Decline in proliferation and immature neuron markers in the human subependymal zone during aging: Relationship to EGF- and FGF-related transcripts. *Frontiers in Aging Neuroscience*, *8*, 274.

Werner, L., Müller-Fielitz, H., Ritzal, M., Werner, T., Rossner, M., & Schwaninger, M. (2012). Involvement of doublecortin-expressing cells in the arcuate nucleus in body weight regulation. *Endocrinology*, *153*, 2655–2664.

Wickings, E. J., & Nieschlag, E. (1980). Seasonality in endocrine and exocrine testicular function of the adult rhesus monkey (Macaca mulatta) maintained in a controlled laboratory environment. *International Journal of Andrology*, *3*, 87–104.

Wilcoxon, J. S., Nadolski, G. J., Samarut, J., Chassande, O., & Redei, E. E. (2007). Behavioral inhibition and impaired spatial learning and memory in hypothyroid mice lacking thyroid hormone receptor α. *Behavioural Brain Research*, *177*, 109–116.

Winner, B., Cooper-Kuhn, C. M., Aigner, R., Winkler, J., & Kuhn, H. G. (2002). Long-term survival and cell death of newly generated neurons in the adult rat olfactory bulb. *The European Journal of Neuroscience*, *16*, 1681–1689.

Wirz-Justice, A., Graw, P., Kräuchi, K., Sarrafzadeh, A., English, J., Arendt, J., et al. (1996). 'Natural' light treatment of seasonal affective disorder. *Journal of Affective Disorders*, *37*, 109–120.

Wolff, J. O., & Durr, D. S. (1986). Winter nesting behavior of Peromyscus leucopus and Peromyscus maniculatus. *Journal of Mammalogy*, *67*, 409–412.

Wood, S., & Loudon, A. (2014). Clocks for all seasons: Unwinding the roles and mechanisms of circadian and interval timers in the hypothalamus and pituitary. *The Journal of Endocrinology*, *222*(2), R39–R59.

Woon, F. L., Sood, S., & Hedges, D. W. (2010). Hippocampal volume deficits associated with exposure to psychological trauma and posttraumatic stress disorder in adults: A meta-analysis. *Progress in Neuro-Psychopharmacology & Biological Psychiatry*, *34*, 1181–1188.

Workman, J. L., Bowers, S. L., & Nelson, R. J. (2009). Enrichment and photoperiod interact to affect spatial learning and hippocampal dendritic morphology in white-footed mice (Peromyscus leucopus). *The European Journal of Neuroscience*, *29*, 161–170.

Wu, Y.-H., Ursinus, J., Zhou, J.-N., Scheer, F. A. J. L., Ai-Min, B., Jockers, R., et al. (2013). Alterations of melatonin receptors MT1 and MT2 in the hypothalamic suprachiasmatic nucleus during depression. *Journal of Affective Disorders*, *148*, 357–367.

Xing, Y. L., Röth, P. T., Stratton, J. A. S., Chuang, B. H. A., Danne, J., Ellis, S. L., et al. (2014). Adult neural precursor cells from the subventricular zone contribute significantly to oligodendrocyte regeneration and remyelination. *The Journal of Neuroscience*, *34*(42), 14128–14146.

Xu, Y., Tamamaki, N., Noda, T., Kimura, K., Itokazu, Y., Matsumoto, N., et al. (2005). Neurogenesis in the ependymal layer of the adult rat 3rd ventricle. *Experimental Neurology*, *192*, 251–264.

Yao, W., Liu, W., Deng, K., Wang, Z., Wang, D.-H., & Zhang, X.-Y. (2019). GnRH expression and cell proliferation are associated with seasonal breeding and food hoarding in Mongolian gerbils (Meriones unguiculatus). *Hormones and Behavior, 112*, 42–53.

Yeung, M. S. Y., Zdunek, S., Bergmann, O., Bernard, S., Salehpour, M., Alkass, K., et al. (2014). Dynamics of oligodendrocyte generation and myelination in the human brain. *Cell, 159*, 766–774.

Yoo, S., Cha, D., Kim, S., Jiang, L., Cooke, P., Adebesin, M., et al. (2020). Tanycyte ablation in the arcuate nucleus and median eminence increases obesity susceptibility by increasing body fat content in male mice. *Glia, 68*(10), 1987–2000.

Yoshimura, T., Yasuo, S., Watanabe, M., Iigo, M., Yamamura, T., Hirunagi, K., et al. (2003). Light-induced hormone conversion of T4 to T3 regulates photoperiodic response of gonads in birds. *Nature, 426*, 178–181.

Young, R. A., Rajatanavin, R., Braverman, L. E., & Tennant, B. U. D. C. (1986). Seasonal changes in serum thyroid hormone binding proteins in the woodchuck (Marmota monax). *Endocrinology, 119*, 967–971.

Zhang, L., Cooper-Kuhn, C. M., Nannmark, U., Blomgren, K., & Kuhn, H. G. (2010). Stimulatory effects of thyroid hormone on brain angiogenesis in vivo and in vitro. *Journal of Cerebral Blood Flow and Metabolism, 30*, 323–335.

Zilkha-Falb, R., Kaushansky, N., & Ben-Nun, A. (2020). The median eminence, A new oligodendrogenic niche in the adult mouse brain. *Stem Cell Reports, 14*, 1076–1092.

Zoeller, R. T., & Rovet, J. (2004). Timing of thyroid hormone action in the developing brain: Clinical observations and experimental findings. *Journal of Neuroendocrinology, 16*, 809–818.

Zou, J., Wang, W., Pan, Y. W., Lu, S., & Xia, Z. (2015). Methods to measure olfactory behavior in mice. *Current Protocols in Toxicology, 63*, 11.18.1–11.18.21.

CHAPTER SIX

Insights from single cell studies of human pancreatic islets and stem cell-derived islet cells to guide functional beta cell maturation *in vitro*

Natasha Hui Jin Ng[a], Claire Wen Ying Neo[a,b], Shirley Suet Lee Ding[a], and Adrian Kee Keong Teo[a,b,c,*]

[a]Stem Cells and Diabetes Laboratory, Institute of Molecular and Cell Biology (IMCB), A*STAR, Proteos, Singapore, Singapore
[b]Department of Medicine, Yong Loo Lin School of Medicine, National University of Singapore, Singapore, Singapore
[c]Department of Biochemistry, Yong Loo Lin School of Medicine, National University of Singapore, Singapore, Singapore
*Corresponding author: e-mail addresses: ateo@imcb.a-star.edu.sg; drainteo@gmail.com

Contents

Abstract

There is now a sizeable number of single cell transcriptomics studies performed on human and rodent pancreatic islets that have shed light on the unique gene signatures and level of heterogeneity within each individual islet cell type. Following closely from

Vitamins and Hormones, Volume 116
ISSN 0083-6729
https://doi.org/10.1016/bs.vh.2021.02.011

these studies, there is also rapidly-growing activity on characterizing islet-like cells derived from *in vitro* differentiation of human pluripotent stem cells (hPSCs) at the single cell level. The overall consensus across the studies so far suggests that the first few stages of differentiation are largely uniform, whereas during pancreatic endocrine commitment, cell trajectories start to diverge, resulting in multiple end-stage pancreatic cells that include progenitor-like, endocrine and non-endocrine cells. Comprehensive transcriptional profiling is important for understanding how and why islet cells, especially the insulin-secreting beta cells, exist in subpopulations that differ in maturity, proliferation rate, sensitivity to stress, and insulin secretion function. For hPSC-derived beta cells to be used confidently for cell therapy, optimal differentiation and thorough characterization is required. The key questions to address are—What is the trajectory of differentiation? Is heterogeneity a natural occurrence or is it a consequence of imperfect differentiation protocols? Can lessons be drawn from the extensive single cell transcriptomic data to help guide maturation of beta cells *in vitro*? This book chapter seeks to address some of these questions, and facilitate ongoing efforts in improving the beta cell differentiation pipeline or enriching for desired beta cell populations following differentiation, to make way for better mechanistic studies and future clinical translation.

1. Introduction

Since the advent of single cell sequencing technologies a decade ago, we are now able to collect gene expression profiles at single cell resolution, with the appreciation that individual cells within a given cell population can exhibit differences in transcriptional signature and functional states. Specifically, the pancreatic islets, which are central to glucose homeostasis, are made up of different cell types that each have a unique set of cellular markers to direct a specialized function. Single cell analyses performed on pancreatic islets from both humans and rodents have not only enabled the precise, in-depth characterization of the different cell populations contained within the islet, but also revealed new knowledge of the extent of cellular heterogeneity even within each islet cell type. The pancreatic beta cells are of particular scrutiny due to their fundamental role in releasing insulin in response to nutrient intake, to achieve glucose homeostasis. Single cell RNA-Sequencing (scRNA-Seq) studies have shed light on the existence of different beta cell subpopulations, identified markers that can be used to detect and isolate beta cells, and pinpointed alterations in beta cell expression profiles in the context of diabetes, including that of type 2 diabetes (T2D).

In more recent years, scRNA-Seq methodologies have also been used by stem cell scientists to characterize the differentiation process of human pluripotent stem cells (hPSCs) into islet cells. Several protocols have been established to derive a renewable resource of pancreatic beta cells from hPSCs, through mimicking pancreas development in an *in vitro* environment (Pagliuca et al., 2014; Rezania et al., 2014; Russ et al., 2015; Veres et al., 2019). Using conventional techniques such as flow cytometry and immunohistochemistry (IHC) staining, which are essentially analyses of single cells, we know that differentiating cells do not necessarily respond equally to environmental signals (in this case media conditions), and as a result, are not all molecularly or functionally identical. Therefore, single cell transcriptomics can be used to evaluate the level of heterogeneity during beta cell differentiation, at least at the transcript level, in a global and unbiased manner. The scRNA-Seq studies on hPSC-derived beta-like cells can firstly be used to identify genes and pathways involved in beta cell development that can be modulated to achieve better differentiation outcomes, and secondly to identify markers that can be used to enrich for specific subpopulations of cells that are more mature or have a superior functional signature. As current *in vitro* differentiation experiments still give rise to beta cells that resemble functionally-immature cells with limited glucose responsiveness, it is the hope that these studies can shed light on the mechanisms involved in beta cell maturation and identify ways to develop hPSC-derived cells that more closely resemble *bona fide* human beta cells.

In this book chapter, we will cover the current understanding of human pancreas development and relate that to how beta cells are being developed in the lab. We then summarize the scRNA-Seq studies in primary human pancreatic islets and hPSC-derived islet cells, and place the findings in the perspective of beta cell development and maturation. The insights gathered may support the ongoing research in beta cell regeneration. Of note, we recommend reading a recent review article which provided a thorough discussion on the technical limitations of scRNA-Seq (Mawla & Huising, 2019), that emphasized the need to interpret study results with caution while advocating validation using multiple means.

2. Human pancreatic islet development and (dys) function

2.1 Human pancreas development

The human pancreas consists of exocrine and endocrine compartments, with the exocrine portion constituting the majority of total pancreatic mass. The endocrine compartment is primarily made up of the islets of Langerhans, which account for only ~2% of the adult human pancreas, and are comprised of the insulin-producing beta cells and other hormone-producing cells such as alpha, delta, epsilon, and pancreatic polypeptide cells (Mastracci & Sussel, 2012) (Fig. 1). The endocrine cell types are uniformly distributed in the human islet, favoring heterologous cell-cell contacts. This cellular arrangement has been linked to implications for islet cell function (Cabrera et al., 2006). Insulin secreted from beta cells and glucagon secreted from alpha cells are the primary regulators of systemic glucose homeostasis. A subset of only 1%–10% of the beta cell population, now coined as "hub" cells (Johnston et al., 2016), was recently reported to be dominant regulators of insulin release dynamics across the islet in response to glucose. A better understanding of the developmental processes that govern endocrine cell formation and compartmentalization will therefore help to enhance our current attempts to generate hPSC-derived beta cells that better resemble those found in primary human islets.

Taking reference from how the human endocrine pancreas develops (based on very limited studies of human embryonic and fetal tissue) will help to inform us on the necessary steps to recapitulate this process *in vitro*. Human pancreas organogenesis begins from the definitive endoderm at Carnegie stage (CS) 9 (approximately 22–26 days post-conception) which invaginates to form the foregut endoderm that lies adjacent to the notochord. Notochord patterning of the dorsal foregut permits the development of the dorsal and ventral pancreatic buds by excluding sonic hedgehog (SHH) signaling (Jennings, Berry, Strutt, Gerrard, & Hanley, 2015). At CS 13, the pancreatic buds express the fate-specifying transcription factors PDX1, SOX9 and GATA4, all of which are essential for early human pancreatic development (Jennings et al., 2013). The multipotent pancreatic progenitors subsequently grow and expand, mediated by WNT signaling and YAP-TEAD-dependent mechanisms (Cebola et al., 2015), after which proliferation rate declines. During this period, differences in the fates of the NKX6.1+ pancreatic progenitor populations also arise, whereby peripheral "tip" cells are directed towards acinar cell fate and eventually lose NKX6.1, whereas central "trunk" cells are directed towards duct or endocrine cell fates (Jennings et al., 2015). Unlike that observed during mouse embryonic development, pancreatic endocrine differentiation in humans do not proceed through two temporal waves (known as the primary and secondary transitions), characterized by the expression of Neurogenin 3 (encoded by *NEUROG3*), a pro-endocrine basic-helix-loop-helix (bHLH) transcription factor (Jennings et al., 2013). Instead, NEUROG3, which is the key driver of endocrine commitment, arises towards the end of embryogenesis

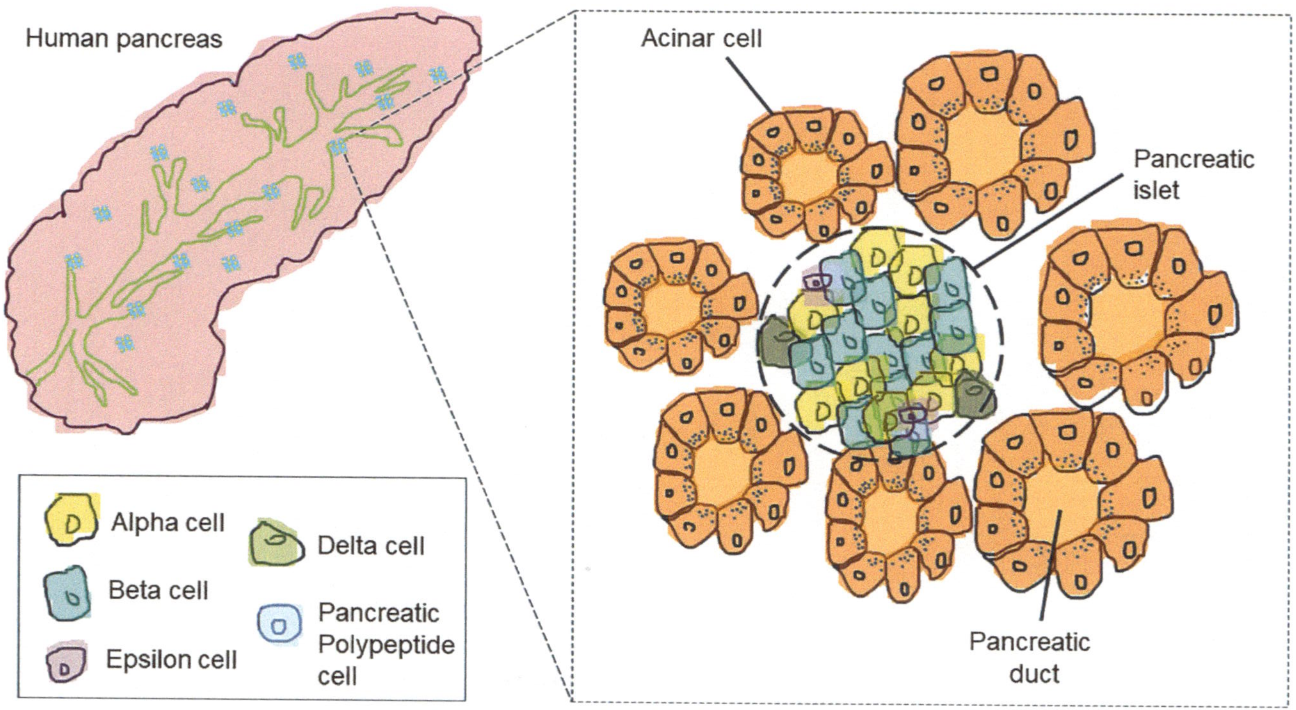

Fig. 1 Anatomical organization of the pancreas. The human pancreas consists of both exocrine and endocrine compartments. The endocrine pancreas is organized into islets of Langerhans that are made up of multiple different types of hormone-producing cells. The exocrine pancreas, which is composed of acinar and ductal epithelial cells, accounts for >95% of the adult pancreatic mass.

(CS 21–22) in a single phase, and peaks right after the embryonic period. At ~8 weeks post conception (wpc), the first signs of fetal beta cells appear, though some insulin positivity can already be detected before that. Other islet cell types that display unique hormone and transcription factor expression profiles subsequently form, and may only be fully resolved from their acinar (GATA4-retaining) and ductal (SOX9-retaining) counterparts at ~14 wpc (Jennings et al., 2013). Consistent with that observed in the mouse pancreas, NEUROG3 expression in the human endocrine pancreas is transient and is switched off before birth (Salisbury et al., 2014).

Although current beta cell differentiation protocols have benefitted greatly from the understanding of human pancreas development, and can now generate insulin-expressing beta-like cells, the expression of insulin and canonical beta cell transcription factors alone is not sufficiently indicative of a mature and functional beta cell. A glucose-stimulated insulin secretion (GSIS) functional assay is typically also required to demonstrate functionality. In addition, increased expression of several categories of functional beta cell markers, as shown in Fig. 2, may also indirectly signify characteristics of beta cells that are functionally mature (Nair, Tzanakakis, & Hebrok, 2020).

2.2 Beta cell replacement for treating diabetes

In the context of disease, type 1 diabetes mellitus (T1D) and T2D arise as the pancreatic beta cells are unable to meet the demand of insulin secretion due

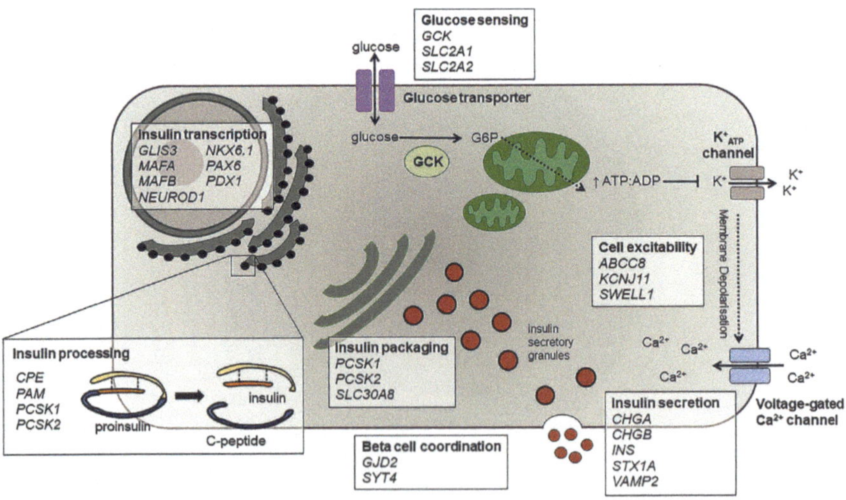

Fig. 2 Features and gene markers that are associated with a functionally mature beta cell phenotype.

to either an almost complete loss, or a deficit of functional beta cells in T1D and T2D respectively. In T2D, this is further compounded by peripheral insulin resistance. The mechanisms underlying the loss of beta cell mass and/or function are being intensively studied, and include both environmental stress and genetic susceptibility factors. Exogenous insulin administration through daily injections is a standard therapy required to achieve glucose control in T1D patients and some late-stage T2D patients. However, patients on life-long insulin therapy often suffer from discomfort, poor compliance, and also risk of hypoglycemia. Thus, alternative therapeutic strategies to replenish the loss of beta cell mass or restore beta cell function are needed to tackle the disease in the long run. Therefore, ongoing efforts to replace the beta cells lost in T1D, either through primary human islet transplantation (an approved therapy in Canada) or stem cell-derived islet cell transplantation (undergoing clinical trials), have made the pancreatic islets a key focus of recent regenerative medicine efforts (Tan, Lau, Tan, Khoo, & Teo, 2020).

Currently, transplantation of human islets from cadaveric donors represents a promising approach for the restoration of beta cell mass in T1D patients (Ng, Tan, Koh, & Teo, 2019). Through long-term follow-up (up to 10 years) of islet transplantation outcomes in a group of T1D patients, it was reported that a majority of subjects continued to show persistent production of insulin/C-peptide (Brennan et al., 2016). However, only a small subset of these patients achieved partial to complete insulin independence. The beneficial effects of islet transplantation appear to remain short-term as islet graft function deteriorates over time, reverting patients to full insulin dependency. Additionally, donor islets are limited in supply and also require concurrent use of life-long immunosuppression (McCall & Shapiro, 2012), thus making islet transplantation a potentially unsustainable approach. In the search for comparable islet surrogates, hPSCs provide a renewable source of insulin-secreting beta cells that can potentially be used to combat diabetes, and also aid the investigation of mechanisms underlying pancreatic development and diabetes pathophysiology.

3. Modeling human pancreas development using human pluripotent stem cells

hPSC-derived beta cells are an important source of human beta-like cells for both investigative studies and regenerative medicine applications. Over the past 6 years, new and improved protocols that have been developed are capable of generating functional insulin-secreting beta-like cells that express PDX1, NKX6.1, and other mature beta cell markers in a highly scalable manner (Pagliuca et al., 2014; Rezania et al., 2014; Russ et al., 2015; Veres et al., 2019). Various research groups are however still performing continuous optimizations to improve the efficiency of differentiation at each stage along the pancreatic lineage. The stepwise differentiation involves sequential activation and repression of signaling cues through the use of recombinant growth factors and small molecules to guide hPSCs toward the formation of beta-like cells (Loo, Lau, Jasmen, Lim, & Teo, 2018) (Tables 1 and 2). As summarized in Fig. 3, the journey begins with the induction of the definitive endoderm through activation of the Wnt/β-catenin and transforming growth factor β (TGF-β) signaling pathways (Scheibner, Bakhti, Bastidas-Ponce, & Lickert, 2019), generation of the foregut endoderm through FGF signaling, followed by pancreatic specification contributed by repression of Wnt/β-catenin and SHH signaling, BMP inhibition and activation of retinoic acid (RA) signaling. The pancreatic progenitors, which express PDX1, SOX9 and lack NKX2.2, appear to mimic human pancreas development (Jennings et al., 2013; Rezania et al., 2014; Russ et al., 2015). The cells are then directed towards the endocrine program through a combination of activating MAPK/ERK signaling, inhibiting BMP, Notch and TGF-β signaling, and also introducing the thyroid hormone triiodothyronine (T3) (Fig. 3). Endocrine differentiation from the PDX1 + NKX6.1 + NEUROG3 + endocrine progenitor population ultimately gives rise to fetal-like beta-like cells that express the crucial beta cell transcription factor MAFA and are monohormonal for insulin, and also bihormonal cells that are positive for both insulin and glucagon (Pagliuca et al., 2014; Rezania et al., 2014; Russ et al., 2015). The hPSC-derived beta-like cells generally showed modest insulin secretion in response to glucose stimulation, though in some reports they are sufficient for restoring glucose homeostasis in diabetic rodents upon xenotransplantation. In many cases, further *in vivo* maturation of these hPSC-derived beta-like cells is required to achieve full functionality. It is therefore clear that there are still challenges in achieving fully functional beta cell maturation *in vitro*.

More recent studies have demonstrated greater success in generating functional beta-like cells *in vitro* through re-introduction of Activin A to modulate TGF-β signaling at Stage 4 (pancreatic endoderm), and removal of serum and exogenous signaling factors during the final stage (Stage 6) of the differentiation (Velazco-Cruz et al., 2019; Veres et al., 2019) (Table 1). The authors who were involved in the initial protocol development now

Table 1 Comparison of several well-established beta cell differentiation protocols and the small molecules used during differentiation.

Stage	Ramond et al. (2018)	Velazco-Cruz et al. (2019)	Nair et al. (2019)	Veres et al. (2019)
Stage 1 (Definitive endoderm)	GDF8 GSK	Activin A CHIR99021		
Stage 2 (Primitive gut tube)	FGF7 (KGF)			
Stage 3 (Posterior foregut)	FGF7 (KGF) RA SANT1 LDN193189			
	TPB	PdbU Y27632		
Stage 4 (Pancreatic progenitors)	FGF7 (KGF) RA SANT1			
	Ascorbic acid TPB LDN193189	Y27632 Activin A		
Stage 5 (Endocrine progenitors)	SANT1 RA Alk5i T3			
	LDN193189	XXI Betacellulin		
Stage 6 (Beta-like cells)	Alk5i T3			
	LDN193189 GSiXX N-Cys AXL inh	CMRL supplemented only		

GDF8, Growth Differentiation Factor 8; GSK, Glycogen Synthase Kinase; CHIR99021, GSK3 inhibitor; FGF7, Fibroblast Growth Factor 7; KGF, Keratinocyte Growth Factor, TPB; protein kinase C activator; PdbU, Phorbol 12,13-dibutyrate; Y27632, Rho-associated coiled-coil containing protein kinase (ROCK) inhibitor; RA, retinoic acid; SANT1, hedgehog signaling pathway antagonist; LDN193189, BMP pathway inhibitor; XXI, γ-Secretase inhibitor; ALK5i, TGF-β RI kinase inhibitor; T3, 3,3′,5-Triiodo-L-thyronine; GSiXX, γ-Secretase inhibitor; N-Cys, *N*-acetyl cysteine; AXL inh, AXL inhibitor.

Table 2 Growth factors and small molecules used during differentiation of hPSCs into beta cells, their molecular targets and biological effects.

Factor	Primary function	Reported effects on differentiation
Activin A	TGF-β agonist	Support cell proliferation and drive definitive endoderm formation
CHIR99021	Activator of WNT signaling	Support cell proliferation and drive definitive endoderm formation
Fibroblast Growth Factor 7 (FGF7)/ Keratinocyte Growth Factor (KGF)	FGFR2b agonist	Improves survival, proliferation of cells and expression of PDX1 and NKX6.1
Retinoic Acid (RA)	RARα,β,γ agonist	High concentration induces PDX1 expression during Stage 3; low concentration maintains PDX1 and induces/maintains NKX6.1 expression during Stages 4 and 5
SANT1	Smoothened antagonist—inhibits Hedgehog signaling	Improves PDX1 and NKX6.1 expression
LDN193189 (LDN)	BMP type 1 receptor inhibitor	Improves PDX1 expression
Phorbol 12,13-dibutyrate (PdbU)	Protein kinase C activator	Improves PDX1 expression
Y-27632	Rho-associated coiled-coil containing protein kinase (ROCK) inhibitor	Increases insulin activity, regulates glucose homeostasis and promotes beta cell maturation
XXI	γ-Secretase inhibitor—inhibits Notch signaling	Induces endocrine differentiation

Alk5 Receptor Inhibitor II (Alk5i)	Alk5 inhibitor—inhibits TGF-β/Activin signaling	Induces endocrine differentiation, maintains NKX6.1 expression in endocrine cells and improves function of stem cell-derived beta cells
Triiodothyronine (T3)	Thyroid receptor agonist	Increases insulin expression in NKX6.1-expressing endocrine cells and improves function of stem cell-derived beta cells
Betacellulin	Erbb1 (EGFR) and Erbb4 agonist	Improves maintenance of NKX6.1 expression in endocrine cells
High glucose	NA	Increases endocrine differentiation and insulin expression

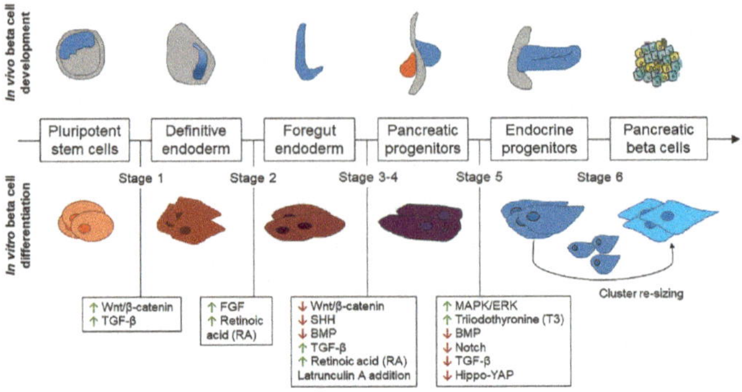

Fig. 3 Illustration of the key stages of pancreatic beta cell differentiation *in vivo* (top) and *in vitro* (bottom). The signaling pathways that are activated/upregulated or repressed/downregulated to drive each stage of the differentiation process, and other relevant strategies to achieve better differentiation efficiency based on the established protocols in Table 1 are highlighted.

suggest that Alk5i, a TGF-β inhibitor, is necessary at Stage 5 (endocrine progenitors) but limits beta cell functional maturation at Stage 6 (Velazco-Cruz et al., 2019). Interestingly, recent protocols also eliminated the use of T3, which was reported earlier to drive the induction of mature beta cell markers (Millman et al., 2016; Pagliuca et al., 2014; Rezania et al., 2014), in the final stage of differentiation (Velazco-Cruz et al., 2019; Veres et al., 2019). Besides adjusting media conditions, re-aggregation and re-sizing of cell clusters was also found to enrich for beta cells and enhance beta cell function, likely due to improved compartmentalization of endocrine cell populations and increased mitochondrial metabolic activity (Nair et al., 2019; Velazco-Cruz et al., 2019; Veres et al., 2019). Other recent developments include modulation of actin polymerization (using latrunculin A treatment) to influence NEUROG3-mediated endocrine differentiation (Hogrebe, Augsornworawat, Maxwell, Velazco-Cruz, & Millman, 2020), and targeting of the Hippo pathway through downregulation of YAP to control proliferation and enhance endocrine progenitor differentiation (Rosado-Olivieri, Anderson, Kenty, & Melton, 2019). Though current methods often suffer from several drawbacks that hinder the clinical translation of the technology, including heterogeneous generation of progenitors at differing developmental stages and generation of polyhormonal and/or non-endocrine cells, the rapid advancements in this field to further fine-tune differentiation protocols and/or devise effective purification strategies are steps in the right direction.

Single cell technologies have at the same time opened up a new avenue to study both primary pancreatic islet tissue and hPSC-derived islet cells. Single cell analyses facilitate the characterization of cellular heterogeneity within the human tissues *via* unbiased transcriptional profiling of hundreds of thousands of single cells. Multiple groups have now performed scRNA-Seq on juvenile and adult human islets, including non-diabetic and diabetic human islets, fetal pancreatic tissue, rodent islets, as well as various cells from different developmental stages during *in vitro* beta cell differentiation. A summary of the insights into islet cell biology using these approaches, and a discussion of their potential to guide functional beta cell maturation *in vitro*, will be a key focus of the next few sections.

4. Insights into human pancreas biology from single cell transcriptomics studies of pancreatic islets

In the past 4 years alone, there have been at least 25 studies dedicated to using scRNA-Seq technologies to characterize primary pancreatic islet tissue, including both human and rodent tissues, from both healthy and diabetic donors (Table 3). The scRNA-Seq platforms used are typically fluorescence-activated cell sorting (FACS)-based, microfluidics-based (Fluidigm C1), or droplet-based (InDrops, 10 × Genomics), that all aim to achieve single cell separation. The abundance of large, high resolution, transcriptional (and some protein level) data collected over the past few years presents an excellent opportunity to consolidate these findings. Besides scRNA-Seq, we also share complementary efforts employing mass cytometry that characterize cells based on a large panel of selected protein markers (Table 4). Globally, it appears that cell clustering is primarily governed by cell type-specific transcriptome signatures, above and beyond any inter-individual/donor differences or T2D status. Many studies also stated that the enriched genes that were identified within the beta and alpha cell populations corroborated with other studies. While this may appear consistent at first, upon closer look, the level of overlap found is surprisingly limited (Mawla & Huising, 2019). One concern is that the high level of technical noise within scRNA-Seq studies could render it difficult to identify true biological signals. Our efforts here will attempt to consolidate current findings and present the repeatedly-observed gene expression profiles to guide future hPSC-derived beta cell work (Table 5).

Table 3 Summary of scRNA-Seq studies of primary pancreatic tissue in humans and mice.

Study reference	Tissue	Number of non-diabetic (ND) and diabetic adult tissues (unless otherwise specified), or mouse strain	Single cell capture technology	Total number of cells sequenced	Number of cells used in final profiling	Number of reads/cell
Human tissues						
Li et al. (2016)	Islets	1 ND	FACS	70	64	12.7M
Wang et al. (2016)	Islets	3 ND, 1 T1D, 2 T2D 2 ND (children)	Fluidigm C1	635	430	2.2M
Segerstolpe et al. (2016)	Islets	6 ND, 4 T2D	FACS	3386	2209	0.75M
Xin et al. (2016)	Islets	12 ND, 6 T2D	Fluidigm C1	3709	1492	0.95M
Baron et al. (2016)	Islets	4 ND	inDrop	~10,000	8629	0.1M
Muraro et al. (2016)	Islets	4 ND	FACS	3072	Not disclosed	0.041M
Lawlor et al. (2017)	Islets	5 ND, 3 T2D	Fluidigm C1	1050	638	3M
Dorajoo et al. (2017)	Islets	3 ND	Fluidigm C1	448	276	3.68M
Enge et al. (2017)	Islets	8 ND	FACS	2680	2544	0.93M
Xin et al. (2018)	Islets	12 ND	10 × Genomics	20,811	19,174	Not disclosed
Tyler et al. (2019)	Islets	3 ND	Fluidigm C1	247	185	2.84M

Fang et al. (2019)	Islets	6 ND, 3 T2D	Drop-Seq	39,905	28,026	Not disclosed
Marquina-Sanchez et al. (2020)	Islets	4 ND	10 × Genomics	127,032	103,303	Not disclosed
Camunas-Soler et al. (2020)	Islets	21 ND, 7 T2D, 3 T1D	Patch-Seq and FACS	5204	4887	1–2M
Mouse tissues						
Baron et al. (2016)	Islets	C57BL/6 and ICR	inDrop	~2000	1886	0.1M
Xin, Okamoto, et al. (2016)	Islets	C57BL/6 (3 months and 26 months)	Fluidigm C1	207	Not disclosed	1.4M
Xin et al. (2016)	Islets	C57BL/6	Fluidigm C1	622	341	1M
Zeng et al. (2017)	Beta cells	mIns1-H2B-mCherry on P1, P7, P14, P21, P28	FACS, Fluidigm C1	401	387	4.3M
Qiu et al. (2017)	Alpha and beta cells	Ins1-RFP and Gcg-Cre; Rosa-RFP on E17.5, P0, P3, P9, P15, P18, P60	FACS	891	866	1–3M
Krentz et al. (2018)	Embryonic pancreata	Neurog3-Cre; Rosa26mT/mG on E15.5, E18.5	10 × Genomics	14,525	13,531	> 0.05M
Li et al. (2018)	Pdx1-expressing cells in dorsal/ventral pancreatic regions	Pdx1-GFP on E9.5, E10.5	FACS	232	Not disclosed	1.7M

Continued

Table 3 Summary of scRNA-Seq studies of primary pancreatic tissue in humans and mice.—cont'd

Study reference	Tissue	Number of non-diabetic (ND) and diabetic adult tissues (unless otherwise specified), or mouse strain	Single cell capture technology	Total number of cells sequenced	Number of cells used in final profiling	Number of reads/cell
Byrnes et al. (2018)	Embryonic pancreata	Swiss Webster, Fev-Cre; ROSA26mTmG on E12.5, E14.5, E17.5	10 × Genomics	93,800	Not disclosed	Not disclosed
Scavuzzo et al. (2018)	Embryonic pancreata	ICR, Ngn3-eGFP on E14.5, E16.5	Droplet-based, Fluidigm C1	Not disclosed	Not disclosed	Not disclosed
Bastidas-Ponce et al. (2019)	Embryonic pancreata	Ngn3-Venus fusion on E12.5, E13.5, E14.5, E15.5	FACS, 10 × Genomics	36,351	Not disclosed	Not disclosed
Sharon et al. (2019)	Embryonic pancreata	Ngn3-eGFP on E13.5, E14.5, E15.5	FACS	440	Not disclosed	Not disclosed
Marquina-Sanchez et al. (2020)	Islets	Not disclosed	10 × Genomics	33,913	24,729	Not disclosed

Table 4 Summary of non-RNA-Seq-based single cell studies of primary pancreatic tissue in humans and rodents.

Study reference	Tissue	Number of non-diabetic (ND) and diabetic tissues	Technology platform
Wang et al. (2016)	Human islets	17 ND, 3 T2D	Mass cytometry (24 antibodies)
Jansson, Comi, Rubakhin, and Sweedler (2016)	Rat islets (3506 cells)	Sprague–Dawley	MALDI mass spectrometry
Ramond et al. (2018)	Human fetal pancreas at 9 weeks of development (698 cells)	3 ND	FACS and single cell qPCR
Wang et al. (2019)	Human pancreas sections (2191 islets)	6 ND, 12 T1D	Imaging mass cytometry (33 antibodies)
Damond et al. (2019)	Human pancreas sections (1581 islets)	4 ND, 4 T1D (recent onset), 4 T1D (long-standing)	Imaging mass cytometry (35 antibodies)
Rai et al. (2020)	Human islets (1690 cells)	1 ND	ATAC-Seq
Camunas-Soler et al. (2020)	Human islets (1369 cells)	21 ND, 7 T2D, 3 T1D	Patch-Seq (single cell patch clamp measurement)

4.1 Identification of beta cell heterogeneity and gene signatures

As seen in Table 3, most of the scRNA-Seq studies have focused on healthy islets, while a handful compared islets from both healthy and T2D donors (Fang et al., 2019; Lawlor et al., 2017; Segerstolpe et al., 2016; Wang, Schug, et al., 2016; Xin, Kim, Okamoto, et al., 2016). Besides identifying the predominant endocrine cell types in the scRNA-Seq datasets, a few studies were able to identify rarer cell types or other non-endocrine cells such as acinar, ductal cells, stellate cells and even resident immune cells, based on specific gene markers. The ability to pick up rare cell populations is in part due to differences in sample size (number of cells analyzed) and sequencing depth across studies. In fact, one study detected up to 14 different cell types in or surrounding the pancreatic islets, albeit with a sequencing depth of

Table 5 Genes identified to be beta cell-specific/enriched based on 10 human islet scRNA-Seq studies.

Beta cell-specific/enriched genes identified in at least 6 out of 10 studies	Beta cell-specific/enriched genes identified in only 5 out of 10 studies
ADCYAP1	*AP3B1*
BMP5	*C1orf127*
CAPN13	*CDKN1C*
CASR	*GPM6A*
CYYR1	*GPSM1*
DLK1	*GSN*
ENTPD3	*HSP90AB1*
G6PC2	*HSPD1*
HADH	*IGF2*
IAPP	*KLHDC8A*
INS (all studies)	*RASGRF1*
LRRTM3	*ROBO1*
MAFA	*SAMD11*
MEG3	*SIX3*
NPTX2	*SLC25A34*
PDX1	*SLC30A8*
PFKFB2	*SLC6A6*
RBP4	*SRD5A1*
RGS16	*TFCP2L1*
ROBO2	*TGFBR3*
SCD5	*TMED6*
SORL1	*TMEM150C*
SYT13	
TMEM37	
TSPAN1	
WSCD2	

References: Baron et al. (2016); Dorajoo et al. (2017); Fang et al. (2019); Lawlor et al. (2017); Li et al. (2016); Muraro et al. (2016); Segerstolpe et al. (2016); Wang, Schug, et al. (2016); Xin et al. (2018); Xin, Kim, Okamoto, et al. (2016).

~100,000 reads per cell (Baron et al., 2016), which is considerably lower than most others. A handful of reports also identified islet cell subpopulations, especially within the alpha, beta and exocrine cells, through detailed investigation of the clustering of cells within each cell type. Other studies that did not find evidence of subpopulations within individual islet cell types had suggested that human islet cells are in fact rather homogeneous (Lawlor et al., 2017; Xin, Kim, Okamoto, et al., 2016).

4.1.1 Beta cell heterogeneity

Firstly, multiple beta cell subpopulations have been identified that are governed by differences in expression of genes that are so-called markers of beta cell heterogeneity. Segerstople et al. identified five beta cell sub-clusters with generally similar levels of *INS* that were marked by their expression pattern of five genes, namely *RBP4*, *FFAR4/GPR120*, *ID1*, *ID2* and *ID3* (Segerstolpe et al., 2016). Xin et al. found four beta cell subpopulations, one of which expressed generally lower levels of *INS* and displayed enrichment in genes involved in the endoplasmic reticulum (ER) stress response and protein folding, such as *DDIT3*, *FTL*, *HSPA5* and *SQSTM1*, but with little evidence for differences in apoptosis and de-differentiation genes (Xin et al., 2018). Pseudo-time analysis showed that the beta cells could be ordered based on their levels of *INS* expression and enrichment of unfolded protein response (UPR) genes. In beta cells with low *INS* and a high UPR signature, the superoxide dismutase genes *SOD1* and *SOD2*, genes involved in glutathione and thioredoxin production and other NRF2-dependent anti-oxidant defense mechanisms were particularly enriched (Xin et al., 2018). These cells also had more proliferative signatures and high expression of energy metabolism genes including those involved in glycolysis, the tricarboxylic acid (TCA) cycle, electron transport chain (ETC) and pentose phosphate pathway (PPP). The authors postulated a phenomenon in beta cells that demonstrated possible transition from high INS production to low INS production activity that is coupled with UPR-mediated stress recovery, proliferation and high energy consumption (Xin et al., 2018). In this case, it is also plausible that INS production, which is a major activity in the beta cells, is being tuned by the level of stress response in the cells. Another scRNA-Seq study also described significant beta cell heterogeneity due to varied ER stress pathway genes such as *DDIT3*, *HERPUD1*, and *HSPA5*, as well as beta cell functional genes (Baron et al., 2016). The expression of heat shock genes appears to be consistently identified as a marker of cellular heterogeneity (Fang et al., 2019). A recent meta-analysis evaluated the reported beta cell heterogeneity genes across five human islet scRNA-Seq studies (Baron et al., 2016; Lawlor et al., 2017; Muraro et al., 2016; Segerstolpe et al., 2016; Xin, Kim, Okamoto, et al., 2016) and identified a total of 24 genes that were commonly driving heterogeneous expression in the beta cells from healthy donors (Mawla & Huising, 2019). These included *NPY*, a known beta cell heterogeneity marker, *HERPUD1*, *RBP4*, *DLK1*, and also *GCG*. When beta cells from diabetic donors were also included, the beta cell markers *G6PC2*, *MAFA*, and non-beta cell endocrine markers *SST*, *PPY* and *PRSS1* were also found in the expanded list.

4.1.2 Beta cell-specific gene signatures

The results across the various scRNA-Seq studies also differed in the number of beta cell signature genes, or genes reported to be differentially expressed in beta cells compared to other cells (Table 5). For instance, one study identified a conservative collection of 61 beta cell signature genes (Lawlor et al., 2017), whereas another identified 953 genes (Muraro et al., 2016). Various studies have found the Notch pathway component gene, *DLK1*, also a T1D-associated gene, to be largely restricted to human beta cells (Baron et al., 2016; Li et al., 2016). The specificity of *DLK1* to beta cells is, however, not observed in mouse islets, as *Dlk1* is also found in glucagon-expressing cells in mice (Li et al., 2016). The transcription factors encoded by *SIX2* and *SIX3* have also been identified by multiple studies as novel beta cell marker genes (Baron et al., 2016; Muraro et al., 2016; Segerstolpe et al., 2016). Other apparently highly-specific beta cell transcription factors included *BMI1* (Li et al., 2016), *OLIG1* (Baron et al., 2016), and *ESR1* (Fang et al., 2019), though these were not found in other studies. Establishing the role of these transcriptional regulators could aid in developing and maintaining beta cell identity. Other previously unreported beta cell-enriched genes include *PFKFB2* (thought to regulate insulin secretion), and a group of genes implicated in ER and oxidative stress (*SRNX1*, *SQSTM1*, *FTH1P3*, *FTH1*, *FTL*) that also appeared to be dictating beta cell heterogeneity (Muraro et al., 2016). Beta cell-specific genes involved in the glucose-stimulated insulin secretion (GSIS) pathway, such as *GPD1* (part of the glycerol-3-phosphate shuttle) and *SLC25A34* (mitochondrial solute transporter) were also highlighted (Lawlor et al., 2017). Additionally, cell type-specific surface markers can be used to enrich for, or exclude, specific cell populations using FACS. For instance, CD24 and CD44 were shown to be restricted to acinar and ductal cell clusters and hence negative selection for CD24/CD44 can be used to enrich for endocrine over exocrine cells (Muraro et al., 2016). These strategies would potentially help to improve purification protocols, such as in the context of a mixed population of hPSC-derived islet cells. We have consolidated a consensus list of beta cell-specific/enriched genes based on available information from up to 10 human islet scRNA-Seq studies (Baron et al., 2016; Dorajoo et al., 2017; Fang et al., 2019; Lawlor et al., 2017; Li et al., 2016; Muraro et al., 2016; Segerstolpe et al., 2016; Wang, Schug, et al., 2016; Xin et al., 2018; Xin, Kim, Okamoto, et al., 2016) (Table 5).

4.1.3 Transcriptomic alterations in T2D

In comparing beta cell transcriptomes between non-diabetic and diabetic individuals, a number of genes were found to be altered in T2D. For instance, *FXYD2* was downregulated in T2D, while *GPD2* (involved in mitochondrial metabolism), *DLK1* and *LEPROTL1* were upregulated (Segerstolpe et al., 2016). Perturbations in these genes were suggested to contribute to diabetes pathogenesis. Gene set enrichment analysis also showed that mitochondrial energy metabolism and aerobic respiration was significantly downregulated in most cell types in T2D, while hypoxia, apoptosis, diabetic nephropathy and cytokine signaling pathways were upregulated (Fang et al., 2019; Segerstolpe et al., 2016). In the diabetic state, both beta and alpha cells expressed a more immature gene signature, due to an alteration of cyclins and cell cycle genes such as *CDKN2B*, *BARD1*, and *JUNB*, which could contribute to partial de-differentiation and transcriptional dysregulation in T2D (Wang, Schug, et al., 2016). A different study had however reported no evidence that reflected de-differentiation signatures in T2D (Lawlor et al., 2017). A recent review integrated the gene lists from three scRNA-Seq studies (Lawlor et al., 2017; Segerstolpe et al., 2016; Xin, Kim, Okamoto, et al., 2016) and found poor consensus in the differentially expressed genes in T2D (Wang & Kaestner, 2019). One of these studies also reported that most of the differentially expressed genes in T2D seemed to have no clear link to islet cell growth and function, making it challenging to determine the biological significance of these findings (Xin, Kim, Okamoto, et al., 2016). Overall, it appears that more scRNA-Seq data from T2D samples are needed, and more importantly, study samples should have well-established clinical parameters and appropriate controls to enable more robust comparisons.

4.1.4 Transcriptomic alterations during aging

On top of investigating gene alterations in islets from non-diabetic and diabetic donors, other studies have focused on identifying transcriptomic changes in islets that occur with age. One such study which compared islet cells from young and adult healthy donors found that some beta cell signature genes expressed in adult cells were not expressed in juvenile cells in the same cell type-specific manner (Wang, Schug, et al., 2016). The authors also found that juvenile beta cells "mis-express" adult alpha cell genes, suggesting a beta cell profile that is not fully mature. Another study by Enge et al. tackled this question by evaluating pancreatic cells from eight healthy donors classified into three different age groups (juvenile, young adult and middle-aged) (Enge et al., 2017). They found that transcriptional noise increased with age, potentially due to age-dependent transcriptional instability, that can affect endocrine function. The mutational signatures in endocrine cells that were found to be most correlated with age involved transcription (*TCEB2*), protein synthesis (*RPL36*) and modulation of reactive oxygen species (*ROMO1*). Specifically, the authors showed that both alpha and beta cells not only co-expressed more *INS* and *GCG* transcripts respectively with age, but also stress response genes such as *FOSB*, *HSPA1A* and *JUND* (Enge et al., 2017).

4.1.5 Transcriptomic changes during early pancreas development

Single cell studies on developing human pancreatic tissues are very limited but can provide much needed insight on the human developmental processes *in vivo*. Ramond et al. investigated the human fetal pancreas at 9 wpc, after pancreatic endocrine cell differentiation starts, to characterize the cell populations present during endocrine development (Ramond et al., 2018). Instead of scRNA-Seq, they performed single cell profiling by FACS and qPCR, and were able to identify cells at three different maturation stages, representing (1) pancreatic progenitors, (2) early endocrine progenitors and (3) early endocrine cells. This could help to bring us a step closer to understanding the trajectory of human endocrine development, which is not possible in studies of mature adult islets. The presence of *EGFR*, *ERBB3*, *FGFR2* and *SPROUTY2* in pancreatic progenitors suggested that EGF and FGF pathways are involved in the maintenance and proliferation of the progenitor cells. The pancreatic progenitor population was also enriched for genes linked to cell migration, such as *FZD5*, *TCF7L2*, *GNDF* and *PDGFC*. In contrast, during the endocrine progenitor stage (marked by *NEUROG3*), cell division and cell cycle progression may be blocked due to the upregulation of *CDKN1C* and *RIPPLY3*, followed by the onset of hormone transcripts (Fig. 3). The endocrine cell population displayed a clear upregulation of multiple genes involved in secretory machinery as well as activation of Activin signaling (marked by *ACVR1C*) (Ramond et al., 2018). Importantly, pseudo-time ordering was used to map the developmental trajectory across the three cell populations into three main branches. One branch off the main trunk represented endocrine progenitors, a second branch represented polyhormonal endocrine cells, and the last branch further diverged into two endocrine populations that are likely to mimic the path to beta and alpha/delta cells, respectively. The authors did find rare cells at 10 wpc pancreas sections that co-expressed multiple hormones at the protein level (Ramond et al., 2018).

4.1.6 Other non-RNA-Seq single cell studies

Besides scRNA-Seq, other single cell platforms appear to be the next major focus in single cell analyses in the years to come. Single cell mass cytometry has already progressed the field from measuring transcript level data to protein data for understanding cellular function. Though this is done at much lower throughput, these efforts are important as transcript changes or differences may not necessarily be mirrored at the protein level. Protein-based analysis using mass cytometry benefits from low background due to the use of antibodies conjugated to rare earth metals instead of fluorophores. The first of these studies in dissociated human islets used ~24 antibodies for selected markers, including KI67, a marker of proliferation (Wang, Golson, et al., 2016). Due to significant inter-individual variability, the authors analyzed and presented data by a per donor basis. The authors identified three main clusters of beta cells in multiple donors, two of which contained cells that were KI67+. One of these clusters also had a higher expression of a suite of other proliferative markers such as CD44, CD49F, PDGFRA, pERK1/2, pSTAT3 and pSTAT5. They further reported that beta cells had a lower basal rate of proliferation compared to alpha cells, and that proliferation generally appeared to decrease with age across all endocrine cell types (Wang, Golson, et al., 2016). A more recent study from the same group followed up with 33 markers on pancreas slides that allowed the preservation of spatial resolution and visualization of the head, body and tail regions of the pancreas (Wang et al., 2019). They found that the head region had lower percentages of beta and alpha cells, and higher percentages of pancreatic polypeptide cells compared to the body and tail regions. They also assessed pancreas sections from both healthy and T1D donors, and reported that T1D donors had, as expected, few C-peptide+ cells. However, they detected a cell population with no PDX1, but expressing intermediate NKX6.1 levels and GCG, which seemed to be enriched in T1D tissues (Wang et al., 2019). These could be dysregulated/trans-differentiating alpha cells or dedifferentiated beta cells. Another recent single cell mass cytometry study focused on mapping T1D progression by studying the expression pattern of 35 biomarkers in pancreas tissue obtained from recent onset T1D (<0.5 years) and long term T1D donors (≥ 8 years) (Damond et al., 2019). It showed that in long-term T1D individuals, the pancreata are almost devoid of all beta cells, whereas in recent disease onset, individuals may still retain ~40% to nearly 100% of its fraction of beta cells with mostly preserved islet architecture. Despite that, recent onset T1D cells seem to possess an altered beta cell phenotype as measured by changes in INS, PIN, IAPP and PTPRN expression

(Damond et al., 2019). Consistent with Wang et al., they purported that beta cell destruction in long-standing T1D is preceded by a detectable loss of beta cell markers or beta cell dysfunction (Damond et al., 2019; Wang et al., 2019). Mass cytometry remains a very powerful and complementary tool, and studies like these are likely to continue opening up new ways of exploring islet biology and disease pathology.

4.2 Channeling undesired alpha cells towards beta cell fate

Although both beta cells and alpha cells have contrasting and distinctive functions in regulating blood glucose homeostasis, both cell types originate from a common endocrine precursor (Gradwohl, Dierich, LeMeur, & Guillemot, 2000) and express overlapping transcriptomes and epigenomes (Bramswig et al., 2013). Close examination of existing scRNA-Seq data on alpha cells will provide us with insights into genes that are unique to alpha cells, or that overlap with beta cells, to help us better understand the alpha cell-like signatures that may arise from *in vitro* beta cell differentiation protocols. Furthermore, this knowledge may highlight key alpha cell pathways to repress, to more effectively direct the differentiation of endocrine progenitors towards the beta cell fate. Based on scRNA-Seq studies of human pancreatic islets (Baron et al., 2016; Dominguez Gutierrez et al., 2018; Dorajoo et al., 2017; Fang et al., 2019; Lawlor et al., 2017; Li et al., 2016; Muraro et al., 2016; Segerstolpe et al., 2016; Wang, Schug, et al., 2016; Xin, Kim, Okamoto, et al., 2016), we have identified several commonly-reported alpha cell-specific/enriched genes (Table 6). A common thread among several studies relate to the detection of a subset of alpha cells with proliferative signatures, that suggest a rare alpha cell population exhibiting high potential for self-renewal (Fang et al., 2019; Segerstolpe et al., 2016; Wang, Golson, et al., 2016; Wang, Schug, et al., 2016).

Due to the similarities in developmental origins between alpha cells and beta cells, alpha cells also harbor the plasticity to transdifferentiate into beta cells (Ye, Robertson, Hesselson, Stainier, & Anderson, 2015). As such, leveraging on this transdifferentiation potential is an alternative route to generate more beta cells. The *ARX* transcription factor that is uniquely expressed in alpha cells, is an alpha cell lineage-defining factor. Repression of *ARX* directly (Chakravarthy et al., 2017; Courtney et al., 2013), through overexpression of *PAX4* (Al-Hasani et al., 2013; Collombat et al., 2009), or through long term γ-aminobutyric acid (GABA) administration (Ben-Othman et al., 2017) was sufficient to convert mouse alpha cells into functional beta cells.

Table 6 Genes identified to be alpha cell-specific/enriched based on 10 human islet scRNA-Seq studies.

Alpha cell-specific/enriched genes identified in at least 6 out of 10 studies	Alpha cell-specific/enriched genes identified in only 5 out of 10 studies
ARRDC4	*ASPH*
ARX	*CD36*
CLU	*CD99L2*
COTL1	*CFC1*
DPP4	*CRH*
ELL2	*CRYBA2*
F10	*FAM110B*
FABP5	*FEV*
FAM84A	*FXYD3*
FAP	*FXYD6*
FSTL5	*GADD45G*
FXYD5	*GRIA3*
GC	*KCNJ3*
GCG (all studies)	*KLHL41*
GLS	*LDB2*
GPX3	*MBOAT2*
HIGD1A	*MRC1*
HLA-B	*NPNT*
HLA-E	*PCSK2*
HMGB3	*PLIN3*
IRX2	*PPP2R2B*
KCTD12	*PRKCE*
LOXL4	*PTPRT*
MUC13	*RAB3C*
NAA20	*VGF*
PALLD	*YIPF2*
PAPPA2	
PDK4	
PLCE1	
POPDC3	
RGS4	
RNASE4	
SERPINE2	
SLC38A4	
SLC7A2	
SMOC1	
SPOCK3	
SPTSSB	
TM4SF4	
TMEM176A	
TMEM176B	
TMEM236	
TTR (all studies)	

References: Baron et al. (2016); Dominguez Gutierrez et al. (2018); Dorajoo et al. (2017); Fang et al. (2019); Lawlor et al. (2017); Li et al. (2016); Muraro et al. (2016); Segerstolpe et al. (2016); Wang, Schug, et al. (2016); Xin, Kim, Okamoto, et al. (2016).

Additionally, two recent studies demonstrated that artemisinins (artemether), belonging to a class of antimalarial drugs, can inhibit ARX and induce the expression of insulin and other beta cell genes in mouse and human alpha cells, while downregulating the alpha cell-specific genes, based on scRNA-Seq data in human islets (Li et al., 2017; Marquina-Sanchez et al., 2020). However, functional data of artemether treatment on primary islets performed by another group showed detrimental effects on overall islet function instead, calling into question the suitability of artemether to be used as a diabetes treatment strategy (van der Meulen et al., 2018). Overall, while ARX is known to be important for alpha cell identity, more thorough follow-up studies are needed to investigate the pathways to achieve targeted alpha to beta cell conversion.

As such, identifying genes and pathways uniquely expressed by alpha cells could possibly pinpoint specific pathways that may be manipulated to increase the percentage of beta cells obtained *via* hPSC differentiation, or devise better transdifferentiation strategies. Nevertheless, the potential for the transdifferentiated beta cells to revert back to an alpha cell signature, and the long-term functionality of the transdifferentiated beta cells compared to *bona fide* beta cells remain to be considered.

4.3 Learning from studies of rodent development

Several studies have examined a wide variety of rodent islets at varying stages of development in hopes of shedding light on the key genes and pathways essential for beta cell maturation, at least in rodents (Table 3). To gain a comprehensive understanding of the beta cell maturation process, Qiu et al. performed a bulk beta cell transcriptomics analysis together with single cell analyses of rodent beta cells at different developmental time points, and reported a difference in gene signatures between immature and mature beta cells (Qiu et al., 2017). Immature beta cells displayed upregulation of cell cycle pathways, enrichment of regulatory cell signaling pathways (such as TGFβ, Notch and MAPK pathways) and expressed more genes involved in cell migration, adhesion and differentiation compared to their mature counterparts. Mature beta cells demonstrated upregulation of genes involved in metabolism, hormone secretion and transportation (Qiu et al., 2017). This study has provided a resource for understanding the key processes involved in lineage maturation, key players in regulating proliferation, and also highlighted transcription factors enriched in immature or mature beta cell clusters that require further investigation.

Although many studies have leveraged on rodent pancreatic cells to study islet development and maturation, it is important to be mindful of the differences between human and rodent islets (Santosa, Low, Pek, & Teo, 2015). For example, *Ucn3*, which encodes urocortin 3, a peptide hormone of the corticotropin-releasing factor family, is expressed specifically and abundantly in mature rodent beta cells (Blum et al., 2012; van der Meulen & Huising, 2014). However, in humans, *UCN3* is not beta cell-specific and is also found in other endocrine cells (van der Meulen & Huising, 2014). Therefore, *UCN3* alone is insufficient as a human beta cell marker. Other mature beta cell markers must be used concurrently to identify or isolate mature human beta cells. Similarly, although there is a switch from *Mafb* to *Mafa* expression as rodent beta cells mature, *MAFB* remains to be expressed with *MAFA* in mature human beta cells (Benner et al., 2014). These findings emphasize species-specific differences in islet gene expression.

Globally, although most alpha and beta cells exhibit similar expression patterns in both mouse and human islets based on scRNA-Seq data, there are also many exceptions. Genes that have been found to be highly expressed in human but not in mouse beta cells included *RBP4, NPTX2, HIGD1A, RGS16, SOX4, BMP5, TSPAN1* and *TFF3*, while genes robustly expressed in mouse but not in human beta cells included *GSTM5, S100A1, ADH1C, FAM151A, PRSS53, COX6A2, BACE2, SYTL4* and *FRZB* (Xin, Kim, Okamoto, et al., 2016). In another example, Li et al. compared the expression patterns of the human alpha cell gene *GC* and the human beta cell gene *DLK1* in human and mouse islets or pancreas sections, and reported an opposing expression pattern between the two species at both transcript and protein levels (Li et al., 2016). Together, these studies suggest that cell type-specific markers identified from rodent islet data will need to be thoroughly validated in human cell models (Santosa et al., 2015).

5. scRNA-Seq data from human beta cell differentiation inform gaps in functional maturation

Finally, there is no doubt that single cell technologies will provide an unbiased way to characterize the cells that are generated from beta cell differentiation protocols, to obtain a deeper understanding of the cell types that emerge from each stage of differentiation (Table 7). Thus far, most studies have only been able to achieve ~30%–50% beta cells from hPSCs, suggesting a continued need to improve the efficiency of current differentiation protocols. It will be important to know how the current *in vitro*

Table 7 Summary of single cell RNA-based studies performed on hPSC-derived islet cells.

Study reference	Cell line	Time point of cells analyzed	Single cell capture technology	Total number of cells analyzed/sequenced	Number of cells used in final profiling	Number of reads/cell
Petersen et al. (2017)	NEUROG3-eGFP hESC	Stage 4 day 1 (S4d1), S4d3, S5d3, S6d2, S6d7, S7d7	FACS, qPCR	546	517	NA
Krentz et al. (2018)	CyT49 NEUROG3-2A-eGFP hESC	Stage 6 day 1	10 × Genomics	4462	Not disclosed	Not disclosed
Ramond et al. (2018)	SA121 hESC and iPSCs AD2.1 and AD3.1	Stage 5 day 1	FACS, qPCR	264	264	NA
Balboa et al. (2018)	Patient-derived iPSCs for INS C96R and INS corrected	Stage 7	InDrop	2287	2171	0.038M
Veres et al. (2019)	iPSC, HUES8 hESC	Stages 3, 4, 5, 6	InDrop	85,444	Not disclosed	Not disclosed
Sharon et al. (2019)	iPSC, HUES8 hESC	ESCs, Stages 1, 2, 3, 4.3, 4, 5.3, 5, 6.2, 6.7, 6.14	Fluidigm C1	949	773	0.89M
Maxwell et al. (2020)	Patient-derived iPSCs for WFS1 unedited and corrected	Stage 6	10 × Genomics	10,855	Not disclosed	50,996

differentiation process compares with fetal beta cell development *in vivo*, so as to devise ways to improve the protocols further.

Petersen et al. was the first to report single cell gene expression profiling of pancreatic cells derived from human embryonic stem cells (hESCs) (Petersen et al., 2017), using single cell qPCR. The authors characterized each stage of cells from Stages 4 to 7, capturing the pancreatic progenitors, early and late endocrine progenitors, and subsequently endocrine cells. They identified an early *NKX6.1+MNX1+* pancreatic progenitor population that gave rise to subsequent *NKX6.1+NEUROG3+* endocrine progenitors that expressed other beta cell markers. They also identified two other *NEUROG3+* endocrine progenitor populations that did not express *NKX6.1* and several pancreatic progenitor markers. The divergence of differentiation trajectories appeared to occur from Stage 4, resulting in endocrine populations at later stages that displayed heterogeneity. For instance, *NKX6.1* can be acquired downstream of *NEUROG3* (in both *NEUROG3+* and *NEUROG3−* cells) during the endocrine progenitor stage and still give rise to monohormonal cells at the final stage. Cells that do not acquire *NKX6.1* instead appear to become polyhormonal cells expressing *INS*, *GCG* and *ARX*. Reminiscent of previous findings in primary human islets, the authors also showed that higher levels of *GCG* transcripts did not always translate to the same observations at the protein level (Petersen et al., 2017). The eventual population of endocrine cells separated into four groups—early immature beta-like cells that express *NEUROG3*, immature-beta like cells with low *NEUROG3*, beta-like cells, and polyhormonal cells. A subset of these *in vitro* generated cells mimicked the transcriptional identity of primary human beta cells. However, they expressed higher levels of *KRT19*, *FEV* and *PRDM16*, and lower levels *GPM6A*, *MAFA*, *HDAC9*, *IAPP*, and *HADH* than the primary beta cells, reflecting some level of immaturity. When the single cell profiles of hPSC-derived cells and human fetal pancreatic cells were compared in a follow-up study (Ramond et al., 2018), the authors found that even though similar cell populations were generated both *in vitro* and *in vivo*, expression profiles were not all the same. The genes *CDX2*, *GPM6A*, *CDKN1A*, *BMP5* and *ZNF453* appear to be mis-expressed in the hPSC-derived pancreatic progenitors compared to those found in the human fetal pancreas. In addition, the *in vitro*-derived cells continued to co-express multiple hormones at the transcriptional level more than that seen in the fetal pancreas, and did not reach the levels of expression for beta cell maturity genes as seen in adult human islets (Ramond et al., 2018). This study has provided a reference for evaluating the quality of pancreatic cells produced *in vitro*.

An scRNA-Seq study by Krentz et al. focused on profiling the transcriptome of human endocrine progenitors using a NEUROG3-eGFP reporter hESC line, targeting the time point at which the number of GFP+ cells peaks (Krentz et al., 2018). They identified five cell types: endocrine progenitors, polyhormonal endocrine cells, duct cells, liver cells and an unknown cell type. The three endocrine progenitor clusters observed expressed *NEUROD1*, a downstream target of *NEUROG3*, but were differentiated from one another by the expression of *NEUROG3* (endocrine progenitor 1), serotonin production genes such as *TPH1* and *FEV* (endocrine progenitor 2), and *GAST* (endocrine progenitor 3). The authors also described three endocrine clusters that expressed signatures indicating fates toward beta cells (*ERO1B* and *SLC30A8*), alpha cells (*GCG* and *PEMT*) and delta cells (*SST* and *HHEX*). This study further compared the hESC-derived endocrine progenitors with primary mouse and human tissue. They found that about half of the hESC-derived endocrine progenitor clusters resembled mouse endocrine progenitor cells, while a third of hESC-derived endocrine clusters mapped to beta cells in the mouse embryonic pancreas. When compared against human islet single cell data, a vast majority of the endocrine clusters mapped to beta cells, followed by some alpha and delta cells (Krentz et al., 2018). This study also highlighted the generation of a cluster of liver-like cells that are expressing *AFP*, which could represent unintended generation of other endodermal cell types given that both the liver and pancreas are derived from the foregut endoderm. Overall, it was suggested that methods to stabilize NEUROG3 protein could help to reduce formation of such off-target cells.

Another study also identified early and late endocrine progenitors as two distinct progenitor branches, with one marking early endocrine progenitors that expressed *HES1*, *PAX6*, *PROX1*, and genes related to cell adhesion, migration and TGF-β signaling, whereas the other marked cells resembled a more advanced, granulated, late endocrine progenitor identity with high levels of *NEUROD1*, *CHGA* and *FEV* (Balboa et al., 2018). The investigators also tapped on the scRNA-Seq pipeline for disease modeling, to determine the impact of a permanent neonatal diabetes mellitus (PNDM)-causing INS missense mutation (C96R) on the endocrine progenitor and beta cell populations. Since current beta cell differentiations only yield a fraction of INS+ cells, single cell transcriptomics enables selective investigation of the phenotype of only the INS+ cells. The study found that the INS C96R-expressing mutant beta cells displayed upregulation of chaperone genes and many components of the ER-associated degradation pathway, and downregulation of genes associated with beta cell proliferation and function, indicating that the mutation causes ER stress and UPR activation (Balboa et al., 2018). Another recent disease modeling study of Wolfram syndrome (caused by mutations in *WFS1*) showed that the patient-derived iPSCs generated mostly pancreatic exocrine or non-pancreatic cells, due to the loss of *WFS1*, affecting cell fate specification (Maxwell et al., 2020). Within the small fraction of *WFS1* mutant beta cells that were found (~10%), though the expression of a number of beta cell transcripts (except for *INS*) appeared unperturbed, there was an observable upregulation of UPR, mitochondrial stress and apoptotic markers. Therefore, the use of single cell techniques in disease modeling studies can help to determine the transcriptional changes due to specific gene mutations at the single cell level.

More recent reports utilizing a similar differentiation protocol have now comprehensively characterized hPSC-derived beta-like cells across up to 10 consecutive time points during differentiation (Sharon et al., 2019; Veres et al., 2019). It appears to be widely acknowledged that the first three stages of differentiation generally result in a largely uniform population of cells, with heterogeneity only observed from Stage 4 onwards. Compared to previous studies, the protocol involved a 3D suspension culture throughout differentiation. The authors then identified two groups of early progenitors—replicating pancreatic progenitors and *NKX6.1*+ progenitors at Stages 3 and 4 respectively. They also identified three groups of *CHGA*+ endocrine cells, namely pancreatic beta cells, alpha cells, and a population of enterochromaffin (EC) cells that expressed *CHGA*, *TPH1*, *LMX1A* and *SLC18A1* (Veres et al., 2019). Both beta-like cells and EC cells are likely to have emerged from the same endocrine induction events at Stage 5, and thus represented a late bifurcation event along the lineage of beta cell generation. They also found a *SOX9*+ non-endocrine cell population arising from progenitors that failed to progress through endocrine induction, and that retained expression of cell cycle-associated genes and appeared to be differentiating towards pancreatic exocrine fates. Only a small population of *SST*+ and *HHEX*+ cells were found. Using pseudo-time analysis, they found markers of beta cell maturation that increased over time such as *HOPX*, *NEFM* and *SIX2*, but not *MAFA* and *SIX3*, genes which are known to be expressed in mature beta cells. In general, the hPSC-derived beta cells expressed key beta cell markers and profiles similar to that in human islets, with minimal signatures of cell replication. In an attempt to improve the differentiation process, the authors tested a few protocol modifications from Stage 4 onward, that resulted in overall changes in cell population ratios rather than gene signatures of the detected cell types. They also identified pathways that were still active in the non-endocrine populations at the end of differentiation, providing hints to the pathways that could be suppressed to allow for proper endocrine development. Another method that improved efficiency entailed reaggregation of cells after Stage 5 to deplete the non-endocrine cells and obtain purer endocrine cell clusters. Without such comprehensive multi-stage transcriptomic profiling, one would have mistaken the hPSC-derived EC cells as beta cells, given the expression of some common endocrine genes.

Based on the study by Sharon et al., the Stage 4 cells could be divided into two main clusters representing epithelial cord-like progenitors (that expressed epithelial cord markers and cell cycle-related genes) and endocrine cells (that expressed endocrine markers and reduced cell proliferation gene expression) (Sharon et al., 2019). Among the endocrine population, multiple distinct cell types were observed – endocrine precursors marked by *NEUROG3*, *SOX11*, *RUNX1T1* and low hormone expression, alpha cells marked by *GCG*, *ETV1* and *ARX*, beta cells marked by *INS*, *PDX1*, *NKX6.1*, *LMO1* and low *GCG*, and finally non-pancreatic endocrine cells again resembling enteroendocrine populations that had low *ISL1* but expressed *CHGA* serotonin genes such as *TPH1*, and non-islet genes such as *ADH6* and *LMX1A*. Unlike what may be observed during embryonic development, *in vitro* beta cell differentiation gave rise to two waves of endocrine precursor formation, each during Stages 4 and 5. Alpha cells are formed in the first wave, followed by beta cells in the second wave. *CPEP+* and *NKX6.1+* beta cells appeared almost exclusively from Stage 5 onwards. To investigate what governed the timing of endocrine differentiation, the authors looked into the different pathway genes expressed in the epithelial progenitor cells and the endocrine cells, and found that WNT inhibition as marked by the expression of WNT inhibitor *APC* appeared to be necessary for proper endocrine differentiation. To test this experimentally, the authors treated Stage 4 and 5 cells with the WNT-tankyrase inhibitor IWR1-endo or XAV939, and saw an increase in the expression of endocrine markers with concurrent downregulation of progenitor markers. They also reported more than 50% increase in the proportion of *CPEP+NKX6.1+* beta cells (Sharon et al., 2019).

Multiple groups have now identified an off-target population resembling intestinal EC cells, that expressed serotonin production machinery genes (Balboa et al., 2018; Krentz et al., 2018; Sharon et al., 2019; Veres et al., 2019). ScRNA-Seq comparisons across differentiated cell populations can help to identify the gaps in cell maturation, and provide clues to the manipulations that could be done at the relevant stage. These studies could also advise on markers that can be used to enrich for the desired cell populations, or to deplete the off-target cells. For instance, Veres et al. also identified a potential cell surface marker for beta cell enrichment such as CD49a, even though in adult human islets, CD49a is not specific to beta cells (Veres et al., 2019).

6. Discussion

From characterizing human and rodent, fetal and adult pancreatic tissues, to mapping the *in vitro* differentiation of islet cells from hPSCs, single cell technologies such as scRNA-Seq can help us better understand cellular heterogeneity and the trajectory of beta cell development. Based on the findings from human and rodent islet tissues, it is clear that individual islet cell populations exhibit molecular heterogeneity that may correlate with changes in islet cell identity, survival, metabolic signature, stress responses, and function. There is increasing evidence that beta cells co-exist in different subpopulations that may play specialized roles (such as in the case of "hub" cells) or that exhibit different levels of stress responses that are tightly linked to cell function (such as insulin production). However, to establish biological significance, experimental validation of the beta cell heterogeneity genes that have been identified needs to be performed, to determine if they are functional drivers or simply markers of cell heterogeneity, the former of which may be targeted to induce functional maturation in immature cell types (Liu & Hebrok, 2017). Despite the large number of scRNA-Seq studies on the pancreatic islets, the level of congruence in terms of beta cell heterogeneity genes and beta cell gene signatures across studies remains surprisingly little. This could be due to differences in the single cell capture and sequencing technology, but also by variabilities in islet source, quality of islet preparation and length of *ex vivo* culture duration (Ng et al., 2019). Nonetheless, the 48 beta cell-specific/enriched genes that we have identified to be consistently found across ~50% of 10 human islet scRNA-Seq studies (Table 5) represent both well-established (*e.g., INS, G6PC2, IAPP, PDX1, MAFA*) and relatively less well-known beta cell targets (*e.g., SORL1, AP3B1, GSN, HSPD1, KLHDC8A, SAMD11*) that could play a part in regulating beta cell function, or be used to identify beta cell populations, warranting further follow-up studies.

In comparing the findings from human pancreatic tissue and hPSC-derived cells, it is encouraging that current *in vitro* differentiation protocols are able to generate cell sub-types that resemble those observed in primary tissue, at least at the transcript level. Some key differences seem to lie in the trajectory of endocrine commitment, which appear to occur in multiple phases (as opposed to a single phase *in vivo* during human pancreas development), that may result in a heterogeneous pool of endocrine cells that arose from different progenitors at different time points. This could also contribute to the generation of immature and off-target cells. The generation of bihormonal cells and immature beta cells is, to some extent, however, likely to be a normal phenomenon during fetal pancreas development, and may not necessarily represent an *in vitro* differentiation artifact. Therefore, the insights gathered from these single cell, time point analyses could help to pinpoint specific pathways to target at certain time points during the differentiation to drive functional maturation of beta cells. The pathway of differentiation is relatively homogenous up until the pancreatic progenitor stage, after which cell lineages start to diverge during endocrine differentiation at Stages 4 and 5. As endocrine markers start emerging during the later stages of differentiation, any modifications may need to be done earlier. Although most protocols now generate three-dimensional beta cell-like clusters that look like primary human islets in size and morphology, whether an *in vitro* differentiation environment can sufficiently recapitulate other fine developmental cues provided in an *in vivo* environment remains to be determined.

Despite the potential that scRNA-Seq promises, there are a number of limitations that must be acknowledged. Some studies had excluded cells with a mixed signature, or cells expressing high levels of two or more hormones, which is a stringent criterion needed to exclude potential cell doublets (Lawlor et al., 2017). The bihormonal cells detected (such as alpha cells expressing *INS* or beta cells expressing *GCG*) could arise from cross-contamination during the processing steps. However, this exclusion could also have excluded cells of truly mixed signatures, such as precursor or de-/trans-differentiating cells (Teo et al., 2018). The dissociation of cells also introduces selection bias to specific types of cells that persist from the extraction process to analysis. A recent review emphasized the importance of considering the sample preparation process when interpreting the results in each study (Bonnycastle et al., 2020). Single cell analyses that require single cell dissociation, such as scRNA-Seq, also do not take into account the cell-cell interactions within the islet that influence cell behavior and function. Therefore, analyses that do not require a dissociation step, such as mass cytometry or fluorescence imaging, would better facilitate the study of cells in their "native" states while also providing positional information. On the other hand, these protein studies are significantly more limiting in the number of markers that can be used and studied, and may be less suitable for discovery efforts unlike scRNA-Seq. Another recent review extensively discussed the pitfalls of scRNA-Seq pertaining to reliable detection and interpretation of gene expression results across single cells (Mawla & Huising, 2019). The authors found that the endogenous level of expression of a gene directly influenced its own propensity to be detected in single cells across several studies, which may limit the identification of common heterogeneous beta cell genes for meaningful biological interpretation. The "underdetection" of genes in samples is therefore common in scRNA-Seq studies which tend to operate at low levels of detection.

In light of the various limitations, it is important to increase the number of cells sequenced, achieve sufficient depth of sequencing, conduct more independent studies to validate findings and also employ other independent complementary techniques (to generate genomic, protein and functional data), much akin to the approach of genome-wide association studies. Thankfully, many of these are currently being tackled. For instance, sample sizes have increased and analytical methods have improved in a short span of a few years. A study by Tyler et al. that used a different analysis method for scRNA-Seq data suggested that the network structure from scRNA-Seq data is actually reproducible across different labs and platforms (Tyler et al., 2019). Given the large amount of published data so far, it would be useful to perform meta-analyses of all single cell transcriptomic studies, as well as systematic comparisons between human islet studies and stem cell differentiation studies to identify the trends that are robust and consistent. In addition, defining standard criteria for calling cell identities and consensus sets of transcriptional profiles for each islet cell type may help to harmonize cell clustering efforts in future studies. There are also emerging single cell chromatin profiling studies using ATAC-Seq in pancreatic islets that can define regulatory signatures in a cell type- and cell state-specific manner (Chiou et al., 2019; Rai et al., 2020), as well as the first report that combined scRNA-Seq with electrophysiological measurements of endocrine cell exocytosis and ion channel activity to correlate transcriptomic signatures with functional activity (Camunas-Soler et al., 2020) (Table 4). We believe that in due course, these studies would be similarly performed in hPSC-derived cells. Overall, it is clear that the identification of pancreatic subpopulations and cell state-specific genes, through whichever means, will lead to the discovery and clearer definition of human pancreas development and biology, and that this would aid the ongoing efforts to improve the quality and efficiency of islet/beta cell differentiation for both research and clinical applications.

Acknowledgments

The authors thank members of the Teo laboratory for the critical reading of this manuscript. N.H.J.N. is supported by NMRC OFYIRG18may040 and the 1st A*STAR Career Development Award (CDA). C.W.Y.N. is supported by the NUS Research Scholarship. A.K.K.T. is supported by the Institute of Molecular and Cell Biology (IMCB), A*STAR, NMRC Open Fund-Young Individual Research Grant (OF-YIRG) OFYIRG16may014, A*STAR ETPL Gap Funding ETPL/18-GAP005-R20H, Lee Foundation Grant SHTX/LFG/002/2018, Skin Innovation Grant SIG18011, NMRC OF-LCG/DYNAMO, FY2019 SingHealth Duke-NUS Surgery Academic Clinical Programme Research Support Programme Grant, Precision Medicine and Personalised Therapeutics Joint Research Grant 2019, Industry Alignment Fund—Industry Collaboration Project (IAF-ICP) I1901E0049 and the 2nd A*STAR-AMED Joint Grant Call 192B9002.

Declaration of interests

N.H.J.N. and A.K.K.T. are founders and shareholders of BetaLife Pte Ltd.

References

Al-Hasani, K., Pfeifer, A., Courtney, M., Ben-Othman, N., Gjernes, E., Vieira, A., et al. (2013). Adult duct-lining cells can reprogram into beta-like cells able to counter repeated cycles of toxin-induced diabetes. *Developmental Cell*, *26*(1), 86–100. https://doi.org/10.1016/j.devcel.2013.05.018.

Balboa, D., Saarimaki-Vire, J., Borshagovski, D., Survila, M., Lindholm, P., Galli, E., et al. (2018). Insulin mutations impair beta-cell development in a patient-derived iPSC model of neonatal diabetes. *Elife*, *7*, e38519. https://doi.org/10.7554/eLife.38519.

Baron, M., Veres, A., Wolock, S. L., Faust, A. L., Gaujoux, R., Vetere, A., et al. (2016). A single-cell transcriptomic map of the human and mouse pancreas reveals inter- and intra-cell population structure. *Cell Systems*, *3*(4), 346–360. e344. https://doi.org/10.1016/j.cels.2016.08.011.

Bastidas-Ponce, A., Tritschler, S., Dony, L., Scheibner, K., Tarquis-Medina, M., Salinno, C., et al. (2019). Comprehensive single cell mRNA profiling reveals a detailed roadmap for pancreatic endocrinogenesis. *Development*, *146*(12). https://doi.org/10.1242/dev.173849.

Benner, C., van der Meulen, T., Caceres, E., Tigyi, K., Donaldson, C. J., & Huising, M. O. (2014). The transcriptional landscape of mouse beta cells compared to human beta cells reveals notable species differences in long non-coding RNA and protein-coding gene expression. *BMC Genomics*, *15*, 620. https://doi.org/10.1186/1471-2164-15-620.

Ben-Othman, N., Vieira, A., Courtney, M., Record, F., Gjernes, E., Avolio, F., et al. (2017). Long-term GABA administration induces alpha cell-mediated Beta-like cell neogenesis. *Cell*, *168*(1–2), 73–85. e11. https://doi.org/10.1016/j.cell.2016.11.002.

Blum, B., Hrvatin, S., Schuetz, C., Bonal, C., Rezania, A., & Melton, D. A. (2012). Functional beta-cell maturation is marked by an increased glucose threshold and by expression of urocortin 3. *Nature Biotechnology*, *30*(3), 261–264. https://doi.org/10.1038/nbt.2141.

Bonnycastle, L. L., Gildea, D. E., Yan, T., Narisu, N., Swift, A. J., Wolfsberg, T. G., et al. (2020). Single-cell transcriptomics from human pancreatic islets: Sample preparation matters. *Biology Methods and Protocols*, *5*(1), bpz019. https://doi.org/10.1093/biomethods/bpz019.

Bramswig, N. C., Everett, L. J., Schug, J., Dorrell, C., Liu, C., Luo, Y., et al. (2013). Epigenomic plasticity enables human pancreatic alpha to beta cell reprogramming. *The Journal of Clinical Investigation*, *123*(3), 1275–1284. https://doi.org/10.1172/JCI66514.

Brennan, D. C., Kopetskie, H. A., Sayre, P. H., Alejandro, R., Cagliero, E., Shapiro, A. M., et al. (2016). Long-term follow-up of the Edmonton protocol of islet transplantation in the United States. *American Journal of Transplantation*, *16*(2), 509–517. https://doi.org/10.1111/ajt.13458.

Byrnes, L. E., Wong, D. M., Subramaniam, M., Meyer, N. P., Gilchrist, C. L., Knox, S. M., et al. (2018). Lineage dynamics of murine pancreatic development at single-cell resolution. *Nature Communications*, *9*(1), 3922. https://doi.org/10.1038/s41467-018-06176-3.

Cabrera, O., Berman, D. M., Kenyon, N. S., Ricordi, C., Berggren, P. O., & Caicedo, A. (2006). The unique cytoarchitecture of human pancreatic islets has implications for islet cell function. *Proceedings of the National Academy of Sciences of the United States of America*, *103*(7), 2334–2339. https://doi.org/10.1073/pnas.0510790103.

Camunas-Soler, J., Dai, X. Q., Hang, Y., Bautista, A., Lyon, J., Suzuki, K., et al. (2020). Patch-Seq links single-cell transcriptomes to human islet dysfunction in diabetes. *Cell Metabolism*, *31*(5), 1017–1031. e1014. https://doi.org/10.1016/j.cmet.2020.04.005.

Cebola, I., Rodriguez-Segui, S. A., Cho, C. H., Bessa, J., Rovira, M., Luengo, M., et al. (2015). TEAD and YAP regulate the enhancer network of human embryonic pancreatic progenitors. *Nature Cell Biology*, *17*(5), 615–626. https://doi.org/10.1038/ncb3160.

Chakravarthy, H., Gu, X., Enge, M., Dai, X., Wang, Y., Damond, N., et al. (2017). Converting adult pancreatic islet alpha cells into beta cells by targeting both Dnmt1 and Arx. *Cell Metabolism*, *25*(3), 622–634. https://doi.org/10.1016/j.cmet.2017.01.009.

Chiou, J., Zeng, C., Cheng, Z., Han, J. Y., Schlichting, M., Huang, S., et al. (2019). Single cell chromatin accessibility reveals pancreatic islet cell type- and state-specific regulatory programs of diabetes risk. *bioRxiv*, 693671. https://doi.org/10.1101/693671.

Collombat, P., Xu, X., Ravassard, P., Sosa-Pineda, B., Dussaud, S., Billestrup, N., et al. (2009). The ectopic expression of Pax4 in the mouse pancreas converts progenitor cells into alpha and subsequently beta cells. *Cell*, *138*(3), 449–462. https://doi.org/10.1016/j.cell.2009.05.035.

Courtney, M., Gjernes, E., Druelle, N., Ravaud, C., Vieira, A., Ben-Othman, N., et al. (2013). The inactivation of Arx in pancreatic alpha-cells triggers their neogenesis and conversion into functional beta-like cells. *PLoS Genetics*, *9*(10), e1003934. https://doi.org/10.1371/journal.pgen.1003934.

Damond, N., Engler, S., Zanotelli, V. R. T., Schapiro, D., Wasserfall, C. H., Kusmartseva, I., et al. (2019). A map of human type 1 diabetes progression by imaging mass cytometry. *Cell Metabolism*, *29*(3), 755–768. e755. https://doi.org/10.1016/j.cmet.2018.11.014.

Dominguez Gutierrez, G., Xin, Y., Okamoto, H., Kim, J., Lee, A. H., Ni, M., et al. (2018). Gene signature of proliferating human pancreatic alpha cells. *Endocrinology*, *159*(9), 3177–3186. https://doi.org/10.1210/en.2018-00469.

Dorajoo, R., Ali, Y., Tay, V. S. Y., Kang, J., Samydurai, S., Liu, J., et al. (2017). Single-cell transcriptomics of East-Asian pancreatic islets cells. *Scientific Reports*, 7(1), 5024. https://doi.org/10.1038/s41598-017-05266-4.

Enge, M., Arda, H. E., Mignardi, M., Beausang, J., Bottino, R., Kim, S. K., et al. (2017). Single-cell analysis of human pancreas reveals transcriptional signatures of aging and somatic mutation patterns. *Cell*, *171*(2), 321–330. e314. https://doi.org/10.1016/j.cell.2017.09.004.

Fang, Z., Weng, C., Li, H., Tao, R., Mai, W., Liu, X., et al. (2019). Single-cell heterogeneity analysis and CRISPR screen identify key beta-cell-specific disease genes. *Cell Reports*, *26*(11), 3132–3144. e3137. https://doi.org/10.1016/j.celrep.2019.02.043.

Gradwohl, G., Dierich, A., LeMeur, M., & Guillemot, F. (2000). Neurogenin3 is required for the development of the four endocrine cell lineages of the pancreas. *Proceedings of the National Academy of Sciences of the United States of America*, *97*(4), 1607–1611. https://doi.org/10.1073/pnas.97.4.1607.

Hogrebe, N. J., Augsornworawat, P., Maxwell, K. G., Velazco-Cruz, L., & Millman, J. R. (2020). Targeting the cytoskeleton to direct pancreatic differentiation of human pluripotent stem cells. *Nature Biotechnology*, *38*(4), 460–470. https://doi.org/10.1038/s41587-020-0430-6.

Jansson, E. T., Comi, T. J., Rubakhin, S. S., & Sweedler, J. V. (2016). Single cell peptide heterogeneity of rat islets of Langerhans. *ACS Chemical Biology*, *11*(9), 2588–2595. https://doi.org/10.1021/acschembio.6b00602.

Jennings, R. E., Berry, A. A., Kirkwood-Wilson, R., Roberts, N. A., Hearn, T., Salisbury, R. J., et al. (2013). Development of the human pancreas from foregut to endocrine commitment. *Diabetes*, *62*(10), 3514–3522. https://doi.org/10.2337/db12-1479.

Jennings, R. E., Berry, A. A., Strutt, J. P., Gerrard, D. T., & Hanley, N. A. (2015). Human pancreas development. *Development*, *142*(18), 3126–3137. https://doi.org/10.1242/dev.120063.

Johnston, N. R., Mitchell, R. K., Haythorne, E., Pessoa, M. P., Semplici, F., Ferrer, J., et al. (2016). Beta cell hubs dictate pancreatic islet responses to glucose. *Cell Metabolism*, *24*(3), 389–401. https://doi.org/10.1016/j.cmet.2016.06.020.

Krentz, N. A. J., Lee, M. Y. Y., Xu, E. E., Sproul, S. L. J., Maslova, A., Sasaki, S., et al. (2018). Single-cell transcriptome profiling of mouse and hESC-derived pancreatic progenitors. *Stem Cell Reports*, *11*(6), 1551–1564. https://doi.org/10.1016/j.stemcr.2018.11.008.

Lawlor, N., George, J., Bolisetty, M., Kursawe, R., Sun, L., Sivakamasundari, V., et al. (2017). Single-cell transcriptomes identify human islet cell signatures and reveal cell-type-specific expression changes in type 2 diabetes. *Genome Research*, *27*(2), 208–222. https://doi.org/10.1101/gr.212720.116.

Li, J., Casteels, T., Frogne, T., Ingvorsen, C., Honore, C., Courtney, M., et al. (2017). Artemisinins target GABAA receptor signaling and impair alpha cell identity. *Cell*, *168*(1–2), 86–100. e115. https://doi.org/10.1016/j.cell.2016.11.010.

Li, J., Klughammer, J., Farlik, M., Penz, T., Spittler, A., Barbieux, C., et al. (2016). Single-cell transcriptomes reveal characteristic features of human pancreatic islet cell types. *EMBO Reports*, *17*(2), 178–187. https://doi.org/10.15252/embr.201540946.

Li, L. C., Qiu, W. L., Zhang, Y. W., Xu, Z. R., Xiao, Y. N., Hou, C., et al. (2018). Single-cell transcriptomic analyses reveal distinct dorsal/ventral pancreatic programs. *EMBO Reports*, *19*(10), e46148. https://doi.org/10.15252/embr.201846148.

Liu, J. S., & Hebrok, M. (2017). All mixed up: Defining roles for beta-cell subtypes in mature islets. *Genes & Development*, *31*(3), 228–240. https://doi.org/10.1101/gad.294389.116.

Loo, L. S. W., Lau, H. H., Jasmen, J. B., Lim, C. S., & Teo, A. K. K. (2018). An arduous journey from human pluripotent stem cells to functional pancreatic beta cells. *Diabetes, Obesity & Metabolism*, *20*(1), 3–13. https://doi.org/10.1111/dom.12996.

Marquina-Sanchez, B., Fortelny, N., Farlik, M., Vieira, A., Collombat, P., Bock, C., et al. (2020). Single-cell RNA-seq with spike-in cells enables accurate quantification of cell-specific drug effects in pancreatic islets. *Genome Biology*, *21*(1), 106. https://doi.org/10.1186/s13059-020-02006-2.

Mastracci, T. L., & Sussel, L. (2012). The endocrine pancreas: Insights into development, differentiation and diabetes. *Wiley Interdisciplinary Reviews. Membrane Transport and Signaling*, *1*(5), 609–628. https://doi.org/10.1002/wdev.44.

Mawla, A. M., & Huising, M. O. (2019). Navigating the depths and avoiding the shallows of pancreatic islet cell transcriptomes. *Diabetes*, *68*(7), 1380–1393. https://doi.org/10.2337/dbi18-0019.

Maxwell, K. G., Augsornworawat, P., Velazco-Cruz, L., Kim, M. H., Asada, R., Hogrebe, N. J., et al. (2020). Gene-edited human stem cell-derived beta cells from a patient with monogenic diabetes reverse preexisting diabetes in mice. *Science Translational Medicine*, *12*(540), eaax9106. https://doi.org/10.1126/scitranslmed.aax9106.

McCall, M., & Shapiro, A. M. (2012). Update on islet transplantation. *Cold Spring Harbor Perspectives in Medicine*, *2*(7), a007823. https://doi.org/10.1101/cshperspect.a007823.

Millman, J. R., Xie, C., Van Dervort, A., Gurtler, M., Pagliuca, F. W., & Melton, D. A. (2016). Generation of stem cell-derived beta-cells from patients with type 1 diabetes. *Nature Communications*, 7, 11463. https://doi.org/10.1038/ncomms11463.

Muraro, M. J., Dharmadhikari, G., Grun, D., Groen, N., Dielen, T., Jansen, E., et al. (2016). A single-cell transcriptome atlas of the human pancreas. *Cell Systems*, *3*(4), 385–394. e383. https://doi.org/10.1016/j.cels.2016.09.002.

Nair, G. G., Liu, J. S., Russ, H. A., Tran, S., Saxton, M. S., Chen, R., et al. (2019). Recapitulating endocrine cell clustering in culture promotes maturation of human stem-cell-derived beta cells. *Nature Cell Biology*, *21*(2), 263–274. https://doi.org/10.1038/s41556-018-0271-4.

Nair, G. G., Tzanakakis, E. S., & Hebrok, M. (2020). Emerging routes to the generation of functional beta-cells for diabetes mellitus cell therapy. *Nature Reviews. Endocrinology*, *16*(9), 506–518. https://doi.org/10.1038/s41574-020-0375-3.

Ng, N. H. J., Tan, W. X., Koh, Y. X., & Teo, A. K. K. (2019). Human islet isolation and distribution efforts for clinical and basic research. *OBM Transplantation*, *3*(2), 31.

Pagliuca, F. W., Millman, J. R., Gurtler, M., Segel, M., Van Dervort, A., Ryu, J. H., et al. (2014). Generation of functional human pancreatic beta cells in vitro. *Cell*, *159*(2), 428–439. https://doi.org/10.1016/j.cell.2014.09.040.

Petersen, M. B. K., Azad, A., Ingvorsen, C., Hess, K., Hansson, M., Grapin-Botton, A., et al. (2017). Single-cell gene expression analysis of a human ESC model of pancreatic endocrine development reveals different paths to beta-cell differentiation. *Stem Cell Reports*, *9*(4), 1246–1261. https://doi.org/10.1016/j.stemcr.2017.08.009.

Qiu, W. L., Zhang, Y. W., Feng, Y., Li, L. C., Yang, L., & Xu, C. R. (2017). Deciphering pancreatic islet beta cell and alpha cell maturation pathways and characteristic features at the single-cell level. *Cell Metabolism*, *25*(5), 1194–1205. e1194. https://doi.org/10.1016/j.cmet.2017.04.003.

Rai, V., Quang, D. X., Erdos, M. R., Cusanovich, D. A., Daza, R. M., Narisu, N., et al. (2020). Single-cell ATAC-Seq in human pancreatic islets and deep learning upscaling of rare cells reveals cell-specific type 2 diabetes regulatory signatures. *Molecular Metabolism*, *32*, 109–121. https://doi.org/10.1016/j.molmet.2019.12.006.

Ramond, C., Beydag-Tasoz, B. S., Azad, A., van de Bunt, M., Petersen, M. B. K., Beer, N. L., et al. (2018). Understanding human fetal pancreas development using subpopulation sorting, RNA sequencing and single-cell profiling. *Development*, *145*(16). https://doi.org/10.1242/dev.165480. dev165480.

Rezania, A., Bruin, J. E., Arora, P., Rubin, A., Batushansky, I., Asadi, A., et al. (2014). Reversal of diabetes with insulin-producing cells derived in vitro from human pluripotent stem cells. *Nature Biotechnology*, *32*(11), 1121–1133. https://doi.org/10.1038/nbt.3033.

Rosado-Olivieri, E. A., Anderson, K., Kenty, J. H., & Melton, D. A. (2019). YAP inhibition enhances the differentiation of functional stem cell-derived insulin-producing beta cells. *Nature Communications*, *10*(1), 1464. https://doi.org/10.1038/s41467-019-09404-6.

Russ, H. A., Parent, A. V., Ringler, J. J., Hennings, T. G., Nair, G. G., Shveygert, M., et al. (2015). Controlled induction of human pancreatic progenitors produces functional

beta-like cells in vitro. *EMBO Journal, 34*(13), 1759–1772. https://doi.org/10.15252/embj.201591058.
Salisbury, R. J., Blaylock, J., Berry, A. A., Jennings, R. E., De Krijger, R., Piper Hanley, K., et al. (2014). The window period of NEUROGENIN3 during human gestation. *Islets, 6*(3), e954436. https://doi.org/10.4161/19382014.2014.954436.
Santosa, M. M., Low, B. S., Pek, N. M., & Teo, A. K. (2015). Knowledge gaps in rodent pancreas biology: Taking human pluripotent stem cell-derived pancreatic beta cells into our own hands. *Frontiers in Endocrinology (Lausanne), 6*, 194. https://doi.org/10.3389/fendo.2015.00194.
Scavuzzo, M. A., Hill, M. C., Chmielowiec, J., Yang, D., Teaw, J., Sheng, K., et al. (2018). Endocrine lineage biases arise in temporally distinct endocrine progenitors during pancreatic morphogenesis. *Nature Communications, 9*(1), 3356. https://doi.org/10.1038/s41467-018-05740-1.
Scheibner, K., Bakhti, M., Bastidas-Ponce, A., & Lickert, H. (2019). Wnt signaling: Implications in endoderm development and pancreas organogenesis. *Current Opinion in Cell Biology, 61*, 48–55. https://doi.org/10.1016/j.ceb.2019.07.002.
Segerstolpe, A., Palasantza, A., Eliasson, P., Andersson, E. M., Andreasson, A. C., Sun, X., et al. (2016). Single-cell transcriptome profiling of human pancreatic islets in health and type 2 diabetes. *Cell Metabolism, 24*(4), 593–607. https://doi.org/10.1016/j.cmet.2016.08.020.
Sharon, N., Vanderhooft, J., Straubhaar, J., Mueller, J., Chawla, R., Zhou, Q., et al. (2019). Wnt signaling separates the progenitor and endocrine compartments during pancreas development. *Cell Reports, 27*(8), 2281–2291. e2285. https://doi.org/10.1016/j.celrep.2019.04.083.
Tan, W. X., Lau, H. H., Tan, N. S., Khoo, C. M., & Teo, K. K. A. (2020). Considerations in using human pluripotent stem cell-derived pancreatic beta cells to treat type 1 diabetes. In *Vol. 3. Recent advances in iPSCs for therapy* (pp. 173–203). Academic Press.
Teo, A. K. K., Lim, C. S., Cheow, L. F., Kin, T., Shapiro, J. A., Kang, N. Y., et al. (2018). Single-cell analyses of human islet cells reveal de-differentiation signatures. *Cell Death Discovery, 4*, 14. https://doi.org/10.1038/s41420-017-0014-5.
Tyler, S. R., Rotti, P. G., Sun, X., Yi, Y., Xie, W., Winter, M. C., et al. (2019). PyMINEr finds gene and autocrine-paracrine networks from human islet scRNA-Seq. *Cell Reports, 26*(7), 1951–1964. e1958. https://doi.org/10.1016/j.celrep.2019.01.063.
van der Meulen, T., & Huising, M. O. (2014). Maturation of stem cell-derived beta-cells guided by the expression of urocortin 3. *The Review of Diabetic Studies, 11*(1), 115–132. https://doi.org/10.1900/RDS.2014.11.115.
van der Meulen, T., Lee, S., Noordeloos, E., Donaldson, C. J., Adams, M. W., Noguchi, G. M., et al. (2018). Artemether does not turn alpha cells into beta cells. *Cell Metabolism, 27*(1), 218–225. e214. https://doi.org/10.1016/j.cmet.2017.10.002.
Velazco-Cruz, L., Song, J., Maxwell, K. G., Goedegebuure, M. M., Augsornworawat, P., Hogrebe, N. J., et al. (2019). Acquisition of dynamic function in human stem cell-derived beta cells. *Stem Cell Reports, 12*(2), 351–365. https://doi.org/10.1016/j.stemcr.2018.12.012.
Veres, A., Faust, A. L., Bushnell, H. L., Engquist, E. N., Kenty, J. H., Harb, G., et al. (2019). Charting cellular identity during human in vitro beta-cell differentiation. *Nature, 569*(7756), 368–373. https://doi.org/10.1038/s41586-019-1168-5.
Wang, Y. J., Golson, M. L., Schug, J., Traum, D., Liu, C., Vivek, K., et al. (2016). Single-cell mass cytometry analysis of the human endocrine pancreas. *Cell Metabolism, 24*(4), 616–626. https://doi.org/10.1016/j.cmet.2016.09.007.
Wang, Y. J., & Kaestner, K. H. (2019). Single-cell RNA-Seq of the pancreatic islets—A promise not yet fulfilled? *Cell Metabolism, 29*(3), 539–544. https://doi.org/10.1016/j.cmet.2018.11.016.

Wang, Y. J., Schug, J., Won, K. J., Liu, C., Naji, A., Avrahami, D., et al. (2016). Single-cell transcriptomics of the human endocrine pancreas. *Diabetes*, *65*(10), 3028–3038. https://doi.org/10.2337/db16-0405.

Wang, Y. J., Traum, D., Schug, J., Gao, L., Liu, C., HPAP Consortium, et al. (2019). Multiplexed in situ imaging mass cytometry analysis of the human endocrine pancreas and immune system in type 1 diabetes. *Cell Metabolism*, *29*(3), 769–783. e764. https://doi.org/10.1016/j.cmet.2019.01.003.

Xin, Y., Dominguez Gutierrez, G., Okamoto, H., Kim, J., Lee, A. H., Adler, C., et al. (2018). Pseudotime ordering of single human beta-cells reveals states of insulin production and unfolded protein response. *Diabetes*, *67*(9), 1783–1794. https://doi.org/10.2337/db18-0365.

Xin, Y., Kim, J., Ni, M., Wei, Y., Okamoto, H., Lee, J., et al. (2016). Use of the Fluidigm C1 platform for RNA sequencing of single mouse pancreatic islet cells. *Proceedings of the National Academy of Sciences of the United States of America*, *113*(12), 3293–3298. https://doi.org/10.1073/pnas.1602306113.

Xin, Y., Kim, J., Okamoto, H., Ni, M., Wei, Y., Adler, C., et al. (2016). RNA sequencing of single human islet cells reveals type 2 diabetes genes. *Cell Metabolism*, *24*(4), 608–615. https://doi.org/10.1016/j.cmet.2016.08.018.

Xin, Y., Okamoto, H., Kim, J., Ni, M., Adler, C., Cavino, K., et al. (2016). Single-cell RNAseq reveals that pancreatic beta-cells from very old male mice have a young gene signature. *Endocrinology*, *157*(9), 3431–3438. https://doi.org/10.1210/en.2016-1235.

Ye, L., Robertson, M. A., Hesselson, D., Stainier, D. Y., & Anderson, R. M. (2015). Glucagon is essential for alpha cell transdifferentiation and beta cell neogenesis. *Development*, *142*(8), 1407–1417. https://doi.org/10.1242/dev.117911.

Zeng, C., Mulas, F., Sui, Y., Guan, T., Miller, N., Tan, Y., et al. (2017). Pseudotemporal ordering of single cells reveals metabolic control of postnatal beta cell proliferation. *Cell Metabolism*, *25*(5), 1160–1175. e1111. https://doi.org/10.1016/j.cmet.2017.04.014.

CHAPTER SEVEN

Generation of high yield insulin-producing cells (IPCs) from various sources of stem cells

Adele Soltani[a], Samaneh Khazaei[b,c], Seyedeh Mahsoumeh Mirtaghi[d,e], Seyedeh Fatemeh Ahmadi[d,e], Mohammad Rahmati[f], Reyhaneh Nassiri Mansour[d,g], Mohammad Foad Abazari[h], and Seyed Ehsan Enderami[e,i,*]

[a]Department of Clinical Biochemistry, Faculty of Medical Sciences, Tarbiat Modares University, Tehran, Iran
[b]Department of Hematology and Cell Therapy, Faculty of Medical Sciences, Tarbiat Modares University, Tehran, Iran
[c]Department of Cell and Molecular Biology, Faculty of Biological Science and Technology, Isfahan University, Isfahan, Iran
[d]Student Research Committee, Mazandaran University of Medical Sciences, Sari, Iran
[e]Department of Medical Biotechnology, School of Advanced Technologies in Medicine, Mazandaran University of Medical Sciences, Sari, Iran
[f]Department of Medical Biotechnology, Faculty of Paramedicine, Guilan University of Medical Sciences, Rasht, Iran
[g]Department of Tissue Engineering, School of Advanced Technologies in Medicine, Mazandaran University of Medical Sciences, Sari, Iran
[h]Research Center for Clinical Virology, Tehran University of Medical Sciences, Tehran, Iran
[i]Molecular and Cell biology Research Center, Faculty of Medicine, Mazandaran University of Medical Sciences, Sari, Iran
*Corresponding author: e-mail address: ehsan.enderami@gmail.com

Contents

Vitamins and Hormones, Volume 116
ISSN 0083-6729
https://doi.org/10.1016/bs.vh.2021.02.006

Abstract

Type 1 diabetes mellitus occurs when beta cell mass is reduced to less than 20% of the normal level due to immune system destruction of beta cell resulting in an inability to secrete enough insulin. The prevalence of diabetes is expanding according to the American Diabetes Association and the World Health Organization (WHO), foretold to exceed 350 million by 2030. The current treatment does not cure many of the serious complications associated with the disease such as neuropathy, nephropathy, dyslipidemia, retinopathy and cardiovascular disease. Whole pancreas or isolated pancreatic islet transplantation as an alternative therapy can prevent or reduce some of the complications of diabetes. However, the shortage of matched organ or islets cells donor and alloimmune responses limit this therapeutic strategy. Recently, several reports have raised extremely promising results to use different sources of stem cells to differentiate insulin-producing cells and focus on the expansion of these alternative sources. Stem cells, due to their potential for multiple differentiation and self-renewal can differentiate into all cell types, including insulin-producing cells (IPCs). Generation of new beta cells can be achieved from various stem cell sources, including embryonic stem cells (ESCs), adult stem cells, such as mesenchymal stem cells (MSCs) and induced pluripotent stem cells (iPSCs). Thus, this chapter discusses on the assistance of cellular reprogramming of various stem cells as candidates for the generation of IPCs using transcription factors/miRNA, cytokines/small molecules and tissue engineering.

1. Generation of IPCs from iPSCs

1.1 Generation of IPCs from iPSCs with growth factors/cytokines involved

The induced pluripotent stem cells (iPSCs) are a group of cells that could be generated from differentiated somatic cells by nuclear reprogramming of embryonic and adult fibroblasts using plasmids transfecting. Initial production of iPSCs was done utilizing the retroviral-mediated of Oct3/4, Sox2, Klf4, and c-Myc (Kaji et al., 2009; Kumar, Delgado, Mendez, & Joddar, 2019). Then, non-viral methodologies were reported using repeated transfection of plasmids containing Oct3/4, Sox2, Klf4, and c-Myc resulted in the generation of iPS cells without evidence of plasmid integration (Okita, Hong, Takahashi, & Yamanaka, 2010).

The iPS cells generated from somatic cells are expected to use a replacement in cell-based therapy. The iPSCs as a suitable source of pluripotent stem cells are resolved problems that pertain to embryonic cells restrictive ethical and legal issues. So, the patient received iPSCs would not suffer immune rejection since they are autologous, like the use of embryonic cells that have limited to established clones (Shen, Cheng, Han, Mu, & Han, 2013;

Yamanaka, 2012). However, the yield of their generation is low. Moreover, the generated iPSCs show infinite proliferation activity and form tumors upon transplantation due to viral transfection of transcription factors require the use of genomes (Calafiore & Basta, 2015). These genomes can result in mutations and prevent the usual function of iPSCs and their ability to differentiate. Self-renew infinitely of iPSCs and their potential for differentiation into any type of cell of the three germ layers such as insulin-producing cells (IPCs) were favorited for cell-based therapy in diabetic patients (Tateishi et al., 2008). It was demonstrated that iPSC represents a valuable source for cell base therapy because of their ability to differentiate into beta-cell type in vitro (Csobonyeiova, Polak, & Danisovic, 2020). The initial step of differentiation is definitive endoderm formation followed by pancreatic epithelial cell induction and the ultimate differentiation into the beta-cells (Fig. 1) (Kroon et al., 2008). Production of beta-like cells derived iPSCs was also achieved by pancreatic developmental stages using growth

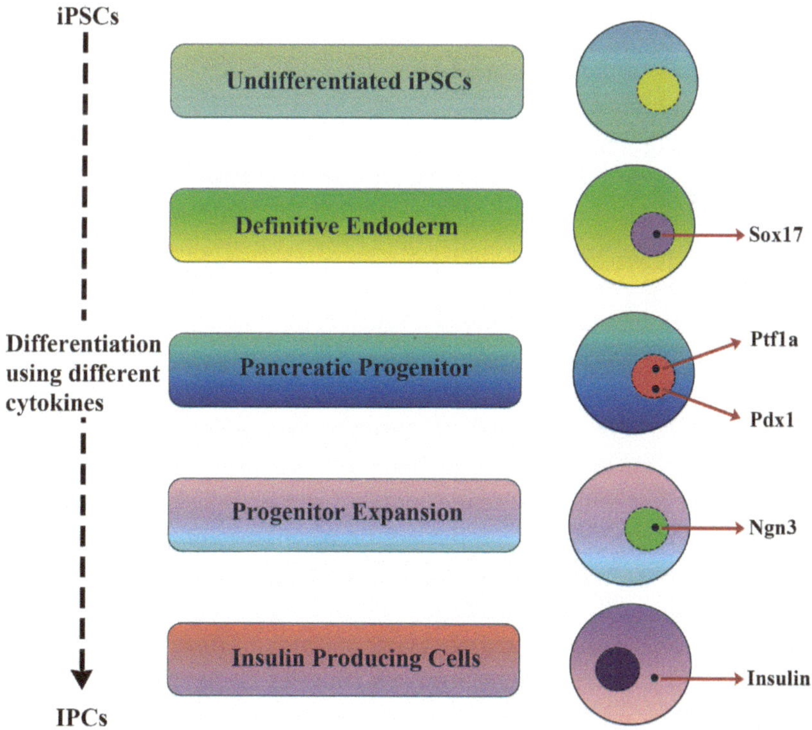

Fig. 1 Schematic representation of the stepwise differentiation of induced pluripotent stem (iPSC) cells to insulin-producing cells.

factors, small molecules, cytokines and vectors carrying different genes in vitro (Raikwar et al., 2015). There are numerous published protocols about the production of beta-like-cells challenging process.

The first published report showed the differentiation of iPSCs derived from fibroblasts into insulin-producing cells by mimicking the process of pancreatic development using a four-stage differentiation protocol. During this protocol iPSCs have also been most recently differentiated into C-peptide and glucagon-positive cells through stimulation with a combination of sodium butyrate and activin A with additional growth factors (Tateishi et al., 2008).

One of the published reports showed the differentiation of iPSCs derived from mouse fibroblasts into insulin-producing cells using a three-stage differentiation protocol, containing laminin, insulin, selenic acid, nicotinamide, transferrin, progesterone, and KO serum, based on embryoid bodies formation (Alipio et al., 2010).

The first study to produce iPSCs from a large animal model of T1D was reported by Zhu et al. Generating IPCs from monkey iPSC by a four-stage protocol demonstrated that the cells could secrete insulin in response to glucose stimulation, in addition, reducing the blood glucose levels of 50% of diabetic mice (Zhu et al., 2011).

During a differentiation protocol to induce beta-cells derived from Non-obese diabetic (NOD) mouse pancreatic epithelial cells and embryonic fibroblasts, it was found transplantation of IPC with pancreatic epithelial cells origins could normalize blood glucose levels in diabetic mice. Their chemically defined stepwise protocol was concluded bovine serum albumin (BSA), activin A and insulin-transferrin-selenium, in summary (Jeon et al., 2012).

In a published protocol by Maehr et al. differentiation of iPSCs into IPCs was conducted using WNT3A, activin A for DE formation, and then treatment with FGF10, retinoic acid, and cyclopamine for the formation of gut tube endoderm and pancreas progenitors. These produced cells were able to respond to glucose stimulation and secrete C-peptide (Maehr et al., 2009).

The laboratory of Rezania reported the generation of functional mature beta-cells with the optimized protocol. Their protocol was included thyroid hormones, vitamin C, transforming growth factor, receptor inhibitors, and protein kinase C pathway activators. However, ultimate cells limited in insulin secretory kinase and calcium dynamics in comparison with natural islets (Rezania et al., 2014).

A five-step protocol for direct differentiation of human iPSCs without EBs formation was reported by Rajaei group. They had shown that the usage of phosphodiesterase inhibitor IBMX and small molecule SB431542 had high efficacy in the differentiation process. At the 5th stage, the maturity of beta cells was confirmed by expression of, NKX2-2, PDX1, NKX6-1, and FOXA2 in addition to the presence of insulin secretory granules in cytoplasm of the generated beta cells (Rajaei, Shamsara, Amirabad, Massumi, & Sanati, 2017).

The laboratory of Melton reported a differentiation protocol that can produce glucose-responsive cells with insulin-secreting ability. This protocol involves the use of WNT pathway activator, retinoic acid, activin, TGF, and Gama secretase inhibitors. The produced cells by them were able to expose C-peptide and Nxk6.1. These cells secreted insulin into the serum of diabetic mice shortly after transplantation and then ameliorate hyperglycemia in mice (Pagliuca et al., 2014).

The other protocol consists of treatment of human iPSC with noggin combined with epidermal growth factor (EGF) and nicotinamide was reported by Nostro group. These factors were used to induce NKX6.1 expression via inhibition of bone morphogenetic protein (BMP). The generated insulin-positive cells not only are able to respond to glucose but also secrete insulin following transplantation in vivo (Nostro et al., 2015; Pellegrini et al., 2015).

Enderami et al. described an experimental method using a stepwise differentiation protocol for the differentiation of hiPS cells into IPCs that optimized with adding platelet-rich plasma (PRP). Their data demonstrated that the use of PRP as an effective bioactive substance play a synergistic role for pancreatic differentiation during this optimized multi-step protocol for cell-based replacement therapy to treat T1DM by generating of IPCs. The result of this study demonstrated that PRP led to a change in the cell cycle in undifferentiated iPSCs. This process occurred by the signaling MAPK pathway activating, followed by cyclinD-CDK4 activity and prevent apoptosis phase after G1-s phase exiting (Enderami et al., 2017).

The other study reported about generation of the insulin-producing cells from human iPSCs using recombinant growth factors that mimic the essential signaling for pancreatic development. The protocol was the inclusion of pancreatic endoderm inducer indolactam V and glucagon-like peptide-1 (GLP-1) in pancreatogenic cocktail regimens facilitates generation of glucose-responsive islet-like cells from a human iPSCs source. This platform for islet-like cell generation allow modeling of patient-specific disease pathogenesis, and ultimately lead to personalized approaches for diabetes cell therapy with insulin-producing cells from human iPS cells (Thatava et al., 2011).

A developed stepwise protocol to generate insulin-producing cell with 85% efficiency from iPSCs were reported by Zhu et al. The generated $PDX1^+$ cells could secrete insulin in response to glucose stimulation and after transplantation into mouse models of diabetes; these insulin-producing cells could decrease blood glucose levels in approximately 50% of streptozotocin (STZ)-treated diabetic mice. This four-stage protocol initiates with activin A and wingless-related MMTV integration site 3A (WNT3A) treatment of the iPSCs. In step 2, the cells were exposed to retinoic acid and keratinocyte growth factor for 6 days. Following this, in the next step, the cells were treated with EGF and SB431542. In the final step, the cells were treated with hepatocyte growth factor, IGF1, exendin-4 and nicotinamide for 7 days (Zhu et al., 2011).

1.2 Generation of IPCs from iPSCs with miRNAs/genes involved

Generation of insulin-producing cells (IPCs) from induced pluripotent stem cells by means of induction of certain transgene expression can be a promising approach in the field of type 1 diabetes treatment (Walczak, Drozd, Stoczynska-Fidelus, Rieske, & Grzela, 2016).

MicroRNAs are a class of small noncoding ribonucleic acid that regulates gene expression through post-transcriptional mechanisms. MicroRNAs can regulate iPSCs fate decisions and pancreatic beta-like cell differentiation through direct repression of their target genes (Soltani, Jafarian, & Allameh, 2019).

In this regard, Li et al. differentiated human iPSCs into IPCs in a stepwise process that iPSCs were transduced with lentiviral vectors carrying miR-181c-5p precursor. The differentiated cells significantly increased the endodermal markers FOXA2, CXCR4, SOX17, and GATA4 and pancreatic endocrine-specific gene expression, including NKX6.1, PDX1, MAFA and Insulin. The miR-181c-5p directly targets the 3′UTR of smad7 and TGIF2 mRNA, which are known to be endogenous repressors of TGF-β-smad2/3 signaling. On the other hand, induction of miR-181c-5p allows for the increased expression of the genes involved in the maturation and formation of IPCs. In addition, the inhibition of TGIF2 and smad7 facilitates the phosphorylation of smad2/3, thus favoring the pancreatic beta cell phenotype specification.

Furthermore, the functional evaluation of generated cells was performed by in vivo transplanting differentiated cells under the kidney capsule of mice.

The results showed that the glucose stimulated C-peptide secretion elevated and the fasting blood glucose levels reduced in miR-181c-5p overexpression group (Li et al., 2020).

Shaer et al. differentiated iPSCs through in vitro chemical transfection by microRNA-7 to obtain functional insulin like clusters in a short time. The level of microRNA-7 increased by 107-fold in iPSCs 24 h after chemical transfection, whereas the expression level of microRNA-7 was low in iPS cells before the transfection.

The clusters were characterized by the expression of pancreatic specific transcription factors and were stained for insulin and neurogenin-3 proteins. The cells were stained with DTZ and secreted insulin in response to high glucose concentration in the culture medium. During their one step protocol, the iPS colonies without MEF were cultured for transfection application (Shaer, Azarpira, Karimi, Soleimani, & Dehghan, 2016).

Walczak et al. (2016) coupled two strategies for the generation of insulin-producing cells. During the first approach, iPSCs were transduced with Tet-On 3G transactivation gene by lentiviral system. These cells were again transduced with lentiviral particles carrying NKX6.1 and PDX1-VP16. The second approach includes differentiation of initial stem cells by activation of TGF or inhibition of GSK3 signaling pathways by small chemical molecules and protein ligands, such as B27, NS21 supplements, retinoic acid, SB431542, nicotinamide and dorsomorphin. This study demonstrated that over-expression of NKX6.1 and PDX1 genes, which encode transcription factors indispensable for pancreatic development enhances insulin production at the stage of maturation of IPSC.

Generated cells were shown molecular markers of pancreatic cells and secreted insulin into the culture medium. Furthermore, cells expressing both NKX6.1 and PDX1-VP16 in the course of the differentiation process secreted a higher level of insulin (Walczak et al., 2016).

Lahmy et al. laboratory presented a growth factor-free approach to differentiate human iPSCs into insulin-producing cells by overexpressing lentiviral carrying miR-375. During this protocol, infected cells were aggregated to produce embryoid bodies (EBs), in the absence of MEF in "hanging drops." Ultimately, miR-375 transduced cells have similar characteristics to pancreatic beta cells and they were shown in vitro functionality by secretion insulin in a glucose-dependent manner. They have demonstrated the application of miRNAs in the protocols without any other stimulator to the generation of IPCs for cell replacement therapy in diabetes type 1 patients (Lahmy et al., 2014).

1.3 Generation of IPCs from iPSCs with 3D condition involved

In recent years, various methods, such as the use of differentiation medium and gene transfer, have been used to differentiate iPSCs into insulin-producing cell-based therapy for T1DM. These strategies are time consuming and expensive (Enderami et al., 2017; Piran et al., 2017). Many researchers have been tried to obtain new opportunities from tissue engineering and regenerative medicine to solve these problems. Tissue engineering through the scaffolds made of biodegradable and biocompatible polymers can help overcome the limitations of current beta cell production protocols by providing tissue microenvironments outside the body (Dawson, Mapili, Erickson, Taqvi, & Roy, 2008; Goh et al., 2013). The scaffolds made of biomaterials quickly adapt to the human body and do not stimulate the immune response. These scaffolds provide a three dimensional (3D) environment that enhances the proliferation and differentiation of stem cell types into pancreatic beta cells by mimicking in vivo environment; support the cell-cell contacts and cell-extracellular matrix (ECM) interactions and form the tissue (Aloysious & Nair, 2014; Ravi, Paramesh, Kaviya, Anuradha, & Solomon, 2015). Similar to the ECM, electrospinning is the other technique for the preparation of very thin polymer fibers as 3D construction, which can be easily applied in a lab. The culture of cells on 3D fibrous scaffolds by imitating in vivo conditions could increase the efficiency of making beta-like cells. So, simulation of the ECM plays a vital role in the differentiation and function of pancreatic cells (Benton, George, Kleinman, & Arnaoutova, 2009; Zhang, Lim, Ramakrishna, & Huang, 2005).

There are numerous studies have tried to utilize various polymers based scaffold for the promotion of beta-like cells differentiation.

Enderami et al. study have investigated the effect of the poly-L-lactic acid and polyvinyl alcohol (PLLA/PVA) scaffold as a substrate for the differentiation of iPSCs into higher functional beta-like cells. To evaluate the effect of PLLA/PVA scaffold on iPSCs during pancreatic differentiation, they seeded the iPSC colonies in tissue culture plates (TCPs) and PLLA/PVA scaffold with the same pancreatic induction medium and further assessed cell morphology and function. During the differentiation process, in both groups, morphology of cells gradually changed and finally, hiPSC-IPCs converted spherical-shaped cell that was the typical shape of islets of Langerhans. The results of this study showed that these IPCs had beta cells pancreatic characteristic in the mRNA and protein levels and also were seen to respond to glucose stimulation test (Enderami, Kehtari, et al., 2018).

In the other study, the differentiation potential of the iPSCs into IPCs while cultured on poly (lactic-*co*-glycolic) acid (PLGA)/polyethylene glycol were presented by Abazari et al. (Enderami, Kehtari, et al.) nanofibrous scaffold as a 3D substrate, compared to mesenchymal stem cells cultures on tissue culture polystyrene (TCPS) as a 2D substrate. Biocompatibility and morphological assays showed that the scaffold had a nanofibrous structure without cytotoxicity.

The typical cell's morphology of IPC and survival rate of the human iPSCs cultured on the scaffold were significantly higher than differentiated MSCs. Examination of the pancreatic related gene and insulin and C-peptide secretion assays confirmed the major efficacy of this 3D culture systems because of helping to the fusion of the iPS cells in each other, which induced by fibrous scaffolds that are similar to the in vivo microenvironment of cells in the body (Abazari et al., 2020).

Mansour et al. showed that PES-collagen is a great potential to be used for pancreatic tissue engineering and regenerative medicine applications. They have improved the biological activity of Polyethersulfone (PES) nanofibers by coating them with collagen, while iPSCs culture on the surface of them to generate beta-like cells, because of the abundant protein present in the islets extra-matrix in the body is collagen to further simulate the pancreatic ECM in vitro. Their results were demonstrated a higher expression level of pancreatic specific markers in generated IPCs on PES-collagen nanofibers in comparison with those generated on culture plate dishes as a control group during the study period.

Also, it was shown that the coating of the collagen on the surface of the PES scaffold resulted in improved IPCs differentiation efficacy of the iPSCs and it can be introduced as a more appropriate scaffold in comparison with intact PES nanofibers for use in pancreatic tissue engineering applications. In addition, IPCs produced in this study were demonstrated as functional cells with a response to the glucose challenges test (Mansour et al., 2018).

The other study of Abazari group showed that the polycaprolactone and polyvinyl alcohol (PCL/PVA) scaffold as a 3D culture have positive effects on the viability and expression of pancreatic-specific genes during differentiation condition for iPSCs and generates the IPCs. PCL/PVA-based scaffold may indicate biological properties and give suitable synthetic ECM for the improvement of beta-like cell differentiation. PCL is biodegradable and biocompatible polyester with high mechanical stability and hydrophobicity characteristic which has been intensively studied as a good candidate for fabrication of 3D culture. About PVA, this polymer is a biodegradable polymer with low toxicity, hydrophilic and inexpensive polymer with good mechanical and thermal stability (Xue et al., 2014; Zarekhalili, Bahrami, Ranjbar-Mohammadi, & Milan, 2017). In other word, the connection between beta-like cells and the extracellular matrix was provided by PCL/PVA scaffold for maintaining pancreatic cells spherical architecture, function and survival. In the present study, to evaluate the effect of this scaffold on iPSCs during pancreatic differentiation, they seeded the iPSCs in nanofibrous scaffolds. In order to induce IPCs, they used a cocktail of growth factors such as bFGF, EGF, activin-A and betacellulin. EGF and bFGF play an essential role in cellular proliferation, survival and differentiation. EGF and bFGF, have been shown to be useful in IPCs differentiation. The result suggests that the nanofibrous scaffold efficiently mimics the role of ECM by preparing a positive environment for IPCs. This condition obtainable by the nanofibrous scaffold is due to the excellent porosity, presence of biocompatible and biodegradable polymer, and lack of toxic agents (Abazari et al., 2018).

2. Generation of IPCs from ESCs

2.1 Generation of IPCs from ESCs with growth factors/cytokines involved

Embryonic stem cells (ESCs) originate from the inner cell mass of the blastocyst. Despite morphological differences and the need for different growth factors, human and mouse embryonic stem cells are similar at the molecular level (Sato et al., 2003). Among different stem cell types, ESCs are pluripotent that means they have the capacity to differentiate into three primary germ layers; ectoderm, endoderm and mesoderm and produce all of the cell types in the body (Park et al., 2008). Also, they have an unlimited capacity for self-renewal (Yao et al., 2006). These two properties of ESCs make them an exciting candidate for regenerative medicine and biomedical research. There are three different basic methods of the differentiation of ESCs: (1) three-dimensional aggregates of ESCs, which are known as embryoid bodies (EBs). EBs have spherical structures and are an intermediate stage that has undergone an early stage in ESCs differentiation, which can differentiate into specific cell types according to environmental conditions and various factors. (2) The culture of ESCs on extracellular matrix (ECM) proteins. The ECMs are consists of two groups: natural ECM such as collagen and fibrin and artificial ECM such as silk (Murry & Keller, 2008). (3) The culture of ESCs on supportive stromal layers. Although it is not yet clear which method is better to use (Kubo et al., 2004).

As mentioned above, ECSs have the ability to differentiate into all types of cells, including insulin-producing cells (IPCs) (Soria, Skoudy, & Martin, 2001). In embryonic pancreas development, there are five main stages: definitive endoderm (DE), primitive gut tube (PG), posterior foregut (PF), endocrine precursor (EP) cells and IPCs (Kroon et al., 2008). The most important key in the production of IPCs from ESCs is the recapitulation of essential signals in pancreatic development in embryogenesis (Lumelsky et al., 2001). In this regard, various studies by using chemical factors and small molecules have been conducted.

In the early study, IPCs were produced from hESCs in a five-step protocol (D'Amour et al., 2006). In stage 1, hESCs were differentiated through mesendoderm to DE using activin A and Wnt3a. SOX17, FOXA2 and CXCR4 expression were detected in produced cells (McLean et al., 2007). This stage is critical for the efficient production of IPCs from ESCs. In the next stage, by removing activin A and adding fibroblast growth factor 10 (FGF10) and cyclopamine (SHH signaling inhibitor), the primitive gut was formed and gut-tube markers expression such as HNF1B and HNF4A was upregulated (Lau, Kawahira, & Hebrok, 2006). In stage 3, in addition to cyclopamine and FGF10, retinoic acid (RA) was added to the medium (Chen et al., 2004). The presence of RA increased PDX1 and HNF6 expression, while HNF1B and HNF4A were still expressed. Simultaneous expression of these genes are indicative of posterior foregut (Jacquemin, Lemaigre, & Rousseau, 2003). In stage 4, PDX1-expressing posterior foregut endoderm cells become pancreatic endoderm and endocrine precursor. In addition to the PDX1, NKX6.1, NGN3, PAX4 and NKX2.2 also express in pancreatic endoderm cells (Wilson, Scheel, & German, 2003). On the 15th day of differentiation, endocrine cells expressing the pancreatic hormones insulin, somatostatin, glucagon and ghrelin are produced. Moreover, the expression of INS, CGC and GHRL can be detected in production cells. (Fig. 2).

Using this protocol, ESCs-derived IPCs contain many cell components that are critical to β-cell function. But these cells secrete very small C-peptide in response to glucose. This feature is similar to immature fetal β-cell. Therefore, future research will focus on further differentiation of cells into mature pancreatic islet cells.

For this purpose, Jiang et al. differentiated mesenchymal stem cells into insulin-producing cells using a four-step protocol over 36 days (Jiang et al., 2007). In stage 1, using sodium butyrate and activin A, definitive endoderm is generated from hESCs that they co-express CXCR4 and SOX17, and CXCR4 and FOXA2. In stage 2, cellular aggregates are formed under

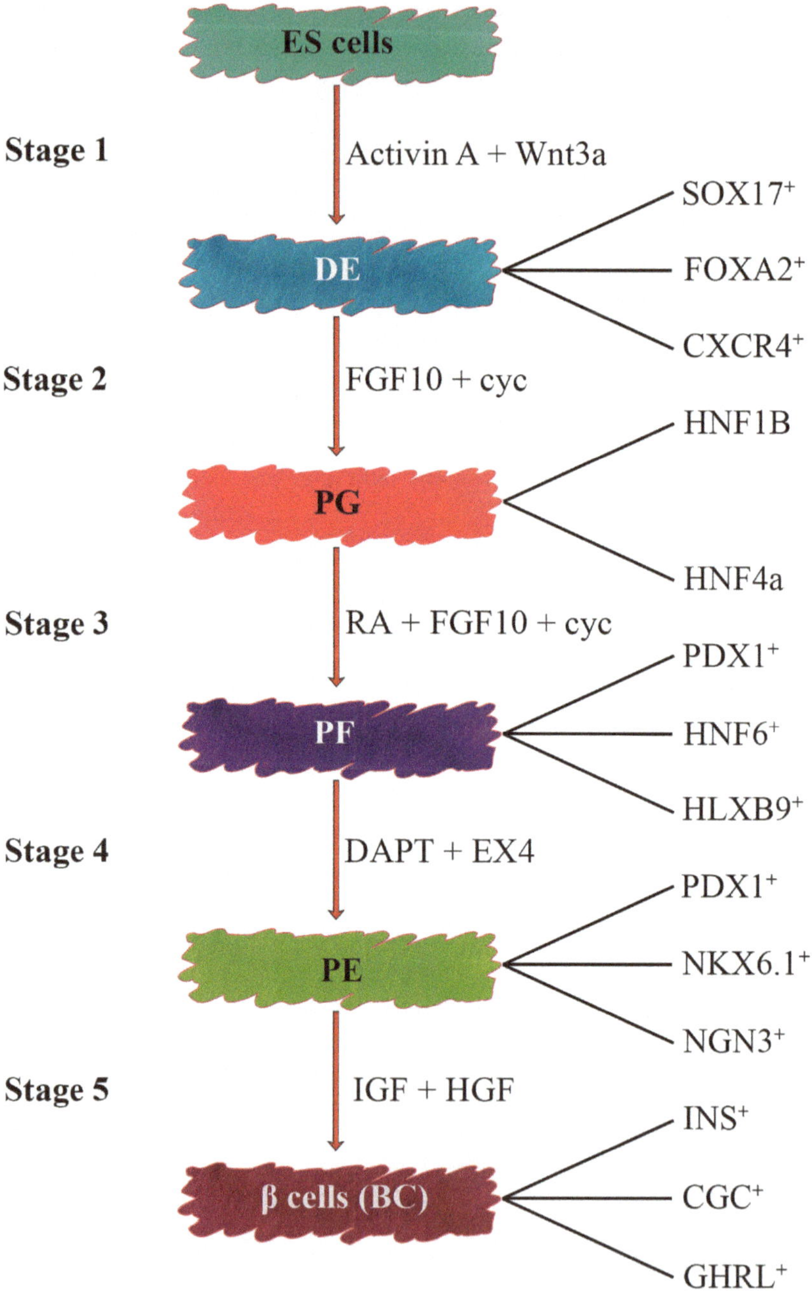

Fig. 2 Schematic representation of the steps involved in ESCs differentiation toward β-cell fate and the factors involved in this process. ESC, embryonic stem cell; DE, definitive endoderm; PG, primitive gut; PF, posterior foregut, PE, pancreatic endoderm; BC, β-cell.

low-adherent culture conditions. Subsequently, these aggregates are grown in medium containing epidermal growth factor (EGF), basic fibroblast growth factors (bFGF) and noggin. So, Pdx1-expressing pancreatic endoderm is formed (Rossi, Dunn, Hogan, & Zaret, 2001). In stage 3, the bFGF is removed from the medium and the aggregate cells become more differentiated and formed pancreatic endocrine (Deutsch, Jung, Zheng, Lóra, & Zaret, 2001). In this stage, expression of Ptf1a and Ngn3 is detected. Finally, aggregate cells are transferred into a medium containing IGFII and nicotinamide and IPCs are generated (Otonkoski, Beattie, Mally, Ricordi, & Hayek, 1993). The hESCs-derived IPCs population contained ductal, exocrine and endocrine pancreas. Also, these cells express insulin, glucagon and somatostatin hormone. Most importantly, differentiated cells produce C-peptide in response to glucose stimulation.

In another protocol, Pezzolla et al. used resveratrol (RSV) to improve ESCs differentiation conditions (Pezzolla et al., 2015). There are two advantages to using RSV: first, improving the maturation process. Second, increased insulin secretion in response to glucose levels (Fiori et al., 2013). RSV activates both AMPK and PI3K/AKT signaling pathway, thereby increase PDX1 expression. So, β-cell maturation and insulin secretion ameliorate. In this study, a combination of activin A and Wnt3 is used to definitive endoderm generation. As in the previous protocol, SOX17 and FOXA2 are expressed at this stage. In the next step, the primitive gut then posterior foregut is formed from definitive endoderm using basic bFGFs, sonic hedgehog (SHH) inhibitor, bone morphogenic proteins (BMPs) and retinoic acid. bFGF and RA induce PDX1 expression in foregut-derivate endoderm. By adding noggin and cyclopamine, pancreatic differentiation progress and pancreatic endoderm is formed and early pancreatic progenitor cells express such as ISL1, NGN3 and PDX1. In the last stage, resveratrol is used to maturation the PE into endocrine precursors (EP) and β-like cells.

The early stages of pancreatic differentiation, including the formation of ED, gut-tube like cells and pancreatic progenitors, can be efficiently induced in vitro. But differentiation of progenitor cells to IPCs is less efficient and produces a mixed population of polyhormonal endocrine cells (Guo et al., 2013). The protocols mentioned so far all have this problem and using these protocols, ESCs are differentiated into polyhormonal cells which, in addition to insulin, express other hormone and more importantly, they don't express key beta cell transcription factors. To address this issue, Russ et al. performed a detailed stepwise protocol of pancreatic progenitor generation and endocrine induction (Russ et al., 2015).

Two points are essential in the production of functional beta-like cells that were considered by Russ et al.: first, the production of progenitor cells that co-express PDX1 and NKX6.1 simultaneously. Second, the timing of Neurog3 expression. If Neurog3 is not expressed at the right time, alpha cells are produced instead of beta cells (Johansson et al., 2007).

In this protocol, as in previous protocols, activin A and Wnt3a are used to form DE. Instead of using a combination of retinoic acid, cyclopamine and noggin (referred to as RCN), only retinoic acid is used to produce PDX1$^+$ cells. In the next step, EGF and keratinocyte growth factor (KGF) is used to induce NKX6.1 expression and PDX1$^+$/NKX6.1$^+$ progenitor cells are produced (Hua et al., 2013). Endocrine differentiation in PDX1$^+$/NKX6.1$^+$ is initiated by using a cocktail of factors consisting of TBP (T), ALK inhibitor (A), Noggin (N) and KGF (K). One day after treatment with TANK, NEUROG expression is induced and endocrine differentiation initiates. Using this method, NEUROG is transiently expressed and glucose-responsive insulin-expressing beta-like cells are formed. At the same time, only a few polyhormonal cells were observed (~3%). Therefore, using this protocol, glucose-responsive beta-like cells can be produced with high efficiency.

2.2 Generation of IPCs from ESCs with miRNAs/genes involved

It is not cost-effective to use different growth factors and cytokines to induce differentiation in stem cells (Chen et al., 2009). For this reason, one of the alternative methods is to use genetic manipulation.

In early research, PDX-1 (a crucial transcription factor for pancreatic development) was used to differentiated mESCs from IPCs (Miyazaki, Yamato, & Miyazaki, 2004; Zhu, Liu, Zhou, & Ikeda, 2017). In this study, Pdx-1 expression was controlled by Tet-off system integrated into the ROSA26 locus. Pdx-1-transfected ESCs formed embryoid bodies (EBs) on day 5. The EB cells were transferred to an attached culture system. IPCs were produced within 21 days according to previous protocols and by various factors such as EGF, KGF, bFGF and nicotinamide (Hua et al., 2013). In differentiated cells the expression of insulin 2, somatostatin, Kir6.2, Pax6 and HNF6 was increased. Although Pdx-1 expression increased insulin 2 gene expression, the amount of insulin secreted into the medium was much less than that from pancreatic β-cells. In addition, differentiated cells were not able to secret insulin in response to glucose. Also, these cells after transplantation into streptozotocin-induced diabetic mice could not lower glucose.

Similar studies were performed with other transcription factors, including Nkx2.2 and Pax4, but none of the results were acceptable, and the differentiated cells were very different from beta cells (Blyszczuk et al., 2003; Shiroi et al., 2005). In pancreatic development not only transcription factors and different signaling pathways are involved, but also microRNAs (miRNA) have an important role in differentiation. Hence, the role of miRNAs in pancreatic development has been studied in various studies. Some of the miRNAs involved in pancreatic development are miR-375, miR-200a/c, miR-376, miR-148, miR-195, miR-26a, miR-7, miR-16 and miR-96 (López-Beas et al., 2018). Each of these miRNAs has a specific function in the pancreas development. For example, studies on hESCs have shown that miR-7 and miR-375 are progressively upregulated in the early stages of beta-cell generation (Wei et al., 2013). Moreover, miR-15a and miR-16a show increased expression in endocrine induction. Expression of miR-29b and miR-148a reach peak levels at the endocrine proliferation stage (Kaspi, Pasvolsky, & Hornstein, 2014). Among these, miR-375 and miR-7 are vital regulators of pancreatic development in humans and mice. Therefore, most studies have focused on two of these miRNA.

The miR-375 is a candidate miRNA for ESCs differentiation to IPCs because it is expressed exclusively in the pancreas. Also, miR-375 is necessary for glucose homeostasis. Thus, miR-375 is essential not only for the islet pancreas development, but also for the function of the mature islet (Wei et al., 2013). In a study by Lahmy et al. hESCs were transduced by pCDH-375 lentiviral vector and human embryoid bodies (hEBs) were formed, which had similar characteristics to mature islets (Lahmy et al., 2016). Morphological changes were observed in pCDH-375 hESCs and islet-like clusters were formed at the end of the experiment. In addition, Double immunofluorescence staining showed that differentiated cells secreted insulin and C-peptide. Also, DTZ staining detected IPCs within clusters. The expression pattern of genes in differentiated cells was similar to the expression pattern at the time of pancreatic development. On day 7, Foxa2 reached to maximum expression level, indicating definitive endoderm induction. Also, HNF4α and PDX1 were expressed from day 7 until the end of induction. From day 14 to day 21, NKX6.1 and PAX6 were expressed, suggesting pancreatic specialization. Taken together, in this study via miR-375 overexpression in hESCs, IPCs were produced that expressed beta cell markers. Also, these cells secreted insulin in response to glucose.

miR-7 is one of the most important miRNAs in the development of the pancreas and is the most abundant endocrine miRNA. Increased miR-7 expression at weeks 14–18 induces PDX-1 expression as well as other transcription factors involved in endocrine cell differentiation (Nieto et al., 2012). In a study by López-Beas et al. the hESC line HS181 was transfected by miR-7 mimic and maintained for 24 h in pancreatic differentiation medium in order to form cell clusters and to mature the beta cell-like phenotype. With increasing miR-7 expression, the expression of pancreatic markers, including PDX-1, FOXA2 and insulin was increased. Also, up-regulation of miR-7 caused secretion of insulin from hESC-derived IPCs in a glucose-induced manner. Thus, miR-7 is not only essential for the differentiation of beta cells and DE formation, but also is necessary to regulate maturation and functional aspects of IPCs.

2.3 Generation of IPCs from ESCs with 3D condition involved

The protocols mentioned so far for differentiating ESCs into IPCs were all in a 2-dimensional (2D) environment. Because activated and inactivated signaling pathways are different between 2D and 3D environments, the cells produced in the 2D environment are phenotypically immature pancreatic endocrine (Jin, Yao, Krisanarungson, Haukas, & Ye, 2012). Cells in cellular organisms are surrounded by a 3D network consisting of a complex of different non-living materials, which is called the extracellular matrix (ECM). Cell-cell and cell-matrix contact in a 3D environment improves ESCs differentiation and the producing cells are phenotypically more similar to mature cells. Therefore, most studies have focused on simulating physiological that conditions occurred in vivo in the laboratory. Various studies have shown that the composition of ECM proteins in scaffold, promotes differentiation of ESCs into different cell types. For instance, vitronectin differentiates hESC into oligodendrocytes during neurogenesis (Gil et al., 2009), while the addition of fibronectin differentiates them into hepatocyte-like cells (Rodríguez-Juan et al., 2009).

Collagen is one of the main components of the ECM. In a study, Jin et al. showed that collagen improved the differentiation of ESCs into IPCs (Jin, Wang, Wang, & Ye, 2014). In this study, it was shown that hESCs proliferate in the collagen scaffold exponentially. Also, by calcein-PI live/dead cells, staining revealed that 98% of cells were viable. Therefore, it clear that collagen allows hESCs to proliferate and cause them to be viable in the environment.

The results of hESCs culture in a 3D environment compared to 2D were as follows:

(1) Expression of three gene markers (Foxa2, CXCR4 and Sox17) in differentiated cells in the 3D culture shows a four- to fivefold increase compared to 2D culture. (2) Cells in the clusters formed in 3D culture expressed 50–60% more insulin than clusters formed in 2D culture. In addition, in the 3D culture environment, insulin^{+} cells were at the center of the clusters, similar to adult islet cell clusters in the body. While insulin^{+} cells in the 2D culture environments were scattered around the clusters. (3) IPCs secreted more insulin in response to glucose in a 3D culture environment than in a 2D environment. (4) IPCs in the collagen scaffold produced a high C-peptide level, whereas little C-peptide were synthesized in the IPCs formed in the 2D culture. Generally, the collagen scaffold improves the differentiation of hESCs into IPCs and leads to mature cells' formation.

Organoids are 3D assemblies that contain multiple cell types, organized similarly to the cells in a specific tissue (Yin et al., 2016). The production of an in vitro organoid system requires key components of both an organ-specific cell source as well as a platform to induce lineage specification and self-organization. In this regard, Candiello et al. used a new synthetic hydrogel called Amikagel (Grandhi, Potta, Nitiyanandan, Deshpande, & Rege, 2017) (generated by amikacin polymerization with PEG-diglycidylether at 40 °C) to make islets organoids (Candiello et al., 2018). Also, Matrigel was used as a control 3D scaffold. This study consisted of two steps: spheroid formation and islet organoids formation. In the first step, human embryonic stem cell derived pancreatic progenitor cells (hESC-PP) aggregate spontaneously in Amikagel and form spheroid. In the second step, spheroids are co-cultured with HUVEC to form heterogeneous organoids (Fig. 3).

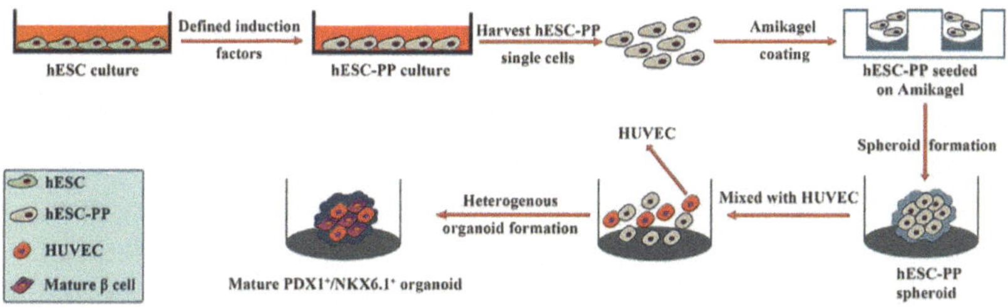

Fig. 3 Schematic of heterogeneous organoid formation process on Amikagel. hESCs-PP are generated from hESCs. Then, after hESCs-PP seeding on Amikagel, spheroids are formed. Finally, with the addition of HUVEC, heterogeneous organoids are formed.

Various reports have suggested that HUVEC induce insulin gene and protein expression in hESC-PP cells via paracrine signals (Jaramillo, Mathew, Mamiya, Goh, & Banerjee, 2015). The hESC-PP spheroids had a significant increase of cell populations co-expressing NKX6.1 and PDX1. Nearly 70% of aggregate cell populations in the Amikagel expressed both NKX6.1 and PDX1, while only 16% of the cells were $NKX6.1^{+}$ and $PDX1^{+}$ in the Matrigel. One of the hallmarks of differentiated cells is an increase in their population in the G1 phase (Jaramillo, Mathew, Task, Barner, & Banerjee, 2014). Flow cytometry analysis revealed that 70% of hESC-PP spheroids were in the G1 phase, a sign of cell maturation. Also, 30% of the hESC-PP population was positive for C-peptide protein, which was able to secrete insulin in response to glucose. Taken together, Amikagel provides conditions that cause spontaneous aggregation of differentiating hPSCs and ultimately, production of the spheroid. Subsequently, heterogeneous organoids can be produced with hESC-PP spheroids and supporting cells such as HUVEC, which improve the differentiation conditions and maturation of cells.

3. Generation of IPCs from MSCs

3.1 Generation of IPCs from MSCs with growth factors/cytokines involved

The field of regenerative medicine prepares great chances for the formation of beta cells from different stem cell sources for cell therapy. In spite that insulin-producing cells (IPC) can be generated from a diversity of stem cell type such as induced pluripotent stem cells (iPSCs) and embryonic stem cells (ESC), Mesenchymal stem cells (MSCs) are a noteworthy option for cell therapy for type 1 diabetes (Zhang et al., 2017), because of their considerable features including easy isolation, immune privilege and highly plastic stem cells, clinical translation to generate autologous cells, induced revascularization and angiogenesis of transplanted cells and has fewer ethical problem compared with iPSCs and ESCs (Watt et al., 2013). In addition mesenchymal stem cells are the most pivotal members in adult stem cells family and are the best candidate for cell therapy because of these properties: differentiation multipotency, self-renewal capability, long-term ex vivo proliferation, immunomodulatory potential, and can differentiate into ectodermal, mesodermal, and ectodermal tissues (Krampera, Franchini, Pizzolo, & Aprili, 2007). The MSCs can be obtained from two sources: perinatal and adult tissues. Perinatal include the amniotic fluid, Wharton's jelly, umbilical cord blood, and placenta. Adult sources include bone marrow (BM), adipose tissue, dental pulp and endometrial polyps (Kalaszczynska & Ferdyn, 2015; Marino et al., 2019). Among these sources, bone marrow and adipose tissue are the most studied.

Mesenchymal stem cells are useful in regenerative therapy in view of these cells secreting various growth factors, cytokines and chemokines in response to environmental cues and secrete them in culture media (Somal et al., 2016). Most of these molecules are considered as paracrine agents; these molecules can become active or inactive adjacent cells by their influence on intracellular pathways without influencing other cells of the body (Yang et al., 2013). In this regard, various studies by using chemical factors and small molecules have been conducted.

The first published report showed the differentiation of MSCs derived from adipose cells (ADSCs) into insulin-producing cells by a new 2-step acceleration protocol. In this 2-step method, by using histone deacetylase inhibitor, the pro-endocrine pancreatic lineage was promoted. Differentiation of ADSCs into IPCs was significantly accelerated in comparison with conventional methods without the histone deacetylase inhibitor (Ikemoto et al., 2018).

One of the published reports by Ikemoto et al. showed that the differentiation of bone marrow derived mesenchymal stem cells (BMSCs) into IPCs could be improved by knockdown or repression of some factors and molecule. The result in this study demonstrates that by suppression of chicken ovalbumin upstream promoter transcriptional factor 1 (COUP-TF1) that bound to direct repeats with 1 spacer (DR1) element in the mouse insulin 2 (Ins2) promotor and overexpression of MafA, this lead to induce differentiation of BMSCs into IPCs. In addition, transplantation of these IPCs into streptozotocin-induced diabetic mice caused a reduction in blood glucose levels (Ikemoto et al., 2018).

Allahverdi et al. explain that differentiation of MSCs into IPCs could be improved by genetic manipulation of them. Lentiviral vectors are useful tools for genetic manipulation. In this experiment, lentiviral vectors are used to deliver the pancreatic and duodenal homeobox factor 1 (PDX1) gene into MSCs in order to IPCs generation, BMSCs can efficiently differentiate into IPCs by PDX1 overexpression (Allahverdi et al., 2015). Boroujeni and Aleyasin added B27, nicotinamide and basic fibroblast growth factor (bFGF) to the culture medium in order to induce differentiation of human adipose tissue-derived stem cells (hADSCs) into IPCs (Boroujeni & Aleyasin, 2013). The combination of some small molecules can improve their function, Gabr et al. suggest that synergistic action between nicotinamide and betacellulin has improved the differentiation of BMSCs into IPCs (Gabr et al., 2018). In the present study, the role of PAX4, a transcriptional factor, was investigated in the differentiation of mesenchymal stem cells into IPCs. The effects of PAX4 and PDX1 on differentiation of MSCs into IPCs in vitro by using adenoviruses carrying PAX4 and PDX1 was evaluated. PAX4 regulates the expression of Nkx6.1, MafA, insulin, and GLUT2, in PDX1-positive MSCs to promote their differentiation. The finding data displayed that PAX4 promoted the differentiation of PDX1-positive MSCs to IPCs (Xu et al., 2017).

3.2 Generation of IPCs from MSCs with miRNAs/genes involved

Generation of insulin-producing cell from mesenchymal stem cells with key roles of microRNAs (miRNAs) can be a promising cell-based approach for insulin replacement therapy (Zhao et al., 2020). MicroRNAs are a large family of highly conserved small non-coding RNAs have been recognized as notable regulators of gene expression at post transcriptional level (Jafarian et al., 2015). In addition they play a crucial role in the regulation of mesenchymal stem cells fate, differentiation and self-renewal (Shenoy & Blelloch, 2014). Direct differentiation of MSCs to IPC also depend on significant changes in miRNA expression (Bai, Gao, Zhang, Yang, & Guan, 2017). There is a large group of miRNAs with an established role in beta cell biology (Filios & Shalev, 2015). In this regard, Jafarian et al. provided for the pancreatic differentiation of human bone marrow mesenchymal stem cells (hBM-MSCs) by up-regulation of mir-375 and downregulation of mir-9 by lentiviruses containing mir-375 and anti-mir-9, investigated whether mir-375 and/or mir-9 lentiviruses infected cells were able to produce, store and secret insulin. The differentiated cells significantly increased the endodermal markers SOX17, HNF-3 beta/foxA2 and pancreatic endocrine-specific gene expression, including PDX1, Ngn3, insulin, glucagon, somatostatin, pancreatic polypeptide and B cells specific genes like Nkx2.2 and GLUT2. Despite insulin gene and protein expression and even a useful factor for in vitro differentiation of hMSCs into islet-like cellular aggregate, there is no response to glucose in derived IPC by alone mir-375. However, downregulation of mir-9 in MSCs did not have any effect on differentiation and insulin secretion but after induction into islet-like aggregates containing mir-375 and mir-9, could increase cellular response to different concentration of glucose because of increasing of OCT-2 protein expression that leads to a decrease in the level of its target gene, granuphilin, that has a negative role in insulin secretion (Jafarian et al., 2015).

The second study explains a procedure for expansion of bone marrow mesenchymal stem cells in in vitro. Their differentiation to insulin-producing cells was reported by Zalzman, Anker-Kitai, and Efrat, a transcription factor that plays important roles in pancreas development and B-cells gene expression is duodenal homeobox 1 (PDX1) (Zalzman, Anker-Kitai, & Efrat, 2005). BM-MSCs has been achieved from 14 donors; both sexes, ages 19–70 was cultured and infected with a retrovirus expressing rat pdx1 cDNA. Then evaluated the extent of phenotypic change induced by stable expression of PDX1. Analyses data in the level of RNA and protein demonstrate the expression of all four islet hormones. On the other hand, the cells lacked expression of NEUROD, a key transcription factor. Finding results show that PDX1 induces a deep change in gene expression in an adult human mesodermal tissue cultured in vitro and can increase their differentiation toward B-cell phenotype. Cells transplantation under the renal capsule into STZ-diabetic SCID mice resulted in more differentiation (induction of NEUROD1), and reduced blood glucose levels (Karnieli, Izhar-Prato, Bulvik, & Efrat, 2007).

In a published protocol by Pang group, the generation of IPCs from bone marrow mesenchymal stem cells could be used as a therapeutic option for diabetic patients. In this study, four kinds of cells were used: normal mouse pancreatic B cells, non-B cells as a control, bone marrow mesenchymal stem cells (BM-MSCs), and adipose-derived stem cells (ADSCs), in the following next-generation sequencing (NGS) used to detect the miRNA expression profile of these cells. Data shows the expression of insulin in B cells is higher than others, and compare with the above cells. Among all of the target genes of miRNA, four novel miRNA candidates (13_7354-5p, 3_17977-5p, 7_25147-5p, and 7_25719-5p) were more expressed in B cells were demonstrate to target genes that prevent differentiation of MSCs to IPCs. They found that the potentially novel mouse miRNA 13_7354-5p is a mouse pancreatic B cell-specific miRNA. By using mirDeep2.0, found that chr13_novelMir7354-5p, a novel candidate, notably promoted the differentiation of BM-MSCs into insulin-producing cells in vitro. IPC in the 13_7354-5p group expressed pdx1, NeuroD1, somatostatin, IAPP, and glucagon on day 56 of differentiation. In addition, chr13_novelMir7354-5p transfected BM-MSCs transplanted under left renal capsule of streptozotocin (STZ) treated diabetic mice; an obvious reduction in blood glucose was appear. Moreover, bioinformatic analyses show that Notch1 and Rbpj are direct targets of chr13_novelMir7354-5p. This study presents a new treatment strategy to generate insulin-producing cells from stem cells (Zhao et al., 2020). Guan and his colleagues explain the potent production of IPCs from nestin-positive umbilical cord-derived mesenchymal stem cells (N-UCMSCs) following overexpression of mir-375 and mir-26a.

microRNA expression profile in N-UCMSCs with a sequencing method was examined. mir375 cannot induce the differentiation of N-UCMSCs to IPC alone. In this regard mir375 and mir26a co-transfected into N-UMSCs. The result show that by downregulating target genes such as mtpn, sox6, bhlhe22 and ccnd1; mir375 and mir26a can induce IPC differentiation from N-UCMSCs. Small interfering RNAs were also used to knock down these genes in N-UCMSCs to induce the formation of IPCs (Bai, Gao, Li, et al., 2017).

In another protocol, Bai et al. used RNA-seq and qPCR to confirm the result of miRNA expression profiles of MSCs and IPCs, so as to screen for miRNA and their targets that function during the formation of IPC from MSCs. The result demonstrates that mir-34c display transient upregulation at an early stage of the generation of IPCs derived from MSCs. In another part of the study, by using bioinformatic tools, found that mir-34c and its targets play a crucial role in the generation of IPCs. So, to recognize the function of mir-34c in the maturation of IPC from MSCs, in stepwise protocol, by overexpressing mir-34c and expressing small interfering RNAs of its targets, it was found that mir-34c play as a bidirectional switch in the generation of IPC derived from MSCs. Mir-34c can increase proinsulin synthesis by downregulating its targets, including PDE7B, PDGFRA, MAP2K1. But when mir-34c enduringly dysregulate such expression, it suppressed the expression of other targets, like ACSL4 and SAR1A, caused a reduction of insulin secretion in IPCs derived from MSCs. They suggested that endogenous miRNAs involved in the generation of IPCs from stem cells show promise for advancing the development of an effective cell transplant therapy for diabetes (Bai, Gao, Li, et al., 2017).

3.3 Generation of IPCs from MSCs with 3D condition involved

Nowadays, cell-based therapy is proposed as an alternative treatment for diabetics in which islets or endocrine cells containing beta-cells are implanted in the body. Thus, they restore the normal function of the pancreas. During many studies, various protocols have been developed culturing and differentiating MSCs into IPC; it includes culturing these cells in two-dimensional or three-dimensional media culture, using multiple scaffolds and different biomaterials, and further into optimizing these scaffolds. In other words, the use of tissue engineering, it consists of cells, growth factor and scaffolds (Enderami et al., 2017; Khorsandi, Nejad-Dehbashi, Ahangarpour, & Hashemitabar, 2015). The purpose of using scaffolds, one of the important parts of pancreatic tissue engineering, is to create a supporting structure and provide the 3D conditions. It provides the biological materials needed for tissue differentiation, growth and formation (Aloysious & Nair, 2014). The cells may experience a series of physical and chemical damage during isolation, and their intercellular interactions are destroyed, which negatively affects the survival and long-term culture of cells. Thus, these cell-cell and cell-matrix interactions help to improve cell function and survival. These interactions are important for insulin secretion and the compact nature of the islets and the coordination between cells demonstrate the importance of a 3D environment on beta-cell function (Hunckler & García, 2020). The 3D culture by imitation of natural tissue and cell-extracellular matrix (ECM) allows cell-cell interactions. The 3D culture by simulating of in vivo environment and provides conditions similar to the body niche and uses different materials and methods; it provides the requirements for the 3D growth of cells and facilitates differentiation (Wang et al., 2019). The 3D scaffolds increase the effectiveness of cell therapy in the treatment of diabetes. IPCs produced by these environments, are more viable than 2D due to the addition of extracellular molecules, and cells produce more insulin and their survival increase during the differentiation period (Amer, Mahoney, & Bryant, 2014). The ECM is responsible for biochemical and structural support, which execute both signaling and repairing the tissue, and it also provides controlled environmental support. Some of these ECMs regulate cytokine sensitivity, and some of them are involved in insulin production and survival (Llacua, Faas, & de Vos, 2018). Many 3D culture systems have been developed to differentiate stem cells. When cells are implanted in these 3D scaffolds, spheroids are formed, which are 3D structure and very similar to the structure of the pancreas. Overall, the 3D culture system is beneficial for pancreatic differentiation (Ichihara, Utoh, Yamada, Shimizu, & Uchigata, 2016; Toyoda et al., 2015). The scaffold used for the 3D culture, it generally has two types, natural and synthetic. After the process of cell lysis, the decellularized tissues turn into structures that are similar to scaffold that it naturally has the 3D infrastructure and growth factors and they are used in tissue engineering as a scaffold, this type of scaffolding is one of the most suitable scaffolds due to its natural nature, it also has a positive effect on the differentiation of stem cells into beta-cells and the maturation of these cells (Wang, Wang, Zhang, Qiang, & Luo, 2017; Zhou et al., 2016). Taken into consideration the fact that, transplanted cells are prone to immune system attacks, encapsulating is one way to protect these cells. The importance of biomaterials has been proven not only in encapsulation and safety protection, but also improving the transplant site, and angiogenesis. A wide range of natural and synthetic biomaterials is used in vivo and in vitro, and they are also used in the construction of scaffold and encapsulation (Ernst, Wang, & Ma, 2018; Gattás-Asfura, Abuid, Labrada, & Stabler, 2019; Song et al., 2019). Among the natural biomaterials used in tissue engineering is collagen (Yap et al., 2013), laminin, gelatin (Echave, Burgo, Pedraz, & Orive, 2017), alginate (de Vos, Lazarjani, Poncelet, & Faas, 2014), silk (Enderami et al., 2020) and fibrin-based biomaterials (Riopel, Trinder, & Wang, 2015). Hence, synthetic biomaterials are PLA, PLLA, PGA, which are among the polyesters (Liu et al., 2016) and the group of polyvinyl are PVA, PVP (Nyitray, Chavez, & Desai, 2014), and PEG which are the most abundant biomaterials in tissue engineering (Spicer, 2020). In the following, we review a number of studies on the effect of the 3D media culture on the survival and differentiation of mesenchyme stem cells into insulin-producing cells and examine their effect on the function of these cells.

Ojaghi et al. adipose tissue-derived stem cells (ADSCs) cultured in a poly-L-lactic acid/polyvinyl alcohol (PLLA/PVA) nanofiber scaffold and differentiated into IPCs (Ojaghi et al., 2019). After isolation, ADSCs were seeded on nanofiber scaffolds and then the morphology of these cells was examined after 21 days and spheroid formation was confirmed. Then, gene expression was assessed at the mRNA and protein levels. The results demonstrated the expression of the cultured sample in the 3D scaffold increased in comparison to the control and 2D groups. The results also showed that the response to glucose and the secretion of insulin and glucagon by generated beta-cells in the 3D scaffold increased significantly and this study confirmed that PLLA/PVA scaffolds are a suitable option for pancreatic tissue engineering.

In the study of Abazari et al. the differentiation of MSCs and induced pluripotent stem cells (iPSCs) into the beta-cell in the 2D and 3D media cultures were compared. In this study, poly (lactic-*co*-glycolic) acid (PLGA)/polyethylene glycol (PEG) nanofibrous scaffold was used for the 3D culture and tissue culture polystyrene (TCPS) was employed as a substrate for the 2D culture. Results of morphology displayed both cells formed a cluster in the 3D scaffold. They used insulin and C-peptide secretion in response to glucose to evaluate cell function in the 3D scaffold compared to the 2D environment. The data of ELISA assay indicated iPSCs had more functionality than MSCs. The results of gene and protein expression showed that pancreatic markers significantly increased in the 3D culture (in both cells); meanwhile, iPSCs expressed more protein markers than MSCs (Abazari et al., 2020).

In another study, differentiation of adipose-derived mesenchyme stem cells (ADSCs) into IPC in the collagen/hyaluronic acid scaffold was evaluated by Khorsandi, Khodadadi, Nejad-Dehbashi, and Saremy (2015). The results of the study showed collagen and HA had positive effects on cell differentiation and proliferation and used as a scaffold in pancreatic tissue engineering (Tang, Vickers, Hsu, & Spector, 2007; Wang, Zhao, Jiang, & Ding, 2009). In this study, ADSCs were extracted from adipose tissue and placed on a Col/HA scaffold after fabrication. Thanks to flow cytometry test, the expression of insulin protein was measured and it was found that in the 3D environment, expression was significantly higher than other groups. PCR results also proved that the 3D scaffold had positive results on the expression of pancreatic genes, including pdx1, glucagon and insulin, and gene expression significantly increased in this group. In the 3D group, the response to glucose or functionality was better than the 2D media and control groups.

Enderami et al. added platelet-rich plasma (PRP) into the differentiation media along with the PVA scaffold to improve pancreatic differentiation protocol (Enderami, Soleimani, Mortazavi, Nadri, & Salimi, 2018). In this study, adipose-derived stem cells (ADSCs) were differentiated into IPCs on PVA scaffold. PVA polymer has several features such as hydrophilic, flexible biocompatible and biodegradable. The results of the MTT test showed survival in seeded cells on PVA significantly increased in comparison to the 2D groups. The gene and protein expression of differentiated cells in the 3D groups was significantly increased compared to the 2D groups. IPCs in the 3D groups were successful in response to high glucose concentration and were significantly different from the other groups.

Another study, Barati et al. trabecular meshwork derived stem cells were differentiated into IPCs in the microfluidic encapsulation. They used microfluidic encapsulation to protect IPCs against the host immune system. They made an alginate microfiber from a microfluid device and they utilized it to differentiate TM-MSCs. Morphological features confirmed cluster formation in the differentiation media. High expression of pancreatic genes was observed in the TM-MSCs-derived IPCs in the 3D alginate microfiber and it was significantly higher than the 2D culture. At the protein level, pancreatic markers were expressed, such as PDX-1. The study also performed an in vivo phase and it was transplanted encapsulated and non-encapsulated IPCs into the diabetic rat. The results showed that, encapsulated IPCs were able to decrease glucose blood levels in a few months (Barati et al., 2019).

4. Conclusion

Diabetes is a disease that occurs basically due to insulin deficiency, and hyperglycemia. The employment of stem cells for the diabetes treatment toward insulin-producing cells can promise therapeutic purposes for treatment of diabetes. There has been a continuous search for (Mesenchymal stem cells) MSCs, (Embryonic stem cells) ESCs and (induced Pluripotent stem cells) iPSCs as stem cell sources with optimal properties, such as broad accessibility, differentiation efficiency, low complexity safety and post-transplantation of procedures involved. In addition to different sources of stem cells, there are a number of approaches that have been commonly observed in many induction protocols for the generation of IPCs. The need to formulate novel methods for the differentiation of functional beta-cell is a priority. The regeneration of insulin-producing cells can be made possible by controlling the regulation of different factors. Usage of cocktail/induction medium, gene/miRNA delivered and three-dimensional conditions can be trigger proliferation, differentiation, and insulin expression in a population of stem cells (Fig. 4). These extrinsic factors have been shown to promote IPCs direct differentiation and increase insulin content of beta-like cells both in vitro and in vivo. Each of the strategies to generate IPCs has its competency and drawbacks. So, it is not possible to define which one will be victorious in different induced methods exactly.

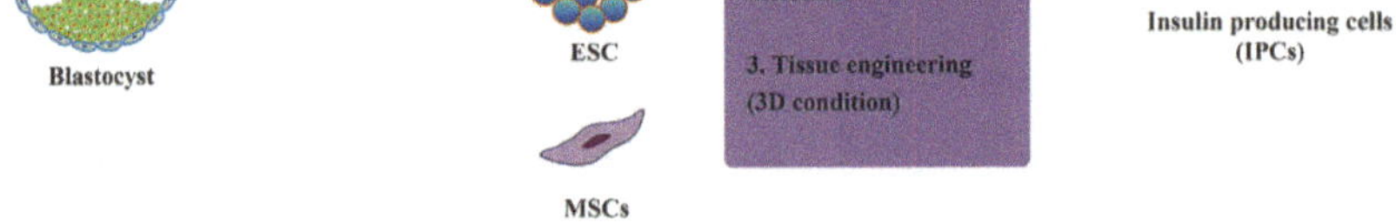

Fig. 4 A variety of candidate stem cells with the potential to develop into insulin-secreting cells. Embryonic stem cells (ESC), induced pluripotent stem cells (iPSC) and mesenchymal stem cells (MSCs) are being tested as cell sources. Current strategies for the development of stem cell-based therapies for diabetes are shown in this figure.

References

Abazari, M. F., Nasiri, N., Nejati, F., Zare Karizi, S., Amini Faskhodi, M., Saburi, E., et al. (2020). Comparison of human-induced pluripotent stem cells and mesenchymal stem cell differentiation potential to insulin producing cells in 2D and 3D culture systems in vitro. *Journal of Cellular Physiology*, *235*(5), 4239–4246.

Abazari, M. F., Soleimanifar, F., Aleagha, M. N., Torabinejad, S., Nasiri, N., Khamisipour, G., et al. (2018). PCL/PVA nanofibrous scaffold improve insulin-producing cells generation from human induced pluripotent. *Stem Cells*, *671*, 50–57.

Alipio, Z., Liao, W., Roemer, E. J., Waner, M., Fink, L. M., Ward, D. C., et al. (2010). Reversal of hyperglycemia in diabetic mouse models using induced-pluripotent stem (iPS)-derived pancreatic β-like cells. *Proceedings of the National Academy of Sciences of the United States of America*, *107*(30), 13426–13431.

Allahverdi, A., Abroun, S., Jafarian, A., Soleimani, M., Taghikhani, M., & Eskandari, F. (2015). Differentiation of human mesenchymal stem cells into insulin producing cells by using a lentiviral vector carrying PDX1. *Cell Journal (Yakhteh)*, *17*(2), 231.

Aloysious, N., & Nair, P. D. (2014). Enhanced survival and function of islet-like clusters differentiated from adipose stem cells on a three-dimensional natural polymeric scaffold: An in vitro study. *Tissue Engineering Part A*, *20*(9–10), 1508–1522.

Amer, L. D., Mahoney, M. J., & Bryant, S. J. (2014). Tissue engineering approaches to cell-based type 1 diabetes therapy. *Tissue Engineering, Part B: Reviews*, *20*(5), 455–467.

Bai, C., Gao, Y., Li, X., Wang, K., Xiong, H., Shan, Z., et al. (2017). MicroRNAs can effectively induce formation of insulin-producing cells from mesenchymal stem cells. *Journal of Tissue Engineering and Regenerative Medicine*, *11*(12), 3457–3468.

Bai, C., Gao, Y., Zhang, X., Yang, W., & Guan, W. (2017). MicroRNA-34c acts as a bidirectional switch in the maturation of insulin-producing cells derived from mesenchymal stem cells. *Oncotarget*, *8*(63), 106844.

Barati, G., Nadri, S., Hajian, R., Rahmani, A., Mostafavi, H., Mortazavi, Y., et al. (2019). Differentiation of microfluidic-encapsulated trabecular meshwork mesenchymal stem cells into insulin producing cells and their impact on diabetic rats. *Journal of Cellular Physiology*, *234*(5), 6801–6809.

Benton, G., George, J., Kleinman, H. K., & Arnaoutova, I. P. (2009). Advancing science and technology via 3D culture on basement membrane matrix. *Journal of Cellular Physiology*, *221*(1), 18–25.

Blyszczuk, P., Czyz, J., Kania, G., Wagner, M., Roll, U., St-Onge, L., et al. (2003). Expression of Pax4 in embryonic stem cells promotes differentiation of nestin-positive progenitor and insulin-producing cells. *Proceedings of the National Academy of Sciences of the United States of America*, *100*(3), 998–1003.

Boroujeni, Z. N., & Aleyasin, A. (2013). Insulin producing cells established using non-integrated lentiviral vector harboring PDX1 gene. *World Journal of Stem Cells*, *5*(4), 217.

Calafiore, R., & Basta, G. (2015). Stem cells for the cell and molecular therapy of type 1 diabetes mellitus (T1D): The gap between dream and reality. *American Journal of Stem Cells*, *4*(1), 22.

Candiello, J., Grandhi, T. S. P., Goh, S. K., Vaidya, V., Lemmon-Kishi, M., Eliato, K. R., et al. (2018). 3D heterogeneous islet organoid generation from human embryonic stem cells using a novel engineered hydrogel platform. *Biomaterials*, *177*, 27–39.

Chen, S., Borowiak, M., Fox, J. L., Maehr, R., Osafune, K., Davidow, L., et al. (2009). A small molecule that directs differentiation of human ESCs into the pancreatic lineage. *Nature Chemical Biology*, *5*(4), 258–265.

Chen, Y., Pan, F. C., Brandes, N., Afelik, S., Sölter, M., & Pieler, T. (2004). Retinoic acid signaling is essential for pancreas development and promotes endocrine at the expense of exocrine cell differentiation in Xenopus. *Developmental Biology*, *271*(1), 144–160.

Csobonyeiova, M., Polak, S., & Danisovic, L. (2020). Generation of pancreatic β-cells from iPSCs and their potential for type 1 diabetes mellitus replacement therapy and modelling. *Experimental and Clinical Endocrinology and Diabetes*, *128*(05), 339–346.

D'Amour, K. A., Bang, A. G., Eliazer, S., Kelly, O. G., Agulnick, A. D., Smart, N. G., et al. (2006). Production of pancreatic hormone–expressing endocrine cells from human embryonic stem cells. *Nature Biotechnology*, *24*(11), 1392–1401.

Dawson, E., Mapili, G., Erickson, K., Taqvi, S., & Roy, K. (2008). Biomaterials for stem cell differentiation. *Advanced Drug Delivery Reviews*, *60*(2), 215–228.

de Vos, P., Lazarjani, H. A., Poncelet, D., & Faas, M. M. (2014). Polymers in cell encapsulation from an enveloped cell perspective. *Advanced Drug Delivery Reviews*, *67*, 15–34.

Deutsch, G., Jung, J., Zheng, M., Lóra, J., & Zaret, K. S. (2001). A bipotential precursor population for pancreas and liver within the embryonic endoderm. *Development*, *128*(6), 871–881.

Echave, C. M., Burgo, S. L., Pedraz, L. J., & Orive, G. (2017). Gelatin as biomaterial for tissue engineering. *Current Pharmaceutical Design*, *23*(24), 3567–3584.

Enderami, S. E., Ahmadi, S. F., Mansour, R. N., Abediankenari, S., Ranjbaran, H., Mossahebi-Mohammadi, M., et al. (2020). Electrospun silk nanofibers improve differentiation potential of human induced pluripotent stem cells to insulin producing cells. *Materials Science and Engineering: C*, *108*, 110398.

Enderami, S. E., Kehtari, M., Abazari, M. F., Ghoraeian, P., Nouri Aleagha, M., Soleimanifar, F., et al. (2018). Generation of insulin-producing cells from human induced pluripotent stem cells on PLLA/PVA nanofiber scaffold. *Artificial Cells, Nanomedicine, and Biotechnology*, *46*(Suppl. 1), 1062–1069.

Enderami, S. E., Mortazavi, Y., Soleimani, M., Nadri, S., Biglari, A., & Mansour, R. N. (2017). Generation of insulin-producing cells from human-induced pluripotent stem cells using a stepwise differentiation protocol optimized with platelet-rich plasma. *Journal of Cellular Physiology*, *232*(10), 2878–2886.

Enderami, S. E., Soleimani, M., Mortazavi, Y., Nadri, S., & Salimi, A. (2018). Generation of insulin-producing cells from human adipose-derived mesenchymal stem cells on PVA scaffold by optimized differentiation protocol. *Journal of Cellular Physiology*, *233*(5), 4327–4337.

Ernst, A. U., Wang, L.-H., & Ma, M. (2018). Islet encapsulation. *Journal of Materials Chemistry B*, *6*(42), 6705–6722.

Filios, S. R., & Shalev, A. (2015). β-cell microRNAs: Small but powerful. *Diabetes*, *64*(11), 3631–3644.

Fiori, J. L., Shin, Y.-K., Kim, W., Krzysik-Walker, S. M., González-Mariscal, I., Carlson, O. D., et al. (2013). Resveratrol prevents β-cell dedifferentiation in nonhuman primates given a high-fat/high-sugar diet. *Diabetes*, *62*(10), 3500–3513.

Gabr, M. M., Zakaria, M. M., Refaie, A. F., Ismail, A. M., Khater, S. M., Ashamallah, S. A., et al. (2018). Insulin-producing cells from adult human bone marrow mesenchymal stromal cells could control chemically induced diabetes in dogs: A preliminary study. *Cell Transplantation*, *27*(6), 937–947.

Gattás-Asfura, K., Abuid, N., Labrada, I., & Stabler, C. (2019). Promoting dendrimer self-assembly enhances covalent layer-by-layer encapsulation of pancreatic islets. *ACS Biomaterials Science & Engineering*, *6*(5), 2641–2651.

Gil, J.-E., Woo, D.-H., Shim, J.-H., Kim, S.-E., You, H.-J., Park, S.-H., et al. (2009). Vitronectin promotes oligodendrocyte differentiation during neurogenesis of human embryonic stem cells. *FEBS Letters*, *583*(3), 561–567.

Goh, S.-K., Bertera, S., Olsen, P., Candiello, J. E., Halfter, W., Uechi, G., et al. (2013). Perfusion-decellularized pancreas as a natural 3D scaffold for pancreatic tissue and whole organ engineering. *Biomaterials*, *34*(28), 6760–6772.

Grandhi, T. S. P., Potta, T., Nitiyanandan, R., Deshpande, I., & Rege, K. (2017). Chemomechanically engineered 3D organotypic platforms of bladder cancer dormancy and reactivation. *Biomaterials*, *142*, 171–185.

Guo, S., Dai, C., Guo, M., Taylor, B., Harmon, J. S., Sander, M., et al. (2013). Inactivation of specific β cell transcription factors in type 2 diabetes. *The Journal of Clinical Investigation*, *123*(8), 3305–3316.

Hua, H., Shang, L., Martinez, H., Freeby, M., Gallagher, M. P., Ludwig, T., et al. (2013). iPSC-derived β cells model diabetes due to glucokinase deficiency. *The Journal of Clinical Investigation*, *123*(7), 3146–3153.

Hunckler, M. D., & García, A. J. (2020). Engineered biomaterials for enhanced function of insulin-secreting β-cell organoids. *Advanced Functional Materials*, *30*, 2000134. https://doi.org/10.1002/adfm.202000134.

Ichihara, Y., Utoh, R., Yamada, M., Shimizu, T., & Uchigata, Y. (2016). Size effect of engineered islets prepared using microfabricated wells on islet cell function and arrangement. *Heliyon*, *2*(6), e00129.

Ikemoto, T., Feng, R., Shimada, M., Saito, Y., Iwahashi, S., Morine, Y., et al. (2018). A new 2-step acceleration protocol using a histone deacetylase inhibitor to generate insulin-producing cells from adipose-derived mesenchymal stem cells. *Pancreas*, *47*(4), 477.

Jacquemin, P., Lemaigre, F. P., & Rousseau, G. G. (2003). The Onecut transcription factor HNF-6 (OC-1) is required for timely specification of the pancreas and acts upstream of Pdx-1 in the specification cascade. *Developmental Biology*, *258*(1), 105–116.

Jafarian, A., Taghikani, M., Abroun, S., Allahverdi, A., Lamei, M., Lakpour, N., et al. (2015). The generation of insulin producing cells from human mesenchymal stem cells by MiR-375 and anti-MiR-9. *PLoS One*, *10*(6), e0128650.

Jaramillo, M., Mathew, S., Mamiya, H., Goh, S. K., & Banerjee, I. (2015). Endothelial cells mediate islet-specific maturation of human embryonic stem cell-derived pancreatic progenitor cells. *Tissue Engineering Part A*, *21*(1–2), 14–25.

Jaramillo, M., Mathew, S., Task, K., Barner, S., & Banerjee, I. (2014). Potential for pancreatic maturation of differentiating human embryonic stem cells is sensitive to the specific pathway of definitive endoderm commitment. *PLoS One*, *9*(4), e94307.

Jeon, K., Lim, H., Kim, J.-H., Thuan, N. V., Park, S. H., Lim, Y.-M., et al. (2012). Differentiation and transplantation of functional pancreatic beta cells generated from induced pluripotent stem cells derived from a type 1 diabetes mouse model. *Stem Cells and Development*, *21*(14), 2642–2655.

Jiang, J., Au, M., Lu, K., Eshpeter, A., Korbutt, G., Fisk, G., et al. (2007). Generation of insulin-producing islet-like clusters from human embryonic stem cells. *Stem Cells*, *25*(8), 1940–1953.

Jin, S., Wang, X., Wang, W., & Ye, K. (2014). Bioinspired 3D scaffold for inducing differentiation of embryonic stem cells. In *Handbook of biomimetics and bioinspiration: 3 Tissue models* (pp. 1241–1256). World Scientific.

Jin, S., Yao, H., Krisanarungson, P., Haukas, A., & Ye, K. (2012). Porous membrane substrates offer better niches to enhance the wnt signaling and promote human embryonic stem cell growth and differentiation. *Tissue Engineering Part A*, *18*(13–14), 1419–1430.

Johansson, K. A., Dursun, U., Jordan, N., Gu, G., Beermann, F., Gradwohl, G., et al. (2007). Temporal control of neurogenin3 activity in pancreas progenitors reveals competence windows for the generation of different endocrine cell types. *Developmental Cell*, *12*(3), 457–465.

Kaji, K., Norrby, K., Paca, A., Mileikovsky, M., Mohseni, P., & Woltjen, K. (2009). Virus-free induction of pluripotency and subsequent excision of reprogramming factors. *Nature*, *458*(7239), 771–775.

Kalaszczynska, I., & Ferdyn, K. (2015). Wharton's jelly derived mesenchymal stem cells: Future of regenerative medicine? Recent findings and clinical significance. *BioMed Research International, 2015*. https://doi.org/10.1155/2015/430847.

Karnieli, O., Izhar-Prato, Y., Bulvik, S., & Efrat, S. (2007). Generation of insulin-producing cells from human bone marrow mesenchymal stem cells by genetic manipulation. *Stem Cells, 25*(11), 2837–2844.

Kaspi, H., Pasvolsky, R., & Hornstein, E. (2014). Could microRNAs contribute to the maintenance of β cell identity? *Trends in Endocrinology and Metabolism, 25*(6), 285–292.

Khorsandi, L., Khodadadi, A., Nejad-Dehbashi, F., & Saremy, S. (2015). Three-dimensional differentiation of adipose-derived mesenchymal stem cells into insulin-producing cells. *Cell and Tissue Research, 361*(3), 745–753.

Khorsandi, L., Nejad-Dehbashi, F., Ahangarpour, A., & Hashemitabar, M. (2015). Three-dimensional differentiation of bone marrow-derived mesenchymal stem cells into insulin-producing cells. *Tissue and Cell, 47*(1), 66–72.

Krampera, M., Franchini, M., Pizzolo, G., & Aprili, G. (2007). Mesenchymal stem cells: From biology to clinical use. *Blood Transfusion, 5*(3), 120.

Kroon, E., Martinson, L. A., Kadoya, K., Bang, A. G., Kelly, O. G., Eliazer, S., et al. (2008). Pancreatic endoderm derived from human embryonic stem cells generates glucose-responsive insulin-secreting cells in vivo. *Nature Biotechnology, 26*(4), 443–452.

Kubo, A., Shinozaki, K., Shannon, J. M., Kouskoff, V., Kennedy, M., Woo, S., et al. (2004). Development of definitive endoderm from embryonic stem cells in culture. *Development, 131*(7), 1651–1662.

Kumar, S. A., Delgado, M., Mendez, V. E., & Joddar, B. (2019). Applications of stem cells and bioprinting for potential treatment of diabetes. *World Journal of Stem Cells, 11*(1), 13.

Lahmy, R., Soleimani, M., Sanati, M. H., Behmanesh, M., Kouhkan, F., & Mobarra, N. (2014). MiRNA-375 promotes beta pancreatic differentiation in human induced pluripotent stem (hiPS) cells. *Molecular Biology Reports, 41*(4), 2055–2066.

Lahmy, R., Soleimani, M., Sanati, M. H., Behmanesh, M., Kouhkan, F., & Mobarra, N. (2016). Pancreatic islet differentiation of human embryonic stem cells by microRNA overexpression. *Journal of Tissue Engineering and Regenerative Medicine, 10*(6), 527–534.

Lau, J., Kawahira, H., & Hebrok, M. (2006). Hedgehog signaling in pancreas development and disease. *Cellular and Molecular Life Sciences: CMLS, 63*(6), 642–652.

Li, N., Jiang, D., He, Q., He, F., Li, Y., Deng, C., et al. (2020). microRNA-181c-5p promotes the formation of insulin-producing cells from human induced pluripotent stem cells by targeting smad7 and TGIF2. *Cell Death & Disease, 11*(6), 1–12.

Liu, J. M., Zhang, J., Zhang, X., Hlavaty, K. A., Ricci, C. F., Leonard, J. N., et al. (2016). Transforming growth factor-beta 1 delivery from microporous scaffolds decreases inflammation post-implant and enhances function of transplanted islets. *Biomaterials, 80*, 11–19.

Llacua, L. A., Faas, M. M., & de Vos, P. (2018). Extracellular matrix molecules and their potential contribution to the function of transplanted pancreatic islets. *Diabetologia, 61*(6), 1261–1272.

López-Beas, J., Capilla-González, V., Aguilera, Y., Mellado, N., Lachaud, C. C., Martín, F., et al. (2018). miR-7 modulates hESC differentiation into insulin-producing beta-like cells and contributes to cell maturation. *Molecular Therapy- -Nucleic Acids, 12*, 463–477.

Lumelsky, N., Blondel, O., Laeng, P., Velasco, I., Ravin, R., & McKay, R. (2001). Differentiation of embryonic stem cells to insulin-secreting structures similar to pancreatic islets. *Science, 292*(5520), 1389–1394.

Maehr, R., Chen, S., Snitow, M., Ludwig, T., Yagasaki, L., Goland, R., et al. (2009). Generation of pluripotent stem cells from patients with type 1 diabetes. *Proceedings of the National Academy of Sciences of the United States of America, 106*(37), 15768–15773.

Mansour, R. N., Soleimanifar, F., Abazari, M. F., Torabinejad, S., Ardeshirylajimi, A., Ghoraeian, P., et al. (2018). Collagen coated electrospun polyethersulfon nanofibers improved insulin producing cells differentiation potential of human induced pluripotent stem cells. *Artificial Cells, Nanomedicine, and Biotechnology*, *46*(Suppl. 3), S734–S739.

Marino, L., Castaldi, M. A., Rosamilio, R., Ragni, E., Vitolo, R., Fulgione, C., et al. (2019). Mesenchymal stem cells from the Wharton's Jelly of the human umbilical cord: Biological properties and therapeutic potential. *International Journal of Stem Cells*, *12*(2), 218.

McLean, A. B., D'Amour, K. A., Jones, K. L., Krishnamoorthy, M., Kulik, M. J., Reynolds, D. M., et al. (2007). Activin a efficiently specifies definitive endoderm from human embryonic stem cells only when phosphatidylinositol 3-kinase signaling is suppressed. *Stem Cells*, *25*(1), 29–38.

Miyazaki, S., Yamato, E., & Miyazaki, J.-i. (2004). Regulated expression of pdx-1 promotes in vitro differentiation of insulin-producing cells from embryonic stem cells. *Diabetes*, *53*(4), 1030–1037.

Murry, C. E., & Keller, G. (2008). Differentiation of embryonic stem cells to clinically relevant populations: Lessons from embryonic development. *Cell*, *132*(4), 661–680.

Nieto, M., Hevia, P., Garcia, E., Klein, D., Alvarez-Cubela, S., Bravo-Egana, V., et al. (2012). Antisense miR-7 impairs insulin expression in developing pancreas and in cultured pancreatic buds. *Cell Transplantation*, *21*(8), 1761–1774.

Nostro, M. C., Sarangi, F., Yang, C., Holland, A., Elefanty, A. G., Stanley, E. G., et al. (2015). Efficient generation of NKX6-1 + pancreatic progenitors from multiple human pluripotent stem cell lines. *Stem Cell Reports*, *4*(4), 591–604.

Nyitray, C. E., Chavez, M. G., & Desai, T. A. (2014). Compliant 3D microenvironment improves β-cell cluster insulin expression through mechanosensing and β-catenin signaling. *Tissue Engineering Part A*, *20*(13–14), 1888–1895.

Ojaghi, M., Soleimanifar, F., Kazemi, A., Ghollasi, M., Soleimani, M., Nasoohi, N., et al. (2019). Electrospun poly-l-lactic acid/polyvinyl alcohol nanofibers improved insulin-producing cell differentiation potential of human adipose-derived mesenchymal stem cells. *Journal of Cellular Biochemistry*, *120*(6), 9917–9926.

Okita, K., Hong, H., Takahashi, K., & Yamanaka, S. (2010). Generation of mouse-induced pluripotent stem cells with plasmid vectors. *Nature Protocols*, *5*(3), 418.

Otonkoski, T., Beattie, G. M., Mally, M. I., Ricordi, C., & Hayek, A. (1993). Nicotinamide is a potent inducer of endocrine differentiation in cultured human fetal pancreatic cells. *The Journal of Clinical Investigation*, *92*(3), 1459–1466.

Pagliuca, F. W., Millman, J. R., Gürtler, M., Segel, M., Van Dervort, A., Ryu, J. H., et al. (2014). Generation of functional human pancreatic β cells in vitro. *Cell*, *159*(2), 428–439.

Park, I.-H., Zhao, R., West, J. A., Yabuuchi, A., Huo, H., Ince, T. A., et al. (2008). Reprogramming of human somatic cells to pluripotency with defined factors. *Nature*, *451*(7175), 141–146.

Pellegrini, S., Ungaro, F., Mercalli, A., Melzi, R., Sebastiani, G., Dotta, F., et al. (2015). Human induced pluripotent stem cells differentiate into insulin-producing cells able to engraft in vivo. *Acta Diabetologica*, *52*(6), 1025–1035.

Pezzolla, D., López-Beas, J., Lachaud, C. C., Domínguez-Rodríguez, A., Smani, T., Hmadcha, A., et al. (2015). Resveratrol ameliorates the maturation process of β-cell-like cells obtained from an optimized differentiation protocol of human embryonic stem cells. *PLoS One*, *10*(3), e0119904.

Piran, M., Enderami, S. E., Piran, M., Sedeh, H. S., Seyedjafari, E., & Ardeshirylajimi, A. (2017). Insulin producing cells generation by overexpression of miR-375 in adipose-derived mesenchymal stem cells from diabetic patients. *Biologicals*, *46*, 23–28.

Raikwar, S. P., Kim, E.-M., Sivitz, W. I., Allamargot, C., Thedens, D. R., & Zavazava, N. (2015). Human iPS cell-derived insulin producing cells form vascularized organoids under the kidney capsules of diabetic mice. *PLoS One*, *10*(1), e0116582.

Rajaei, B., Shamsara, M., Amirabad, L. M., Massumi, M., & Sanati, M. H. (2017). Pancreatic endoderm-derived from diabetic patient-specific induced pluripotent stem cell generates glucose-responsive insulin-secreting cells. *Journal of Cellular Physiology*, *232*(10), 2616–2625.

Ravi, M., Paramesh, V., Kaviya, S., Anuradha, E., & Solomon, F. P. (2015). 3D cell culture systems: Advantages and applications. *Journal of Cellular Physiology*, *230*(1), 16–26.

Rezania, A., Bruin, J. E., Arora, P., Rubin, A., Batushansky, I., Asadi, A., et al. (2014). Reversal of diabetes with insulin-producing cells derived in vitro from human pluripotent stem cells. *Nature Biotechnology*, *32*(11), 1121.

Riopel, M., Trinder, M., & Wang, R. (2015). Fibrin, a scaffold material for islet transplantation and pancreatic endocrine tissue engineering. *Tissue Engineering, Part B: Reviews*, *21*(1), 34–44.

Rodríguez-Juan, C., de la Torre, P., García-Ruiz, I., Díaz-Sanjuán, T., Muñoz-Yagüe, T., Gómez-Izquierdo, E., et al. (2009). Fibronectin increases survival of rat hepatic stellate cells-a novel profibrogenic mechanism of fibronectin. *Cellular Physiology and Biochemistry*, *24*(3–4), 271–282.

Rossi, J. M., Dunn, N. R., Hogan, B. L., & Zaret, K. S. (2001). Distinct mesodermal signals, including BMPs from the septum transversum mesenchyme, are required in combination for hepatogenesis from the endoderm. *Genes & Development*, *15*(15), 1998–2009.

Russ, H. A., Parent, A. V., Ringler, J. J., Hennings, T. G., Nair, G. G., Shveygert, M., et al. (2015). Controlled induction of human pancreatic progenitors produces functional beta-like cells in vitro. *The EMBO Journal*, *34*(13), 1759–1772.

Sato, N., Sanjuan, I. M., Heke, M., Uchida, M., Naef, F., & Brivanlou, A. H. (2003). Molecular signature of human embryonic stem cells and its comparison with the mouse. *Developmental Biology*, *260*(2), 404–413.

Shaer, A., Azarpira, N., Karimi, M. H., Soleimani, M., & Dehghan, S. (2016). Differentiation of human-induced pluripotent stem cells into insulin-producing clusters by microRNA-7. *Experimental and Clinical Transplantation: Official Journal of the Middle East Society for Organ Transplantation*, *14*(5), 555.

Shen, J., Cheng, Y., Han, Q., Mu, Y., & Han, W. (2013). Generating insulin-producing cells for diabetic therapy: Existing strategies and new development. *Ageing Research Reviews*, *12*(2), 469–478.

Shenoy, A., & Blelloch, R. H. (2014). Regulation of microRNA function in somatic stem cell proliferation and differentiation. *Nature Reviews Molecular Cell Biology*, *15*(9), 565–576.

Shiroi, A., Ueda, S., Ouji, Y., Saito, K., Moriya, K., Sugie, Y., et al. (2005). Differentiation of embryonic stem cells into insulin-producing cells promoted by Nkx2. 2 gene transfer. *World Journal of Gastroenterology*, *11*(27), 4161.

Soltani, A., Jafarian, A., & Allameh, A. (2019). The predominant microRNAs in β-cell clusters for insulin regulation and diabetic control. *Current Drug Targets*, *21*, 722–734.

Somal, A., Bhat, I. A., Pandey, S., Panda, B. S., Thakur, N., Sarkar, M., et al. (2016). A comparative study of growth kinetics, in vitro differentiation potential and molecular characterization of fetal adnexa derived caprine mesenchymal stem cells. *PLoS One*, *11*(6), e0156821.

Song, W., Chiu, A., Wang, L.-H., Schwartz, R. E., Li, B., Bouklas, N., et al. (2019). Engineering transferrable microvascular meshes for subcutaneous islet transplantation. *Nature Communications*, *10*(1), 1–12.

Soria, B., Skoudy, A., & Martin, F. (2001). From stem cells to beta cells: New strategies in cell therapy of diabetes mellitus. *Diabetologia*, *44*(4), 407–415.

Spicer, C. D. (2020). Hydrogel scaffolds for tissue engineering: The importance of polymer choice. *Polymer Chemistry*, *11*(2), 184–219.
Tang, S., Vickers, S. M., Hsu, H. P., & Spector, M. (2007). Fabrication and characterization of porous hyaluronic acid–collagen composite scaffolds. *Journal of Biomedical Materials Research Part A*, *82*(2), 323–335.
Tateishi, K., He, J., Taranova, O., Liang, G., D'Alessio, A. C., & Zhang, Y. (2008). Generation of insulin-secreting islet-like clusters from human skin fibroblasts. *The Journal of Biological Chemistry*, *283*(46), 31601–31607.
Thatava, T., Nelson, T. J., Edukulla, R., Sakuma, T., Ohmine, S., Tonne, J. M., et al. (2011). Indolactam V/GLP-1-mediated differentiation of human iPS cells into glucose-responsive insulin-secreting progeny. *Gene Therapy*, *18*(3), 283–293.
Toyoda, T., Mae, S.-I., Tanaka, H., Kondo, Y., Funato, M., Hosokawa, Y., et al. (2015). Cell aggregation optimizes the differentiation of human ESCs and iPSCs into pancreatic bud-like progenitor cells. *Stem Cell Research*, *14*(2), 185–197.
Walczak, M. P., Drozd, A. M., Stoczynska-Fidelus, E., Rieske, P., & Grzela, D. P. (2016). Directed differentiation of human iPSC into insulin producing cells is improved by induced expression of PDX1 and NKX6. 1 factors in IPC progenitors. *Journal of Translational Medicine*, *14*(1), 341.
Wang, C., Feng, N., Chang, F., Wang, J., Yuan, B., Cheng, Y., et al. (2019). Injectable cholesterol-enhanced stereocomplex polylactide thermogel loading chondrocytes for optimized cartilage regeneration. *Advanced Healthcare Materials*, *8*(14), 1900312.
Wang, X., Wang, K., Zhang, W., Qiang, M., & Luo, Y. (2017). A bilaminated decellularized scaffold for islet transplantation: Structure, properties and functions in diabetic mice. *Biomaterials*, *138*, 80–90.
Wang, L.-X., Zhao, H., Jiang, B., & Ding, Y. (2009). Adhesion and growth of human periodontal ligament cells on hyaluronic acid/collagen scaffold. *Hua xi kou qiang yi xue za zhi= Huaxi kouqiang yixue zazhi= West China Journal of Stomatology*, *27*(2), 220.
Watt, S. M., Gullo, F., van der Garde, M., Markeson, D., Camicia, R., Khoo, C. P., et al. (2013). The angiogenic properties of mesenchymal stem/stromal cells and their therapeutic potential. *British Medical Bulletin*, *108*(1), 25–53.
Wei, R., Yang, J., Liu, G.-Q., Gao, M.-J., Hou, W.-F., Zhang, L., et al. (2013). Dynamic expression of microRNAs during the differentiation of human embryonic stem cells into insulin-producing cells. *Gene*, *518*(2), 246–255.
Wilson, M. E., Scheel, D., & German, M. S. (2003). Gene expression cascades in pancreatic development. *Mechanisms of Development*, *120*(1), 65–80.
Xu, L., Xu, C., Zhou, S., Liu, X., Wang, J., Liu, X., et al. (2017). PAX4 promotes PDX1-induced differentiation of mesenchymal stem cells into insulin-secreting cells. *American Journal of Translational Research*, *9*(3), 874.
Xue, J., He, M., Liu, H., Niu, Y., Crawford, A., Coates, P. D., et al. (2014). Drug loaded homogeneous electrospun PCL/gelatin hybrid nanofiber structures for anti-infective tissue regeneration membranes. *Biomaterials*, *35*(34), 9395–9405.
Yamanaka, S. (2012). Induced pluripotent stem cells: Past, present, and future. *Cell Stem Cell*, *10*(6), 678–684.
Yang, D., Wang, W., Li, L., Peng, Y., Chen, P., Huang, H., et al. (2013). The relative contribution of paracine effect versus direct differentiation on adipose-derived stem cell transplantation mediated cardiac repair. *PLoS One*, *8*(3), e59020.
Yao, S., Chen, S., Clark, J., Hao, E., Beattie, G. M., Hayek, A., et al. (2006). Long-term self-renewal and directed differentiation of human embryonic stem cells in chemically defined conditions. *Proceedings of the National Academy of Sciences of the United States of America*, *103*(18), 6907–6912.
Yap, W. T., Salvay, D. M., Silliman, M. A., Zhang, X., Bannon, Z. G., Kaufman, D. B., et al. (2013). Collagen IV-modified scaffolds improve islet survival and function and reduce time to euglycemia. *Tissue Engineering Part A*, *19*(21–22), 2361–2372.

Yin, X., Mead, B. E., Safaee, H., Langer, R., Karp, J. M., & Levy, O. (2016). Stem cell organoid engineering. *Cell Stem Cell*, *18*(1), 25.
Zalzman, M., Anker-Kitai, L., & Efrat, S. (2005). Differentiation of human liver-derived, insulin-producing cells toward the β-cell phenotype. *Diabetes*, *54*(9), 2568–2575.
Zarekhalili, Z., Bahrami, S. H., Ranjbar-Mohammadi, M., & Milan, P. B. (2017). Fabrication and characterization of PVA/Gum tragacanth/PCL hybrid nanofibrous scaffolds for skin substitutes. *International Journal of Biological Macromolecules*, *94*, 679–690.
Zhang, T., Li, X.-H., Zhang, D.-B., Liu, X.-Y., Zhao, F., Lin, X.-W., et al. (2017). Repression of COUP-TFI improves bone marrow-derived mesenchymal stem cell differentiation into insulin-producing cells. *Molecular Therapy- -Nucleic Acids*, *8*, 220–231.
Zhang, Y., Lim, C. T., Ramakrishna, S., & Huang, Z.-M. (2005). Recent development of polymer nanofibers for biomedical and biotechnological applications. *Journal of Materials Science. Materials in Medicine*, *16*(10), 933–946.
Zhao, F., Liu, X., Wang, Z., Lang, H., Zhang, T., Wang, R., et al. (2020). Novel mouse miRNA Chr13_novelMiR7354-5p improves bone-marrow-derived mesenchymal stem cell differentiation into insulin-producing cells. *Molecular Therapy--Nucleic Acids*, *19*, 1110–1122.
Zhou, P., Guo, Y., Huang, Y., Zhu, M., Fan, X., Wang, L., et al. (2016). The dynamic three-dimensional culture of islet-like clusters in decellularized liver scaffolds. *Cell and Tissue Research*, *365*(1), 157–171.
Zhu, Y., Liu, Q., Zhou, Z., & Ikeda, Y. (2017). PDX1, neurogenin-3, and MAFA: critical transcription regulators for beta cell development and regeneration. *Stem Cell Research & Therapy*, *8*(1), 1–7.
Zhu, F., Zhang, P., Zhang, D., Sui, X., Yin, M., Xiang, T., et al. (2011). Generation of pancreatic insulin-producing cells from rhesus monkey induced pluripotent stem cells. *Diabetologia*, *54*(9), 2325.

CHAPTER EIGHT

The development of adult intestinal stem cells: Insights from studies on thyroid hormone-dependent anuran metamorphosis

Yun-Bo Shi*, Yuki Shibata, Yuta Tanizaki, and Liezhen Fu
Section on Molecular Morphogenesis, Eunice Kennedy Shriver National Institute of Child Health and Human Development (NICHD), National Institutes of Health (NIH), Bethesda, MD, United States
*Corresponding author: e-mail address: shi@helix.nih.gov

Contents

Abstract

Vertebrates organ development often takes place in two phases: initial formation and subsequent maturation into the adult form. This is exemplified by the intestine. In mouse, the intestine at birth has villus, where most differentiated epithelial cells are located, but lacks any crypts, where adult intestinal stem cells reside. The crypt is formed during the first 3 weeks after birth when plasma thyroid hormone (T3) levels are high. Similarly, in anurans, the intestine undergoes drastic remodeling into the adult form during metamorphosis in a process completely dependent on T3. Studies on *Xenopus* metamorphosis have revealed important clues on the formation of the adult intestine during metamorphosis. Here we will review our current understanding on how T3 induces the degeneration of larval epithelium and *de novo* formation of adult intestinal stem cells. We will also discuss the mechanistic conservations in intestinal development between anurans and mammals.

Vitamins and Hormones, Volume 116
ISSN 0083-6729
https://doi.org/10.1016/bs.vh.2021.02.010

1. Introduction

Thyroid hormone (T3) is critical for not only the physiology and function of many adult organs but also their development, especially the maturation into the adult form during vertebrate postembryonic development when plasma T3 level peaks. This period corresponds to several weeks or months around birth in mouse and human, respectively (Shi, 1999; Tata, 1993). The maternal dependence of newborns/neonates and uterus-enclosed nature of the embryos in mammals has made it difficult to study how T3 functions during postembryonic development or how adult organs are formed during this period. Interestingly, this period resembles amphibian metamorphosis. In anurans *Xenopus laevis*, T3 is low or not detectable before stage 54, about 4 weeks old and the onset of metamorphosis, reaches the peak level around stage 62 (about 7 weeks old), and drops to a lower, adult level by the end of metamorphosis (about 8 weeks old or stage 66) (Leloup & Buscaglia, 1977; Nieuwkoop & Faber, 1965). The high levels of T3 during this period triggers the transformations of essentially every organ/tissue in the tadpole. These changes include *de novo* formation of adult organs such as limbs, total degeneration of larval specific organs such as the tail and gills, and drastic remodeling of most organs, including the intestine, from the larval forms into their adult forms (Dodd & Dodd, 1976; Shi, 1999; Tata, 1993). Many of these changes resemble what happen during mammalian postembryonic development (Shi, 1999; Tata, 1993). Importantly, anuran metamorphosis can be easily manipulated by controlling the availability of T3 to tadpoles or even organ or primary cell cultures, making anuran metamorphosis a highly valuable model for studying T3 action and organ maturation during postembryonic development in vertebrates. This is further enhanced by the recent development of transgenic and gene-editing technologies suitable for genetic studies in the highly related anuran species *Xenopus laevis* and *tropicalis* (Blitz, Biesinger, Xie, & Cho, 2013; Fu, Buchholz, & Shi, 2002; Kroll & Amaya, 1996; Lei, Guo, Deng, Chen, & Zhao, 2013; Lei et al., 2012; Nakade et al., 2014; Nakayama et al., 2013; Shi et al., 2015; Wang et al., 2015; Young et al., 2011).

2. T3 and intestinal stem cell development

The vertebrate intestine has long been used as a model system for investigating the property and function of adult stem cells in vertebrates. Organ-specific adult stem cells are essential for organ homeostasis and function (Bao, Shi, & Shi, 2020; Clevers, 2013; Shi, Hasebe, Fu, Fujimoto, & Ishizuya-Oka, 2011; Sirakov, Kress, Nadjar, & Plateroti, 2014; Sun, Fu, & Shi, 2014; Sun & Shi, 2012; van der Flier & Clevers, 2009). In adult mammalian intestine, epithelial stem cells are located at the bottom of the crypts (Fig. 1). After proliferation, their daughter cells migrate along the crypt-villus axis and differentiate into different types of epithelial cells. The differentiated epithelial cells eventually die via programmed cell death, mostly at the villus tip. Such a self-renewing cycle occurs once every 1–6 days in adult mammals (MacDonald, Trier, & Everett, 1964; Toner, Carr, & Wyburn, 1971; van der Flier & Clevers, 2009) and every 2 weeks in the anuran *Xenopus laevis* (McAvoy & Dixon, 1977).

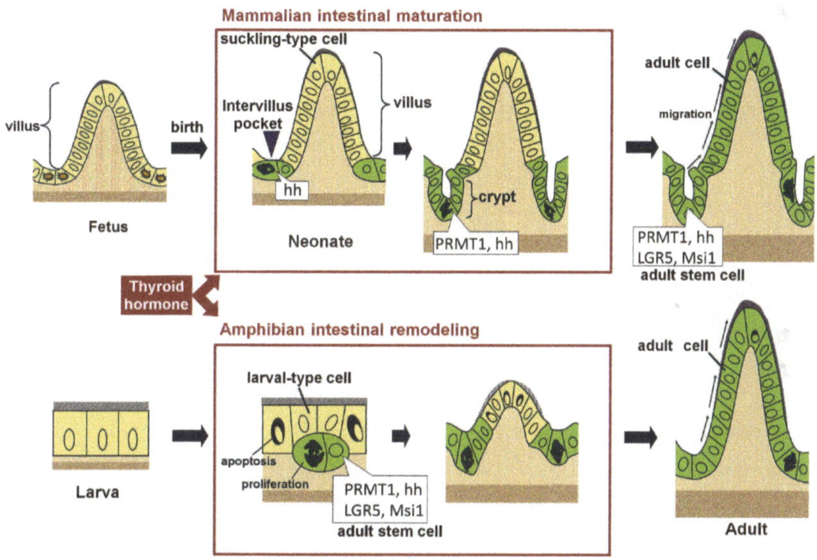

Fig. 1 Intestinal maturation takes place during postembryonic development in vertebrates. Both mouse intestinal maturation (upper panel) and frog intestinal remodeling during *Xenopus* metamorphosis (lower panel) involve the formation of adult epithelial stem cells. Some of the pre-existing epithelial cells develop into adult stem cells during the period when plasma thyroid hormone (T3) levels are high. In mouse, intervillus, sonic hedgehog (hh)-expressing epithelial cells develop into stem cells that express both hh and protein arginine methyltransferase 1 (PRMT1) in the newly formed intestinal crypts within the first 3 weeks after birth. The *Xenopus* adult intestinal stem cells are formed during metamorphosis through dedifferentiation of some larval epithelial cells to also express high levels of PRMT1 and Shh. The adult stem cell markers LGR5 and Msi1 are expressed in the developing adult intestinal stem cells at the climax of metamorphosis, although they have not been analyzed in the maturing neonatal mouse intestine. *Modified after Ishizuya-Oka, A., & Shi, Y.B. (2011). Evolutionary insights into postembryonic development of adult intestinal stem cells.* Cell & Bioscience *1, 37.*

While there have been extensive studies on the adult intestinal stem cells in mammals, revealing interesting properties of and molecular pathways important for adult intestinal stem cells (Sancho, Eduard Batlle, & Clevers, 2004; van der Flier & Clevers, 2009), much less is known about their development. Recent studies have suggested that the formation of mouse adult intestinal stem cells occurs during the neonatal period when plasma thyroid hormone (T3) level peaks (Fig. 1) (Harper, Mould, Andrews, Bikoff, & Robertson, 2011; Matsuda & Shi, 2010; Muncan et al., 2011; Sun & Shi, 2012). In addition, mutations in T3 receptors (TRs) and deficiency in T3 or TR in mouse cause intestinal defects, including reduced cell proliferation (Bao et al., 2019; Flamant et al., 2002; Kress, Rezza, Nadjar, Samarut, & Plateroti, 2009; Plateroti et al., 1999, 2001; Plateroti, Kress, Mori, & Samarut, 2006). Thus, it is very likely that T3 regulates adult intestinal stem cell development during intestinal maturation.

Similarly, the adult frog intestine is formed during T3-dependent amphibian metamorphosis (Dodd & Dodd, 1976; Shi, 1999; Tata, 1993). In premetamorphic tadpoles, the intestine consists of mainly a monolayer of larval epithelial cells with a single epithelial fold, the typhlosole, surrounded by thin layers of connective tissue and muscles (Figs. 1 and 2) (Shi & Ishizuya-Oka, 1996; Sterling, Fu, Matsuura, & Shi, 2012). As T3 level rises during metamorphosis, most larval epithelial cells undergo apoptosis but some, through yet unknown mechanisms, undergoes dedifferentiation to be highly proliferative cells expressing well-known adult intestinal stem cell markers including leucine-rich repeat-containing G-protein coupled receptor 5 (Lgr5) and Musashi-1 (Msi-1) (Figs. 1 and 2) (Ishizuya-Oka, Shimizu, Sakakibara, Okano, & Ueda, 2003; Okada, Wen, Miller, Su, & Shi, 2015; Wen, Hasebe, Miller, Ishizuya-Oka, & Shi, 2015). By the end of metamorphosis, the proliferating cells differentiate to form a multiply folded adult epithelium surrounded by thick layers of connective tissue and muscles. Accompanying these changes is drastic reduction in the length of the intestine (Ishizuya-Oka et al., 2009; Schreiber, Cai, & Brown, 2005; Shi et al., 2011; Shi & Ishizuya-Oka, 1996; Sterling et al., 2012). In the adult frog, the intestinal stem cells are localized in the trough of the epithelial fold while cell death occurs mainly at the crest of the fold, resembling those occurring in the mammalian crypt-villus unit (Ishizuya-Oka & Shi, 2011; Shi et al., 2011; Shi & Ishizuya-Oka, 1996). Importantly, like all other events during anuran metamorphosis, intestinal remodeling, including adult epithelial stem cell formation, requires T3 signaling. Blocking the synthesis of

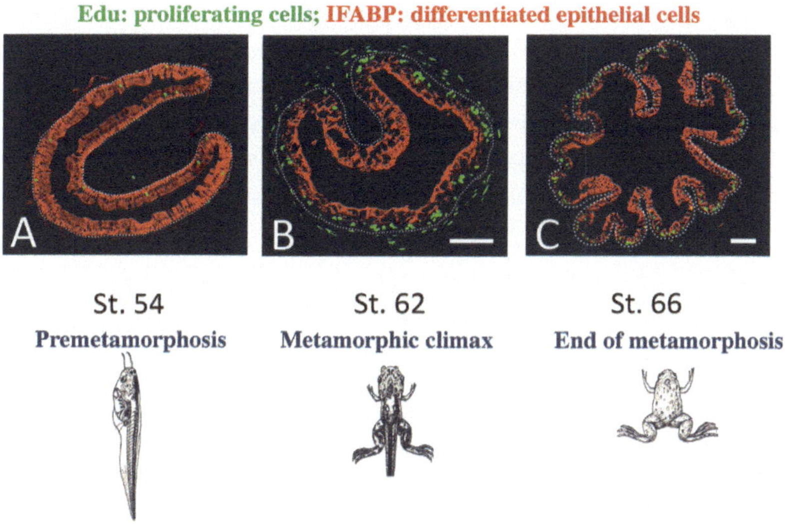

Fig. 2 Proliferating adult epithelial stem cells are formed *de novo* at the climax of *Xenopus* metamorphosis. Edu (5-ethynyl-2′-deoxyuridine, labeling newly synthesized DNA) were injected into tadpoles at premetamorphic stage 54 (A), metamorphic climax (B, stage 62), or the end of metamorphosis (C, stage 66) to label proliferating cells. One hour later, the animals were sacrificed for preparation of intestinal cross-sections, which were then double-stained for EdU and differentiated intestinal epithelial cell marker IFABP (intestinal fatty acid binding protein) by immunohistochemistry. The epithelium-mesenchyme boundary is marked by dotted lines. Note that at the climax, high levels of EdU-labeled, IFABP-negative proliferating cells were present as clusters between the connective tissue and differentiated cells expressing IFABP (B). After metamorphosis, the proliferating cells were in the troughs where IFABP was low (C), mimicking that in the adult mammalian crypt. *Modified after Okada, M., Wen, L., Miller, T.C., Su, D., & Shi, Y.B. (2015). Molecular and cytological analyses reveal distinct transformations of intestinal epithelial cells during Xenopus metamorphosis.* Cell & Bioscience *5, 74.*

endogenous T3 in tadpoles inhibits metamorphosis, including intestinal remodeling (Gilbert, Tata, & Atkinson, 1996; Shi, 1999). On the other hand, T3-treatment of premetamorphic tadpoles or even intestinal organ cultures results in premature intestinal metamorphosis, including larval epithelial cell death and *de novo* formation of adult stem cells (Shi & Ishizuya-Oka, 1996). Thus, T3 controls the *de novo* formation of the adult epithelial stem cells, which originate organ-autonomously within the larval intestine. Furthermore, T3 treatment of recombinant organ cultures made of wild type intestinal epithelium and transgenic non-epithelial tissues (the rest of the intestine) expressing GFP results in adult epithelium lacking GFP expression while GFP positive adult stem cells are produced if the transgenic epithelium expressing GFP is used in combination with either wild type or transgenic non-epithelial tissues in the organ cultures, demonstrating the epithelial origin of the adult stem cells (Ishizuya-Oka et al., 2009). Given the lack of epithelial stem cells in the tadpole intestine and the differentiated larval epithelial cells are capable of proliferation (Ishizuya-Oka & Shi, 2008; Shi & Ishizuya-Oka, 1996), T3 appears to induce the dedifferentiation of some larval epithelial cells into adult stem cells.

3. Gene regulation by TR and its roles during anuran metamorphosis

T3 is believed to function mainly through its genomic effects via binding to T3 receptors (TRs), although it also has non-genomic effects, at least in cell cultures (Buchholz, Paul, Fu, & Shi, 2006; Davis & Davis, 1996; Evans, 1988; Laudet & Gronemeyer, 2002; Lazar, 1993; Shi, 1999; Tsai & O'Malley, 1994; Yen, 2001). All vertebrates have two TR genes, TRα and TRβ. Extensive biochemical and molecular studies *in vitro* and in cell cultures have shown that both TRs can activate or repress the transcription of T3-inducible genes in the presence or absence of T3 respectively. TRs do so by mainly functioning as heterodimers with 9-cis retinoic acid receptors (RXRs) to bind to specific DNA sequences called T3-response elements (TREs) within target genes (Lazar, 1993; Mangelsdorf et al., 1995; Tsai & O'Malley, 1994; Wong et al., 1998; Wong, Shi, & Wolffe, 1995, 1997; Yen, 2001). Unliganded TR can recruit corepressor complexes containing histone deacetylases to the target genes to repress their expression (Chen & Evans, 1995; Horlein et al., 1995; Jones, Sachs, Rouse, Wade, & Shi, 2001; Jones & Shi, 2003; Li et al., 2002; Perissi, Jepsen, Glass, & Rosenfeld, 2010; Stewart, Li, & Wong, 2005; Stewart, Tomita, Shi, & Wong, 2006; Tomita, Buchholz, & Shi, 2004; Yoon et al., 2003), while liganded TRs recruit coactivator complexes such as those containing histone acetyltransferases and histone methyltransferases to activate transcription, likely via epigenetic modifications of histones and chromatin remodeling (Bulynko & O'Malley, 2011; Demarest et al., 2002; O'Malley, Malovannaya, & Qin, 2012; Sheppard, Harries, Hussain, Bevan, & Heery, 2001; Shi, Matsuura, Fujimoto, Wen, & Fu, 2012; Yen, 2001). These properties have led to a dual function model for TRs during amphibian metamorphosis (Sachs et al., 2000). That is, during premetamorphosis, TRs are mainly in the unliganded state when T3 is absent or at low levels, and thus recruit corepressors to repress T3-inducible genes. This helps to prevent precocious initiation of metamorphosis to ensure a proper growth period prior to metamorphosis. When T3 levels rise during metamorphosis, T3 binds to TR, which in turn releases the corepressors and recruit coactivators, leading to the de-repression and further activation of the T3-inducible genes, thus triggering metamorphic transformations of different tissues and organ.

With the development of chromatin immunoprecipitation (ChIP) assays and genetic tools for *in vivo* studies, we and others have tested this dual function model and investigated the roles of TRs during *Xenopus* metamorphosis. Indeed, TR and RXR have been found to be associated with T3-inducible genes in pre- and metamorphosing *Xenopus laevis* and *tropicalis* tadpoles (Sachs & Shi, 2000; Wang, Matsuda, & Shi, 2008) and recruits corepressors or coactivators to modify histones and remodel chromatin in the absence or presence of T3, respectively (Bilesimo et al., 2011; Bulynko & O'Malley, 2011; Demarest et al., 2002; Grimaldi, Buisine, Miller, Shi, & Sachs, 2013; Havis, Sachs, & Demeneix, 2003; Heimeier, Hsia, & Shi, 2008; Huang, Li, Sachs, Cole, & Wong, 2003; Matsuda, Paul, Choi, Hasebe, & Shi, 2009; Matsuda, Paul, Choi, & Shi, 2007; Matsuura, Fujimoto, Fu, & Shi, 2012; O'Malley et al., 2012; Paul, Buchholz, Fu, & Shi, 2005, 2007; Paul, Fu, Buchholz, & Shi, 2005; Sachs, Amano, Rouse, & Shi, 2001; Sachs, Amano, & Shi, 2001; Sachs et al., 2002; Sachs & Shi, 2000; Sheppard et al., 2001; Shi et al., 2012; Tomita et al., 2004; Wang et al., 2008; Wen, Fu, & Shi, 2017; Yen, 2001). More importantly, genetic studies have now demonstrated that TRs are necessary and sufficient for T3-dependent amphibian metamorphosis. First, transgenic expression of dominant negative mutant TRs that cannot bind to T3 inhibits *Xenopus laevis* metamorphosis while heat shock-inducible transgenic expression of a dominant positive TR that does not bind to T3 but functions like liganded TR mimics T3 induction of metamorphosis in wild type tadpoles, indicating that TR is sufficient to mediate most if not all effects of T3 during amphibian metamorphosis (Bagamasbad, Howdeshell, Sachs, Demeneix, & Denver, 2008; Brown & Cai, 2007; Buchholz et al., 2006; Buchholz, Hsia, Fu, & Shi, 2003; Buchholz, Tomita, Fu, Paul, & Shi, 2004; Denver, Hu, Scanlan, & Furlow, 2009; Nakajima & Yaoita, 2003; Sachs et al., 2000; Schreiber, Das, Huang, Marsh-Armstrong, & Brown, 2001; Schreiber, Mukhi, & Brown, 2009; Shi, 1994, 2009). More recently, with the adaptation of gene-editing technologies to knockout endogenous genes in the diploid *Xenopus tropicalis*, we and others have shown that both TRα and TRβ are important for metamorphosis (Buchholz & Shi, 2018; Choi, Ishizuya-Oka, & Buchholz, 2017; Choi et al., 2015; Nakajima, Tazawa, & Shi, 2019; Nakajima, Tazawa, & Yaoita, 2018; Sachs, 2015; Sakane et al., 2018; Shibata, Tanizaki, & Shi, 2020; Shibata,

Wen, Okada, & Shi, 2020; Wen et al., 2017; Wen & Shi, 2015, 2016; Yen, 2015). While animals lacking either TRα or TRβ can complete metamorphosis and develop into mature frogs, they have various developmental abnormalities during metamorphosis, suggesting that TRα and TRβ have distinct roles in coordinating tissue transformation during metamorphosis but nonetheless can compensate for each other during development. When both TR genes are knocked out, the resulting TR double knockout tadpoles initiate metamorphosis earlier (at younger age) and subsequently progress much slower through metamorphosis (Shibata, Wen, et al., 2020). Thus, consistent with the dual function model, de-repression of the T3-inducible genes leads to premature initiation of metamorphosis and that TRs are important for metamorphic progression. Interestingly and surprisingly, tadpoles lacking both TRα and TRβ can develop up to the climax of metamorphosis (stage 61, Fig. 3) (Shibata, Wen, et al., 2020), suggesting that TRs are not required for the initiation of metamorphosis and de-repression of T3 target genes due to the loss of TRs is sufficient for many of the metamorphic processes.

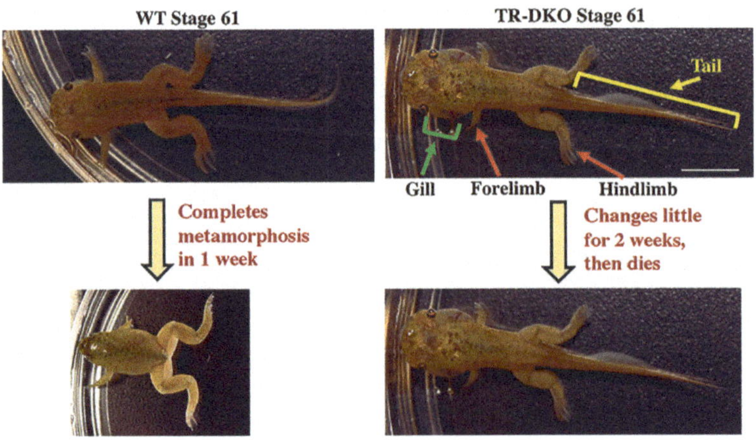

Fig. 3 TRs are not required for the initiation of metamorphosis but essential for animal survival and the completion of metamorphosis. TRα and TRβ double knockout (TR-DKO) *Xenopus tropicalis* tadpoles can develop to the climax stage of 61. Compared to wild type (WT) tadpoles at stage 61, the tadpoles lacking any TR have larger gills, suggesting inhibition of gill resorption due to the TR knockout. Interestingly, these TR-DKO tadpoles are stalled at stage 61 for 2 weeks or so and then die while the WT tadpoles at stage 61 need only about 1 week to complete the remaining metamorphic changes to become tailless froglets. *See Shibata, Y., Wen, L., Okada, M., & Shi, Y.B. (2020). Organ-specific requirements for thyroid hormone receptor ensure temporal coordination of tissue-specific transformations and completion of xenopus metamorphosis.* Thyroid *30, 300–313 for more details.*

On the other hand, compared to wild type tadpoles at stage 61, the tadpoles lacking any TR have larger gills at stage 61 (Fig. 3), suggesting inhibition of gill resorption due to the TR knockout. Furthermore, these TR double knockout tadpoles are stalled at stage 61 for 2 weeks or so and then die, in contrast to the wild type tadpoles at stage 61, which need only about 1 week to complete the remaining metamorphic changes to become tailless froglets. Further analyses suggest that the development of many of the adult tissues/organs can occur or even takes place earlier in the TR double knockout tadpoles compared to the wild type animals, while the resorption of many larval tissues/organs is inhibited due to the lack of TR (Shibata, Wen, et al., 2020). These finding suggest that de-repression of T3 target genes is sufficient for adult organ development while larval tissue degeneration requires gene activation by liganded TR.

4. An essential role of TR in adult intestinal stem cell development

Recent gene knockout studies in *Xenopus tropicalis* indicate that both TRα and TRβ are important for intestinal metamorphosis (Choi et al., 2015, 2017; Nakajima et al., 2018; Sakane et al., 2018; Shibata, Tanizaki, & Shi, 2020; Shibata, Wen, et al., 2020; Wen & Shi, 2015, 2016; Wen, Shibata, et al., 2017). TRα knockout delays intestinal remodeling (Choi et al., 2017; Wen, Shibata, et al., 2017), while there are only relatively subtle differences in intestinal remodeling between TRβ knockout tadpoles and wild type siblings during natural metamorphosis (Nakajima et al., 2018; Sakane et al., 2018), despite a strong upregulation of TRβ expression during intestinal metamorphosis. These findings suggest that there is a strong compensation between TRα and TRβ during the lengthy natural metamorphosis. This was supported by the recent analysis of intestinal remodeling in TRβ knockout during T3-induced metamorphosis (by treating premetamorphic tadpoles at stage 54 with exogenous but physiological levels of T3), where the larval epithelial cell death and adult intestinal stem cell formation occur within 2–3 days, much shorter than the 2–3 weeks required for the premetamorphic tadpoles at stage 54 to reach stages 60–62 when larval cell death and adult epithelial stem cell formation take place during the natural metamorphosis (Shibata, Tanizaki, & Shi, 2020). Consistent with the reduction in T3-induction of TR target genes in the intestine in TRβ knockout tadpoles, T3-induced reduction in intestinal length, larval epithelial cell death, and adult epithelial stem cell proliferation are all delayed/inhibited in the TRβ knockout tadpoles compared to the wild type siblings. These findings not only reveal a critical role of TRβ for intestinal metamorphosis but also demonstrate that the compensation for the loss of TRβ by TRα is too slow to prevent the defects caused by TRβ knockout during the fast intestinal remodeling induced by T3 treatment.

As described above, TRα and TRβ double knockout animals died at the climax (stage 61) of metamorphosis. Analyses of the intestine have revealed that unlike the wild type intestine at stage 61, little larval epithelial cell death or adult epithelial stem cell proliferation is present in the intestine of TRα and TRβ double knockout animals at stage 61, although adult epithelial folds are formed precociously (Shibata, Y. and Shi, Y.-B., unpublished observation). Furthermore, T3 treatment of premetamorphic TRα and TRβ double knockout animals does not induce any metamorphic changes, including the reduction of intestinal length, accompanied by the lack of the induction of T3 target genes (Shibata, Wen, et al., 2020). Thus, TRs are essential for adult epithelial stem cell development during intestinal remodeling.

A role of TR in adult stem cell development during intestinal metamorphosis has also been supported by recombinant organ culture studies involving animals containing a heat shock-inducible transgenic dominant positive TR (dpTR) that cannot bind to T3 but functions as a constitutively activated TR (Buchholz et al., 2004; Hasebe, Buchholz, Shi, & Ishizuya-Oka, 2011). By recombining wild type intestinal epithelium with transgenic non-epithelium or vice versa, T3 target genes can be selectively activated in either the epithelium or non-epithelium by heat shock treatment of the organ cultures. Analyses of such recombinant organ cultures have shown that dpTR expression in both epithelium and non-epithelium can induce larval epithelial degeneration and development of the adult intestine in the absence of T3, indicating that TR is sufficient for mediating all the effects of T3 for intestinal remodeling (Hasebe et al., 2011). Interestingly, dpTR expression in either the epithelium or non-epithelium can induce larval epithelial cell death, suggesting T3 can induce epithelial apoptosis both cell autonomously and via cell-cell interaction. On the other hand, dpTR in the non-epithelium fails to induce any adult stem cell development while its expression in the epithelium alone leads to the dedifferentiation of some larval epithelial cells, i.e., activating sonic hedgehog gene expression while repressing the expression of the differentiation marker gene encoding intestinal fatty acid binding protein (Hasebe et al., 2011). Interestingly, such dedifferentiated cells lack the expression of Msi-1, a well-known adult intestinal stem cell marker, indicating that adult stem cell development requires T3 signaling in both the epithelium and the non-epithelium (Hasebe et al., 2011). Thus, T3 signaling in the epithelium induces some larval epithelial cells to dedifferentiate and such cells develop further into adult stem cells if T3 signaling is also present in the non-epithelium, which presumably functions to ensure a proper stem cell niche for the developing stem cells (Hasebe et al., 2011; Ishizuya-Oka & Hasebe, 2013; Ishizuya-Oka & Shimozawa, 1992; Schreiber et al., 2009).

5. TR target genes in intestinal stem cell development

TR regulates target gene transcription. Thus, a key to understand how T3 induces the formation of adult stem cells is to identify and functionally characterize TR target genes in the intestine, especially within the epithelium. Over the years, different methods have been used to identify many T3-regulated genes in *Xenopus laevis* intestine and the involvement of many such genes in intestinal remodeling has been implicated by their expression profiles during metamorphosis (Amano & Yoshizato, 1998; Buchholz, Heimeier, Das, Washington, & Shi, 2007; Fu et al., 2017; Hasebe, Fujimoto, Kajita, & Ishizuya-Oka, 2017; Hasebe, Kajita, Fu, Shi, & Ishizuya-Oka, 2012; Heimeier, Das, Buchholz, Fiorentino, & Shi, 2010; Ishizuya-Oka et al., 2001; Luu et al., 2013; Luu, Fu, Fujimoto, & Shi, 2017; Miller et al., 2013; Na, Fu, Luu, & Shi, 2020; Okada, Miller, Fu, & Shi, 2015; Okada, Miller, Wen, & Shi, 2017; Okada & Shi, 2018a, 2018b; Shi & Brown, 1993; Sun et al., 2013; Sun, Fu, Wen, & Shi, 2014; Wen, Fu, & Shi, 2017). Perhaps not surprisingly, many genes in signaling pathways known to be important for stem cell proliferation and function are induced by T3 during intestinal remodeling. Among them include hedgehog pathway (Hasebe et al., 2012; Hasebe, Kajita, Shi, & Ishizuya-Oka, 2008; Ishizuya-Oka, Ueda, Inokuchi, et al., 2001; Stolow & Shi, 1995; Wen et al., 2015), wnt signaling (Hasebe, Fujimoto, Kajita, & Ishizuya-Oka, 2016; Ishizuya-Oka, Kajita, & Hasebe, 2014; Shi, 2018), Notch pathway (Hasebe et al., 2017), and BMP signaling (Ishizuya-Oka et al., 2001; Ishizuya-Oka, Hasebe, Shimizu, Suzuki, & Ueda, 2006), etc.

If and how such T3-regulated genes participate in adult intestinal stem cell development largely remain to be investigated. Limited genetic studies for some of the genes support their involvement. For example, transgenic overexpression of protein arginine methyltransferase 1 (PRMT1), a known TR coactivator that is upregulated during intestinal remodeling

(Fujimoto, Matsuura, Hu-Wang, Lu, & Shi, 2012; Matsuda et al., 2009), leads to increased intestinal stem cells during metamorphosis while antisense morpholino oligonucleotide-mediated knockdown of endogenous PRMT1 reduces proliferating epithelial stem cells (Matsuda & Shi, 2010). PRMT1 likely exerts its effect on intestinal stem cells by functioning as a TR coactivator to increase T3 signaling (Matsuda et al., 2009), although PRMT1 may also act as a coactivator for other transcription factors during stem cell development and/or proliferation or via methylation of non-histone proteins. Similarly, oncogene ectopic viral integration site 1 (EVI) and its variant myelodysplastic syndrome 1 (MDS)/EVI are highly induced by T3 at the climax of intestinal remodeling in *Xenopus laevis* and *Xenopus tropicalis* (Miller et al., 2013; Okada & Shi, 2018b). Knocking out the genes by targeting a common region with gene editing causes severe inhibition of tadpole growth and development during metamorphosis and results in partial lethality at metamorphic climax (Okada & Shi, 2018b). Importantly, the knockout tadpoles have reduced adult intestinal epithelial stem cell proliferation at the end of metamorphosis (for the few surviving animals) or during T3-induced metamorphosis (Okada & Shi, 2018b), revealing a novel function of EVI and/or MDS/EVI during vertebrate development.

A surprising finding among the T3-induced genes having an effect on intestinal stem cell development is Mad1, an antagonist of c-Myc in the Myc/Mad/Max network (Okada et al., 2017). Both Mad1 and c-Myc heterodimerize with Max and bind to the same target genes with opposing effect. While c-Myc is a well-known oncogene that activate target gene transcription and promote cell proliferation, Mad represses the expression of c-Myc target genes and is associated with quiescence or cell differentiation. In addition, c-Myc overexpression can induce cell death (Amati & Land, 1994; Dang, 1999; Koskinen & Alitalo, 1993; Kuchino, Asai, & Kitanaka, 1996; McMahon, 2014; Nieminen, Partanen, & Klefstrom, 2007; Packham & Cleveland, 1995; Pelengaris, Rudolph, & Littlewood, 2000; Shi et al., 1992; Thompson, 1998), while Mad is often anti-apoptotic (Amati & Land, 1994; Grandori, Cowley, James, & Eisenman, 2000; Luscher, 2012; McArthur et al., 1998). Consistent with these opposing roles, Mad1 and c-Myc are expressed in distinct epithelial cells during intestinal metamorphosis, with c-Myc expressed in the proliferating adult stem cells while Mad1 in the apoptotic larval epithelial cells (Fig. 4), implicating a surprising role of Mad1 in developmental cell death (Okada et al., 2017). Indeed, knocking out Mad1 inhibits/delays larval epithelial cell death during

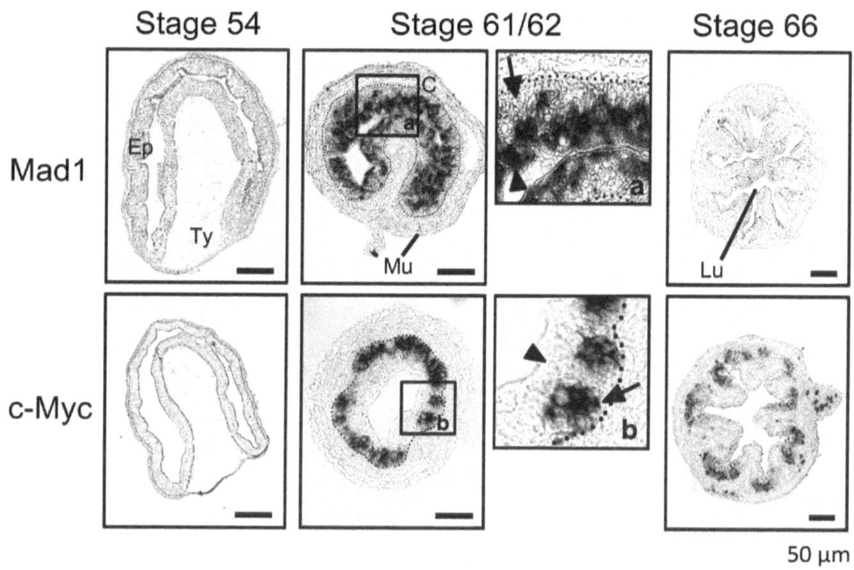

Fig. 4 Distinct expression patterns for the oncogene c-Myc and its antagonist Mad1 in the intestinal epithelium during metamorphosis. *In situ* hybridization analysis of Mad1 and c-Myc mRNAs was carried out on intestinal cross-sections of *Xenopus laevis* premetamorphic tadpoles (stage 54), metamorphosing tadpoles (stages 61/62 or climax), and post metamorphic froglets (stage 66). Panels a and b are higher magnification photos of the boxed region at stage 61/62 for Mad1 and c-Myc, respectively. Note that both Mad1 and c-Myc have little expression during premetamorphosis but are strongly activated at the climax (stages 61/62) with Mad expressed in the dying larval epithelial cells (arrowheads) while c-Myc expressed in the epithelial region close to the connective tissue (arrows). The dotted lines mark the approximate boundary between the epithelium (Ep) and the connective tissue (C). Scale bar, 50 μm. Lu, lumen; Mu, muscle; Ty, typhlosole. *Modified after Okada, M., Miller, T.C., Wen, L., & Shi, Y.B. (2017). A balance of Mad and Myc expression dictates larval cell apoptosis and adult stem cell development during Xenopus intestinal metamorphosis.* Cell Death & Disease *8, e2787.*

T3-induced intestinal metamorphosis but unexpectedly, the knockout tadpoles have increased adult epithelial stem cell proliferation during T3-induced metamorphosis (Okada et al., 2017). While the mechanism remains to be determined, one likely scenario is that Mad1 knockout inhibits/delays larval epithelial cell death, allowing more larval epithelial cells or giving larval epithelial cells, which otherwise would undergo T3-induced apoptosis, more time to undergo dedifferentiation into adult epithelial stem cells. Alternatively, Mad1 may be expressed at low levels, not detectable by *in situ* hybridization as shown in Fig. 4, in developing adult stem cells that express c-Myc. The removal of this low level of Mad1 in the knockout animals thus alters the balance of Mad-Myc, leading to increased stem cell proliferation during T3-induced metamorphosis.

The involvement of Myc/Mad/Max network in both larval epithelial cell death and adult stem cell formation and/or proliferation is also consistent with recent RNA-seq analysis of genes regulated by T3 in the intestine of premetamorphic *Xenopus tropicalis* tadpoles after 1 day T3 treatment (Tanizaki, Shibata, Zhang, & Shi, 2020). T3 treatment of premetamorphic tadpoles at stage 54 is known to induce larval epithelial cell death after 2 days and formation and proliferation of adult intestinal stem cells after 3 days. Interestingly, gene ontology and biological pathway analyses of the T3-regulated genes revealed that the gene ontology terms and signaling pathways related to cell cycle regulation were highly enriched among the regulated genes after 1 day of T3 treatment, which was too short for T3 to induce any larval cell death or stem cell formation/proliferation in the tadpole intestine. By comparing the RNA-seq data of the wild type TRα knockout tadpole intestine, it was found that the endogenous TRα was important for the regulation of the cell cycle program. The intestine in premetamorphic tadpoles are made of predominantly larval epithelial cells, surrounded by thin layers of connective tissue and muscles (Fig. 2). Furthermore, most cell death and proliferation during metamorphosis occur in the epithelium (Fig. 2) (Okada, Wen, et al., 2015). The observed activation of the cell cycle program after 1 day of T3 treatment presumably reflected the changes in gene expression within the larval epithelium. The findings from the RNA-seq analyses thus suggest a novel model for T3 function during intestinal remodeling. That is, T3, mainly via TRα, which is expressed at a much higher level compared to TRβ in premetamorphic tadpole intestine (Wang et al., 2008), induces the cell cycle program in the larval epithelial cells. This, in turn, may facilitate the differentiated larval intestinal epithelial cells to initiate cellular changes toward two alternative fates: death via apoptosis or dedifferentiation to become adult stem cells. Clearly, it would be interesting to test this model.

6. Conservations between *Xenopus* intestinal metamorphosis and mammalian intestinal maturation

As mentioned above, mouse intestinal maturation occurs within the first 3 weeks or so after birth when plasma T3 concentration peaks, just like during amphibian metamorphosis. In addition, T3 or TR deficiency leads to adult intestinal defects in mouse, including reduced proliferating crypt cells (Flamant et al., 2002; Kress et al., 2009; Plateroti et al., 1999, 2001, 2006). Furthermore, human patients with TRα mutations have constipations, implicating intestinal defects, unlike patients with TRβ mutations (Bochukova et al., 2012; Moran & Chatterjee, 2015; Refetoff, Weiss, & Usala, 1993; van Mullem et al., 2012), supporting a specific role of TRα in the intestine. Consistently, a knockin mutation in the mouse TRα gene mimicking human TRα mutant patients also results in constipations in the adult mice (Bao et al., 2019). More importantly, analyses of the adult intestine in the knockin mutant have revealed drastically reduced cell proliferation in the crypts and altered epithelial cell differentiation (Bao et al., 2019), although only minor effect on the morphology of the villus-crypt axis due to compensatory reduction in epithelial cell death on the villus (Bao et al., 2019, 2020). All these point to a role of T3 and TR in intestinal maturation and adult stem cell function.

Interestingly, PRMT1, a gene that is induced by T3 and important for intestinal stem cell formation/proliferation (see above), is also upregulated during intestinal maturation in mouse and zebrafish when T3 levels are high (Matsuda & Shi, 2010). There is little or no PRMT1 mRNA in the zebrafish larval intestine or mouse neonatal intestine. During intestinal maturation when T3 levels peaks (Brown, 1997; Friedrichsen et al., 2003), PRMT1 is upregulated specifically in the bottom of the developing epithelial fold in zebrafish intestine and the developing crypt in mouse intestine, suggesting that conserved roles of T3 and PRMT1 for the formation and/or proliferation of the adult stem cells in vertebrates. Further support for the conservation in adult intestinal stem cell development came from genetic studies on B lymphocyte-induced maturation protein 1 (Blimp1) (Harper et al., 2011; Muncan et al., 2011). Blimp1, a transcriptional repressor, is strongly and uniformly expressed in the intestinal epithelium up to newborn mice before crypts are formed. With rising T3 concentration after birth, the crypt begins to form within the inter-villus pockets of the intestine where Blimp1 expression is reduced. As the intestine develops further, Blimp1 expression is repressed in all crypt cells and eventually the rest of the epithelium in the adult mice, likely due to the gradual replacement of the neonatal epithelium by the newly differentiated cells originated from the crypts. These findings suggest that Blimp1 repression is an early event in the epithelial transformation to form the adult stem cells. The distinct expression profiles of PRMT1 and Blimp1 in the neonatal intestine compared to the adult intestinal stem cells in the crypts suggest that like anuran intestinal metamorphosis, mouse intestinal maturation involves *de novo* formation of adult stem cells in a conserved process regulated by T3.

7. Conclusion

The ability to easily manipulate amphibian metamorphosis and the recent advancement in molecular and genetic studies in *Xenopus* have made this process highly valuable model for studying postembryonic organ development in vertebrates (Bilesimo et al., 2011; Brown & Cai, 2007; Buchholz et al., 2006; Buchholz & Shi, 2018; Grimaldi et al., 2013; Matsuura et al., 2012; Shi, 2013; Shi et al., 2012). Important insights have been gained from studying intestinal metamorphosis in *Xenopus laevis* and *Xenopus tropicalis*. Among them are the discovery that adult intestinal epithelial stem cells are formed *de novo* via T3-induced dedifferentiation of some larval epithelial cells and that this process resembles postembryonic intestinal maturation in mouse. Furthermore, a number of novel genes important for adult stem cell development/proliferation have been discovered from the identification and functional analyses of some of T3 target genes during intestinal remodeling. Given the important role of T3 in intestinal maturation and adult intestinal physiology, these novel adult stem cell genes will likely have conserved roles in mammalian intestinal maturation and function. The availability of many T3-induced genes in the developing intestinal stem cells, i.e., candidate stem cell genes, and the recent advancements in genetic analysis in *Xenopus* (Blitz et al., 2013; Lei et al., 2012, 2013; Nakade et al., 2014; Nakayama et al., 2013; Shi et al., 2015; Wang et al., 2015; Young et al., 2011) will ensure intestinal metamorphosis to be as a highly valuable model for further studying adult stem cell development in vertebrates.

Acknowledgment

This work was supported by the intramural Research Program of NICHD, NIH.

References

Amano, T., & Yoshizato, K. (1998). Isolation of genes involved in intestinal remodeling during anuran metamorphosis. *Wound Repair and Regeneration*, *6*, 302–313.

Amati, B., & Land, H. (1994). Myc-Max-Mad: A transcription factor network controlling cell cycle progression, differentiation and death. *Current Opinion in Genetics & Development*, *4*, 102–108.

Bagamasbad, P., Howdeshell, K. L., Sachs, L. M., Demeneix, B. A., & Denver, R. J. (2008). A role for basic transcription element-binding protein 1 (BTEB1) in the autoinduction of thyroid hormone receptor beta. *The Journal of Biological Chemistry*, *283*, 2275–2285.

Bao, L., Roediger, J., Park, S., Fu, L., Shi, B., Cheng, S. Y., et al. (2019). Thyroid hormone receptor alpha mutations lead to epithelial defects in the adult intestine in a mouse model of resistance to thyroid hormone. *Thyroid*, *29*, 439–448.

Bao, L., Shi, B., & Shi, Y. B. (2020). Intestinal homeostasis: A communication between life and death. *Cell & Bioscience*, *10*, 66.

Bilesimo, P., Jolivet, P., Alfama, G., Buisine, N., Le Mevel, S., Havis, E., et al. (2011). Specific histone lysine 4 methylation patterns define TR-binding capacity and differentiate direct T3 responses. *Molecular Endocrinology*, *25*, 225–237.

Blitz, I. L., Biesinger, J., Xie, X., & Cho, K. W. (2013). Biallelic genome modification in F(0) Xenopus tropicalis embryos using the CRISPR/Cas system. *Genesis*, *51*, 827–834.

Bochukova, E., Schoenmakers, N., Agostini, M., Schoenmakers, E., Rajanayagam, O., Keogh, J. M., et al. (2012). A mutation in the thyroid hormone receptor alpha gene. *The New England Journal of Medicine*, *366*, 243–249.

Brown, D. D. (1997). The role of thyroid hormone in zebrafish and axoloft development. *Proceedings of the National Academy of Sciences of the United States of America*, *94*, 13011–13016.

Brown, D. D., & Cai, L. (2007). Amphibian metamorphosis. *Developmental Biology*, *306*, 20–33.

Buchholz, D. R., Heimeier, R. A., Das, B., Washington, T., & Shi, Y.-B. (2007). Pairing morphology with gene expression in thyroid hormone-induced intestinal remodeling and identification of a core set of TH-induced genes across tadpole tissues. *Developmental Biology*, *303*, 576–590.

Buchholz, D. R., Hsia, V. S.-C., Fu, L., & Shi, Y.-B. (2003). A dominant negative thyroid hormone receptor blocks amphibian metamorphosis by retaining corepressors at target genes. *Molecular and Cellular Biology*, *23*, 6750–6758.

Buchholz, D. R., Paul, B. D., Fu, L., & Shi, Y. B. (2006). Molecular and developmental analyses of thyroid hormone receptor function in *Xenopus laevis*, the African clawed frog. *General and Comparative Endocrinology*, *145*, 1–19.

Buchholz, D. R., & Shi, Y. B. (2018). Dual function model revised by thyroid hormone receptor alpha knockout frogs. *General and Comparative Endocrinology*, *265*, 214–218.

Buchholz, D. R., Tomita, A., Fu, L., Paul, B. D., & Shi, Y.-B. (2004). Transgenic analysis reveals that thyroid hormone receptor is sufficient to mediate the thyroid hormone signal in frog metamorphosis. *Molecular and Cellular Biology*, *24*, 9026–9037.

Bulynko, Y. A., & O'Malley, B. W. (2011). Nuclear receptor coactivators: Structural and functional biochemistry. *Biochemistry*, *50*, 313–328.

Chen, J. D., & Evans, R. M. (1995). A transcriptional co-repressor that interacts with nuclear hormone receptors. *Nature*, *377*, 454–457.

Choi, J., Ishizuya-Oka, A., & Buchholz, D. R. (2017). Growth, development, and intestinal remodeling occurs in the absence of thyroid hormone receptor alpha in tadpoles of Xenopus tropicalis. *Endocrinology*, *158*, 1623–1633.

Choi, J., Suzuki, K. I., Sakuma, T., Shewade, L., Yamamoto, T., & Buchholz, D. R. (2015). Unliganded thyroid hormone receptor alpha regulates developmental timing via gene repression as revealed by gene disruption in Xenopus tropicalis. *Endocrinology*, *156*, 735–744.

Clevers, H. (2013). The intestinal crypt, a prototype stem cell compartment. *Cell*, *154*, 274–284.

Dang, C. V. (1999). c-Myc target genes involved in cell growth, apoptosis, and metabolism. *Molecular and Cellular Biology*, *19*, 1–11.

Davis, P. J., & Davis, F. B. (1996). Nongenomic actions of thyroid hormone. *Thyroid*, *6*, 497–504.

Demarest, S. J., Martinez-Yamout, M., Chung, J., Chen, H., Xu, W., Dyson, H. J., et al. (2002). Mutual synergistic folding in recruitment of CBP/p300 by p160 nuclear receptor coactivators. *Nature*, *415*, 549–553.

Denver, R. J., Hu, F., Scanlan, T. S., & Furlow, J. D. (2009). Thyroid hormone receptor subtype specificity for hormone-dependent neurogenesis in *Xenopus laevis*. *Developmental Biology, 326*, 155–168.

Dodd, M. H. I., & Dodd, J. M. (1976). The biology of metamorphosis. In B. Lofts (Ed.), *Physiology of the amphibia* (pp. 467–599). New York: Academic Press.

Evans, R. M. (1988). The steroid and thyroid hormone receptor superfamily. *Science, 240*, 889–895.

Flamant, F., Poguet, A. L., Plateroti, M., Chassande, O., Gauthier, K., Streichenberger, N., et al. (2002). Congenital hypothyroid Pax8(−/−) mutant mice can be rescued by inactivating the TRalpha gene. *Molecular Endocrinology, 16*, 24–32.

Friedrichsen, S., Christ, S., Heuer, H., Schäfer, M. K. H., Mansouri, A., Bauer, K., et al. (2003). Regulation of iodothyronine deiodinases in the Pax8 −/− mouse model of congenital hypothyroidism. *Endocrinology, 144*, 777–784.

Fu, L., Buchholz, D., & Shi, Y. B. (2002). Novel double promoter approach for identification of transgenic animals: A tool for in vivo analysis of gene function and development of gene-based therapies. *Molecular Reproduction and Development, 62*, 470–476.

Fu, L., Das, B., Matsuura, K., Fujimoto, K., Heimeier, R. A., & Shi, Y. B. (2017). Genome-wide identification of thyroid hormone receptor targets in the remodeling intestine during Xenopus tropicalis metamorphosis. *Scientific Reports*, 7, 6414.

Fujimoto, K., Matsuura, K., Hu-Wang, E., Lu, R., & Shi, Y. B. (2012). Thyroid hormone activates protein arginine methyltransferase 1 expression by directly inducing c-Myc transcription during xenopus intestinal stem cell development. *The Journal of Biological Chemistry, 287*, 10039–10050.

Gilbert, L. I., Tata, J. R., & Atkinson, B. G. (1996). *Metamorphosis: Post-embryonic reprogramming of gene expression in amphibian and insect cells*. New York: Academic Press.

Grandori, C., Cowley, S. M., James, L. P., & Eisenman, R. N. (2000). The Myc/Max/Mad network and the transcriptional control of cell behavior. *Annual Review of Cell and Developmental Biology, 16*, 653–699.

Grimaldi, A., Buisine, N., Miller, T., Shi, Y. B., & Sachs, L. M. (2013). Mechanisms of thyroid hormone receptor action during development: Lessons from amphibian studies. *Biochimica et Biophysica Acta, 1830*, 3882–3892.

Harper, J., Mould, A., Andrews, R. M., Bikoff, E. K., & Robertson, E. J. (2011). The transcriptional repressor Blimp1/Prdm1 regulates postnatal reprogramming of intestinal enterocytes. *Proceedings of the National Academy of Sciences of the United States of America, 108*, 10585–10590.

Hasebe, T., Buchholz, D. R., Shi, Y. B., & Ishizuya-Oka, A. (2011). Epithelial-connective tissue interactions induced by thyroid hormone receptor are essential for adult stem cell development in the *Xenopus laevis* intestine. *Stem Cells, 29*, 154–161.

Hasebe, T., Fujimoto, K., Kajita, M., Fu, L., Shi, Y. B., & Ishizuya-Oka, A. (2017). Thyroid hormone-induced activation of notch signaling is required for adult intestinal stem cell development during *Xenopus Laevis* metamorphosis. *Stem Cells, 35*, 1028–1039.

Hasebe, T., Fujimoto, K., Kajita, M., & Ishizuya-Oka, A. (2016). Thyroid hormone activates Wnt/beta-catenin signaling involved in adult epithelial development during intestinal remodeling in *Xenopus laevis*. *Cell and Tissue Research, 365*, 309–318.

Hasebe, T., Fujimoto, K., Kajita, M., & Ishizuya-Oka, A. (2017). Essential roles of thyroid hormone-regulated hyaluronan/CD44 signaling in adult stem cell development during *Xenopus laevis* intestinal remodeling. *Stem Cells, 35*, 2175–2183.

Hasebe, T., Kajita, M., Fu, L., Shi, Y. B., & Ishizuya-Oka, A. (2012). Thyroid hormone-induced sonic hedgehog signal up-regulates its own pathway in a paracrine manner in the *Xenopus laevis* intestine during metamorphosis. *Developmental Dynamics, 241*, 403–414.

Hasebe, T., Kajita, M., Shi, Y. B., & Ishizuya-Oka, A. (2008). Thyroid hormone-up-regulated hedgehog interacting protein is involved in larval-to-adult intestinal remodeling by regulating sonic hedgehog signaling pathway in *Xenopus laevis*. *Developmental Dynamics*, *237*, 3006–3015.

Havis, E., Sachs, L. M., & Demeneix, B. A. (2003). Metamorphic T3-response genes have specific co-regulator requirements. *EMBO Reports*, *4*, 883–888.

Heimeier, R. A., Das, B., Buchholz, D. R., Fiorentino, M., & Shi, Y. B. (2010). Studies on *Xenopus laevis* intestine reveal biological pathways underlying vertebrate gut adaptation from embryo to adult. *Genome Biology*, *11*, R55.

Heimeier, R. A., Hsia, V. S.-C., & Shi, Y.-B. (2008). Participation of BAF57 and BRG1-containing chromatin remodeling complexes in thyroid hormone-dependent gene activation during vertebrate development. *Molecular Endocrinology*, *22*, 1065–1077.

Horlein, A. J., Naar, A. M., Heinzel, T., Torchia, J., Gloss, B., Kurokawa, R., et al. (1995). Ligand-independent repression by the thyroid hormone receptor mediated by a nuclear receptor co-repressor. *Nature*, *377*, 397–404.

Huang, Z.-Q., Li, J., Sachs, L. M., Cole, P. A., & Wong, J. (2003). A role for cofactor–cofactor and cofactor–histone interactions in targeting p300, SWI/SNF and mediator for transcription. *The EMBO Journal*, *22*, 2146–2155.

Ishizuya-Oka, A., & Hasebe, T. (2013). Establishment of intestinal stem cell niche during amphibian metamorphosis. *Current Topics in Developmental Biology*, *103*, 305–327.

Ishizuya-Oka, A., Hasebe, T., Buchholz, D. R., Kajita, M., Fu, L., & Shi, Y. B. (2009). Origin of the adult intestinal stem cells induced by thyroid hormone in *Xenopus laevis*. *The FASEB Journal*, *23*, 2568–2575.

Ishizuya-Oka, A., Hasebe, T., Shimizu, K., Suzuki, K., & Ueda, S. (2006). Shh/BMP-4 signaling pathway is essential for intestinal epithelial development during Xenopus larval-to-adult remodeling. *Developmental Dynamics*, *235*, 3240–3249.

Ishizuya-Oka, A., Kajita, M., & Hasebe, T. (2014). Thyroid hormone-regulated Wnt5a/Ror2 signaling is essential for dedifferentiation of larval epithelial cells into adult stem cells in the *Xenopus laevis* intestine. *PLoS One*, *9*, e107611.

Ishizuya-Oka, A., & Shi, Y. B. (2008). Thyroid hormone regulation of stem cell development during intestinal remodeling. *Molecular and Cellular Endocrinology*, *288*, 71–78.

Ishizuya-Oka, A., & Shi, Y. B. (2011). Evolutionary insights into postembryonic development of adult intestinal stem cells. *Cell & Bioscience*, *1*, 37.

Ishizuya-Oka, A., Shimizu, K., Sakakibara, S., Okano, H., & Ueda, S. (2003). Thyroid hormone-upregulated expression of Musashi-1 is specific for progenitor cells of the adult epithelium during amphibian gastrointestinal remodeling. *Journal of Cell Science*, *116*, 3157–3164.

Ishizuya-Oka, A., & Shimozawa, A. (1992). Connective tissue is involved in adult epithelial development of the small intestine during anuran metamorphosis in vitro. *Roux's Archives of Developmental Biology*, *201*, 322–329.

Ishizuya-Oka, A., Ueda, S., Amano, T., Shimizu, K., Suzuki, K., Ueno, N., et al. (2001). Thyroid-hormone-dependent and fibroblast-specific expression of BMP-4 correlates with adult epithelial development during amphibian intestinal remodeling. *Cell and Tissue Research*, *303*, 187–195.

Ishizuya-Oka, A., Ueda, S., Inokuchi, T., Amano, T., Damjanovski, S., Stolow, M., et al. (2001). Thyroid hormone-induced expression of Sonic hedgehog correlates with adult epithelial development during remodeling of the Xenopus stomach and intestine. *Differentiation*, *69*, 27–37.

Jones, P. L., Sachs, L. M., Rouse, N., Wade, P. A., & Shi, Y. B. (2001). Multiple N-CoR complexes contain distinct histone deacetylases. *The Journal of Biological Chemistry*, *276*, 8807–8811.

Jones, P. L., & Shi, Y.-B. (2003). N-CoR-HDAC corepressor complexes: Roles in transcriptional regulation by nuclear hormone receptors. In J. L. Workman (Ed.), *Current topics in microbiology and immunology: Protein complexes that modify chromatin* (pp. 237–268). Berlin: Springer-Verlag.

Koskinen, P. J., & Alitalo, K. (1993). Role of myc amplification and overexpression in cell growth, differentiation and death. *Seminars in Cancer Biology*, *4*, 3–12.

Kress, E., Rezza, A., Nadjar, J., Samarut, J., & Plateroti, M. (2009). The frizzled-related sFRP2 gene is a target of thyroid hormone receptor alpha1 and activates beta-catenin signaling in mouse intestine. *The Journal of Biological Chemistry*, *284*, 1234–1241.

Kroll, K. L., & Amaya, E. (1996). Transgenic Xenopus embryos from sperm nuclear transplantations reveal FGF signaling requirements during gastrulation. *Development*, *122*, 3173–3183.

Kuchino, Y., Asai, A., & Kitanaka, C. (1996). Myc-mediated apoptosis. *Progress in Molecular and Subcellular Biology*, *16*, 104–129.

Laudet, V., & Gronemeyer, H. (2002). *The nuclear receptor FactsBook*. San Diego: Academic Press.

Lazar, M. A. (1993). Thyroid hormone receptors: Multiple forms, multiple possibilities. *Endocrine Reviews*, *14*, 184–193.

Lei, Y., Guo, X., Deng, Y., Chen, Y., & Zhao, H. (2013). Generation of gene disruptions by transcription activator-like effector nucleases (TALENs) in Xenopus tropicalis embryos. *Cell & Bioscience*, *3*, 21.

Lei, Y., Guo, X., Liu, Y., Cao, Y., Deng, Y., Chen, X., et al. (2012). Efficient targeted gene disruption in Xenopus embryos using engineered transcription activator-like effector nucleases (TALENs). *Proceedings of the National Academy of Sciences of the United States of America*, *109*, 17484–17489.

Leloup, J., & Buscaglia, M. (1977). La triiodothyronine: Hormone de la métamorphose des amphibiens. *Comptes Rendus. Académie des Sciences*, *284*, 2261–2263.

Li, J., Lin, Q., Yoon, H. G., Huang, Z. Q., Strahl, B. D., Allis, C. D., et al. (2002). Involvement of histone methylation and phosphorylation in regulation of transcription by thyroid hormone receptor. *Molecular and Cellular Biology*, *22*, 5688–5697.

Luscher, B. (2012). MAD1 and its life as a MYC antagonist: An update. *European Journal of Cell Biology*, *91*, 506–514.

Luu, N., Fu, L., Fujimoto, K., & Shi, Y. B. (2017). Direct regulation of histidine ammonia-lyase 2 gene by thyroid hormone in the developing adult intestinal stem cells. *Endocrinology*, *158*, 1022–1033.

Luu, N., Wen, L., Fu, L., Fujimoto, K., Shi, Y. B., & Sun, G. (2013). Differential regulation of two histidine ammonia-lyase genes during Xenopus development implicates distinct functions during thyroid hormone-induced formation of adult stem cells. *Cell & Bioscience*, *3*, 43.

MacDonald, W. C., Trier, J. S., & Everett, N. B. (1964). Cell proliferation and migration in the stomach, duodenum, and rectum of man: Radioautographic studies. *Gastroenterology*, *46*, 405–417.

Mangelsdorf, D. J., Thummel, C., Beato, M., Herrlich, P., Schutz, G., Umesono, K., et al. (1995). The nuclear receptor superfamily: The second decade. *Cell*, *83*, 835–839.

Matsuda, H., Paul, B. D., Choi, C. Y., Hasebe, T., & Shi, Y.-B. (2009). Novel functions of protein arginine methyltransferase 1 in thyroid hormone receptor-mediated transcription and in the regulation of metamorphic rate in *Xenopus laevis*. *Molecular and Cellular Biology*, *29*, 745–757.

Matsuda, H., Paul, B. D., Choi, C. Y., & Shi, Y.-B. (2007). Contrasting effects of two alternative splicing forms of coactivator-associated arginine methyltransferase 1 on thyroid hormone receptor-mediated transcription in *Xenopus laevis*. *Molecular Endocrinology*, *21*, 1082–1094.

Matsuda, H., & Shi, Y. B. (2010). An essential and evolutionarily conserved role of protein arginine methyltransferase 1 for adult intestinal stem cells during postembryonic development. *Stem Cells*, *28*, 2073–2083.

Matsuura, K., Fujimoto, K., Fu, L., & Shi, Y.-B. (2012). Liganded thyroid hormone receptor induces nucleosome removal and histone modifications to activate transcription during larval intestinal cell death and adult stem cell development. *Endocrinology*, *153*, 961–972.

McArthur, G. A., Laherty, C. D., Queva, C., Hurlin, P. J., Loo, L., James, L., et al. (1998). The Mad protein family links transcriptional repression to cell differentiation. *Cold Spring Harbor Symposia on Quantitative Biology*, *63*, 423–433.

McAvoy, J. W., & Dixon, K. E. (1977). Cell proliferation and renewal in the small intestinal epithelium of metamorphosing and adult *Xenopus laevis*. *The Journal of Experimental Zoology*, *202*, 129–138.

McMahon, S. B. (2014). MYC and the control of apoptosis. *Cold Spring Harbor Perspectives in Medicine*, *4*, a014407.

Miller, T. C., Sun, G., Hasebe, T., Fu, L., Heimeier, R. A., Das, B., et al. (2013). Tissue-specific upregulation of MDS/EVI gene transcripts in the intestine by thyroid hormone during Xenopus metamorphosis. *PLoS One*, *8*, e55585.

Moran, C., & Chatterjee, K. (2015). Resistance to thyroid hormone due to defective thyroid receptor alpha. *Best Practice & Research. Clinical Endocrinology & Metabolism*, *29*, 647–657.

Muncan, V., Heijmans, J., Krasinski, S. D., Buller, N. V., Wildenberg, M. E., Meisner, S., et al. (2011). Blimp1 regulates the transition of neonatal to adult intestinal epithelium. *Nature Communications*, *2*, 452.

Na, W., Fu, L., Luu, N., & Shi, Y. B. (2020). Direct activation of tRNA methyltransferase-like 1 (Mettl1) gene by thyroid hormone receptor implicates a role in adult intestinal stem cell development and proliferation during Xenopus tropicalis metamorphosis. *Cell & Bioscience*, *10*, 60.

Nakade, S., Tsubota, T., Sakane, Y., Kume, S., Sakamoto, N., Obara, M., et al. (2014). Microhomology-mediated end-joining-dependent integration of donor DNA in cells and animals using TALENs and CRISPR/Cas9. *Nature Communications*, *5*, 5560.

Nakajima, K., Tazawa, I., & Shi, Y. B. (2019). A unique role of thyroid hormone receptor beta in regulating notochord resorption during Xenopus metamorphosis. *General and Comparative Endocrinology*, *277*, 66–72.

Nakajima, K., Tazawa, I., & Yaoita, Y. (2018). Thyroid hormone receptor alpha- and beta-knockout Xenopus tropicalis tadpoles reveal subtype-specific roles during development. *Endocrinology*, *159*, 733–743.

Nakajima, K., & Yaoita, Y. (2003). Dual mechanisms governing muscle cell death in tadpole tail during amphibian metamorphosis. *Developmental Dynamics*, *227*, 246–255.

Nakayama, T., Fish, M. B., Fisher, M., Oomen-Hajagos, J., Thomsen, G. H., & Grainger, R. M. (2013). Simple and efficient CRISPR/Cas9-mediated targeted mutagenesis in Xenopus tropicalis. *Genesis*, *51*, 835–843.

Nieminen, A. I., Partanen, J. I., & Klefstrom, J. (2007). c-Myc blazing a trail of death: Coupling of the mitochondrial and death receptor apoptosis pathways by c-Myc. *Cell Cycle*, *6*, 2464–2472.

Nieuwkoop, P. D., & Faber, J. (1965). *Normal table of Xenopus laevis*. Amsterdam: North Holland Publishing.

Okada, M., Miller, T. C., Fu, L., & Shi, Y. B. (2015). Direct activation of amidohydrolase domain-containing 1 gene by thyroid hormone implicates a role in the formation of adult intestinal stem cells during Xenopus metamorphosis. *Endocrinology*, *156*, 3381–3393.

Okada, M., Miller, T. C., Wen, L., & Shi, Y. B. (2017). A balance of Mad and Myc expression dictates larval cell apoptosis and adult stem cell development during Xenopus intestinal metamorphosis. *Cell Death & Disease*, *8*, e2787.

Okada, M., & Shi, Y. B. (2018a). The balance of two opposing factors Mad and Myc regulates cell fate during tissue remodeling. *Cell & Bioscience*, *8*, 51.

Okada, M., & Shi, Y. B. (2018b). EVI and MDS/EVI are required for adult intestinal stem cell formation during postembryonic vertebrate development. *The FASEB Journal*, *32*, 431–439.

Okada, M., Wen, L., Miller, T. C., Su, D., & Shi, Y. B. (2015). Molecular and cytological analyses reveal distinct transformations of intestinal epithelial cells during Xenopus metamorphosis. *Cell & Bioscience*, *5*, 74.

O'Malley, B. W., Malovannaya, A., & Qin, J. (2012). Minireview: Nuclear receptor and coregulator proteomics—2012 and beyond. *Molecular Endocrinology*, *26*, 1646–1650.

Packham, G., & Cleveland, J. L. (1995). c-Myc and apoptosis. *Biochimica et Biophysica Acta*, *1242*, 11–28.

Paul, B. D., Buchholz, D. R., Fu, L., & Shi, Y.-B. (2005). Tissue- and gene-specific recruitment of steroid receptor coactivator-3 by thyroid hormone receptor during development. *The Journal of Biological Chemistry*, *280*, 27165–27172.

Paul, B. D., Buchholz, D. R., Fu, L., & Shi, Y.-B. (2007). SRC-p300 coactivator complex is required for thyroid hormone induced amphibian metamorphosis. *The Journal of Biological Chemistry*, *282*, 7472–7481.

Paul, B. D., Fu, L., Buchholz, D. R., & Shi, Y.-B. (2005). Coactivator recruitment is essential for liganded thyroid hormone receptor to initiate amphibian metamorphosis. *Molecular and Cellular Biology*, *25*, 5712–5724.

Pelengaris, S., Rudolph, B., & Littlewood, T. (2000). Action of Myc in vivo—Proliferation and apoptosis. *Current Opinion in Genetics & Development*, *10*, 100–105.

Perissi, V., Jepsen, K., Glass, C. K., & Rosenfeld, M. G. (2010). Deconstructing repression: Evolving models of co-repressor action. *Nature Reviews. Genetics*, *11*, 109–123.

Plateroti, M., Chassande, O., Fraichard, A., Gauthier, K., Freund, J. N., Samarut, J., et al. (1999). Involvement of T3Ralpha- and beta-receptor subtypes in mediation of T3 functions during postnatal murine intestinal development. *Gastroenterology*, *116*, 1367–1378.

Plateroti, M., Gauthier, K., Domon-Dell, C., Freund, J. N., Samarut, J., & Chassande, O. (2001). Functional interference between thyroid hormone receptor alpha (TRalpha) and natural truncated TRDeltaalpha isoforms in the control of intestine development. *Molecular and Cellular Biology*, *21*, 4761–4772.

Plateroti, M., Kress, E., Mori, J. I., & Samarut, J. (2006). Thyroid hormone receptor alpha1 directly controls transcription of the beta-catenin gene in intestinal epithelial cells. *Molecular and Cellular Biology*, *26*, 3204–3214.

Refetoff, S., Weiss, R. E., & Usala, S. J. (1993). The syndromes of resistance to thyroid hormone. *Endocrine Reviews*, *14*, 348–399.

Sachs, L. M. (2015). Unliganded thyroid hormone receptor function: Amphibian metamorphosis got TALENs. *Endocrinology*, *156*, 409–410.

Sachs, L. M., Amano, T., Rouse, N., & Shi, Y. B. (2001). Involvement of histone deacetylase at two distinct steps in gene regulation during intestinal development in *Xenopus laevis*. *Developmental Dynamics*, *222*, 280–291.

Sachs, L. M., Amano, T., & Shi, Y. B. (2001). An essential role of histone deacetylases in postembryonic organ transformations in *Xenopus laevis*. *International Journal of Molecular Medicine*, *8*, 595–601.

Sachs, L. M., Damjanovski, S., Jones, P. L., Li, Q., Amano, T., Ueda, S., et al. (2000). Dual functions of thyroid hormone receptors during Xenopus development. *Comparative Biochemistry and Physiology. Part B, Biochemistry & Molecular Biology*, *126*, 199–211.

Sachs, L. M., Jones, P. L., Havis, E., Rouse, N., Demeneix, B. A., & Shi, Y.-B. (2002). N-CoR recruitment by unliganded thyroid hormone receptor in gene repression during *Xenopus laevis* development. *Molecular and Cellular Biology*, *22*, 8527–8538.

Sachs, L. M., & Shi, Y.-B. (2000). Targeted chromatin binding and histone acetylation in vivo by thyroid hormone receptor during amphibian development. *PNAS*, *97*, 13138–13143.

Sakane, Y., Iida, M., Hasebe, T., Fujii, S., Buchholz, D. R., Ishizuya-Oka, A., et al. (2018). Functional analysis of thyroid hormone receptor beta in Xenopus tropicalis founders using CRISPR-Cas. *Biology Open*, *7*, bio030338. https://doi.org/10.1242/bio.030338.

Sancho, E., Eduard Batlle, E., & Clevers, H. (2004). Signaling pathways in intestinal development and cancer. *Annual Review of Cell and Developmental Biology*, *20*, 695–723.

Schreiber, A. M., Cai, L., & Brown, D. D. (2005). Remodeling of the intestine during metamorphosis of *Xenopus laevis*. *Proceedings of the National Academy of Sciences of the United States of America*, *102*, 3720–3725.

Schreiber, A. M., Das, B., Huang, H., Marsh-Armstrong, N., & Brown, D. D. (2001). Diverse developmental programs of *Xenopus laevis* metamorphosis are inhibited by a dominant negative thyroid hormone receptor. *PNAS*, *98*, 10739–10744.

Schreiber, A. M., Mukhi, S., & Brown, D. D. (2009). Cell-cell interactions during remodeling of the intestine at metamorphosis in *Xenopus laevis*. *Developmental Biology*, *331*, 89–98.

Sheppard, H. M., Harries, J. C., Hussain, S., Bevan, C., & Heery, D. M. (2001). Analysis of the steroid receptor coactivator 1 (SRC1)-CREB binding protein interaction interface and its importance for the function of SRC1. *Molecular and Cellular Biology*, *21*, 39–50.

Shi, Y.-B. (1994). Molecular biology of amphibian metamorphosis: A new approach to an old problem. *Trends in Endocrinology and Metabolism*, *5*, 14–20.

Shi, Y.-B. (1999). *Amphibian metamorphosis: From morphology to molecular biology*. New York: John Wiley & Sons, Inc.

Shi, Y.-B. (2009). Dual functions of thyroid hormone receptors in vertebrate development: The roles of histone-modifying cofactor complexes. *Thyroid*, *19*, 987–999.

Shi, Y. B. (2013). Unliganded thyroid hormone receptor regulates metamorphic timing via the recruitment of histone deacetylase complexes. *Current Topics in Developmental Biology*, *105*, 275–297.

Shi, Y. B. (2018). Thyroid hormone and intestinal tumor: A Wnt connection. *Oncotarget*, *9*, 31941.

Shi, Y.-B., & Brown, D. D. (1993). The earliest changes in gene expression in tadpole intestine induced by thyroid hormone. *The Journal of Biological Chemistry*, *268*, 20312–20317.

Shi, Y., Glynn, J. M., Guilbert, L. J., Cotter, T. G., Bissonnette, R. P., & Green, D. R. (1992). Role for c-myc in activation-induced apoptotic cell death in T cell hybridomas. *Science*, *257*, 212–214.

Shi, Y. B., Hasebe, T., Fu, L., Fujimoto, K., & Ishizuya-Oka, A. (2011). The development of the adult intestinal stem cells: Insights from studies on thyroid hormone-dependent amphibian metamorphosis. *Cell & Bioscience*, *1*, 30.

Shi, Y.-B., & Ishizuya-Oka, A. (1996). Biphasic intestinal development in amphibians: Embryogenesis and remodeling during metamorphosis. *Current Topics in Developmental Biology*, *32*, 205–235.

Shi, Y. B., Matsuura, K., Fujimoto, K., Wen, L., & Fu, L. (2012). Thyroid hormone receptor actions on transcription in amphibia: The roles of histone modification and chromatin disruption. *Cell & Bioscience*, *2*, 42.

Shi, Z., Wang, F., Cui, Y., Liu, Z., Guo, X., Zhang, Y., et al. (2015). Heritable CRISPR/Cas9-mediated targeted integration in Xenopus tropicalis. *The FASEB Journal*, *29*, 4914–4923.

Shibata, Y., Tanizaki, Y., & Shi, Y. B. (2020). Thyroid hormone receptor beta is critical for intestinal remodeling during Xenopus tropicalis metamorphosis. *Cell & Bioscience*, *10*, 46.

Shibata, Y., Wen, L., Okada, M., & Shi, Y. B. (2020). Organ-specific requirements for thyroid hormone receptor ensure temporal coordination of tissue-specific transformations and completion of xenopus metamorphosis. *Thyroid*, *30*, 300–313.

Sirakov, M., Kress, E., Nadjar, J., & Plateroti, M. (2014). Thyroid hormones and their nuclear receptors: New players in intestinal epithelium stem cell biology? *Cellular and Molecular Life Sciences*, *71*, 2897–2907.

Sterling, J., Fu, L., Matsuura, K., & Shi, Y.-B. (2012). Cytological and morphological analyses reveal distinct features of intestinal development during Xenopus tropicalis metamorphosis. *PLoS One*, *7*, e47407.

Stewart, M. D., Li, J., & Wong, J. (2005). Relationship between histone H3 lysine 9 methylation, transcription repression, and heterochromatin protein 1 recruitment. *Molecular and Cellular Biology*, *25*, 2525–2538.

Stewart, D., Tomita, A., Shi, Y. B., & Wong, J. (2006). Chromatin immunoprecipitation for studying transcriptional regulation in Xenopus oocytes and tadpoles. *Methods in Molecular Biology*, *322*, 165–181.

Stolow, M. A., & Shi, Y. B. (1995). Xenopus sonic hedgehog as a potential morphogen during embryogenesis and thyroid hormone-dependent metamorphosis. *Nucleic Acids Research*, *23*, 2555–2562.

Sun, G., Fu, L., & Shi, Y.-B. (2014). Epigenetic regulation of thyroid hormone-induced adult intestinal stem cell development during anuran metamorphosis. *Cell & Bioscience*, *4*, 73.

Sun, G., Fu, L., Wen, L., & Shi, Y. B. (2014). Activation of sox3 gene by thyroid hormone in the developing adult intestinal stem cell during Xenopus metamorphosis. *Endocrinology*, *155*, 5024–5032.

Sun, G., Heimeier, R. A., Fu, L., Hasebe, T., Das, B., Ishizuya-Oka, A., et al. (2013). Expression profiling of intestinal tissues implicates tissue-specific genes and pathways essential for thyroid hormone-induced adult stem cell development. *Endocrinology*, *154*, 4396–4407.

Sun, G., & Shi, Y.-B. (2012). Thyroid hormone regulation of adult intestinal stem cell development: Mechanisms and evolutionary conservations. *International Journal of Biological Sciences*, *8*, 1217–1224.

Tanizaki, Y., Shibata, Y., Zhang, H., & Shi, Y.-B. (2020). Analysis of thyroid hormone receptor α knockout tadpoles reveals that the activation of cell cycle program is involved in thyroid hormone-induced larval epithelial cell death and adult intestinal stem cell development during Xenopus tropicalis metamorphosis. *Thyroid*, *31*, 128–142.

Tata, J. R. (1993). Gene expression during metamorphosis: An ideal model for post-embryonic development. *BioEssays*, *15*, 239–248.

Thompson, E. B. (1998). The many roles of c-Myc in apoptosis. *Annual Review of Physiology*, *60*, 575–600.

Tomita, A., Buchholz, D. R., & Shi, Y.-B. (2004). Recruitment of N-CoR/SMRT-TBLR1 corepressor complex by unliganded thyroid hormone receptor for gene repression during frog development. *Molecular and Cellular Biology*, *24*, 3337–3346.

Toner, P. G., Carr, K. E., & Wyburn, G. M. (1971). *The digestive system: An ultrastructural atlas and review*. London: Butterworth.

Tsai, M. J., & O'Malley, B. W. (1994). Molecular mechanisms of action of steroid/thyroid receptor superfamily members. *Annual Review of Biochemistry*, *63*, 451–486.

van der Flier, L. G., & Clevers, H. (2009). Stem cells, self-renewal, and differentiation in the intestinal epithelium. *Annual Review of Physiology*, *71*, 241–260.

van Mullem, A., van Heerebeek, R., Chrysis, D., Visser, E., Medici, M., Andrikoula, M., et al. (2012). Clinical phenotype and mutant TRalpha1. *The New England Journal of Medicine*, *366*, 1451–1453.

Wang, X., Matsuda, H., & Shi, Y.-B. (2008). Developmental regulation and function of thyroid hormone receptors and 9-cis retinoic acid receptors during Xenopus tropicalis metamorphosis. *Endocrinology*, *149*, 5610–5618.

Wang, F., Shi, Z., Cui, Y., Guo, X., Shi, Y. B., & Chen, Y. (2015). Targeted gene disruption in *Xenopus laevis* using CRISPR/Cas9. *Cell & Bioscience*, *5*, 15.

Wen, L., Fu, L., & Shi, Y. B. (2017). Histone methyltransferase Dot1L is a coactivator for thyroid hormone receptor during Xenopus development. *The FASEB Journal*, *31*, 4821–4831.

Wen, L., Hasebe, T., Miller, T. C., Ishizuya-Oka, A., & Shi, Y. B. (2015). A requirement for hedgehog signaling in thyroid hormone-induced postembryonic intestinal remodeling. *Cell & Bioscience*, *5*, 13.

Wen, L., & Shi, Y. B. (2015). Unliganded thyroid hormone receptor alpha controls developmental timing in Xenopus tropicalis. *Endocrinology*, *156*, 721–734.

Wen, L., & Shi, Y. B. (2016). Regulation of growth rate and developmental timing by Xenopus thyroid hormone receptor alpha. *Development, Growth & Differentiation*, *58*, 106–115.

Wen, L., Shibata, Y., Su, D., Fu, L., Luu, N., & Shi, Y.-B. (2017). Thyroid hormone receptor α controls developmental timing and regulates the rate and coordination of tissue specific metamorphosis in Xenopus tropicalis. *Endocrinology*, *158*, 1985–1998.

Wong, J., Patterton, D., Imhof, D., Guschin, D., Shi, Y.-B., & Wolffe, A. P. (1998). Distinct requirements for chromatin assembly in transcriptional repression by thyroid hormone receptor and histone deacetylase. *The EMBO Journal*, *17*, 520–534.

Wong, J., Shi, Y. B., & Wolffe, A. P. (1995). A role for nucleosome assembly in both silencing and activation of the Xenopus TR beta A gene by the thyroid hormone receptor. *Genes & Development*, *9*, 2696–2711.

Wong, J., Shi, Y.-B., & Wolffe, A. P. (1997). Determinants of chromatin disruption and transcriptional regulation instigated by the thyroid hormone receptor: Hormone-regulated chromatin disruption is not sufficient for transcriptional activation. *The EMBO Journal*, *16*, 3158–3171.

Yen, P. M. (2001). Physiological and molecular basis of thyroid hormone action. *Physiological Reviews*, *81*, 1097–1142.

Yen, P. M. (2015). Unliganded TRs regulate growth and developmental timing during early embryogenesis: Evidence for a dual function mechanism of TR action. *Cell & Bioscience*, *5*, 8.

Yoon, H.-G., Chan, D. W., Huang, Z. Q., Li, J., Fondell, J. D., Qin, J., et al. (2003). Purification and functional characterization of the human N-CoR complex: The roles of HDAC3, TBL1 and TBLR1. *The EMBO Journal*, *22*, 1336–1346.

Young, J. J., Cherone, J. M., Doyon, Y., Ankoudinova, I., Faraji, F. M., Lee, A. H., et al. (2011). Efficient targeted gene disruption in the soma and germ line of the frog Xenopus tropicalis using engineered zinc-finger nucleases. *Proceedings of the National Academy of Sciences of the United States of America*, *108*, 7052–7057.

CHAPTER NINE

Regulation of muscle stem cell function

Julia von Maltzahn*
Leibniz-Institute on Aging—Fritz-Lipmann-Institute, Jena, Germany
*Corresponding author: e-mail address: julia.vonmaltzahn@leibniz-fli.de

Contents

Abstract

Regeneration of skeletal muscle is a finely tuned process which is depending on muscle stem cells, a population of stem cells in skeletal muscle which is also termed satellite cells. Muscle stem cells are a prerequisite for regeneration of skeletal muscle. Of note, the muscle stem cell population is heterogeneous and subpopulations can be identified depending on gene expression or phenotypic traits. However, all muscle stem cells express the transcription factor Pax7 and their functionality is tightly controlled by intrinsic signaling pathways and extrinsic signals. The latter ones include signals form the stem cell niche as well as circulating factors such as growth factors and hormones. Among them are Wnt proteins, growth factors like IGF-1 or FGF-2 and hormones such as thyroid hormones and the anti-aging hormone Klotho. A highly orchestrated interplay between those factors and muscle stem cells is important for their full functionality and ultimately regeneration of skeletal muscle as outlined here.

Vitamins and Hormones, Volume 116
ISSN 0083-6729
https://doi.org/10.1016/bs.vh.2021.02.012

1. Introduction

Skeletal muscle is a postmitotic tissue which can remarkably adapt to physiological demands, a process often driven by hormonal changes (von Maltzahn, Bentzinger, & Rudnicki, 2013). In the organism, skeletal muscle serves a multitude of functions including voluntary movements, breathing and postural behavior. Of note, skeletal muscle is one of the tissues with the highest ability to regenerate (Henze, Jung, Ahrens, Steiner, & von Maltzahn, 2020; Schmidt, Schuler, Huttner, von Eyss, & von Maltzahn, 2019). Muscle stem cells, a small population of stem cells residing underneath the basal lamina of muscle fibers, are a prerequisite for regeneration of skeletal muscle (Lepper, Partridge, & Fan, 2011; Murphy, Lawson, Mathew, Hutcheson, & Kardon, 2011; Sambasivan et al., 2011). However, other cell types in skeletal muscle are supporting the regeneration process, among them immune cells, FAPs (fibro-adipogenic-progenitor cells) and PICs (PWI positive interstitial cells) (Bentzinger, Wang, Dumont, & Rudnicki, 2013; Chazaud et al., 2009; Schmidt et al., 2019). Muscle stem cells (MuSCs) were first characterized in 1961 by Alexander Mauro by their location between the plasma membrane of the myofiber and the surrounding basal lamina (Mauro, 1961). At the time they were termed satellite cells, but in recent years the term muscle stem cell has been used more frequently.

2. Muscle stem cells are a heterogeneous population

MuSCs are characterized by the expression of the paired-box transcription factor Pax7 which is essential for MuSC functionality (Gunther et al., 2013; Lepper, Conway, & Fan, 2009; Seale et al., 2000; von Maltzahn, Jones, Parks, & Rudnicki, 2013). Under resting conditions MuSCs are quiescent, a state characterized by the expression of Sprouty-1 (Chang & Rudnicki, 2014; Mauro, 1961; Shea et al., 2010). Although all MuSCs express Pax7, multiple subpopulations of MuSCs have been described so far (Schmidt et al., 2019).

The individual subpopulations of MuSCs can be distinguished by different means, e.g. through gene expression or phenotypic traits like rate of division. One approach for investigating MuSC heterogeneity through gene expression is based on lineage tracing using a Myf5-Cre reporter mouse line (R26R-YFP/Myf5-Cre). In this model, approximately 10% of MuSCs have never expressed *myf5*, as demonstrated by the absence of the reporter YFP (yellow fluorescent protein) in Pax7 positive MuSCs (Kuang, Kuroda, Le Grand, & Rudnicki, 2007). Interestingly, the YFP^- MuSC subpopulation can engraft into the MuSC niche after transplantation while the YFP^+ MuSC subpopulation gives rise to new myofibers but is not able to home to the MuSC niche. Though, upregulation of *myf5* and consequently induction of the YFP reporter gene was shown in daughter cells of the YFP^- MuSC subpopulation further supporting the notion that the $Pax7^+/YFP^-$ MuSC subpopulation is a rare sub-population with stem cell like features, while the $Pax7^+/YFP^+$ MuSC subpopulation represents a more committed progenitor cell population (Kuang et al., 2007). However, following MuSC activation, e.g. after injury, both subpopulations (YFP^+ and YFP^-) proliferate and undergo planar cell divisions giving rise to two identical daughter cells (identical in terms of YFP expression). Multiple signaling pathways are driving the symmetric division of the YFP^- MuSC subpopulation. Those include signaling induced by Wnt7a together with its receptor Fzd7 and the ECM molecule Fibronectin, as well as JAK/STAT signaling (Bentzinger, Wang, von Maltzahn, et al., 2013; Le Grand, Jones, Seale, Scime, & Rudnicki, 2009; Price et al., 2014; Tierney et al., 2014).

Identification of MuSC subpopulations through phenotypic traits can be performed e.g. through investigation of the clonal dynamics of MuSCs following injury. Here, it was described that the clonal complexity of MuSCs is reduced after multiple rounds of injury in a mouse model using a multicolor lineage tracing approach (Tierney, Stec, Rulands, Simons, & Sacco, 2018). Another way of distinguishing MuSC subpopulations depending on phenotypic traits is based on pulse-chase experiments using the Tg:Pax7-GFP mouse line. Here, the authors demonstrated that MuSCs with a relatively high expression of Pax7 ($Pax7^{High}$) are less primed for commitment and need a longer time to undergo the first mitosis compared to MuSCs with a relatively low expression of Pax7 ($Pax7^{Low}$) (Rocheteau, Gayraud-Morel, Siegl-Cachedenier, Blasco, & Tajbakhsh, 2012). Furthermore, those cells show an asymmetric segregation of their DNA. The daughter cell receiving the template strand also maintains expression of the stem cell markers (Shinin, Gayraud-Morel, Gomes, & Tajbakhsh, 2006; Shinin, Gayraud-Morel, & Tajbakhsh, 2009). Another approach to define subpopulations of MuSCs by phenotypic traits identified a MuSC subpopulation termed label-retaining cells (LRCs) through the use of the TetO-H2B-GFP mouse line.

Here, administration of doxycycline identifies cells which are rarely dividing or non-cycling cells by retention of the expression of the H2B-GFP reporter (Chakkalakal et al., 2014).

3. Muscle stem cells and lineage progression during adult myogenesis

Although MuSCs are a heterogeneous population, all MuSCs are quiescent under resting conditions. However, they can promptly activate, e.g. after injury of skeletal muscle (Evano & Tajbakhsh, 2018) (Fig. 1). Activation of MuSCs is followed by proliferation and migration to the site of injury and the expression of MyoD, an early marker for myogenic commitment (Evano & Tajbakhsh, 2018; Schmidt et al., 2019). Those $Pax7^{+}/MyoD^{+}$ myoblasts can then either undergo further differentiation to become myocytes or downregulate MyoD expression and thereby undergo self-renewal

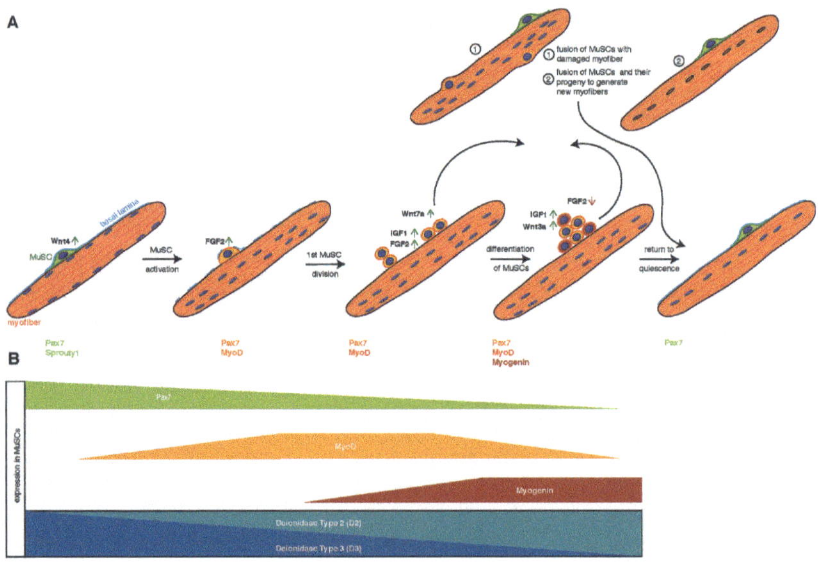

Fig. 1 Activation and differentiation of MuSCs after injury resulting in generation of regenerated or newly formed myofibers. (A) Schematic illustration of the activation and differentiation of MuSCs. MuSCs are quiescent in resting skeletal muscle (in green), they get activated (in orange), e.g. due to injury, start to proliferate, thereby generating differentiating MuSCs (in red), which fuse with damaged myofibers (1) or fuse to form myotubes and mature to become myofibers (2). (B) Expression profile of key modulators of myogenic lineage progression. MuSC: muscle stem cell; FGF-2: fibroblast growth factor 2; IGF-1: insulin growth factor 1.

to replenish the MuSC pool in a Sprouty-1 dependent manner (Olguin & Olwin, 2004; Shea et al., 2010; Zammit et al., 2004). Non-self-renewing myoblasts ($Pax7^+/MyoD^+$) convert into myogenic progenitor cells and further differentiate into myocytes, characterized by the expression of MyoD, myogenin and MRF4 (Bentzinger, Wang, Dumont, et al., 2013; Henze et al., 2020; Hernandez-Hernandez, Garcia-Gonzalez, Brun, & Rudnicki, 2017). This differentiation process during adult myogenesis is accompanied by a transcription factor cascade which highly resembles the cascade during embryonic myogenesis and contains the key myogenic regulatory factors (MyoD, Myf5, Myogenin, MRF4) Those markers are frequently used to determine the state of myogenic differentiation (Bentzinger, Wang, Dumont, et al., 2013; Chal & Pourquie, 2017; Hernandez-Hernandez et al., 2017; Schmidt et al., 2019). Myocytes then fuse and mature into myotubes and myofibers, the functional units of skeletal muscle characterized by the expression of MHC (Myosin heavy chain) (Fig. 1).

4. Regeneration of skeletal muscle

Regeneration of skeletal muscle is usually triggered by injury, e.g. by eccentric exercise, mechanical injury or experimentally via injection of toxins such as cardiotoxin among others (Hardy et al., 2016). The regenerative process following injury can be separated in three phases: the early phase of inflammation, the phase of activation and differentiation of MuSCs and finally the maturation phase, when newly formed myofibers are arising (Schmidt et al., 2019).

The first phase of regeneration is preceded by a short degeneration phase of the muscle tissue. This is due to necrosis accompanied by an increased calcium influx from the sarcoplasmic reticulum of damaged myofibers resulting in degeneration and proteolysis of the damaged muscle tissue (Yin, Price, & Rudnicki, 2013). The necrosis of damaged myofibers then triggers the first phase of regeneration (inflammatory phase). Circulating leukocytes are recruited to the site of injury followed by neutrophils and macrophages (Fielding et al., 1993; Tidball, 1995; Yin et al., 2013). Of note, the macrophage population comprises two distinct populations, the early pro-inflammatory $CD68^+/CD163^-$ macrophages and the late anti-inflammatory $CD68^-/CD163^+$ macrophages (Cantini et al., 2002; Chazaud et al., 2009; Merly, Lescaudron, Rouaud, Crossin, & Gardahaut, 1999). The early pro-inflammatory $CD68^+/CD163^-$ macrophages are important for phagocytosis of damaged tissue parts. Furthermore, they secrete pro-inflammatory cytokines like TNFalpha and IL-1. Late inflammatory cells such as anti-inflammatory macrophages on the other hand secrete cytokines like IL-10 known to facilitate proliferation and differentiation of MuSCs (Arnold et al., 2007; Heredia et al., 2013; Saclier et al., 2013).

The inflammatory phase is followed by the second phase of regeneration marked by activation and differentiation of MuSCs. Activation and differentiation of MuSCs is a highly organized process which is coinciding with changes in cell shape resembling embryonic myogenesis. MuSCs (marked by the expression of Pax7) are activated (characterized by the expression of MyoD) and start to migrate to the site of injury and then either become myogenic progenitor cells or directly fuse with damaged myofibers (Schmidt et al., 2019) (Fig. 1). MuSC migration is controlled by signals from damaged or newly formed myofibers, e.g. Ephrins and Wnt7a control migration of MuSCs and myogenic precursor cells (Arnold et al., 2020; Bentzinger et al., 2014; Stark, Karvas, Siegel, & Cornelison, 2011). Myogenic precursor cells are characterized by the expression of MyoD and Myf5, often this transient amplifying cell population is termed myoblasts (Chang & Rudnicki, 2014; von Maltzahn, Chang, Bentzinger, & Rudnicki, 2012). Myogenic differentiation is driven by the myogenic regulatory factors (MRFs, comprised of Myf5, MyoD, Myogenin and MRF4), which are downstream of Pax7 and Pax3 (Hernandez-Hernandez et al., 2017; Singh & Dilworth, 2013; Soleimani et al., 2012). MyoD drives the expression of Myogenin which then initiates the terminal differentiation process which is accompanied by morphological changes (Hernandez-Hernandez et al., 2017; Singh & Dilworth, 2013; von Maltzahn, Bentzinger, et al., 2013). First, the myogenic progenitor cells become elongated myocytes, which then fuse into multinucleated myotubes. The newly formed—immature—myofibers are characterized by the expression of developmental myosin heavy chain (devMHC) and centrally located myonuclei (Bentzinger, Wang, & Rudnicki, 2012; von Maltzahn, Bentzinger, et al., 2013; Yin et al., 2013). The second phase of regeneration is followed by the maturation phase (phase 3), here myofibers mature and loose expression of devMHC. This is accompanied by the migration of the myonuclei from the center of the myofiber to the periphery (Forcina, Cosentino, & Musaro, 2020).

Regeneration of skeletal muscle is a highly orchestrated process, the interaction of MuSCs with their niche and other supporting cells is essential for proper regeneration. Consequences of perturbations in this interaction can be observed in conditions such as aging resulting in impaired regeneration of skeletal muscle (Munoz-Canoves, Neves, & Sousa-Victor, 2019).

For instance, in the aged Fibronectin levels are reduced in regenerating skeletal muscle hampering the interaction of the niche with MuSCs concomitant with altered Integrin-beta1 activity (Lukjanenko et al., 2016; Rozo, Li, & Fan, 2016). In addition to alterations in the MuSC niche proper activation of intrinsic signaling pathways in MuSCs is required for functional regeneration of skeletal muscle. During aging, different intrinsic signaling pathways display alterations in MuSCs, e.g. changes in p38 and JAK/STAT signaling have been shown to impair MuSC functionality (Henze et al., 2020).

5. Regulation of MuSCs by hormones and other circulating factors

The functionality of MuSCs is not only controlled by multiple intrinsic signaling pathways, but also highly regulated by extrinsic cues as evidenced by parabiosis and transplantation experiments (Conboy et al., 2005; Price et al., 2014; Sousa-Victor et al., 2014).

Different growth factors affect MuSC activation and differentiation. For instance, levels of IGF-1 (insulin growth factor 1) increase during regeneration of skeletal muscle and drive proliferation and differentiation of MuSCs. Other growth factors like HGF (hepatocyte growth factor), FGF-2 and FGF-6 (fibroblast growth factor) specifically promote proliferation of MuSCs and inhibit their differentiation (Bentzinger, von Maltzahn, & Rudnicki, 2010).

5.1 Effect of FGF-2 and IGF-1 on MuSCs

Different fibroblast growth factors (FGFs), especially FGF-2 and FGF-6 induce proliferation of myoblasts in cell culture and inhibit their differentiation (Bentzinger et al., 2010; Pawlikowski, Vogler, Gadek, & Olwin, 2017). These data suggest that FGFs are driving MuSC proliferation and thereby expansion of the MuSC compartment *in vivo*. However, results obtained in regenerating FGF-6 deficient mice are controversial (Bentzinger et al., 2010; Fiore, Sebille, & Birnbaum, 2000; Floss, Arnold, & Braun, 1997; Pawlikowski et al., 2017). On the other hand, the situation for FGF-2 is quite clear. Several groups described that FGF-2 is sequestered from developing and adult myofibers driving MuSC proliferation (Bentzinger et al., 2010; Chakkalakal, Jones, Basson, & Brack, 2012; Pawlikowski et al., 2017). Furthermore, it was demonstrated that blocking antibodies directed against FGF-2 impaired regeneration of skeletal muscle while exogenous FGF-2 improved MuSC divisions and thereby regeneration in dystrophic mice

(Lefaucheur & Sebille, 1995a, 1995b; Pawlikowski et al., 2017). During aging niche FGF-2 signaling is increased due to the release of FGF-2 from myofibers. This coincides with a break of quiescence and decreased expression of Sprouty1 in MuSCs, a natural inhibitor of FGF signaling and a marker for quiescent MuSCs (Chakkalakal et al., 2012).

Insulin-like growth factor 1 (IGF-1) is a growth factor which comprises anabolic functions and has been implicated in hypertrophy, growth and regeneration of skeletal muscle (Bentzinger et al., 2010; Forcina, Miano, Scicchitano, & Musaro, 2019; Glass, 2003, 2005). Multiple cell types in the MuSC niche including MuSCs themselves secrete IGF-1 (Bentzinger et al., 2010; Forcina et al., 2019). IGF-1 promotes regeneration of skeletal muscle by stimulating proliferation and differentiation of myoblasts (Fig. 1). During aging, IGF-1 signaling is altered in skeletal muscle concomitant with a decrease in regenerative capacity (Ascenzi et al., 2019). Of note, administration of IGF-1 or overexpression improves regeneration of skeletal muscle and attenuates impaired regeneration in the aged (Bentzinger et al., 2010; Musaro et al., 2001).

5.2 Regulation of MuSCs by thyroid hormones

Similar to IGF-1 thyroid hormones (TH) affect whole skeletal muscle as well as MuSCs. They regulate the expression of various proteins important for proper muscle development and contractility (Ambrosio, De Stefano, Di Girolamo, & Salvatore, 2017; Milanesi et al., 2016, 2017; Soukup & Smerdu, 2015). Deionidase type 2 (D2) is the main activator of TH at the local level, it converts the inactive pro-hormone T4 into the active form T3. Deionidase 3 (D3) on the other hand is considered to be the main physiological inactivator of TH action (Ambrosio et al., 2017). Cellular entry and efflux of TH are mediated by transmembrane transporters such as the monocarboxylate transporters 8 (MCT8) and 10 (MCT10), organic anion transporting protein 1C1 (OATP1C1), and the L-type aminoacid transporters LAT1 and LAT2 (Bernal, Guadaño-Ferraz, & Morte, 2015; Heuer & Visser, 2009, 2013).

D3 is strongly expressed in proliferating myoblasts, its expression decreases upon differentiation (Dentice et al., 2014) (Fig. 1). Of note, expression of D3 varies in different MuSC populations. For instance, expression of D3 is mainly expressed in $Pax7^{low}$ cells (Dentice et al., 2014), here it most likely prevents their differentiation. High levels of TH result in massive apoptosis of myoblasts, a phenotype reminiscent of D3 deficient mice which are also characterized by impaired regeneration of skeletal muscle (Dentice et al., 2014). In contrast to D3 levels the levels of the activating enzyme D2 are low in MuSCs and myoblasts but increase in the course of differentiation (Ambrosio et al., 2017). This is thought to contribute to the induction of MyoD expression and its downstream myogenic factors (Dentice et al., 2010). Further support for this concept arises from experiments showing that depletion of D2 expression results in impaired myogenic differentiation resembling the MyoD$^{-/-}$ phenotype (Salvatore, Simonides, Dentice, Zavacki, & Larsen, 2014; Yin et al., 2013).

THs tightly regulate MuSC function and regeneration of skeletal muscle (Dentice et al., 2010, 2014; Dentice, Marsili, Zavacki, Larsen, & Salvatore, 2013; Lee, Kim, & Milanesi, 2014; Salvatore et al., 2014). Interestingly, in addition to the activating and inactivating enzymes, the TH transporters Mct8 and Oatp1c1 show a specific expression pattern in MuSCs. While quiescent and activated MuSCs express Mct8, Oatp1c1 is exclusively found in activated MuSCs exhibiting a transient expression (Mayerl et al., 2018). The importance of a tight regulation of TH levels in MuSCs was demonstrated in mice deficient for both TH transporters expressed in MuSCs, namely Mct8 and Oatp1c1. In those mice regeneration of skeletal muscle was delayed, presumably due to an impaired TH uptake which keeps MuSCs for a prolonged period in a proliferative state (Mayerl et al., 2018). This then results in a delayed differentiation and consequently affected regeneration process. These data suggest that the TH transporters Mct8 and Oatp1c1 act as gate-keepers of MuSC activation underscoring the importance of a finely-tuned TH signaling during myogenesis and regeneration of skeletal muscle.

5.3 Wnt signaling regulates MuSC function

During embryonic development Wnt signaling is one of the main drivers of skeletal muscle development, a highly orchestrated process. Furthermore, it is one of the major signaling pathways important for regeneration of skeletal muscle (von Maltzahn et al., 2012). Wnt proteins are a large family of secreted glyco-proteins which are related to the wingless gene in Drosophila (Sethi & Vidal-Puig, 2010). In mammals the Wnt family comprises 19 members. Interestingly, they share homologies in their amino acid sequence but often have fundamentally distinct signaling properties (Clevers, Loh, & Nusse, 2014; Nusse, 2008; von Maltzahn et al., 2011). Importantly, all Wnt proteins contain a signal sequence required for secretion. Additionally, they comprise multiple glycosylation sites and a characteristic distribution of 22 cysteine residues (Nusse, 2008). Wnt proteins bind to Frizzled (Fzd) receptors, a class of seven-transmembrane proteins containing an extracellular cysteine-rich domain (Clevers & Nusse, 2012; Nusse, 2008; Sethi & Vidal-Puig, 2010). The interaction of the Wnt ligand with the Fzd receptor can evoke multiple intracellular responses. Among them the activation of β-catenin/TCF transcriptional complexes, also known as canonical Wnt signaling, is the best described so far (Nusse, 2012).

During regeneration of skeletal muscle multiple Wnt ligands are expressed. In the early phase of regeneration mRNA expression of *Wnt5a*, *Wnt5b* and *Wnt7a* was found, while *Wnt4* mRNA expression was described to decrease (Polesskaya, Seale, & Rudnicki, 2003). *Wnt3a* and *Wnt7b* transcripts were detected during the later phases of regeneration (Brack, Conboy, Conboy, Shen, & Rando, 2008; Polesskaya et al., 2003). Canonical Wnt signaling—mainly through the ligand Wnt3a—drives differentiation of MuSCs in adult skeletal muscle while non-canonical Wnt signaling through the ligand Wnt7a promotes symmetric satellite stem cell divisions in concert with the extracellular matrix protein Fibronectin (Bentzinger et al., 2014; Brack et al., 2008, 2007; Le Grand et al., 2009; Otto et al., 2008) (Fig. 1). Furthermore, Wnt7a improves regeneration of skeletal muscle, migration of MuSCs and myoblasts and induces hypertrophy of myotubes and myofibers (Bentzinger et al., 2014; Le Grand et al., 2009; von Maltzahn et al., 2011; von Maltzahn, Zinoviev, Chang, Bentzinger, & Rudnicki, 2013). A recent publication also demonstrates the importance of R-spondin, a known modulator of canonical Wnt signaling for proper differentiation of myogenic progenitor cells during regeneration (Girardi & Le Grand, 2018; Huels & Sansom, 2017; Lacour et al., 2017). The importance of balanced canonical Wnt signaling for proper regeneration is further supported by a study, which demonstrated that disruption as well as activation of β-catenin in adult MuSCs impairs regeneration of skeletal muscle (Rudolf et al., 2016). Additionally, a recent study also showed that Wnt4 secreted by the myofiber maintains MuSC quiescence via RhoA (Eliazer et al., 2019). With increasing age, canonical Wnt signaling becomes aberrantly active in skeletal muscle resulting in increased fibrosis after injury being one of the major causes of impaired regeneration of skeletal muscle in the elderly (Brack et al., 2007; Girardi & Le Grand, 2018).

5.4 Effect of the anti-aging hormone Klotho on MuSC functionality

Klotho is one of the well-known anti-aging hormones. It is mainly synthesized in the kidney, then secreted into the blood stream and serves as a suppressor of aging in various tissues through multiple mechanisms (Quarles, 2008; Utsugi et al., 2000). Consequently, the amount of Klotho in the serum is decreasing with increasing age (Pedersen, Pedersen, Brasen, & Rasmussen, 2013; Sato et al., 2015).

Klotho hypomorphic mice are characterized by a premature aging phenotype due to an insertion in the promotor region of the *klotho* gene (Kuro-o et al., 1997). Those mice display several aging-associated conditions including aberrations in muscle function which are also observed during natural aging, e.g. reduced muscle strength and running endurance (Phelps, Pettan-Brewer, Ladiges, & Yablonka-Reuveni, 2013). Furthermore, myofiber size and MuSC numbers are significantly decreased in *klotho* hypomorphic mice suggesting that those mice suffer from sarcopenia and impairments in MuSC maintenance as observed during natural aging (Ahrens, Huettemeister, Schmidt, Kaether, & von Maltzahn, 2018; Brack & Munoz-Canoves, 2016; Henze et al., 2020; Hwang & Brack, 2018). Moreover, MuSC function is severely impaired upon loss of *klotho* expression concomitant with a dramatic reduction in the regenerative capacity of skeletal muscle (Ahrens et al., 2018). This is in line with data demonstrating that Klotho expression is upregulated during regeneration of skeletal muscle in the young (Sahu et al., 2018). Of note, addition of recombinant Klotho protein inhibits the aberrant Wnt signaling in aged MuSCs leading to restored MuSC functionality (Ahrens et al., 2018). This suggests that acute Klotho availability can overcome intrinsic changes in aged MuSCs, which have been acquired over life time—at least in short term (Ahrens et al., 2018; Brack & Munoz-Canoves, 2016; Sousa-Victor et al., 2014).

6. Conclusion

The functionality of MuSCs is affected by intrinsic changes but also by changes in the MuSC niche and circulating factors. How intrinsic changes in MuSCs and changes in the niche and the systemic environment are affecting MuSC functionality and how they are interrelated, is one of the scientific questions to be answered in the coming years.

Acknowledgments

This work was supported by a grant from the Deutsche Forschungsgemeinschaft (MA-3975/2-1), the Carl Zeiss foundation and the Deutsche Krebshilfe (DKH-JvM-861005) to JvM.

References

Ahrens, H. E., Huettemeister, J., Schmidt, M., Kaether, C., & von Maltzahn, J. (2018). Klotho expression is a prerequisite for proper muscle stem cell function and regeneration of skeletal muscle. *Skeletal Muscle*, *8*, 20.

Ambrosio, R., De Stefano, M. A., Di Girolamo, D., & Salvatore, D. (2017). Thyroid hormone signaling and deiodinase actions in muscle stem/progenitor cells. *Molecular and Cellular Endocrinology*, *459*, 79–83.

Arnold, L. L., Cecchini, A., Stark, D. A., Ihnat, J., Craigg, R. N., Carter, A., et al. (2020). EphA7 promotes myogenic differentiation via cell-cell contact. *eLife*, *9*, 1–19.

Arnold, L., Henry, A., Poron, F., Baba-Amer, Y., van Rooijen, N., Plonquet, A., et al. (2007). Inflammatory monocytes recruited after skeletal muscle injury switch into antiinflammatory macrophages to support myogenesis. *The Journal of Experimental Medicine*, *204*, 1057–1069.

Ascenzi, F., Barberi, L., Dobrowolny, G., Villa Nova Bacurau, A., Nicoletti, C., Rizzuto, E., et al. (2019). Effects of IGF-1 isoforms on muscle growth and sarcopenia. *Aging Cell*, *18*, e12954.

Bentzinger, C. F., von Maltzahn, J., Dumont, N. A., Stark, D. A., Wang, Y. X., Nhan, K., et al. (2014). Wnt7a stimulates myogenic stem cell motility and engraftment resulting in improved muscle strength. *The Journal of Cell Biology*, *205*, 97–111.

Bentzinger, C. F., von Maltzahn, J., & Rudnicki, M. A. (2010). Extrinsic regulation of satellite cell specification. *Stem Cell Research & Therapy*, *1*, 27.

Bentzinger, C. F., Wang, Y. X., Dumont, N. A., & Rudnicki, M. A. (2013). Cellular dynamics in the muscle satellite cell niche. *EMBO Reports*, *14*, 1062–1072.

Bentzinger, C. F., Wang, Y. X., & Rudnicki, M. A. (2012). Building muscle: Molecular regulation of myogenesis. *Cold Spring Harbor Perspectives in Biology*, *4*, 1–16.

Bentzinger, C. F., Wang, Y. X., von Maltzahn, J., Soleimani, V. D., Yin, H., & Rudnicki, M. A. (2013). Fibronectin regulates Wnt7a signaling and satellite cell expansion. *Cell Stem Cell*, *12*, 75–87.

Bernal, J., Guadaño-Ferraz, A., & Morte, B. (2015). Thyroid hormone transporters—Functions and clinical implications. *Nature Reviews. Endocrinology*, *11*, 506.

Brack, A. S., Conboy, I. M., Conboy, M. J., Shen, J., & Rando, T. A. (2008). A temporal switch from notch to Wnt signaling in muscle stem cells is necessary for normal adult myogenesis. *Cell Stem Cell*, *2*, 50–59.

Brack, A. S., Conboy, M. J., Roy, S., Lee, M., Kuo, C. J., Keller, C., et al. (2007). Increased Wnt signaling during aging alters muscle stem cell fate and increases fibrosis. *Science*, *317*, 807–810.

Brack, A. S., & Munoz-Canoves, P. (2016). The ins and outs of muscle stem cell aging. *Skeletal Muscle*, *6*, 1.

Cantini, M., Giurisato, E., Radu, C., Tiozzo, S., Pampinella, F., Senigaglia, D., et al. (2002). Macrophage-secreted myogenic factors: A promising tool for greatly enhancing the proliferative capacity of myoblasts in vitro and in vivo. *Neurological Sciences*, *23*, 189–194.

Chakkalakal, J. V., Christensen, J., Xiang, W., Tierney, M. T., Boscolo, F. S., Sacco, A., et al. (2014). Early forming label-retaining muscle stem cells require p27kip1 for maintenance of the primitive state. *Development*, *141*, 1649–1659.

Chakkalakal, J. V., Jones, K. M., Basson, M. A., & Brack, A. S. (2012). The aged niche disrupts muscle stem cell quiescence. *Nature, 490*, 355–360.
Chal, J., & Pourquie, O. (2017). Making muscle: Skeletal myogenesis in vivo and in vitro. *Development, 144*, 2104–2122.
Chang, N. C., & Rudnicki, M. A. (2014). Satellite cells: The architects of skeletal muscle. *Current Topics in Developmental Biology, 107*, 161–181.
Chazaud, B., Brigitte, M., Yacoub-Youssef, H., Arnold, L., Gherardi, R., Sonnet, C., et al. (2009). Dual and beneficial roles of macrophages during skeletal muscle regeneration. *Exercise and Sport Sciences Reviews, 37*, 18–22.
Clevers, H., Loh, K. M., & Nusse, R. (2014). Stem cell signaling. An integral program for tissue renewal and regeneration: Wnt signaling and stem cell control. *Science, 346*, 1248012.
Clevers, H., & Nusse, R. (2012). Wnt/beta-catenin signaling and disease. *Cell, 149*, 1192–1205.
Conboy, I. M., Conboy, M. J., Wagers, A. J., Girma, E. R., Weissman, I. L., & Rando, T. A. (2005). Rejuvenation of aged progenitor cells by exposure to a young systemic environment. *Nature, 433*, 760–764.
Dentice, M., Ambrosio, R., Damiano, V., Sibilio, A., Luongo, C., Guardiola, O., et al. (2014). Intracellular inactivation of thyroid hormone is a survival mechanism for muscle stem cell proliferation and lineage progression. *Cell Metabolism, 20*, 1038–1048.
Dentice, M., Marsili, A., Ambrosio, R., Guardiola, O., Sibilio, A., Paik, J. H., et al. (2010). The FoxO3/type 2 deiodinase pathway is required for normal mouse myogenesis and muscle regeneration. *The Journal of Clinical Investigation, 120*, 4021–4030.
Dentice, M., Marsili, A., Zavacki, A., Larsen, P. R., & Salvatore, D. (2013). The deiodinases and the control of intracellular thyroid hormone signaling during cellular differentiation. *Biochimica et Biophysica Acta, 1830*, 3937–3945.
Eliazer, S., Muncie, J. M., Christensen, J., Sun, X., D'Urso, R. S., Weaver, V. M., et al. (2019). Wnt4 from the niche controls the mechano-properties and quiescent state of muscle stem cells. *Cell Stem Cell, 25*, 654–665 e654.
Evano, B., & Tajbakhsh, S. (2018). Skeletal muscle stem cells in comfort and stress. *NPJ Regenerative Medicine, 3*, 24.
Fielding, R. A., Manfredi, T. J., Ding, W., Fiatarone, M. A., Evans, W. J., & Cannon, J. G. (1993). Acute phase response in exercise. III. Neutrophil and IL-1 beta accumulation in skeletal muscle. *The American Journal of Physiology, 265*, R166–R172.
Fiore, F., Sebille, A., & Birnbaum, D. (2000). Skeletal muscle regeneration is not impaired in Fgf6 −/− mutant mice. *Biochemical and Biophysical Research Communications, 272*, 138–143.
Floss, T., Arnold, H. H., & Braun, T. (1997). A role for FGF-6 in skeletal muscle regeneration. *Genes & Development, 11*, 2040–2051.
Forcina, L., Cosentino, M., & Musaro, A. (2020). Mechanisms regulating muscle regeneration: Insights into the interrelated and time-dependent phases of tissue healing. *Cells, 9*, 1297.
Forcina, L., Miano, C., Scicchitano, B. M., & Musaro, A. (2019). Signals from the niche: Insights into the role of IGF-1 and IL-6 in modulating skeletal muscle fibrosis. *Cells, 8*, 232.
Girardi, F., & Le Grand, F. (2018). Wnt signaling in skeletal muscle development and regeneration. *Progress in Molecular Biology and Translational Science, 153*, 157–179.
Glass, D. J. (2003). Signalling pathways that mediate skeletal muscle hypertrophy and atrophy. *Nature Cell Biology, 5*, 87–90.
Glass, D. J. (2005). Skeletal muscle hypertrophy and atrophy signaling pathways. *The International Journal of Biochemistry & Cell Biology, 37*, 1974–1984.
Gunther, S., Kim, J., Kostin, S., Lepper, C., Fan, C. M., & Braun, T. (2013). Myf5-positive satellite cells contribute to Pax7-dependent long-term maintenance of adult muscle stem cells. *Cell Stem Cell, 13*, 590–601.

Hardy, D., Besnard, A., Latil, M., Jouvion, G., Briand, D., Thepenier, C., et al. (2016). Comparative study of injury models for studying muscle regeneration in mice. *PLoS One*, *11*, e0147198.

Henze, H., Jung, M. J., Ahrens, H. E., Steiner, S., & von Maltzahn, J. (2020). Skeletal muscle aging—Stem cells in the spotlight. *Mechanisms of Ageing and Development*, *189*, 111283.

Heredia, J. E., Mukundan, L., Chen, F. M., Mueller, A. A., Deo, R. C., Locksley, R. M., et al. (2013). Type 2 innate signals stimulate fibro/adipogenic progenitors to facilitate muscle regeneration. *Cell*, *153*, 376–388.

Hernandez-Hernandez, J. M., Garcia-Gonzalez, E. G., Brun, C. E., & Rudnicki, M. A. (2017). The myogenic regulatory factors, determinants of muscle development, cell identity and regeneration. *Seminars in Cell & Developmental Biology*, *72*, 10–18.

Heuer, H., & Visser, T. J. (2009). Minireview: Pathophysiological importance of thyroid hormone transporters. *Endocrinology*, *150*, 1078–1083.

Heuer, H., & Visser, T. J. (2013). The pathophysiological consequences of thyroid hormone transporter deficiencies: Insights from mouse models. *Biochimica et Biophysica Acta*, *1830*, 3974–3978.

Huels, D. J., & Sansom, O. J. (2017). R-spondin is more than just Wnt's sidekick. *Developmental Cell*, *41*, 456–458.

Hwang, A. B., & Brack, A. S. (2018). Muscle stem cells and aging. *Current Topics in Developmental Biology*, *126*, 299–322.

Kuang, S., Kuroda, K., Le Grand, F., & Rudnicki, M. A. (2007). Asymmetric self-renewal and commitment of satellite stem cells in muscle. *Cell*, *129*, 999–1010.

Kuro-o, M., Matsumura, Y., Aizawa, H., Kawaguchi, H., Suga, T., Utsugi, T., et al. (1997). Mutation of the mouse klotho gene leads to a syndrome resembling ageing. *Nature*, *390*, 45–51.

Lacour, F., Vezin, E., Bentzinger, C. F., Sincennes, M. C., Giordani, L., Ferry, A., et al. (2017). R-spondin1 controls muscle cell fusion through dual regulation of antagonistic Wnt signaling pathways. *Cell Reports*, *18*, 2320–2330.

Le Grand, F., Jones, A. E., Seale, V., Scime, A., & Rudnicki, M. A. (2009). Wnt7a activates the planar cell polarity pathway to drive the symmetric expansion of satellite stem cells. *Cell Stem Cell*, *4*, 535–547.

Lee, J. W., Kim, N. H., & Milanesi, A. (2014). Thyroid hormone signaling in muscle development, repair and metabolism. *Journal of Endocrinology, Diabetes & Obesity*, *2*, 1046.

Lefaucheur, J. P., & Sebille, A. (1995a). Basic fibroblast growth factor promotes in vivo muscle regeneration in murine muscular dystrophy. *Neuroscience Letters*, *202*, 121–124.

Lefaucheur, J. P., & Sebille, A. (1995b). Muscle regeneration following injury can be modified in vivo by immune neutralization of basic fibroblast growth factor, transforming growth factor beta 1 or insulin-like growth factor I. *Journal of Neuroimmunology*, *57*, 85–91.

Lepper, C., Conway, S. J., & Fan, C. M. (2009). Adult satellite cells and embryonic muscle progenitors have distinct genetic requirements. *Nature*, *460*, 627–631.

Lepper, C., Partridge, T. A., & Fan, C. M. (2011). An absolute requirement for Pax7-positive satellite cells in acute injury-induced skeletal muscle regeneration. *Development*, *138*, 3639–3646.

Lukjanenko, L., Jung, M. J., Hegde, N., Perruisseau-Carrier, C., Migliavacca, E., Rozo, M., et al. (2016). Loss of fibronectin from the aged stem cell niche affects the regenerative capacity of skeletal muscle in mice. *Nature Medicine*, *22*, 897–905.

Mauro, A. (1961). Satellite cell of skeletal muscle fibers. *The Journal of Biophysical and Biochemical Cytology*, *9*, 493–495.

Mayerl, S., Schmidt, M., Doycheva, D., Darras, V. M., Huttner, S. S., Boelen, A., et al. (2018). Thyroid hormone transporters MCT8 and OATP1C1 control skeletal muscle regeneration. *Stem Cell Reports*, *10*, 1959–1974.

Merly, F., Lescaudron, L., Rouaud, T., Crossin, F., & Gardahaut, M. F. (1999). Macrophages enhance muscle satellite cell proliferation and delay their differentiation. *Muscle & Nerve, 22*, 724–732.

Milanesi, A., Lee, J. W., Kim, N. H., Liu, Y. Y., Yang, A., Sedrakyan, S., et al. (2016). Thyroid hormone receptor α plays an essential role in male skeletal muscle myoblast proliferation, differentiation, and response to injury. *Endocrinology, 157*, 4–15.

Milanesi, A., Lee, J. W., Yang, A., Liu, Y. Y., Sedrakyan, S., Cheng, S. Y., et al. (2017). Thyroid hormone receptor α is essential to maintain the satellite cell niche during skeletal muscle injury and sarcopenia of aging. *Thyroid, 27*, 1316–1322.

Munoz-Canoves, P., Neves, J., & Sousa-Victor, P. (2019). Understanding muscle regenerative decline with aging: New approaches to bring back youthfulness to aged stem cells. *The FEBS Journal, 287*, 406–416.

Murphy, M. M., Lawson, J. A., Mathew, S. J., Hutcheson, D. A., & Kardon, G. (2011). Satellite cells, connective tissue fibroblasts and their interactions are crucial for muscle regeneration. *Development, 138*, 3625–3637.

Musaro, A., McCullagh, K., Paul, A., Houghton, L., Dobrowolny, G., Molinaro, M., et al. (2001). Localized Igf-1 transgene expression sustains hypertrophy and regeneration in senescent skeletal muscle. *Nature Genetics, 27*, 195–200.

Nusse, R. (2008). Wnt signaling and stem cell control. *Cell Research, 18*, 523–527.

Nusse, R. (2012). Wnt signaling. *Cold Spring Harbor Perspectives in Biology, 4*, 1–3.

Olguin, H. C., & Olwin, B. B. (2004). Pax-7 up-regulation inhibits myogenesis and cell cycle progression in satellite cells: A potential mechanism for self-renewal. *Developmental Biology, 275*, 375–388.

Otto, A., Schmidt, C., Luke, G., Allen, S., Valasek, P., Muntoni, F., et al. (2008). Canonical Wnt signalling induces satellite-cell proliferation during adult skeletal muscle regeneration. *Journal of Cell Science, 121*, 2939–2950.

Pawlikowski, B., Vogler, T. O., Gadek, K., & Olwin, B. B. (2017). Regulation of skeletal muscle stem cells by fibroblast growth factors. *Developmental Dynamics, 246*, 359–367.

Pedersen, L., Pedersen, S. M., Brasen, C. L., & Rasmussen, L. M. (2013). Soluble serum Klotho levels in healthy subjects. Comparison of two different immunoassays. *Clinical Biochemistry, 46*, 1079–1083.

Phelps, M., Pettan-Brewer, C., Ladiges, W., & Yablonka-Reuveni, Z. (2013). Decline in muscle strength and running endurance in klotho deficient C57BL/6 mice. *Biogerontology, 14*, 729–739.

Polesskaya, A., Seale, P., & Rudnicki, M. A. (2003). Wnt signaling induces the myogenic specification of resident CD45+ adult stem cells during muscle regeneration. *Cell, 113*, 841–852.

Price, F. D., von Maltzahn, J., Bentzinger, C. F., Dumont, N. A., Yin, H., Chang, N. C., et al. (2014). Inhibition of JAK-STAT signaling stimulates adult satellite cell function. *Nature Medicine, 20*, 1174–1181.

Quarles, L. D. (2008). Endocrine functions of bone in mineral metabolism regulation. *The Journal of Clinical Investigation, 118*, 3820–3828.

Rocheteau, P., Gayraud-Morel, B., Siegl-Cachedenier, I., Blasco, M. A., & Tajbakhsh, S. (2012). A subpopulation of adult skeletal muscle stem cells retains all template DNA strands after cell division. *Cell, 148*, 112–125.

Rozo, M., Li, L., & Fan, C. M. (2016). Targeting beta1-integrin signaling enhances regeneration in aged and dystrophic muscle in mice. *Nature Medicine, 22*, 889–896.

Rudolf, A., Schirwis, E., Giordani, L., Parisi, A., Lepper, C., Taketo, M. M., et al. (2016). Beta-catenin activation in muscle progenitor cells regulates tissue repair. *Cell Reports, 15*, 1277–1290.

Saclier, M., Yacoub-Youssef, H., Mackey, A. L., Arnold, L., Ardjoune, H., Magnan, M., et al. (2013). Differentially activated macrophages orchestrate myogenic precursor cell fate during human skeletal muscle regeneration. *Stem Cells, 31*, 384–396.

Sahu, A., Mamiya, H., Shinde, S. N., Cheikhi, A., Winter, L. L., Vo, N. V., et al. (2018). Age-related declines in alpha-Klotho drive progenitor cell mitochondrial dysfunction and impaired muscle regeneration. *Nature Communications, 9*, 4859.

Salvatore, D., Simonides, W. S., Dentice, M., Zavacki, A. M., & Larsen, P. R. (2014). Thyroid hormones and skeletal muscle—New insights and potential implications. *Nature Reviews. Endocrinology, 10*, 206–214.

Sambasivan, R., Yao, R., Kissenpfennig, A., Van Wittenberghe, L., Paldi, A., Gayraud-Morel, B., et al. (2011). Pax7-expressing satellite cells are indispensable for adult skeletal muscle regeneration. *Development, 138*, 3647–3656.

Sato, S., Kawamata, Y., Takahashi, A., Imai, Y., Hanyu, A., Okuma, A., et al. (2015). Ablation of the p16(INK4a) tumour suppressor reverses ageing phenotypes of klotho mice. *Nature Communications, 6*, 7035.

Schmidt, M., Schuler, S. C., Huttner, S. S., von Eyss, B., & von Maltzahn, J. (2019). Adult stem cells at work: Regenerating skeletal muscle. *Cellular and Molecular Life Sciences, 76*, 2559–2570.

Seale, P., Sabourin, L. A., Girgis-Gabardo, A., Mansouri, A., Gruss, P., & Rudnicki, M. A. (2000). Pax7 is required for the specification of myogenic satellite cells. *Cell, 102*, 777–786.

Sethi, J. K., & Vidal-Puig, A. (2010). Wnt signalling and the control of cellular metabolism. *The Biochemical Journal, 427*, 1–17.

Shea, K. L., Xiang, W., LaPorta, V. S., Licht, J. D., Keller, C., Basson, M. A., et al. (2010). Sprouty1 regulates reversible quiescence of a self-renewing adult muscle stem cell pool during regeneration. *Cell Stem Cell, 6*, 117–129.

Shinin, V., Gayraud-Morel, B., Gomes, D., & Tajbakhsh, S. (2006). Asymmetric division and cosegregation of template DNA strands in adult muscle satellite cells. *Nature Cell Biology, 8*, 677–687.

Shinin, V., Gayraud-Morel, B., & Tajbakhsh, S. (2009). Template DNA-strand co-segregation and asymmetric cell division in skeletal muscle stem cells. *Methods in Molecular Biology, 482*, 295–317.

Singh, K., & Dilworth, F. J. (2013). Differential modulation of cell cycle progression distinguishes members of the myogenic regulatory factor family of transcription factors. *The FEBS Journal, 280*, 3991–4003.

Soleimani, V. D., Punch, V. G., Kawabe, Y., Jones, A. E., Palidwor, G. A., Porter, C. J., et al. (2012). Transcriptional dominance of Pax7 in adult myogenesis is due to high-affinity recognition of homeodomain motifs. *Developmental Cell, 22*, 1208–1220.

Soukup, T., & Smerdu, V. (2015). Effect of altered innervation and thyroid hormones on myosin heavy chain expression and fiber type transitions: A mini-review. *Histochemistry and Cell Biology, 143*, 123–130.

Sousa-Victor, P., Gutarra, S., Garcia-Prat, L., Rodriguez-Ubreva, J., Ortet, L., Ruiz-Bonilla, V., et al. (2014). Geriatric muscle stem cells switch reversible quiescence into senescence. *Nature, 506*, 316–321.

Stark, D. A., Karvas, R. M., Siegel, A. L., & Cornelison, D. D. (2011). Eph/ephrin interactions modulate muscle satellite cell motility and patterning. *Development, 138*, 5279–5289.

Tidball, J. G. (1995). Inflammatory cell response to acute muscle injury. *Medicine and Science in Sports and Exercise, 27*, 1022–1032.

Tierney, M. T., Aydogdu, T., Sala, D., Malecova, B., Gatto, S., Puri, P. L., et al. (2014). STAT3 signaling controls satellite cell expansion and skeletal muscle repair. *Nature Medicine, 20*, 1182–1186.

Tierney, M. T., Stec, M. J., Rulands, S., Simons, B. D., & Sacco, A. (2018). Muscle stem cells exhibit distinct clonal dynamics in response to tissue repair and homeostatic aging. *Cell Stem Cell, 22*, 119–127.e113.

Utsugi, T., Ohno, T., Ohyama, Y., Uchiyama, T., Saito, Y., Matsumura, Y., et al. (2000). Decreased insulin production and increased insulin sensitivity in the klotho mutant mouse, a novel animal model for human aging. *Metabolism*, *49*, 1118–1123.

von Maltzahn, J., Bentzinger, C. F., & Rudnicki, M. A. (2011). Wnt7a-Fzd7 signalling directly activates the Akt/mTOR anabolic growth pathway in skeletal muscle. *Nature Cell Biology*, *14*, 186–191.

von Maltzahn, J., Bentzinger, C. F., & Rudnicki, M. A. (2013). *Characteristics of satellite cells and multipotent adult stem cells in the skeletal muscle*. Springer.

von Maltzahn, J., Chang, N. C., Bentzinger, C. F., & Rudnicki, M. A. (2012). Wnt signaling in myogenesis. *Trends in Cell Biology*, *22*, 602–609.

von Maltzahn, J., Jones, A. E., Parks, R. J., & Rudnicki, M. A. (2013). Pax7 is critical for the normal function of satellite cells in adult skeletal muscle. *Proceedings of the National Academy of Sciences of the United States of America*, *110*, 16474–16479.

von Maltzahn, J., Zinoviev, R., Chang, N. C., Bentzinger, C. F., & Rudnicki, M. A. (2013). A truncated Wnt7a retains full biological activity in skeletal muscle. *Nature Communications*, *4*, 2869.

Yin, H., Price, F., & Rudnicki, M. A. (2013). Satellite cells and the muscle stem cell niche. *Physiological Reviews*, *93*, 23–67.

Zammit, P. S., Golding, J. P., Nagata, Y., Hudon, V., Partridge, T. A., & Beauchamp, J. R. (2004). Muscle satellite cells adopt divergent fates: A mechanism for self-renewal? *The Journal of Cell Biology*, *166*, 347–357.

CHAPTER TEN

Estrogen receptor beta and neural development

Ivan Nalvarte[a,*], Mukesh Varshney[a], Jose Inzunza[a], and Jan-Åke Gustafsson[a,b]

[a]Department of Biosciences and Nutrition, Karolinska Institutet, Huddinge, Sweden
[b]Center for Nuclear Receptors and Cell Signaling, University of Houston, Houston, TX, United States
[*]Corresponding author: e-mail address: ivan.nalvarte@ki.se

Contents

Abstract

The female sex hormone estradiol (E2, 17β-estradiol) has important functions in the developing brain. In addition to regulating sexual differentiation of the brain, E2 participates in the development of brain areas involved in functions unrelated to reproduction, such as cognition. E2 signals mainly thorough two estrogen receptors; estrogen receptor alpha (ERα) and beta (ERβ). While ERα has distinct functions for sexual imprinting of the developing brain, ERβ is considered to participate in the development of brain areas related to cognitive function. In this chapter we will focus on ERβ's role during neural development. We will discuss the contributions of sex chromosomal and sex hormonal effects in this process and place it in relation to recent data on ERβ obtained from stem cell models. Finally, we will discuss the lessons learned from mouse and stem cell models in understanding ERβ's role in neural development and how new stem cell models, by addressing the human relevance, may help to advance our progress in this field.

Vitamins and Hormones, Volume 116
ISSN 0083-6729
https://doi.org/10.1016/bs.vh.2021.02.007

1. Sex hormone signaling during early neurodevelopment

The development of the brain requires intricate temporal and spatial orchestration of a multitude of signaling cues. Clearly, a miss-regulation during this process can result in devastating neurodevelopmental consequences and mediate susceptibility to different neuropsychiatric diseases. Interestingly, many diseases of this kind affect females and males differently. For instance, more females are diagnosed with depression and anxiety, while attention deficit hyperactivity disorder (ADHD), autism spectrum disorders (ASD), and early onset schizophrenia are more prevalent in males (Loke, Harley, & Lee, 2015; Pinares-Garcia, Stratikopoulos, Zagato, Loke, & Lee, 2018).

1.1 Sex chromosomal effects

The difference between males and females in the susceptibility and pathology of different neurological diseases suggest that sex chromosomal factors and sex hormones are likely involved in the etiology of these disorders. One sex-specific genetic factor that has received great attention is the *SRY* gene (Sex-determining region on the Y chromosome) that is located on the Y chromosome and regulates male gonadal development and thus fetal testosterone levels (Koopman, Munsterberg, Capel, Vivian, & Lovell-Badge, 1990; Sinclair et al., 1990; Wilson & Davies, 2007). *SRY* is also actively expressed in dopaminergic neurons of the frontal cortex and SNc of the midbrain where it participates in regulating male dopamine biosynthesis machinery (Czech et al., 2012; Dewing et al., 2006; Dewing, Shi, Horvath, & Vilain, 2003). Therefore a dysregulation of *SRY* results in dopaminergic signaling imbalances, which in turn leads to defective control of aggressive and movement behavior ("fight-flight" response) (Lee & Harley, 2012). Furthermore, over- or under-dosage of sex chromosomes (e.g., gain or loss of chromosomal copy numbers or escape from X-chromosome inactivation) may over- or under-activate other important genes involved in neurodevelopment (Pinares-Garcia et al., 2018).

1.2 Sex hormonal effects

In addition to the sex chromosomal effects that underlie neurodevelopmental processes, proper sex hormone levels are essential to neurodevelopment. The masculinization of the brain starts at the second trimester in humans, co-occurring with the testosterone peak produced by the forming male gonads (Wallen, 2005). In contrast, the brain feminization occurs in the absence of fetal testosterone (Baum, 1979). The role of the female sex hormone estradiol (E2, 17β-estradiol) in masculinization and feminization of the brain is a matter of debate. At least in rodents, circulating testosterone is produced by the developing testes from day E18 and is in the fetal brain converted to E2 by aromatase (*CYP19A1* gene). It is thus E2 that is responsible for the rodent brain masculinization and male-specific behavior (Wu et al., 2009). In females, the ovaries do not produce E2 until after birth, thereby surpassing the sensitive period of the developing brain accounting for brain masculinization. In addition, the rodent fetal brain is well-protected from circulating E2 by the plasma protein α-fetoprotein (AFP) which binds E2 and reduces its bioavailability. However, in humans, it is the circulating androgen levels that account for brain masculinization rather than E2 (Wallen, 2005), and human AFP has low affinity to E2 (Swartz & Soloff, 1974). Instead, the human sex hormone-binding globulin (SHBG) may play a similar role to rodent AFP, with the exception that it has higher affinity to testosterone than to E2 (Petra, Woodcock, Orr, Nguyen, & Sui, 2000; Wallen, 2005) and may therefore protect the human fetal female brain to sex hormones. Thus, there are substantial differences in how sex hormones affect the developing human brain compared to the rodent brain.

The increased sensitivity to organizational and activational effects evoked by sex hormones in the fetal brain occurs at day E18 until a few days after birth in rodents, while in humans this period spans between first and second trimester, up until birth. During this period, sex hormones elicit permanent structural and organizational changes to the developing brain in a sex specific manner. Although the molecular mechanisms behind these changes are still far from clear, they include spatial and temporal effects on gene expression to modulate at least neuronal apoptosis, glial differentiation and migration, as well as synaptic plasticity (Davis, Popper, & Gorski, 1996; de Vries & Södersten, 2009; Gore, Martien, Gagnidze, & Pfaff, 2014).

Clearly, an imbalance in sex hormone levels or sex hormone receptor activation can have severe consequences for neurodevelopment that reaches beyond sexual differentiation and reproduction to affect cognitive and social behavior. Although prenatal treatments of rodents with sex hormones have been shown to affect hippocampal-dependent cognitive tasks later in life (Isgor & Sengelaub, 1998; Kight & McCarthy, 2020; Williams, Barnett, & Meck, 1990), information on similar causal influence of sex hormone signaling on the human fetal brain development and human behavior is almost completely missing. Nevertheless, the obvious sex differences in several neurodevelopmental psychiatric diseases point to sex hormones, in addition to genetic factors, to have important roles in the etiology and progression of these diseases.

2. The estrogen receptor beta (ERβ)

E2 has many pleiotropic functions in the body, mainly related to reproduction. The main organ responsible for E2 production are the ovaries of peri- and post-pubertal females of fertile age. However, E2 is also synthesized locally in several other organs of which the brain is a major site for E2 production during development and in adulthood in both males and females (Garcia-Segura, 2008).

E2 signals mainly through two estrogen receptor isoforms, the estrogen receptor alpha (ERα) and estrogen receptor beta (ERβ), that belong to the steroid-activated nuclear receptor (NR) family of transcription factors. Both receptors have typical distinct structurally and functionally conserved NR regions, that include the DNA binding domain (DBD), the ligand binding domain (LBD) and two transcriptional activation function domains (AF-1) and (AF-2), of which the AF-1 domain is constitutively active and the AF-2 is ligand-dependent (Heldring et al., 2007) (Fig. 1). Upon ligand binding the receptor dimerizes and binds to gene regulatory elements (estrogen response elements, EREs) on the DNA to mediate transcription of hundreds of genes involved in proliferation, differentiation, and survival. Although ERα and ERβ show close structural similarities, they display different transcriptional activities that depend on ligand, cell-type, and promoter contexts. In cells that express both ERs the receptors form heterodimers where ERβ mostly opposes the actions of ERα (Matthews & Gustafsson, 2003). In general,

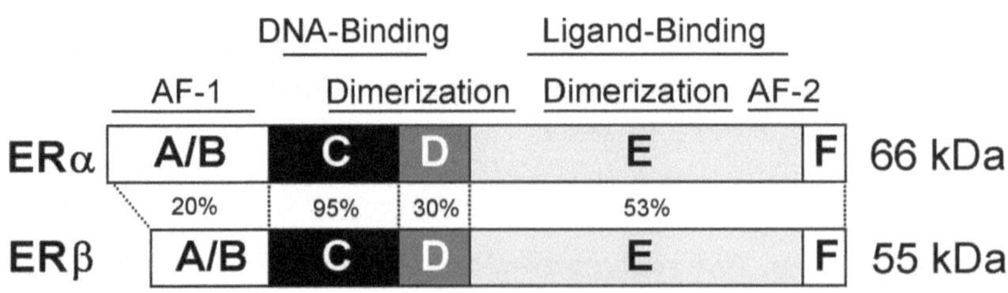

Fig. 1 Comparison between ERα and ERβ. Both ERs share highest sequence similarity in the DNA-binding and ligand-binding domains, and both receptors have a ligand dependent AF-2 domain and a ligand independent, constitutively active, AF-1 domain.

ERα exerts mainly proliferative and pro-survival effects, while ERβ is anti-proliferative and promotes differentiation (Fuentes & Silveyra, 2019; Matthews & Gustafsson, 2003).

3. ERβ in neurodevelopment

3.1 Lessons from rodent models

Both ERα and ERβ expression is highest in the developing fetal rodent brain until day P7. ERα mRNA is found in the rodent brain from E16.5 (Lemmen et al., 1999), while ERβ mRNA is found already at day E10.5 and peaks at E18.5 (Fig. 2) (Wang, Andersson, Warner, & Gustafsson, 2003). After P7, ERα localizes mainly to the hypothalamus, where it participates in reproductive tasks, while ERβ is more dispersed throughout the brain and is the main ER isoform in the hippocampus, serotonergic neurons of the dorsal raphe, interneurons of the cerebral cortex, and in microglia and oligodendrocytes (Gonzalez et al., 2007; Kelly & Qiu, 2010; Liu et al., 2008; Varshney et al., 2021; Wu et al., 2013). Rodent models lacking ERβ display increased oligodendrogliogenesis, defective corticogenesis, and an imbalanced excitatory and inhibitory neurotransmission (Fan, Warner, & Gustafsson, 2006; Fan, Xu, Warner, & Gustafsson, 2010; Varshney et al., 2017; Varshney et al., 2021; Wang et al., 2003; Wang, Andersson, Warner, & Gustafsson, 2001). In addition, loss of ERβ results in a decreased threshold for anxiety behavior in both male and female mice, as well as increased aggressive behavior, increased locomotor activity, and indications of compromised non-selective attention in male mice (Imwalle, Gustafsson, & Rissman, 2005; Nakata et al., 2016; Ogawa et al., 1999; Varshney et al., 2021). It has also been shown that ERβ is needed for proper spatial learning in female mice

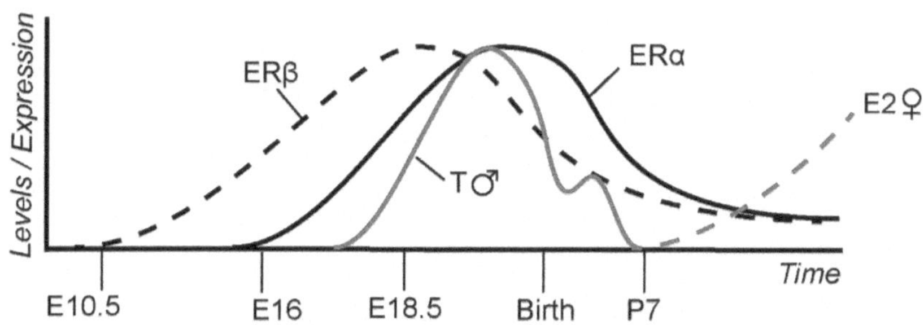

Fig. 2 Comparison between ERα (solid black line) and ERβ (dotted black line) expression and female E2 (dotted gray line) and male testosterone (T, solid gray line) levels in the mouse embryonic and perinatal brain. Axes not drawn to scale.

(Rissman, Heck, Leonard, Shupnik, & Gustafsson, 2002), and ERβ is thus considered to be involved in neurocognitive tasks.

Interestingly, both ERs are expressed in the fetal brain before the rise in prenatal circulating sex hormones (Fig. 2). The developing female ovaries start producing E2 after birth (at day P7 in rodents) (Sugiyama et al., 2009). Before that, the immature ovaries produce the androgen, dihydrotestosterone (DHT), which is rapidly metabolized by 3β-HSD enzyme in the brain to 3β-androstanediol (3β-Adiol) (Guennoun, Fiddes, Gouézou, Lombès, & Baulieu, 1995) (Fig. 3), a potent ERβ selective agonist (Kuiper et al., 1998). Males also produce DHT from the immature testes however at a later time period (after P20 in rodents), resulting in differences in prenatal 3β-Adiol levels between males and females. In contrast to androgens, including DHT, 3β-Adiol is not bound up by SHBG, suggesting that the estrogenic effects of 3β-Adiol may have a more important role in early embryonic brain development than previously thought and that ERβ can be a main mediator of the effects of 3β-Adiol (Sugiyama et al., 2009).

As mentioned above, ERβ is expressed already at E10.5 in mice, before any increase in circulating gonadal androgen levels (Fig. 2), and before the expression of aromatase (that peaks between E15 and E19). This means, that during early embryonic development, ERβ may function independently of estrogenic ligands. However, it also hints that exogenous ligands may modulate its activity during this sensitive neurodevelopmental timeframe. If this holds true in the human setting is not well studied due to the lack of reliable human-relevant models. In this sense, stem cell models may offer invaluable insights into human neurodevelopmental sex hormone signaling.

3.2 Lessons from stem cell models

As mentioned above the generation of ERβ knockout (ERβ−/−) rodent models have anchored the importance of ERβ in neurodevelopment. However, despite more than two decades of studying ERβ−/− rodent models, the mechanistic role of ERβ, and its spatiotemporal dependence on estrogenic ligands, is still largely unknown. This lack in mechanistic knowledge is largely associated with the limitations in studying the fetal rodent brain, which include difficulties in studying spatiotemporal cell-cell-interactions and questionable human relevance.

The advances in stem cell models over the past decade have offered new possibilities to address mechanistic understanding of ERβ's role during neurodevelopment and provide answers whether ERβ may pose as a target

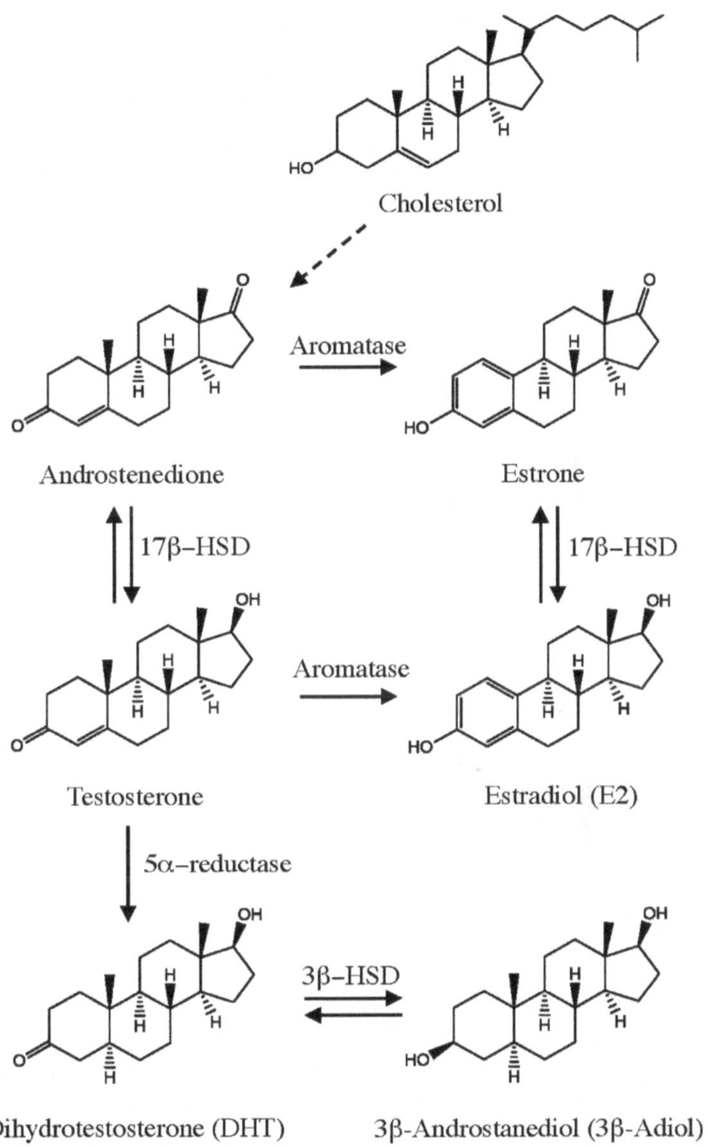

Fig. 3 Metabolism of testosterone into estradiol (E2) and 3β-androstanediol (3β-Adiol). Cholesterol is metabolized in sex hormone producing cells through several steps into androstenedione, which in turn can be converted to testosterone by 17β-HSD (17β-Hydroxysteroid dehydrogenase) or by aromatase (*CYP19A1*) to E2. Testosterone can also be converted to DHT by 5α-reductase, which is rapidly converted by 17β-HSD into 3β-Adiol in the embryonic brain.

for therapeutic interventions in different neurological diseases. Recently, a study using mouse embryonic stem cells derived from ERβ −/− mice showed that ERβ is necessary for proper serotonergic and dopaminergic signaling of midbrain neurons by regulating expression of key factors needed for neuronal differentiation, such as Engrailed1 (*En1*) and Notch1 (*Notch1*) resulting in at least lower dopamine production from midbrain neurons in the absence of ERβ (Varshney et al., 2017). This adds mechanistic knowledge to the in vivo observations that ERβ is involved in maintaining proper dopaminergic and serotonergic signaling, which is reflected in the increased anxiety that ERβ −/− mice display and their lower dopamine levels in several brain regions (Clark et al., 2012; Imwalle et al., 2005; Suzuki et al., 2013; Varshney et al., 2021; Walf, Koonce, & Frye, 2009). Furthermore, by isolating different neural populations from differentiated ERβ −/− stem cells it was shown that important factors involved in GABAergic signaling (e.g., *Gabra4*), oligodendrogenesis (e.g., *Olig2*), and migration, such as reelin (*Reln*) and integrins, were affected by ERβ loss (Varshney et al., 2017), which may reflect the imbalanced excitatory/inhibitory neuronal transmission, precocious oliogodendrogenesis, and impaired corticogenesis, respectively, observed in ERβ −/− mice (Varshney et al., 2021; Wang et al., 2001, 2003).

In addition to the classical ligand-mediated activation of ERβ, it may function independently of ligands through, e.g., phosphorylation, alternative splicing, and coregulator recruitment (Mott et al., 2014; Mott & Pak, 2012; Pinceti, Shults, Rao, Mott, & Pak, 2015). Although such ligand-independent effects of ERβ have been ascribed important functions in the aging female rat brain (Mott et al., 2014), a period of low circulating sex hormone levels, it is far from clear if ERβ has similar ligand independent importance during early development, before any endogenous sex hormone production. Recent work illustrates surprisingly little effects of the selective ERβ agonist LY3201 during neural development despite increased ERβ expression during this differentiation period (Varshney et al., 2017). Out of several factors analyzed, of importance for proper serotonergic and dopaminergic neuron differentiation and maturation, only dopamine synthesis in mature dopaminergic neurons was clearly positively regulated by ligand-mediated ERβ activation (Li et al., 2017; Varshney et al., 2017), while serotonin synthesis or oligodendrocyte maturation/proliferation was not affected by ligand, but was rather affected by the loss of ERβ itself (Varshney et al., 2017). This indicates that ERβ exerts significant ligand-independent effects on neural stem cell differentiation and mature neural function. However, it is not clear how the possible shift from ligand-independent to ligand-dependent activation occurs during neural differentiation. Similarly, the mechanisms behind why some cell types remain insensitive to ERβ ligand activation, but sensitive to ERβ loss, while other cell types respond to ERβ ligand activation, remain to be elucidated. The generation of ever better neural stem cell cultivation and differentiation protocols, as well as the development of new stem cell models such as human brain organoids, harbor great promise in understanding the spatiotemporal effects of sex hormones and sex hormone receptors, such as ERβ, during neurodevelopment and their involvement in neurodevelopmental disorders and diseases.

3.3 Involvement of ERβ in regulating reelin during neurodevelopment

The identification of reelin (mentioned above) as a possible ERβ-regulated gene is highly interesting. Reelin is a large secreted glycoprotein mainly found in the GABAergic interneurons of the post-natal cortex where it participates in synaptogenesis, long-term potentiation, and dendritic outgrowth (Lakatosova & Ostatnikova, 2012). In the developing brain, however, reelin expression is restricted to Cajal-Retzius cells that migrate from the hippocampus and neocortex to the marginal/layer 1 zone of the developing cortex (Lakatosova & Ostatnikova, 2012). In this process, reelin is essential for proper cortical layering by promoting radial glia differentiation, migration, and fiber orientation (Frotscher, 1998; Lakatosova & Ostatnikova, 2012). Interestingly, although ERβ −/− mice appear to have normal Cajal-Retzius cells (Wang et al., 2003), they display an unordered cortical layering (Varshney et al., 2021; Wang et al., 2003), similar to what is seen in mice carrying reelin mutations (i.e., reeler mice) (Hamburgh, 1963), and the finding that ERβ −/− neural stem cells have severely decreased reelin expression (Varshney et al., 2017) puts forward ERβ as a possible key regulator of cortical development, at least in part by regulating neurodevelopmental reelin expression and cell migration. Since reelin has been implicated in several neurodevelopmental disorders or diseases with sex biases, such as dyslexia (Tammimies et al., 2013), autism (Biamonte et al., 2009; Chen et al., 2017; Romano, Cosentino, Laviola, & De Filippis, 2016), and schizophrenia (Fatemi, Earle, & McMenomy, 2000; Potkin et al., 2009), perturbations in sex hormone signaling through ERβ which may affect reelin expression during early neurodevelopment may be one possible mechanism behind the sex differences in these disorders or diseases.

4. Conclusion and perspectives

Many neurological diseases affect men and women differently which is due to sex chromosomal and sex hormonal effects. Since its discovery in the late 90's (Kuiper, Enmark, Pelto-Huikko, Nilsson, & Gustafsson, 1996), ERβ has drawn increasing attention for its importance in neurodevelopment and as a factor that may regulate sex-specific events during this process. Much of our understanding of ERβ's role in neurodevelopment has been gained from invaluable ERβ−/− mouse models, however there are also evident limitations in using mouse models to study human neurodevelopment, such as obvious species differences and limitations in addressing mechanistic detail. Thus, great hope is placed on new complementary human relevant models, such as human stem cell models, to fully understand the importance of ERβ in neurodevelopment. Neural stem cell models have recently illustrated the cell-specific effects of ERβ loss and identified several key developmental genes, such as *Reln*, *En1* and *Notch1* to be targets of ERβ. The recent advances in human brain organoid models in combination with CRISPR/Cas9 technique to regulate ERβ expression will be of great importance to gain deeper mechanistic insights into ERβ functions in the developing and adult human brain, and to bring more understanding to the sex differences observed in several neurodevelopmental diseases, such as ASD, ADHD and schizophrenia.

References

Baum, M. J. (1979). Differentiation of coital behavior in mammals: A comparative analysis. *Neuroscience & Biobehavioral Reviews*, *3*(4), 265–284. 1979/12/01/ https://doi.org/10.1016/0149-7634(79)90013-7.

Biamonte, F., Assenza, G., Marino, R., D'Amelio, M., Panteri, R., Caruso, D., et al. (2009). Interactions between neuroactive steroids and reelin haploinsufficiency in Purkinje cell survival. *Neurobiology of Disease*, *36*(1), 103–115. 2009/10/01/. https://doi.org/10.1016/j.nbd.2009.07.001.

Chen, N., Bao, Y., Xue, Y., Sun, Y., Hu, D., Meng, S., et al. (2017). Meta-analyses of RELN variants in neuropsychiatric disorders. *Behavioural Brain Research*, *332*, 110–119. 2017/08/14/ https://doi.org/10.1016/j.bbr.2017.05.028.

Clark, J. A., Alves, S., Gundlah, C., Rocha, B., Birzin, E. T., Cai, S. J., et al. (2012). Selective estrogen receptor-beta (SERM-beta) compounds modulate raphe nuclei tryptophan hydroxylase-1 (TPH-1) mRNA expression and cause antidepressant-like effects in the forced swim test. *Neuropharmacology*, *63*(6), 1051–1063. https://doi.org/10.1016/j.neuropharm.2012.07.004.

Czech, D. P., Lee, J., Sim, H., Parish, C. L., Vilain, E., & Harley, V. R. (2012). The human testis-determining factor SRY localizes in midbrain dopamine neurons and regulates multiple components of catecholamine synthesis and metabolism. *Journal of Neurochemistry*, *122*(2), 260–271. https://doi.org/10.1111/j.1471-4159.2012.07782.x.

Davis, E. C., Popper, P., & Gorski, R. A. (1996). The role of apoptosis in sexual differentiation of the rat sexually dimorphic nucleus of the preoptic area. *Brain Research*, *734*(1–2), 10–18.

de Vries, G. J., & Södersten, P. (2009). Sex differences in the brain: The relation between structure and function. *Hormones and Behavior*, *55*(5), 589–596. https://doi.org/10.1016/j.yhbeh.2009.03.012.

Dewing, P., Chiang, C. W., Sinchak, K., Sim, H., Fernagut, P. O., Kelly, S., et al. (2006). Direct regulation of adult brain function by the male-specific factor SRY. *Current Biology*, *16*(4), 415–420. https://doi.org/10.1016/j.cub.2006.01.017.

Dewing, P., Shi, T., Horvath, S., & Vilain, E. (2003). Sexually dimorphic gene expression in mouse brain precedes gonadal differentiation. *Brain Research. Molecular Brain Research*, *118*(1–2), 82–90. http://www.ncbi.nlm.nih.gov/pubmed/14559357.

Fan, X., Warner, M., & Gustafsson, J. A. (2006). Estrogen receptor beta expression in the embryonic brain regulates development of calretinin-immunoreactive GABAergic interneurons. *Proceedings of the National Academy of Sciences of the United States of America*, *103*(51), 19338–19343. https://doi.org/10.1073/pnas.0609663103.

Fan, X., Xu, H., Warner, M., & Gustafsson, J. A. (2010). ERbeta in CNS: New roles in development and function. *Progress in Brain Research*, *181*, 233–250. https://doi.org/10.1016/S0079-6123(08)81013-8.

Fatemi, S. H., Earle, J. A., & McMenomy, T. (2000). Reduction in Reelin immunoreactivity in hippocampus of subjects with schizophrenia, bipolar disorder and major depression. *Molecular Psychiatry*, *5*(6), 654–663. 571 https://doi.org/10.1038/sj.mp.4000783.

Frotscher, M. (1998). Cajal—Retzius cells, Reelin, and the formation of layers. *Current Opinion in Neurobiology*, *8*(5), 570–575. 1998/10/01/ https://doi.org/10.1016/S0959-4388(98)80082-2.

Fuentes, N., & Silveyra, P. (2019). Estrogen receptor signaling mechanisms. *Advances in Protein Chemistry and Structural Biology*, *116*, 135–170. https://doi.org/10.1016/bs.apcsb.2019.01.001.

Garcia-Segura, L. M. (2008). Aromatase in the brain: Not just for reproduction anymore. *Journal of Neuroendocrinology*, *20*(6), 705–712. https://doi.org/10.1111/j.1365-2826.2008.01713.x.

Gonzalez, M., Cabrera-Socorro, A., Perez-Garcia, C. G., Fraser, J. D., Lopez, F. J., Alonso, R., et al. (2007). Distribution patterns of estrogen receptor alpha and beta in the human cortex and hippocampus during development and adulthood. *The Journal of Comparative Neurology*, *503*(6), 790–802. https://doi.org/10.1002/cne.21419.

Gore, A. C., Martien, K. M., Gagnidze, K., & Pfaff, D. (2014). Implications of prenatal steroid perturbations for neurodevelopment, behavior, and autism. *Endocrine Reviews*, *35*(6), 961–991. https://doi.org/10.1210/er.2013-1122.

Guennoun, R., Fiddes, R. J., Gouézou, M., Lombès, M., & Baulieu, E. E. (1995). A key enzyme in the biosynthesis of neurosteroids, 3β-hydroxysteroid dehydrogenase/Δ5-Δ4-isomerase (3β-HSD), is expressed in rat brain. *Molecular Brain Research*, *30*(2), 287–300. 1995/06/01/ https://doi.org/10.1016/0169-328X(95)00016-L.

Hamburgh, M. (1963). Analysis of the postnatal developmental effects of "reeler," a neurological mutation in mice. A study in developmental genetics. *Developmental Biology*, *8*, 165–185. https://doi.org/10.1016/0012-1606(63)90040-x.

Heldring, N., Pike, A., Andersson, S., Matthews, J., Cheng, G., Hartman, J., et al. (2007). Estrogen receptors: How do they signal and what are their targets. *Physiological Reviews*, *87*(3), 905–931. https://doi.org/10.1152/physrev.00026.2006.

Imwalle, D. B., Gustafsson, J. A., & Rissman, E. F. (2005). Lack of functional estrogen receptor beta influences anxiety behavior and serotonin content in female mice. *Physiology & Behavior*, *84*(1), 157–163. https://doi.org/10.1016/j.physbeh.2004.11.002.

Isgor, C., & Sengelaub, D. R. (1998). Prenatal gonadal steroids affect adult spatial behavior, CA1 and CA3 pyramidal cell morphology in rats. *Hormones and Behavior*, *34*(2), 183–198. https://doi.org/10.1006/hbeh.1998.1477.

Kelly, M. J., & Qiu, J. (2010). Estrogen signaling in hypothalamic circuits controlling reproduction. *Brain Research*, *1364*, 44–52. https://doi.org/10.1016/j.brainres.2010.08.082.

Kight, K. E., & McCarthy, M. M. (2020). Androgens and the developing hippocampus. *Biology of Sex Differences*, *11*(1), 30. https://doi.org/10.1186/s13293-020-00307-6.

Koopman, P., Munsterberg, A., Capel, B., Vivian, N., & Lovell-Badge, R. (1990). Expression of a candidate sex-determining gene during mouse testis differentiation. *Nature*, *348*(6300), 450–452. https://doi.org/10.1038/348450a0.

Kuiper, G. G., Enmark, E., Pelto-Huikko, M., Nilsson, S., & Gustafsson, J. A. (1996). Cloning of a novel receptor expressed in rat prostate and ovary. *Proceedings of the National Academy of Sciences of the United States of America*, *93*(12), 5925–5930. http://www.ncbi.nlm.nih.gov/pubmed/8650195.

Kuiper, G. G., Lemmen, J. G., Carlsson, B., Corton, J. C., Safe, S. H., van der Saag, P. T., et al. (1998). Interaction of estrogenic chemicals and phytoestrogens with estrogen receptor beta. *Endocrinology*, *139*(10), 4252–4263. https://doi.org/10.1210/endo.139.10.6216.

Lakatosova, S., & Ostatnikova, D. (2012). Reelin and its complex involvement in brain development and function. *The International Journal of Biochemistry & Cell Biology*, *44*(9), 1501–1504. https://doi.org/10.1016/j.biocel.2012.06.002.

Lee, J., & Harley, V. R. (2012). The male fight-flight response: A result of SRY regulation of catecholamines? *BioEssays*, *34*(6), 454–457. https://doi.org/10.1002/bies.201100159.

Lemmen, J. G., Broekhof, J. L., Kuiper, G. G., Gustafsson, J. A., van der Saag, P. T., & van der Burg, B. (1999). Expression of estrogen receptor alpha and beta during mouse embryogenesis. *Mechanisms of Development*, *81*(1–2), 163–167. http://www.ncbi.nlm.nih.gov/pubmed/10330493.

Li, H., Ding, C., Ding, Z. L., Ling, M., Wang, T., Wang, W., et al. (2017). 17β-Oestradiol promotes differentiation of human embryonic stem cells into dopamine neurons via cross-talk between insulin-like growth factors-1 and oestrogen receptor β. *Journal of Cellular and Molecular Medicine*, *21*(8), 1605–1618. https://doi.org/10.1111/jcmm.13090.

Liu, F., Day, M., Muniz, L. C., Bitran, D., Arias, R., Revilla-Sanchez, R., et al. (2008). Activation of estrogen receptor-beta regulates hippocampal synaptic plasticity and improves memory. *Nature Neuroscience*, *11*(3), 334–343. https://doi.org/10.1038/nn2057.

Loke, H., Harley, V., & Lee, J. (2015). Biological factors underlying sex differences in neurological disorders. *The International Journal of Biochemistry & Cell Biology*, *65*, 139–150. https://doi.org/10.1016/j.biocel.2015.05.024.

Matthews, J., & Gustafsson, J. A. (2003). Estrogen signaling: A subtle balance between ER alpha and ER beta. *Molecular Interventions*, *3*(5), 281–292. https://doi.org/10.1124/mi.3.5.281.

Mott, N. N., & Pak, T. R. (2012). Characterisation of human oestrogen receptor beta (ERβ) splice variants in neuronal cells. *Journal of Neuroendocrinology*, *24*(10), 1311–1321. https://doi.org/10.1111/j.1365-2826.2012.02337.x.

Mott, N. N., Pinceti, E., Rao, Y. S., Przybycien-Szymanska, M. M., Prins, S. A., Shults, C. L., et al. (2014). Age-dependent effects of 17β-estradiol on the dynamics of estrogen receptor β (ERβ) protein-protein interactions in the ventral hippocampus. *Molecular & Cellular Proteomics*, *13*(3), 760–779. https://doi.org/10.1074/mcp.M113.031559.

Nakata, M., Sano, K., Musatov, S., Yamaguchi, N., Sakamoto, T., & Ogawa, S. (2016). Effects of prepubertal or adult site-specific knockdown of estrogen receptor beta in the medial preoptic area and medial amygdala on social behaviors in male mice. *eNeuro*, *3*(2), 1–14. Mar-Apr. https://doi.org/10.1523/eneuro.0155-15.2016.

Ogawa, S., Chan, J., Chester, A. E., Gustafsson, J. A., Korach, K. S., & Pfaff, D. W. (1999). Survival of reproductive behaviors in estrogen receptor beta gene-deficient (betaERKO) male and female mice. *Proceedings of the National Academy of Sciences of the United States of America*, *96*(22), 12887–12892. http://www.ncbi.nlm.nih.gov/pubmed/10536018.

Petra, P. H., Woodcock, K. T., Orr, W. R., Nguyen, D. K., & Sui, L. M. (2000). The sex steroid binding protein (SBP or SHBG) of human plasma: Identification of Tyr-57 and Met-107 in the steroid binding site. *The Journal of Steroid Biochemistry and Molecular Biology*, *75*(2–3), 139–145. http://www.ncbi.nlm.nih.gov/pubmed/11226830.

Pinares-Garcia, P., Stratikopoulos, M., Zagato, A., Loke, H., & Lee, J. (2018). Sex: A significant risk factor for neurodevelopmental and neurodegenerative disorders. *Brain Sciences*, *8*(8), 154. https://doi.org/10.3390/brainsci8080154.

Pinceti, E., Shults, C. L., Rao, Y. S., Mott, N. N., & Pak, T. R. (2015). Phosphorylation alters oestrogen receptor β-mediated transcription in neurones. *Journal of Neuroendocrinology*, *27*(12), 861–871. https://doi.org/10.1111/jne.12326.

Potkin, S. G., Turner, J. A., Guffanti, G., Lakatos, A., Fallon, J. H., Nguyen, D. D., et al. (2009). A genome-wide association study of schizophrenia using brain activation as a quantitative phenotype. *Schizophrenia Bulletin*, *35*(1), 96–108. https://doi.org/10.1093/schbul/sbn155.

Rissman, E. F., Heck, A. L., Leonard, J. E., Shupnik, M. A., & Gustafsson, J. A. (2002). Disruption of estrogen receptor beta gene impairs spatial learning in female mice. *Proceedings of the National Academy of Sciences of the United States of America*, *99*(6), 3996–4001. https://doi.org/10.1073/pnas.012032699.

Romano, E., Cosentino, L., Laviola, G., & De Filippis, B. (2016). Genes and sex hormones interaction in neurodevelopmental disorders. *Neuroscience & Biobehavioral Reviews*, *67*, 9–24. 2016/08/01/. https://doi.org/10.1016/j.neubiorev.2016.02.019.

Sinclair, A. H., Berta, P., Palmer, M. S., Hawkins, J. R., Griffiths, B. L., Smith, M. J., et al. (1990). A gene from the human sex-determining region encodes a protein with homology to a conserved DNA-binding motif. *Nature*, *346*(6281), 240–244. https://doi.org/10.1038/346240a0.

Sugiyama, N., Andersson, S., Lathe, R., Fan, X., Alonso-Magdalena, P., Schwend, T., et al. (2009). Spatiotemporal dynamics of the expression of estrogen receptors in the postnatal mouse brain. *Molecular Psychiatry*, *14*(2), 223–232. 117. https://doi.org/10.1038/mp.2008.118.

Suzuki, H., Barros, R. P., Sugiyama, N., Krishnan, V., Yaden, B. C., Kim, H. J., et al. (2013). Involvement of estrogen receptor beta in maintenance of serotonergic neurons of the dorsal raphe. *Molecular Psychiatry*, *18*(6), 674–680. https://doi.org/10.1038/mp.2012.62.

Swartz, S. K., & Soloff, M. S. (1974). The lack of estrogen binding by human alpha-fetoprotein. *The Journal of Clinical Endocrinology and Metabolism*, *39*(3), 589–591. https://doi.org/10.1210/jcem-39-3-589.

Tammimies, K., Vitezic, M., Matsson, H., Le Guyader, S., Burglin, T. R., Ohman, T., et al. (2013). Molecular networks of DYX1C1 gene show connection to neuronal migration genes and cytoskeletal proteins. *Biological Psychiatry*, *73*(6), 583–590. https://doi.org/10.1016/j.biopsych.2012.08.012.

Varshney, M. K., Inzunza, J., Lupu, D., Ganapathy, V., Antonson, P., Ruegg, J., et al. (2017). Role of estrogen receptor beta in neural differentiation of mouse embryonic stem cells. *Proceedings of the National Academy of Sciences of the United States of America*, *114*(48), E10428–E10437. https://doi.org/10.1073/pnas.1714094114.

Varshney, M. K., Yu, N. Y., Katayama, S., Li, X., Liu, T., Wu, W. F., et al. (2021). Motor function deficits in the estrogen receptor beta knockout mouse: Role on excitatory neurotransmission and myelination in the motor cortex. *Neuroendocrinology*, *111*(1–2), 27–44. https://doi.org/10.1159/000506162.

Walf, A. A., Koonce, C. J., & Frye, C. A. (2009). Adult female wildtype, but not oestrogen receptor beta knockout, mice have decreased depression-like behaviour during pro-oestrus and following administration of oestradiol or diarylpropionitrile. *Journal of Psychopharmacology*, *23*(4), 442–450. https://doi.org/10.1177/0269881108089598.

Wallen, K. (2005). Hormonal influences on sexually differentiated behavior in nonhuman primates. *Frontiers in Neuroendocrinology*, *26*(1), 7–26. https://doi.org/10.1016/j.yfrne.2005.02.001.

Wang, L., Andersson, S., Warner, M., & Gustafsson, J. A. (2001). Morphological abnormalities in the brains of estrogen receptor beta knockout mice. *Proceedings of the National Academy of Sciences of the United States of America*, *98*(5), 2792–2796. https://doi.org/10.1073/pnas.041617498.

Wang, L., Andersson, S., Warner, M., & Gustafsson, J. A. (2003). Estrogen receptor (ER) beta knockout mice reveal a role for ERbeta in migration of cortical neurons in the developing brain. *Proceedings of the National Academy of Sciences of the United States of America*, *100*(2), 703–708. https://doi.org/10.1073/pnas.242735799.

Williams, C. L., Barnett, A. M., & Meck, W. H. (1990). Organizational effects of early gonadal secretions on sexual differentiation in spatial memory. *Behavioral Neuroscience*, *104*(1), 84–97. https://doi.org/10.1037//0735-7044.104.1.84.

Wilson, C. A., & Davies, D. C. (2007). The control of sexual differentiation of the reproductive system and brain. *Reproduction*, *133*(2), 331–359. https://doi.org/10.1530/REP-06-0078.

Wu, M. V., Manoli, D. S., Fraser, E. J., Coats, J. K., Tollkuhn, J., Honda, S., et al. (2009). Estrogen masculinizes neural pathways and sex-specific behaviors. *Cell*, *139*(1), 61–72. https://doi.org/10.1016/j.cell.2009.07.036.

Wu, W. F., Tan, X. J., Dai, Y. B., Krishnan, V., Warner, M., & Gustafsson, J. A. (2013). Targeting estrogen receptor beta in microglia and T cells to treat experimental autoimmune encephalomyelitis. *Proceedings of the National Academy of Sciences of the United States of America*, *110*(9), 3543–3548. https://doi.org/10.1073/pnas.1300313110.

CHAPTER ELEVEN

Nuclear receptors linking physiology and germline stem cells in *Drosophila*

Danielle S. Finger, Kaitlin M. Whitehead, Daniel N. Phipps, and Elizabeth T. Ables*
Department of Biology, East Carolina University, Greenville, NC, United States
*Corresponding author: e-mail address: ablese@ecu.edu

Contents

Abstract

Maternal nutrition and physiology are intimately associated with reproductive success in diverse organisms. Despite decades of study, the molecular mechanisms linking maternal diet to the production and quality of oocytes remain poorly defined. Nuclear receptors (NRs) link nutritional signals to cellular responses and are essential for oocyte development. The fruit fly, *Drosophila melanogaster*, is an excellent genetically tractable model to study the relationship between NR signaling and oocyte

Vitamins and Hormones, Volume 116
ISSN 0083-6729
https://doi.org/10.1016/bs.vh.2020.12.008

production. In this review, we explore how NRs in *Drosophila* regulate the earliest stages of oocyte development. Long-recognized as an essential mediator of developmental transitions, we focus on the intrinsic roles of the Ecdysone Receptor and its ligand, ecdysone, in oogenesis. We also review recent studies suggesting broader roles for NRs as regulators of maternal physiology and their impact specifically on oocyte production. We propose that NRs form the molecular basis of a broad physiological surveillance network linking maternal diet with oocyte production. Given the functional conservation between *Drosophila* and humans, continued experimental investigation into the molecular mechanisms by which NRs promote oogenesis will likely aid our understanding of human fertility.

1. *Drosophila* as a model for maternal physiological control of oogenesis

Obesity, metabolic disorders, and diabetes disrupt the body's natural steroid hormone milieu, negatively impacting human health. An important, but often overlooked co-morbidity of metabolic disorders in women is poor reproductive outcome (Gu et al., 2015; Roa & Tena-Sempere, 2014). The quality of preconception maternal diets is tightly correlated with reproductive success (Gaskins & Chavarro, 2018; Hohos & Skaznik-Wikiel, 2017; Luzzo et al., 2012). Maternal obesity is associated with poor oocyte quality, which manifests as increased time to pregnancy, early pregnancy loss, and congenital abnormalities in offspring (Purcell & Moley, 2011; Talmor & Dunphy, 2015). Despite the prevalence of metabolic disorders in the United States, and the clear public health relevance (Panth, Gavarkovs, Tamez, & Mattei, 2018), the molecular mechanisms connecting oocyte quality and production to nutritionally regulated endocrine signals are poorly understood.

Among the many cell-cell communication pathways used by ovarian cells to promote oocyte development, steroid hormones are essential for maintenance of fertility, including follicle assembly, oocyte maturation, and oocyte survival (Grive & Freiman, 2015). Hormones circulating through the blood or hemolymph of an organism create local and long-distance signaling networks by binding to nuclear receptors in target cells, affecting transcriptional change (Weikum, Liu, & Ortlund, 2018). Steroid synthesis is a complex molecular process requiring coordinated activity of multiple substrates and enzymes in gonads and other endocrine tissues (Dallel et al., 2018). Tight spatial and temporal control of steroid synthesis and steroid reception by cognate nuclear receptors must be achieved to facilitate successful oocyte production and fertilization. As such, many female reproductive disorders (such as polycystic ovarian syndrome, endometriosis, and early pregnancy loss) are associated with aberrations in steroid hormone levels or exposure to endocrine disrupting chemicals (Pivonello et al., 2020; Zhang, Wesevich, Chen, Zhang, & Kallen, 2020).

Given the complexity of hormone signaling pathways in mammals, the fruit fly, *Drosophila melanogaster*, has emerged as an excellent genetically tractable model to study the relationship between maternal nutrition and oocyte production. Female fruit flies contain two ovaries made of about 15 ovarioles, each consisting of oocytes in increasing stages of development arranged linearly from anterior to posterior (Fig. 1A and D) (Hinnant, Merkle, & Ables, 2020). The morphology of the ovary enables young female flies to lay 80–90 eggs per day (Drummond-Barbosa & Spradling, 2001). The progressive arrangement of oogenesis also allows researchers to visualize the complete development of oocytes, from their establishment as daughters of a germline stem cell population through ovulation. Female fruit flies are exceptionally sensitive to diet conditions and can be reared on specialized media to manipulate the levels of available nutrients. Moreover, *Drosophila* oogenesis can be monitored on a per-cell basis in vivo using confocal microscopy (Fig. 1A and B) and germ cells can be easily manipulated via a widely available sophisticated genetic toolkit. The ease with which *Drosophila* are reared and their remarkable similarity to human physiology and genetics make fruit flies an incredibly powerful model organism for research.

2. Nuclear receptors as molecular surveyors of physiology

Nuclear receptors (NRs) directly link dietary and hormonal signals to transcriptional responses in target cells and serve as key regulators of development, metabolism, and reproduction (Ables & Drummond-Barbosa, 2017; Bodofsky, Koitz, & Wightman, 2017; Crowder, Seacrist, & Blind, 2017; Dallel et al., 2018; Evans & Mangelsdorf, 2014; King-Jones & Thummel, 2005; Mirth, Nogueira Alves, & Piper, 2019; Pardee, Necakov, & Krause, 2011; Weikum et al., 2018). The nuclear receptor superfamily is one of the largest families of transcription factors, divided into seven subfamilies (NR0-NR6) based on domain conservation across species, including steroid receptors, thyroid hormone receptors, retinoid acid receptors, and retinoid X receptors (Bodofsky et al., 2017; Evans & Mangelsdorf, 2014; King-Jones & Thummel, 2005; Pardee et al., 2011).

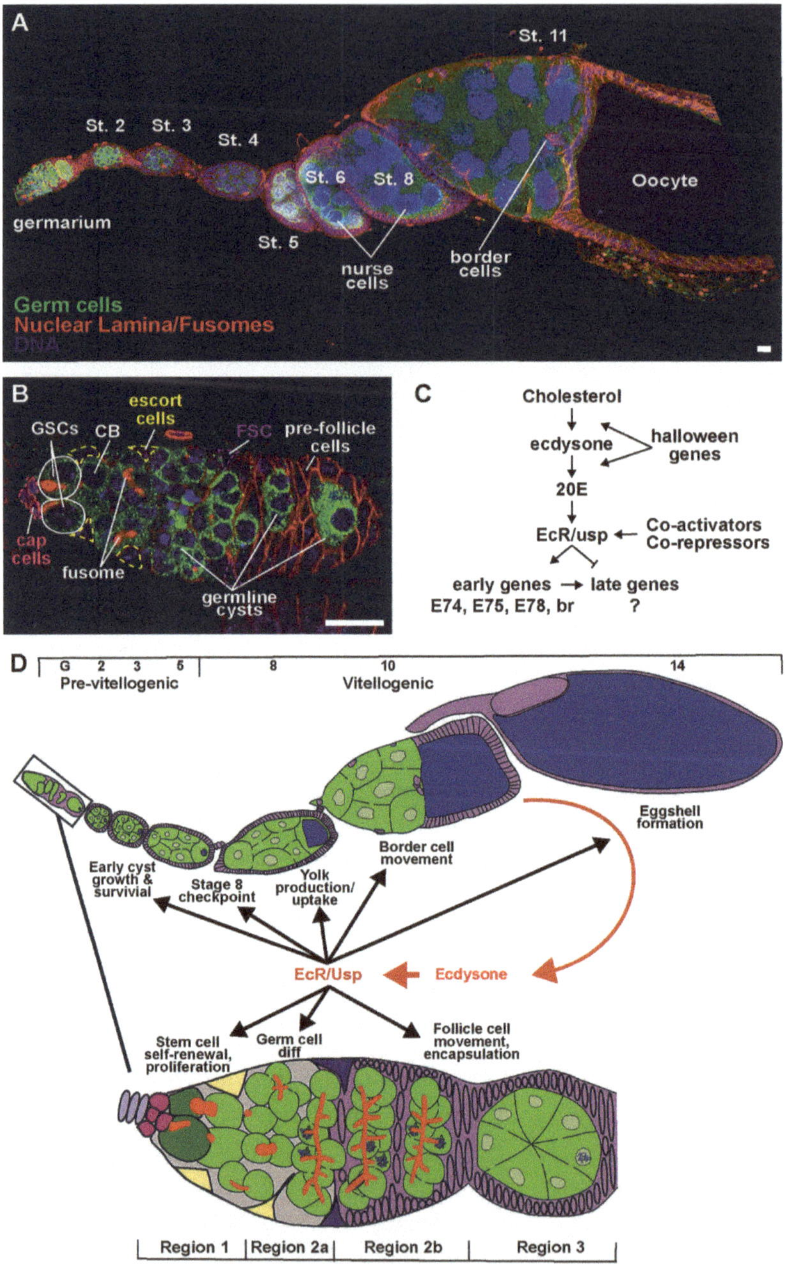

Fig. 1 Ecdysone signaling regulates *Drosophila melanogaster* oogenesis. (A) Maximum intensity projection image of a *Drosophila* ovariole, labeled with anti-Vasa (green; germ cells), anti-Hts (red; fusomes and follicle cell membranes), anti-LamC (red; nuclear envelope of cap cells) and DAPI (blue, DNA). (B) Optical cross section of the *Drosophila* germarium showing GSCs (solid white line) anchored to cap cells (dashed pink line), which make up the GSC niche. Germ cells are characterized by the presence of a fusome (orange), which extends as germ cells divide. Escort cells (yellow dashed line) signal to GSCs to promote differentiation. Follicle stem cells (FSC; purple dashed line) create pre-follicle cells that surround the 16-cell cyst, giving rise to an egg chamber or follicle that leaves the germarium. (C) Schematic of the ecdysone signaling pathway. (D) Summary of ecdysone-regulated processes in the ovariole and germarium. Scale bar = 10 μm.

NRs primarily function as ligand-gated transcription factors. A conserved zinc-finger DNA-binding domain allows NRs to bind specific sequences in the promoter regions of target genes and activate or repress transcription. Small lipophilic ligands, processed from dietary lipids such as cholesterol and fatty acids, bind to a more sequence-diverse C-terminal ligand binding and dimerization domain in the receptor, promoting conformational change when ligand is bound (Bodofsky et al., 2017; Weikum et al., 2018). The ability to directly link transcriptional change with ligands whose concentrations vary with physiology allows organisms to adapt to a wide variety of environmental conditions and stressors. NRs frequently partner with other NRs in heterodimer configurations, and their activity can be modified by the binding of additional co-factors to increase repressor or activator function. Not all NRs, however, have known natural ligands, and at least some can affect transcription in the absence of ligand. Though mammals exhibit exceptional diversity in NRs (encoded by 48 genes in humans), the *Drosophila* genome encodes 18 NR family members, representing each of the known mammalian subfamilies. Unlike humans, the *Drosophila* genome is largely free from duplications, making functional assignments to particular genes easier than in most vertebrate model organisms. Early experimental studies in *Drosophila* took advantage of the distinct hormonally controlled developmental transitions in the insect life cycle, resulting in a wealth of molecular information about steroid hormones and their physiologically relevant NRs and transcriptional responses.

3. Ecdysone signaling as a model for cell-cell communication via nuclear receptors

Perhaps the best characterized hormones in insects are the ecdysteroids, a group of polyhydroxylated steroids essential for development, growth, behavior, and reproduction. *Drosophila* synthesize the predominant hormone, ecdysone, from diet derived cholesterol and other plant sterols (Fig. 1C) (Gilbert, 2004; Gilbert & Warren, 2005; Igarashi, Ogihara, Iga, & Kataoka, 2018; Miller, 2008; Truman, 2019). Loss of the ability to convert cholesterol, such as in flies harboring genetic mutations in the *without children* (*woc*) locus, is lethal but can be rescued with the addition of the cholesterol derivative 7-dehydrocholesterol (Wismar et al., 2000; Warren, Wismar, Subrahmanyam, & Gilbert, 2001). In larvae, ecdysone is synthesized when the neuropeptide prothoracicotropic hormone signals to a specialized organ in the brain, known as the prothoracic gland

(McBrayer et al., 2007; Rewitz, Yamanaka, Gilbert, & O'Connor, 2009). Ecdysone is secreted into the hemolymph, where it is converted into its active form, 20-hydroxyecdysone (20E), which is functionally and structurally analogous to mammalian hormones estrogen, progesterone, and thyroid hormone (Fig. 1C) (Gilbert, Rybczynski, & Warren, 2002; Hoffmeister, Grützmacher, & Dünnebeil, 1967; Mangelsdorf et al., 1995; Petryk et al., 2003; Uryu, Ameku, & Niwa, 2015). Although the prothoracic glands degenerate after metamorphosis, ecdysone is prevalent in adults, albeit at lower titer than in larval stages (Dai, Henrich, & Gilbert, 1991; Richards, 1981; Schwartz, Kelly, Imberski, & Rubenstein, 1985). In adults, ecdysone can be detected in all three body segments and in specific tissues, including the gut, malphigian tubules, testes, and ovaries (Bownes, 1982; Handler, 1982; Schwartz et al., 1985).

3.1 Ecdysone promotes developmental transitions and adult reproductive physiology

Ecdysone was initially identified as a regulator of molting and metamorphosis in arthropods. Pulses of ecdysone precede developmental transitions between life stages and promote critical physiological changes, including germ band retraction and head involution during the first larval instar, larval cuticle development, and programmed cell death and cell remodeling in imaginal discs during metamorphosis (Andres, Fletcher, Karim, & Thummel, 1993; Apple & Fristrom, 1991; Braquart, Bouhin, Quennedey, & Delachambre, 1996; Charles, 2010; Jiang, Baehrecke, & Thummel, 1997; Karlson, 2003; Kozlova & Thummel, 2003; Li & Bender, 2000; Robertson, 1936; Warren et al., 2006). Embryos lacking ecdysone biosynthetic enzymes proceed through embryogenesis due to the availability of maternally deposited transcripts, but fail to make a larval cuticle, yielding a unique ghost-like phenotype (Chavez et al., 2000; Niwa & Niwa, 2014; Uryu et al., 2015). Identification of mutants in large-scale screens sharing this phenotype provided reagents necessary to identify ecdysone biosynthetic enzymes, which were collectively referred to as the "Halloween Genes" (Gilbert, 2004; Niwa & Niwa, 2014).

While the ovary appears to be a major source of ecdysone synthesis in adult females, ecdysone levels in female *Drosophila* are highly dependent on physiological status. Genetic mutant strains that lack ovaries have reduced ecdysone titer in their hemolymph, suggesting that the ovary contributes to the overall ecdysone levels in females (Bainbridge & Bownes, 1988; Tu, Yin, & Tatar, 2002). Ecdysone biosynthesis enzymes, including those encoded by *defective in the avoidance of repellents* (*dare*) and the Halloween genes, are highly expressed in ovarian follicles, suggesting that the ovary is a source of ecdysone in adult *Drosophila* (Freeman, Dobritsa, Gaines, Segraves, & Carlson, 1999; Niwa et al., 2004; Ono et al., 2006; Warren et al., 2004). Ecdysone titer increases in females post-mating in response to the male Sex Peptide protein, transferred via seminal fluid during mating, resulting in higher levels of ecdysone in female than in males (Ameku & Niwa, 2016; Ameku, Yoshinari, Fukuda, & Niwa, 2017; Avila, Sirot, LaFlamme, Rubinstein, & Wolfner, 2011; Carmel, Tram, & Heifetz, 2016; Feng, Palfreyman, Hasemeyer, Talsma, & Dickson, 2014; Hasemeyer, Yapici, Heberlein, & Dickson, 2009; Sieber & Spradling, 2015). Additionally, because ecdysone is synthesized from dietary cholesterol, the nutritional status of the fly greatly impacts ecdysone titer. Ecdysone titer increases 24–48 h after eclosion in well-fed flies, and adult females starved after eclosion fail to accumulate significant levels of ecdysone (Handler, 1982; Terashima, Takaki, Sakurai, & Bownes, 2005). Interestingly, phenotypes seen in ecdysone biosynthesis mutants do not always mimic the phenotypes seen in nutrient deprived flies, suggesting that loss of adequate nutrition does not completely abolish ecdysone titer, but instead significantly lowers it (Carvalho-Santos & Ribeiro, 2018; Terashima et al., 2005).

3.2 Ecdysone exerts cellular effects via conserved nuclear receptors EcR and Usp

Ecdysone functions as a transcriptional regulator by binding to a heterodimeric receptor complex (Fig. 1C) (Billas & Moras, 2005; Thomas, Stunnenberg, & Stewart, 1993; Yao et al., 1993). In *Drosophila*, the receptor complex is encoded by two nuclear receptors: *Ecdysone Receptor* (*EcR*), an ortholog of mammalian farnesoid X and liver X receptors; and *ultraspiracle* (*usp*), an ortholog of mammalian retinoid X receptors (Henrich et al., 1994; King-Jones & Thummel, 2005; Koelle et al., 1991; Oro, McKeown, & Evans, 1990; Yao et al., 1993). Without Usp, EcR can neither bind ecdysone nor its target DNA sequences (Hill, Billas, Bonneton, Graham, & Lawrence, 2013; Yao et al., 1993). Upon forming a heterodimer, the EcR/Usp complex (henceforth referred to as EcR) regulates transcription of target genes via ecdysone response elements (EcREs). In the absence of ecdysone, EcR represses transcription and is bound by co-repressors, including Alien (a COP9 signalosome subunit) and SMRT-related and Ecdysone Receptor Interacting Protein (SMRTER) (Dressel et al., 1999;

Heck et al., 2012; Huang, Lu, Wu, Chien, & Pi, 2014; Schubiger & Truman, 2000). In the presence of ecdysone, cells actively transport ecdysone into the cytoplasm via a membrane-resident transporter belonging to the solute carrier organic anion superfamily (Okamoto et al., 2018). Binding of ecdysone to EcR promotes a conformational change in EcR in which the ligand binding region is internalized into the three-dimensional structure of the protein (Billas et al., 2003; Hu, Cherbas, & Cherbas, 2003; Schubiger, Carré, Antoniewski, & Truman, 2005). Exposure of a new protein surface allows for release of co-repressor proteins and recruitment of transcriptional co-activators, such as the highly conserved Taiman (Tai) and Absent, small, or homeotic discs 2 (Ash2) (Bai, Uehara, & Montell, 2000; Carbonell, Mazo, Serras, & Corominas, 2013; Zhang et al., 2015). Coordinated activity of EcR with nucleosome remodeling complexes, including Putzig, ISWI/NURF, and Kismet, further constrains transcriptional activity at target genes by regulating chromatin accessibility (Badenhorst et al., 2005; Kreher et al., 2017; Kugler, Gehring, Wallkamm, Kruger, & Nagel, 2011; Latcheva, Viveiros, & Marenda, 2019; Uyehara et al., 2017).

Signaling via EcR is required to regulate gene expression in a wide variety of tissues in a spatiotemporally specific manner (Beckstead, Lam, & Thummel, 2005; Gauhar et al., 2009; Gonsalves, Neal, Kehoe, & Westwood, 2011; Li & White, 2003; Shlyueva et al., 2014; Stoiber, Celniker, Cherbas, Brown, & Cherbas, 2016; Uyehara & McKay, 2019). The range of EcR-dependent cellular activities is due, at least in part, to the expression of multiple protein isoforms. The *EcR* locus contains the prototypical genetic structure of the NR superfamily, including conserved ligand-binding and zinc-finger-like DNA-binding domains (Koelle et al., 1991). Although there are seven putative transcripts of *EcR*, there are only three functional protein isoforms, denoted as EcR-A, EcR-B1, and EcR-B2 (Talbot, Swyryd, & Hogness, 1993). Alternative promoter usage results in unique spatiotemporal expression of the three isoforms. Despite the conservation in DNA-binding domains, the EcR isoforms do not appear to be functionally redundant. Each isoform can bind a specific nucleotide sequence, and carefully controlled rescue and mis-expression experiments demonstrated tissue-specificity among the three protein isoforms (Cherbas, Hu, Zhimulev, Belyaeva, & Cherbas, 2003; Davis, Carney, Robertson, & Bender, 2005; Schauer, Callender, Henrich, & Spindler-Barth, 2011; Schubiger, Tomita, Sung, Robinow, & Truman, 2003). In vivo, EcR-A appears to function predominantly as a strong repressor, while both B class isoforms are strong activators (Braun, Azoitei, & Spindler-Barth, 2009;

Dobens, Rudolph, & Berger, 1991; Hu et al., 2003). Unique N-termini of the isoforms also permit differential binding by co-activators and co-repressors to the EcR/Usp complex.

Spatiotemporal specificity of the ecdysone response is also mediated through the transcriptional targets of EcR and Usp. Many of these loci were initially identified more than 45 years ago based on the unique transcriptionally regulated "puffing" of larval salivary polytene chromosomes in response to ecdysone (Ashburner, Chihara, Meltzer, & Richards, 1974; Hill et al., 2013). Initial experiments, followed by more recent whole-genome attempts to catalog the transcriptional response to ecdysone, support a hierarchical model of ecdysone signaling, wherein EcR activation promotes the rapid expression of a small number of targets (Ashburner et al., 1974; Beckstead et al., 2005; Gauhar et al., 2009; Gonsalves et al., 2011; Hill et al., 2013; Li & White, 2003; Shlyueva et al., 2014; Stoiber et al., 2016; Uyehara & McKay, 2019). These so-called early-response genes, including a core group of genes such as *Ecdysone-induced protein 74EF* (*E74*), *Ecdysone-induced protein 75B* (*E75*), and *broad* (*br*), encode transcription factors. These activate a tissue-specific response to ecdysone by regulating a second set of targets (late-response genes). Although EcR and Usp mediate the majority of ecdysone-dependent transcriptional responses, at least six other NR genes are themselves ecdysone-responsive: *Hormone receptor 3* (*Hr3*, also called *Hr46*), *Hormone receptor 4* (*Hr4*), *Hormone receptor-like in 39* (*Hr39*), *E75*, *Ecdysone-induced protein 78C* (*E78*), and *ftz transcription factor 1* (*ftz-f1*) (King-Jones & Thummel, 2005). Intriguingly, while expression of most of these genes is induced by ecdysone, expression of *ftz-f1* is repressed by ecdysone, likely in a feedback loop with ecdysone biosynthesis genes (Parvy et al., 2005; Woodard, Baehrecke, & Thummel, 1994; Yamada et al., 2000). Moreover, some NRs exhibit their own unique ligand activation and can promote cross-regulation between receptors. For example, Ftz-f1 can promote transcriptional activation in the absence of a ligand, but requires Hr3 for full activation (Lam, Jiang, & Thummel, 1997; Lu, Anderson, Zhang, Feng, & Pick, 2013; Musille, Pathak, Lauer, Griffin, & Ortlund, 2013; Yoo et al., 2011). Similarly, while E75 is a classic early response factor, it can also be regulated by small molecules such as nitric oxide (Cáceres et al., 2011; Marvin et al., 2009; Reinking et al., 2005). NRs are dynamically expressed in *Drosophila* tissues and are regulated by a complex web of interactions, placing them in a unique mechanistic position to fine-tune reproduction with metabolic and environmental cues (Fig. 1D) (Palanker et al., 2006; Wilk, Hu, & Krause, 2013).

4. *Drosophila* oogenesis is fueled by stem cells

The development of a viable oocyte requires a coordinated network of somatic and germ cells functioning in unison to support oocyte differentiation, meiotic maturation, and the deposition of maternal mRNA, proteins, and nutrients necessary to sustain embryogenesis post-fertilization. In *Drosophila*, oogenesis begins in the most anterior region of the ovariole, called the germarium (McLaughlin & Bratu, 2015) (Fig. 1B and D). Here, two stem cell populations divide continuously in adult flies to form the cellular precursors of the germline and somatic cells that will ultimately comprise the oocyte. Germline stem cells (GSCs) are housed at the anterior of the germarium and divide asymmetrically to self-renew and produce a differentiating daughter cell called a cystoblast (Hinnant et al., 2020). Following four rounds of mitotic division with incomplete cytokinesis, the cystoblast ultimately differentiates into a 16-cell cyst, wherein one cell becomes the oocyte and the other 15 cells adopt a nurse cell fate. Follicle stem cells (FSCs) reside in the middle of the germarium and produce pre-follicle cells that encapsulate the developing germ cells (Rust & Nystul, 2020). Pre-follicle cells further differentiate into specialized follicle cell populations outside of the germarium, ultimately forming the eggshell, chorion, and associated structures. The differentiated cell daughters of both stem cell populations are essential to form an oocyte that is ready for fertilization, as loss or malfunction of GSCs or FSCs leads to loss of fertility (Lin & Spradling, 1993; Margolis & Spradling, 1995).

While nascent egg chambers are formed in the germarium, oocyte development continues through 14 distinct stages in the vitellarium (Fig. 1A and D) (McLaughlin & Bratu, 2015). Egg chambers bud from the germarium connected by specialized follicle cells called stalk cells. In stages 2–7, egg chambers undergo rapid growth, in part due to developmentally programmed nurse cell endocycling which increases nurse cell DNA content. Follicle cells complete four to five rounds of mitotic divisions to cover the growing germline cyst in an epithelial monolayer, differentiate into a variety of specialized cell types, and switch to endocycling at stage 5 (Duhart, Parsons, & Raftery, 2017). Movement of the follicle cells and subtle rotation of the egg chamber promote elongation (Cetera & Horne-Badovinac, 2015). Stage 8 follicles undergo vitellogenesis, as yolk proteins produced by the fat body and ovarian tissues are loaded into the oocyte (Brennan, Weiner, Goralski, & Mahowald, 1982). Beginning at stage 9, a group of 6–10 follicle cells, called border cells, delaminate from the follicle cell epithelium and migrate between the nurse cells to form the oocyte anterior margin (Peercy & Starz-Gaiano, 2020). At stage 11, concomitant with continued oocyte growth, nurse cells undergo apoptosis and dump their cytoplasmic contents into the oocyte (Quinlan, 2016). During the final stages of egg chamber development, the follicle cells secrete chorion and other proteins necessary to form different layers of the eggshell, helping to protect the oocyte once it is deposited (Osterfield, Berg, & Shvartsman, 2017). At stage 14, the egg chamber is fully developed and ready to be fertilized and deposited (Mahowald & Kambysellis, 1980).

Ovarian stem cells and their differentiating daughters rely on paracrine and endocrine signaling for their development (Ables, Laws, & Drummond-Barbosa, 2012; Armstrong, 2020; Drummond-Barbosa, 2019; Laws & Drummond-Barbosa, 2017). This is perhaps best illustrated by the complex signaling network that guides GSC self-renewal and cystoblast differentiation. GSCs reside in a stem cell niche, composed of adjacent cap cells and escort cells, that provides short range signals required to balance GSC self-renewal vs differentiation (Figs. 1B and 2) (Drummond-Barbosa, 2019; Kahney, Snedeker, & Chen, 2019). GSCs are physically anchored to the cap cells and secrete Bone Morphogenetic Protein (BMP) ligands that regulate GSC self-renewal (Fig. 2) (Song, Zhu, Doan, & Xie, 2002; Wilcockson & Ashe, 2019; Xie & Spradling, 1998). To maintain the undifferentiated fate, GSCs express RNA-binding proteins that promote self-renewal by repressing transcription of the differentiation factor Bag of Marbles (Bam) (D. Chen & McKearin, 2003; Forbes & Lehmann, 1998; Jin et al., 2008; Song et al., 2004; Szakmary, Cox, Wang, & Lin, 2005; Wang & Lin, 2004). Germ cell differentiation and cyst division are thus initiated by the derepression of *bam* transcription. Bam appears to promote differentiation by limiting the production of self-renewal and adhesion factors, and stabilizing CycA expression (Chen, Wang, et al., 2009; Ji et al., 2017; Li, Minor, Park, McKearin, & Maines, 2009; Liu et al., 2017; Perinthottathil & Kim, 2011; Shen, Weng, Yu, & Xie, 2009; Tiwari, Zeitler, Meister, & Wodarz, 2019). Somatic escort cells surrounding the dividing cysts form a second signaling center that promotes germ cell differentiation, posterior movement of the cysts, and follicle cell encapsulation (Banisch, Maimon, Dadosh, & Gilboa, 2017; Decotto & Spradling, 2005; Kirilly, Wang, & Xie, 2011; Morris & Spradling, 2011). Escort cells line the outside of the anterior germarium (regions 1–2) and send long, thin cellular protrusions into the center of the germarium, around the dividing germ cells (Fig. 1B and D) (Banisch et al., 2017; Kirilly et al., 2011; Morris & Spradling, 2011). Escort cells secrete a variety of signaling molecules that

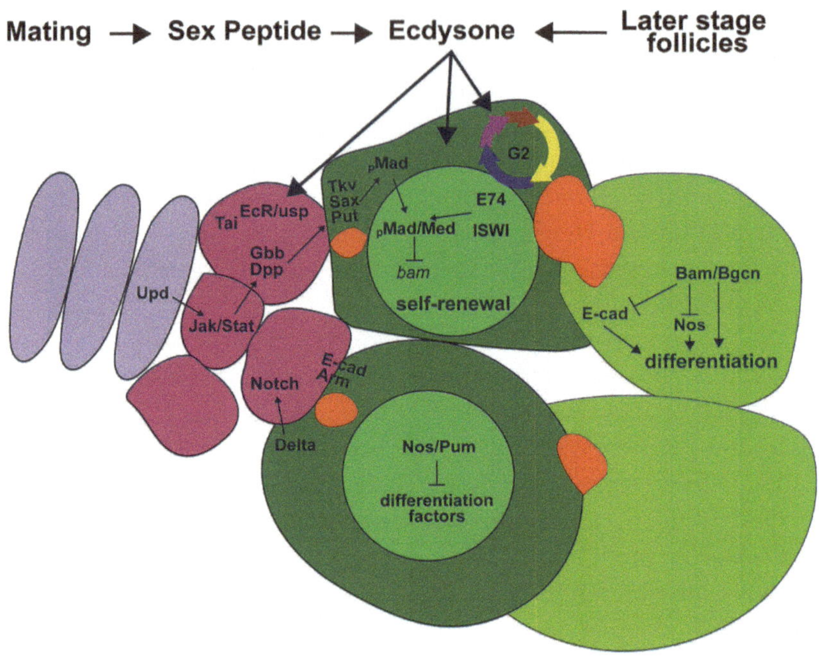

Fig. 2 Signaling pathways that intrinsically regulate GSC self-renewal and cystoblast differentiation. Ecdysone levels are increased due to sex peptide, which is deposited during mating and made in later stage follicles. Ecdysone is received in cap cells (pink) and GSCs (dark green). Terminal filament cells (purple) secrete the ligand Unpaired (Upd) to cap cells to stimulate the Jak/Stat pathway. The Jak/Stat pathway and ecdysone signaling regulate BMP signaling. EcR/Usp binds to Tai to control cap cell function. EcR regulates the BMP signaling pathway to regulate GSC self-renewal. The BMP ligands Gbb and Dpp are secreted from cap cells to the receptors on GSCs. This phosphorylates Mad, which will dimerize with Med and repress *bam* transcription. Ecdysone targets are intrinsically required to maintain GSC self-renewal. In differentiating CBs (light green) bam is transcribed and represses E-cadherin and Nos thereby committing the cell to differentiation.

promote germ cell differentiation. These signals act to sustain the cytoskeletal structure and dynamics of escort cell protrusions and limit the range of BMP signals emanating from cap cells (Antel & Inaba, 2020; Gao et al., 2019).

5. Multifaceted roles of NR signaling in oogenesis

Like its mammalian nuclear receptor counterparts, signaling through EcR provides a crucial link between maternal diet and oogenesis (Ables et al., 2012; Chen, Breen, & Pepling, 2009; Dallel et al., 2018; Drummond et al., 2002; Griswold & Hogarth, 2018; Gu et al., 2015; Martinot et al., 2017; Pepling, 2012). Females produce ecdysone in response to mating and dietary cues, and the ovary is the predominant source of systemic ecdysone in adults

(Ahmed et al., 2020; Ameku & Niwa, 2016; Ameku et al., 2017; Harshman, Loeb, & Johnson, 1999; Mirth et al., 2019; Sieber & Spradling, 2015). In the absence of ecdysone or EcR, oogenesis arrests at four critical developmental processes: germline stem cell (GSC) function, germ cell proliferation, follicle survival, and yolk uptake (Fig. 1D). Here, we highlight the lines of evidence suggesting that *Drosophila* oogenesis is regulated not only by ecdysone signaling, but by a larger network of nutritionally regulated NRs working in both ovarian cells and in peripheral tissues to stimulate oocyte production in response to nutrient availability.

5.1 GSC maintenance and proliferation are both directly and indirectly controlled by EcR signaling

Prior to adulthood, nutritional input guides ovary development and germline stem cell establishment. During early larval development, EcR and Usp repress the differentiation of primordial germ cells and somatic gonadal precursor cells, permitting their proliferation (Gancz, Lengil, & Gilboa, 2011; Hitrik et al., 2016). Later, pulses of ecdysone (likely from the prothoracic gland) promote EcR and Usp activation, providing a timing cue for the formation of stacks of gonadal somatic precursor cells. These so-called terminal filaments will become individual ovarioles in adults (Gancz et al., 2011; Hodin & Riddiford, 1998; Mendes & Mirth, 2016). EcR also promotes allocation of primordial germ cells into nascent somatic niches within the terminal filaments, thereby establishing the number of GSCs in each ovariole at eclosion (Gancz et al., 2011; König, Yatsenko, Weiss, & Shcherbata, 2011; Yatsenko & Shcherbata, 2018). The EcR co-activator Tai is likely involved in larval/pupal ovary development, as loss of *tai* leads to expanded numbers of cap cells and GSCs in newly eclosed females (König et al., 2011). Although most of the targets of EcR/Usp have not been identified in the developing gonad, the ecdysone early gene *br* is necessary for the effects of EcR signaling in this context (Gancz et al., 2011; Hitrik et al., 2016). Additionally, the *Drosophila* NR encoded by *E78* (most similar to vertebrate REV-ERB receptors) is necessary in cap cells prior to adulthood to establish the proper number of GSCs (Ables, Bois, Garcia, & Drummond-Barbosa, 2015). Although other NRs have not been described in ovary development, recent annotation of gene expression profiles for all cell types in the developing ovary will likely aid future experiments geared toward understanding how NRs guide ovary development in response to nutritional cues (Slaidina, Banisch, Gupta, & Lehmann, 2020).

In adult females, EcR signaling is necessary in GSCs for their self-renewal and proliferation (Figs. 2 and 3). Mutants in which ecdysone

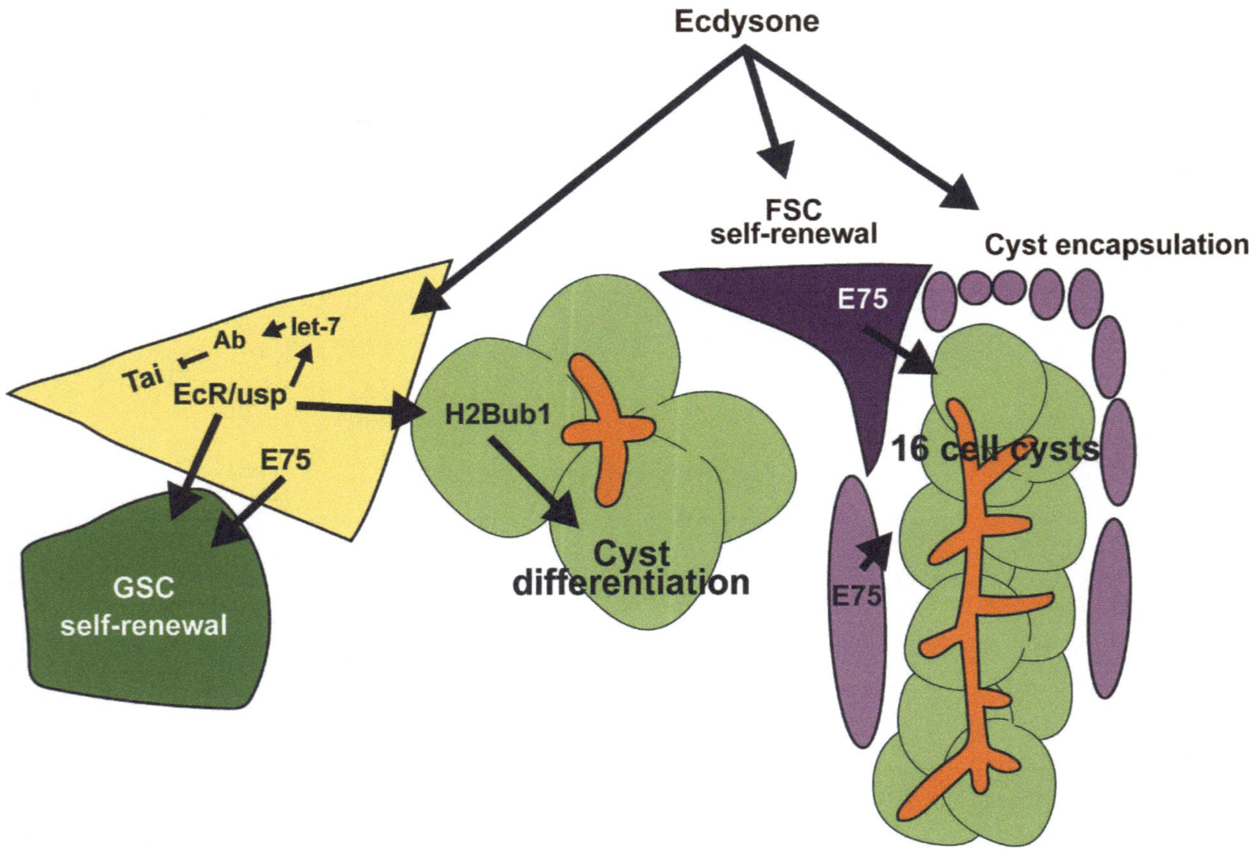

Fig. 3 Ecdysone is required in somatic cells for early germline processes. In the escort cells (yellow), EcR binds to the co-activator Tai to regulate GSC (dark green) self-renewal. Ecdysone functions in a feedback loop with the miRNA let-7 and the repressor Abrupt. Ecdysone regulates the monoubiquitination of the histone H2B (H2Bub1) modification which allows for cysts (light green) to differentiate. Ecdysone signaling in the FSCs (dark purple) and follicle cells (light purple) drives formation and encapsulation of 16 cell cysts.

production (such as the temperature sensitive *ecdysoneless* mutants) or ecdysone reception (such as loss of function of *EcR*) display rapid GSC loss upon switching to a restrictive temperature (Ables & Drummond-Barbosa, 2010; König et al., 2011; Morris & Spradling, 2012). Moreover, a pulse of ecdysone biosynthesis at mating promotes an initial surge of symmetric GSC division, resulting in an overall increased number of GSCs per ovariole (Ameku & Niwa, 2016). While the phenotypes resulting from global loss of ecdysone function are likely a cumulative effect of disrupted signaling in multiple ovarian or peripheral cell types, several lines of evidence suggest that ecdysone is required cell autonomously in the GSCs for self-renewal (Fig. 2) (Ables & Drummond-Barbosa, 2010; Ahmed et al., 2020; Ameku & Niwa, 2016; Ameku et al., 2017; König et al., 2011; Morris & Spradling, 2012; Sieber & Spradling, 2015). GSCs lacking functional *usp* or the early gene *E74* exhibit reduced proliferation and fail to self-renew, likely due to modulation of BMP signaling (Ables & Drummond-Barbosa, 2010; König et al., 2011). Ecdysone also functions with the chromatin remodeling factor ISWI/NURF, an EcR co-activator, to regulate GSC self-renewal, suggesting cell autonomous regulation of GSCs (Ables & Drummond-Barbosa, 2010; Badenhorst et al., 2005). Although *E74* is the only ecdysone early gene known to be required to promote GSC self-renewal and proliferation, other transcriptional targets are likely to promote these processes downstream of EcR (Ables, Hwang, Finger, Hinnant, & Drummond-Barbosa, 2016). Elucidating EcR/Usp and E74 transcriptional targets is a key future direction necessary for understanding how ecdysone directly modulates GSCs.

Ecdysone signaling also regulates GSC self-renewal non-autonomously via somatic escort cells and cap cells (Fig. 3). EcR, Usp, and Taiman are highly expressed in cap and escort cells, and ligand-binding reporters for EcR and Usp indicate that ecdysone signaling is active in these cells (König et al., 2011; Morris & Spradling, 2012). Ecdysone-responsive enhancers in several gene loci, including *E75*, *ftz-f1*, and *br* are also active in cap and escort cells, suggesting a complex signaling network guides escort cell function (McDonald et al., 2019). Loss of *EcR*, *usp*, or *E75* specifically in escort cells results in decreased GSC number (Morris & Spradling, 2012). In contrast, loss of *EcR* or *taiman* in cap cells expanded the number of cap cells, resulting in more GSCs (König et al., 2011). Since escort cells participate non-autonomously in germ cell differentiation by limiting the range of BMP signals (König & Shcherbata, 2015; Luo, Wang, Fan, Liu, & Cai, 2015; Mottier-Pavie, Palacios, Eliazer, Scoggin, & Buszczak, 2016), ecdysone signaling may modulate one of the many paracrine signaling ligands produced by escort cells. Two possible candidates may be Wnt/Wg and/or Epidermal Growth Factor Receptor (EGFR) signaling. In the absence of ecdysone signaling, GSCs do not properly receive BMP signals, EGFR activity is increased, and cell adhesion between germ cells and escort cells is altered (König & Shcherbata, 2015). It is unclear, however, whether these are direct or indirect effects of EcR transcription in escort cells. Additional experiments testing how ecdysone signaling modulates paracrine signals in escort cells are necessary to resolve the molecular mechanism of action.

5.2 Ecdysone signaling mediates germline differentiation, follicle formation and encapsulation

Following separation of the cystoblast from the GSC, the cystoblast progresses through four rounds of mitotic division forming an interconnected cyst, while simultaneously initiating oocyte selection and differentiation (Fig. 1B and D) (Hinnant et al., 2020). Concurrent with oocyte differentiation, individual cysts are packaged into discrete egg chambers, encapsulated by follicle cells. These processes are inextricably intertwined, and include molecular mechanisms maintaining the self-renewal and proliferation of FSCs and their immediate daughters (Rust & Nystul, 2020). A variety of experiments have suggested that ecdysone signaling impacts these processes, perhaps via molecular mechanisms independent of germ cell differentiation. First, loss of ecdysone ligand (*ecdysoneless* mutants) results in fewer dividing cysts and fewer 16-cell cysts, indicating a block to germ cell differentiation (König et al., 2011; Morris & Spradling, 2012). Although inactivation of *E74* in germ cells blocks cyst division, in part due to increased apoptosis, *tai* depletion from escort cells causes a block in cyst differentiation and division, leading to excess single germ cells (Ables & Drummond-Barbosa, 2010; König et al., 2011). The EcR repressor Abrupt regulates this process through a feedback loop with ecdysone (Fig. 3) (König & Shcherbata, 2015; König et al., 2011). Abrupt blocks the ability of Tai to bind to EcR. The ecdysone responsive miRNA, *let-7*, targets *abrupt* transcripts, allowing Tai to bind EcR and increasing ecdysone signaling strength (König & Shcherbata, 2015). Unlike *bam* mutants, which completely block differentiation, loss of EcR signaling leads to a delay of differentiation, accompanied by a change in chromatin state (König et al., 2011; Ohlstein & McKearin, 1997). Ecdysone mutants lack the monoubiquitination of the histone H2B (H2Bub1) modification, which is required for the change from a GSC state to a differentiation state (Karpiuk et al., 2012; König & Shcherbata, 2015).

These cells become temporarily stuck between GSC and cystoblast fates, indicating that ecdysone signaling is required in somatic cells for the committed germ cell differentiation fate.

Loss of ecdysone signaling in escort cells also abrogates cyst formation and encapsulation (Ables & Drummond-Barbosa, 2010; König et al., 2011; Morris & Spradling, 2012). Knock-down of *usp, EcR*, or *E75*, or overexpression of the EcR repressor *Abrupt*, in escort cells and follicle cells resulted in abnormally shaped escort cells and a decrease or absence of membrane extensions (König & Shcherbata, 2015; König et al., 2011; Morris & Spradling, 2012). It is unclear, however, exactly how ecdysone signaling modulates escort cell shape and function, and whether and how this impacts EGFR signaling. Given the unique spatiotemporal specificity of ecdysone signaling, it is also formally possible that ecdysone signaling promotes unique cell activities in posterior escort cells, FSCs, and pre-follicle cells (Fig. 3) (Ables et al., 2016). This could be due to unique combinations of EcR transcriptional targets, or perhaps due to differential availability of the ecdysone ligand. Indeed, knock-down of the ecdysteroidogenic enzymes encoded by *neverland, diembodied*, or *spook* in escort cells (under the control of the *Gal4* driver *c587-Gal4*), is sufficient to block the initial surge of ecdysone production following mating and steroid-dependent midgut growth (Ahmed et al., 2020; Ameku & Niwa, 2016). These results warrant new investigation as to which ovarian cells produce and import ecdysone. Recent characterization of specific reagents for *UAS/Gal4*-mediated CRISPR and RNAi, and ovarian cell transcriptomic signatures, may help distinguish potential roles of ecdysone signaling in these somatic cell types (Hartman et al., 2015; Huang, Sahai-Hernandez, et al., 2014; Jevitt et al., 2020; McDonald et al., 2019; Port et al., 2020; Slaidina et al., 2020).

5.3 Ecdysone is required for continued egg chamber development, survival, and vitellogenesis during mid- and late-stages of oogenesis

The first observed phenotype associated with ecdysone mutants was the loss of vitellogenic egg chambers (Audit-Lamour & Busson, 1981; Buszczak et al., 1999; Carney & Bender, 2000). The few eggs that were laid by females had very thin eggshells with misshapen appendages (Audit-Lamour & Busson, 1981; Hackney, Pucci, Naes, & Dobens, 2007; Oro, McKeown, & Evans, 1992). While injection of ecdysone lead to loss of vitellogenic egg chambers, reduction of ecdysone signaling also abrogated egg chamber development, suggesting that the amount of ecdysone is critical for vitellogenesis. These phenotypes foreshadowed a variety of molecular mechanisms by which ecdysone signaling promotes continued oocyte development outside of the germarium.

After cysts are fully encapsulated, they move outside the germarium as individual egg chambers (Fig. 1A and D). As egg chambers pinch away from the germarium, follicle cells differentiate into stalk cells, pole cells, and main body follicle cells through Notch/Delta and Jak/Stat signaling (Duhart et al., 2017; Osterfield et al., 2017). This establishes egg chamber polarity and subsequent oocyte polarity as the oocyte continues to grow. During vitellogenesis, follicle cells proliferate, grow in size, differentiate, and migrate to specific locations around the oocyte to form the eggshell and exterior structures of the egg chamber, including the micropyle (which allows for sperm to enter the egg), dorsal appendages (which allow for gas exchange), and the operculum (the region from which the larvae emerges at hatching, post-fertilization). Concurrently, the oocyte is loaded with yolk proteins and lipids, especially triacylglycerol (Parra-Peralbo & Culi, 2011; Sieber & Spradling, 2015; Ziegler & Van Antwerpen, 2006). Because oogenesis is closely tied to nutritional status of the female fly, developing egg chambers must bypass a nutritional checkpoint at stage 8 prior to the switch to vitellogenesis. This either allows the egg chamber to proceed to undergo the metabolically demanding process of vitellogenesis or breakdown if nutrients are not favorable for subsequent embryo development (Peterson, Timmons, Mondragon, & McCall, 2015). Insufficient nutrients cause apoptosis and reabsorption of middle stage egg chambers (8–13) and withholding of stage 14 follicles, blocking ovulation.

EcR and several ecdysone responsive genes are highly expressed in both germline and somatic cells during the vitellogenic stages of oocyte development (Ables et al., 2016; Buszczak et al., 1999; McDonald et al., 2019). Loss of *EcR*, *E74*, *E75*, or *E78* all result in egg chamber death, suggesting important roles for ecdysone signaling during early vitellogenesis (Ables et al., 2015; Ables & Drummond-Barbosa, 2010; Ables et al., 2016; Audit-Lamour & Busson, 1981; Buszczak et al., 1999; Carney & Bender, 2000). Ecdysone levels are also the highest during vitellogenic stages, and several ecdysone biosynthesis genes are expressed (Buszczak et al., 1999; Ono et al., 2006; Petryk et al., 2003; Uryu et al., 2015). Interestingly, injection of 20E leads to increased apoptosis of nurse cells in stage 8 follicles and loss of all vitellogenic stages, indicating that the amount of ecdysone is very important for regulation of apoptosis (Soller, Bownes, & Kubli, 1999; Terashima et al., 2005). The ecdysone early genes *br* and *E75* appear to regulate this nutritional checkpoint. When females are nutritionally deprived, *br* isoforms *Z2* and *Z3* are up-regulated, leading to expression of *E75A* which induces apoptosis of cysts, while simultaneously suppressing *E75B*, which is required to inhibit apoptosis (Terashima & Bownes, 2006).

Impairment of EcR and Usp signaling specifically in follicle cells during vitellogenesis results in reduced egg size, thinner eggshell, and malformed dorsal appendages and operculum (Hackney et al., 2007). These phenotypes can be attributed to a variety of molecular mechanisms. First, EcR-B1 in follicle cells promotes epithelial integrity, microvilli formation, and follicle cell shape (Fig. 4) (Hackney et al., 2007; Romani et al., 2009; Romani, Gargiulo, & Cavaliere, 2016). EcR-B1 regulates expression of adherens and basolateral junction proteins to maintain the connections between follicle cells, and between the follicle cells and germ cells (Fig. 4) (Romani et al., 2009). EcR also promotes follicle cell endocycling and expression of eggshell proteins (Bernardi, Romani, Tzertzinis, Gargiulo, & Cavaliere, 2009; Sun, Smith, Armento, & Deng, 2008). Second, follicle cell migration is driven by the interplay of EcR signaling with Jak/Stat signaling (Fig. 5) (Peercy & Starz-Gaiano, 2020). In this context, cell migration occurs in direct response to a pulse of ecdysone synthesis, which dislodges the ecdysone

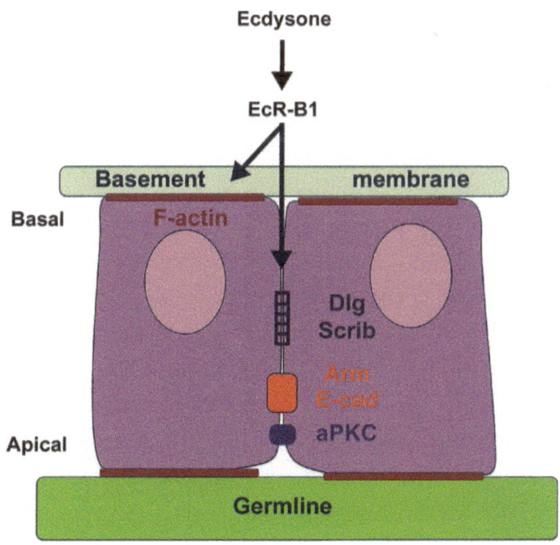

Fig. 4 Follicle cell polarity in mid-stages is driven by ecdysone signaling. EcR-B1 regulates F-actin (maroon) expression, and localization of the adherens junction proteins Arm and E-Cad (red), the septate junction proteins Dlg and Scrib (gray), and aPKC (blue), in mid-stage follicle cells (purple).

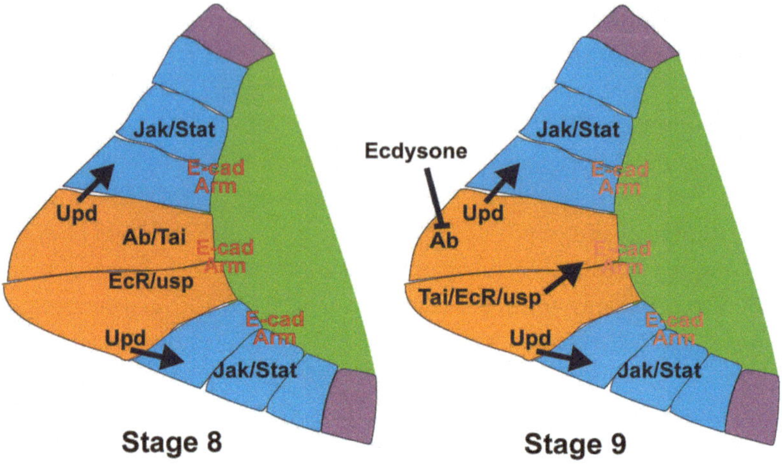

Fig. 5 Ecdysone signaling regulates the timing of border cell movement. Upd ligand in polar cells (orange) establishes future border cells (blue) from follicle cells (light purple). In stage 8 follicles, Abrupt binds Tai and prevents its binding to EcR. This blocks turnover of the cohesion proteins E-cadherin and β-catenin (arm). In stage 9 border cells, ecdysone inhibits Abrupt, allowing Tai to bind to EcR. Tai/EcR creates turnover of E-cadherin and β-catenin, promoting border cell migration.

co-repressor Abrupt (Bai et al., 2000; Buszczak et al., 1999; Jang, Chang, Bai, & Montell, 2009). This allows EcR-A, EcR-B, and Usp to be activated by Tai in border cells, which promotes turnover of the cell adhesion proteins E-cadherin and β-catenin (Bai et al., 2000; Jang et al., 2009) and delamination from the follicle cell epithelium. Border cells then continue to migrate through the nurse cells, sending out projections until they reach the anterior border of the oocyte at stage 10 (Peercy & Starz-Gaiano, 2020). Overexpression of ecdysone signaling leads to early migration of border cells while decreased ecdysone delays migration, indicating that ecdysone helps to regulate timing (Bai et al., 2000; Cherbas et al., 2003; Domanitskaya, Anllo, & Schüpbach, 2014; Jang et al., 2009). At least four other NRs (E75, Hr3, Hr4, and Ftz-f1) also regulate border cell migration (Manning et al., 2017; McDonald et al., 2019; Wang et al., 2020). It is unclear however, whether these NRs functions independently of EcR/Usp, or together in an elaborate transcriptional network. Lastly, Usp and Br are required for dorsal appendage formation and amplification of chorionic genes (Deng & Bownes, 1997; Oro et al., 1992; Osterfield et al., 2017; Tzolovsky, Deng, Schlitt, & Bownes, 1999). In stage 6, Br is expressed in all follicle cells but becomes progressively restricted to two groups of dorsal-lateral-anterior follicle cells that become the cells of the dorsal appendage. Ecdysone signaling and the NR Ftz-f1 are also required for ovulation and reproductive tract development (Knapp, Li, Singh, & Sun, 2020; Knapp & Sun, 2017; Sun & Spradling, 2012).

6. Signaling from peripheral tissues promotes GSC maintenance

Although NRs acting in the ovary are crucial for fertility, a developing area of interest in the field is how NRs located in other tissues can influence oogenesis. Reproduction is energetically costly to females; thus, the maternal metabolic physiology must be tailored to turn specific dietary nutrients into the lipids and proteins necessary to sustain egg development (Armstrong, 2020; Mirth et al., 2019; Sieber & Spradling, 2017). As in humans, maintenance of a maternal physiology capable of sustaining egg production relies on several interrelated organs, including the brain, the fat body (analogous to human liver), and the gut (Droujinine & Perrimon, 2016; Rajan & Perrimon, 2011). Recent studies reveal that a complex interorgan communication network not only links these organs to oogenesis, but also utilize ecdysone produced from the ovary to modulate organ maintenance and function (Ahmed et al., 2020; Ameku & Niwa, 2016; Matsuoka, Armstrong, Sampson, Laws, & Drummond-Barbosa, 2017; Sieber & Spradling, 2015; Weaver & Drummond-Barbosa, 2019).

The fat body, an organ which helps to maintain maternal metabolism, is particularly appealing as a candidate tissue that influences oogenesis. The fat body is composed of adult adipocytes, which store lipids, and oenocytes, the hepatocyte-like cells of insects that produce lipids and other macromolecules (Armstrong, 2020; Arrese & Soulages, 2010). Yolk proteins and the yolk storage protein vitellogenin are produced by the fat body and transported to the oocyte; both of these processes are influenced by ecdysone signaling (Gilbert, Serafin, Watkins, & Richard, 1998; Jowett & Postlethwait, 1980; Schonbaum, Perrino, & Mahowald, 2000; Sieber & Spradling, 2015; Yan & Postlethwait, 1990). EcR promotes the female metabolic state by promoting high levels of whole-body triglycerides and glycogen, likely via effects on the fat body (Sieber & Spradling, 2015). The NR encoded by *seven up* (*svp*), most closely related to mammalian COUP-TFII, is also expressed in adipocytes and essential in adipocytes and oenocytes for ovarian GSC self-renewal and egg chamber survival, respectively (Weaver & Drummond-Barbosa, 2019). Intriguingly, other endocrine signals are also produced by the fat body that regulate oogenesis, notably insulin signaling (Armstrong & Drummond-Barbosa, 2018; Armstrong, Laws, & Drummond-Barbosa, 2014; Matsuoka et al., 2017). These results suggest that peripherally located NRs may act in concert with other endocrine signaling pathways to fine-tune oogenesis in response to maternal diet.

7. Conclusions and open questions

NRs are key regulators of many aspects of development, reproduction, and metabolism. A comprehensive understanding of NR target genes, signaling mechanisms in and between cell types, and communication between organs can provide insight into possible conserved mechanisms between *Drosophila* and humans. Yet many questions remain for future studies. First, what are the specific transcriptional targets of ecdysone signaling in oogenesis, and how are these regulated in a precise spatiotemporal manner? Although ecdysone-mediated processes are well-understood, we are only just beginning to understand how ecdysone biosynthesis, transport, and reception by EcR are achieved in ovarian cells. Addressing this complex question will require sophisticated genetic manipulation in cell-specific studies. Recent transcriptomic profiling of ovarian cells by single cell RNA sequencing may be a tremendous step forward in this area (Jevitt et al., 2020). Second, how does ecdysone signaling promote egg chamber viability and progression into vitellogenesis? Several ecdysone-responsive NRs are clearly expressed in nurse cells during these stages and required for egg chamber survival, but it is unclear whether they may promote an as-yet undescribed nutritional checkpoint in early oogenesis or promote cyst growth. Lastly, are roles of EcR or other NRs unique to female reproduction? Although the role of ecdysone in germline stem cell maintenance has been most commonly associated with oogenesis, recent studies have demonstrated a potential contribution of ecdysone to spermatogenesis in *Drosophila* (Li, Ma, Cherry, & Matunis, 2014). The effects of obesity, metabolic dysfunction, and diabetes on reproductive tissues are not unique to females, and NRs also regulate many aspects of spermatogenesis in both *Drosophila* and humans. Future studies analyzing the role of nutritionally regulated NRs in spermatogenesis are also warranted.

Determining the mechanisms by which hormone signaling facilitates crosstalk between germline and somatic cells could provide valuable insight into possible conserved mechanisms with other species, such as humans. In *Drosophila*, signals from the germline greatly influence follicle cell function and vice versa. It has been well-established that ecdysone signaling in the escort cells influences GSC self-renewal, germ cell differentiation, and follicle cell encapsulation and movement. It is less clear, however, whether ecdysone signaling in germ cells might influence the overlying somatic tissue. Given the parallels between steroid hormone signaling in *Drosophila* and humans, future studies investigating how ecdysone facilitates germline-soma communication could provide a better understanding of the mechanisms of hormone signaling in human oogenesis.

Acknowledgments

We are grateful to all the colleagues whose foundational work has established our field. We apologize to those whose original work could not be cited due to space limitations. Many thanks to members of the Ables lab for critical reading of this manuscript and helpful discussions.

Funding

This work was supported by National Institutes of Health R15 GM117502 (E.T.A.), March of Dimes Basil O'Connor Research Starter Award 5-FY14-62 (E.T.A.), and Sigma Xi Grant-in-Aid of Research (D.S.F.).

References

Ables, E. T., Bois, K. E., Garcia, C. A., & Drummond-Barbosa, D. (2015). Ecdysone response gene E78 controls ovarian germline stem cell niche formation and follicle survival in Drosophila. *Developmental Biology*, *400*(1), 33–42. https://doi.org/10.1016/j.ydbio.2015.01.013.

Ables, E. T., & Drummond-Barbosa, D. (2010). The steroid hormone ecdysone functions with intrinsic chromatin remodeling factors to control female germline stem cells in Drosophila. *Cell Stem Cell*, 7(5), 581–592. https://doi.org/10.1016/j.stem.2010.10.001.

Ables, E. T., & Drummond-Barbosa, D. (2017). Steroid hormones and the physiological regulation of tissue-resident stem cells: Lessons from the drosophila ovary. *Current Stem Cell Reports*, *3*(1), 9–18. https://doi.org/10.1007/s40778-017-0070-z.

Ables, E. T., Hwang, G. H., Finger, D. S., Hinnant, T. D., & Drummond-Barbosa, D. (2016). A genetic mosaic screen reveals ecdysone-responsive genes regulating *Drosophila* oogenesis. *G3: Genes|Genomes|Genetics*, *6*(8), 2629–2642. https://doi.org/10.1534/g3.116.028951.

Ables, E. T., Laws, K. M., & Drummond-Barbosa, D. (2012). Control of adult stem cells in vivo by a dynamic physiological environment: Diet-dependent systemic factors in Drosophila and beyond. *Wiley Interdisciplinary Reviews: Developmental Biology*, *1*(5), 657–674. https://doi.org/10.1002/wdev.48.

Ahmed, S. M. H., Maldera, J. A., Krunic, D., Paiva-Silva, G. O., Pénalva, C., Teleman, A. A., et al. (2020). Fitness trade-offs incurred by ovary-to-gut steroid signalling in Drosophila. *Nature*, *584*(7821), 415–419. https://doi.org/10.1038/s41586-020-2462-y.

Ameku, T., & Niwa, R. (2016). Mating-induced increase in germline stem cells via the neuroendocrine system in female Drosophila. *PLoS Genetics*, *12*(6), e1006123. https://doi.org/10.1371/journal.pgen.1006123.

Ameku, T., Yoshinari, Y., Fukuda, R., & Niwa, R. (2017). Ovarian ecdysteroid biosynthesis and female germline stem cells. *Fly (Austin)*, *11*(3), 185–193. https://doi.org/10.1080/19336934.2017.1291472.

Andres, A. J., Fletcher, J. C., Karim, F. D., & Thummel, C. S. (1993). Molecular analysis of the initiation of insect metamorphosis: A comparative study of Drosophila ecdysteroid-regulated transcription. *Developmental Biology*, *160*(2), 388–404.

Antel, M., & Inaba, M. (2020). Modulation of cell-cell interactions in Drosophila oocyte development. *Cells*, *9*(2), 274. https://doi.org/10.3390/cells9020274.

Apple, R., & Fristrom, J. W. (1991). 20-Hydroxyecdysone is required for, and negatively regulates, transcription of Drosophila pupal cuticle protein genes. *Developmental Biology*, *146*(2), 569–582.

Armstrong, A. R. (2020). Drosophila melanogaster as a model for nutrient regulation of ovarian function. *Reproduction*, *159*(2), R69–R82.

Armstrong, A. R., & Drummond-Barbosa, D. (2018). Insulin signaling acts in adult adipocytes via GSK-3β and independently of FOXO to control Drosophila female germline stem cell numbers. *Developmental Biology*, *440*(1), 31–39. https://doi.org/10.1016/j.ydbio.2018.04.028.

Armstrong, A. R., Laws, K. M., & Drummond-Barbosa, D. (2014). Adipocyte amino acid sensing controls adult germline stem cell number via the amino acid response pathway and independently of Target of Rapamycin signaling in Drosophila. *Development*, *141*(23), 4479–4488. https://doi.org/10.1242/dev.116467.

Arrese, E. L., & Soulages, J. L. (2010). Insect fat body: Energy, metabolism, and regulation. *Annual Review of Entomology*, *55*, 207–225. https://doi.org/10.1146/annurev-ento-112408-085356.

Ashburner, M., Chihara, C., Meltzer, P., & Richards, G. (1974). Temporal control of puffing activity in polytene chromosomes. *Cold Spring Harbor Symposia on Quantitative Biology*, *38*, 655–662.

Audit-Lamour, C., & Busson, D. (1981). Oögenesis defects in the ecd-1 mutant of Drosophila melanogaster, deficient in ecdysteroid at high temperature. *Journal of Insect Physiology*, *27*(12), 829–837.

Avila, F. W., Sirot, L. K., LaFlamme, B. A., Rubinstein, C. D., & Wolfner, M. F. (2011). Insect seminal fluid proteins: Identification and function. *Annual Review of Entomology*, *56*, 21–40. https://doi.org/10.1146/annurev-ento-120709-144823.

Badenhorst, P., Xiao, H., Cherbas, L., Kwon, S. Y., Voas, M., Rebay, I., et al. (2005). The Drosophila nucleosome remodeling factor NURF is required for ecdysteroid signaling and metamorphosis. *Genes & Development*, *19*(21), 2540–2545. https://doi.org/10.1101/gad.1342605.

Bai, J., Uehara, Y., & Montell, D. J. (2000). Regulation of invasive cell behavior by taiman, a Drosophila protein related to AIB1, a steroid receptor coactivator amplified in breast cancer. *Cell*, *103*(7), 1047–1058.

Bainbridge, S. P., & Bownes, M. (1988). Ecdysteroid titers during Drosophila metamorphosis. *Insect Biochemistry*, *18*(2), 185–197.

Banisch, T. U., Maimon, I., Dadosh, T., & Gilboa, L. (2017). Escort cells generate a dynamic compartment for germline stem cell differentiation via combined Stat and Erk signalling. *Development*, *144*, 1937–1947. https://doi.org/10.1242/dev.143727.

Beckstead, R. B., Lam, G., & Thummel, C. S. (2005). The genomic response to 20-hydroxyecdysone at the onset of Drosophila metamorphosis. *Genome Biology*, *6*(12), R99.

Bernardi, F., Romani, P., Tzertzinis, G., Gargiulo, G., & Cavaliere, V. (2009). EcR-B1 and Usp nuclear hormone receptors regulate expression of the VM32E eggshell gene during Drosophila oogenesis. *Developmental Biology*, *328*(2), 541–551. https://doi.org/10.1016/j.ydbio.2009.01.013.

Billas, I. M., Iwema, T., Garnier, J. M., Mitschler, A., Rochel, N., & Moras, D. (2003). Structural adaptability in the ligand-binding pocket of the ecdysone hormone receptor. *Nature, 426*(6962), 91–96. https://doi.org/10.1038/nature02112.

Billas, I. M., & Moras, D. (2005). Ligand-binding pocket of the ecdysone receptor. *Vitamins and Hormones, 73*, 101–129. https://doi.org/10.1016/s0083-6729(05)73004-1.

Bodofsky, S., Koitz, F., & Wightman, B. (2017). Conserved and exapted functions of nuclear receptors in animal development. *Nuclear Receptor Research, 4*, 101305. https://doi.org/10.11131/2017/101305.

Bownes, M. (1982). The role of 20-hydroxy-ecdysone in yolk-polypeptide synthesis by male and female fat bodies of Drosophila melanogaster. *Journal of Insect Physiology, 28*(4), 317–328.

Braquart, C., Bouhin, H., Quennedey, A., & Delachambre, J. (1996). Up-regulation of an adult cuticular gene by 20-hydroxyecdysone in insect metamorphosing epidermis cultured in vitro. *European Journal of Biochemistry, 240*(2), 336–341.

Braun, S., Azoitei, A., & Spindler-Barth, M. (2009). DNA-binding properties of Drosophila ecdysone receptor isoforms and their modification by the heterodimerization partner ultraspiracle. *Archives of Insect Biochemistry and Physiology, 72*(3), 172–191. https://doi.org/10.1002/arch.20328.

Brennan, M. D., Weiner, A. J., Goralski, T. J., & Mahowald, A. P. (1982). The follicle cells are a major site of vitellogenin synthesis in Drosophila melanogaster. *Developmental Biology, 89*(1), 225–236. https://doi.org/10.1016/0012-1606(82)90309-8.

Buszczak, M., Freeman, M. R., Carlson, J. R., Bender, M., Cooley, L., & Segraves, W. A. (1999). Ecdysone response genes govern egg chamber development during mid-oogenesis in Drosophila. *Development, 126*(20), 4581–4589.

Cáceres, L., Necakov, A. S., Schwartz, C., Kimber, S., Roberts, I. J., & Krause, H. M. (2011). Nitric oxide coordinates metabolism, growth, and development via the nuclear receptor E75. *Genes & Development, 25*(14), 1476–1485. https://doi.org/10.1101/gad.2064111.

Carbonell, A., Mazo, A., Serras, F., & Corominas, M. (2013). Ash2 acts as an ecdysone receptor coactivator by stabilizing the histone methyltransferase Trr. *Molecular Biology of the Cell, 24*(3), 361–372. https://doi.org/10.1091/mbc.E12-04-0267.

Carmel, I., Tram, U., & Heifetz, Y. (2016). Mating induces developmental changes in the insect female reproductive tract. *Current Opinion in Insect Science, 13*, 106–113. https://doi.org/10.1016/j.cois.2016.03.002.

Carney, G. E., & Bender, M. (2000). The Drosophila ecdysone receptor (EcR) gene is required maternally for normal oogenesis. *Genetics, 154*(3), 1203–1211.

Carvalho-Santos, Z., & Ribeiro, C. (2018). Gonadal ecdysone titers are modulated by protein availability but do not impact protein appetite. *Journal of Insect Physiology, 106*(Pt. 1), 30–35. https://doi.org/10.1016/j.jinsphys.2017.08.006.

Cetera, M., & Horne-Badovinac, S. (2015). Round and round gets you somewhere: Collective cell migration and planar polarity in elongating Drosophila egg chambers. *Current Opinion in Genetics & Development, 32*, 10–15. https://doi.org/10.1016/j.gde.2015.01.003.

Charles, J. P. (2010). The regulation of expression of insect cuticle protein genes. *Insect Biochemistry and Molecular Biology, 40*(3), 205–213. https://doi.org/10.1016/j.ibmb.2009.12.005.

Chavez, V. M., Marques, G., Delbecque, J. P., Kobayashi, K., Hollingsworth, M., Burr, J., et al. (2000). The Drosophila disembodied gene controls late embryonic morphogenesis and codes for a cytochrome P450 enzyme that regulates embryonic ecdysone levels. *Development, 127*(19), 4115–4126.

Chen, Y., Breen, K., & Pepling, M. E. (2009). Estrogen can signal through multiple pathways to regulate oocyte cyst breakdown and primordial follicle assembly in the neonatal mouse ovary. *Journal of Endocrinology, 202*(3), 407.

Chen, D., & McKearin, D. (2003). Dpp signaling silences bam transcription directly to establish asymmetric divisions of germline stem cells. *Current Biology*, *13*, 1786–1791. https://doi.org/10.1016/j.cub.2003.09.033.

Chen, D., Wang, Q., Huang, H., Xia, L., Jiang, X., Kan, L., et al. (2009). Effete-mediated degradation of Cyclin A is essential for the maintenance of germline stem cells in Drosophila. *Development*, *136*(24), 4133–4142. https://doi.org/10.1242/dev.039032.

Cherbas, L., Hu, X., Zhimulev, I., Belyaeva, E., & Cherbas, P. (2003). EcR isoforms in Drosophila: Testing tissue-specific requirements by targeted blockade and rescue. *Development*, *130*(2), 271–284.

Crowder, M. K., Seacrist, C. D., & Blind, R. D. (2017). Phospholipid regulation of the nuclear receptor superfamily. *Advances in Biological Regulation*, *63*, 6–14. https://doi.org/10.1016/j.jbior.2016.10.006.

Dai, J. D., Henrich, V. C., & Gilbert, L. I. (1991). An ultrastructural analysis of the ecdysoneless (l(3)ecd1ts) ring gland during the third larval instar of Drosophila melanogaster. *Cell and Tissue Research*, *265*(3), 435–445.

Dallel, S., Tauveron, I., Brugnon, F., Baron, S., Lobaccaro, J. M. A., & Maqdasy, S. (2018). Liver X receptors: A possible link between lipid disorders and female infertility. *International Journal of Molecular Sciences*, *19*(8), 2177. https://doi.org/10.3390/ijms19082177.

Davis, M. B., Carney, G. E., Robertson, A. E., & Bender, M. (2005). Phenotypic analysis of EcR-A mutants suggests that EcR isoforms have unique functions during Drosophila development. *Developmental Biology*, *282*(2), 385–396. https://doi.org/10.1016/j.ydbio.2005.03.019.

Decotto, E., & Spradling, A. C. (2005). The Drosophila ovarian and testis stem cell niches: Similar somatic stem cells and signals. *Developmental Cell*, *9*(4), 501–510.

Deng, W.-M., & Bownes, M. (1997). Two signalling pathways specify localised expression of the broad-complex in Drosophila eggshell patterning and morphogenesis. *Development*, *124*(22), 4639–4647.

Dobens, L., Rudolph, K., & Berger, E. M. (1991). Ecdysterone regulatory elements function as both transcriptional activators and repressors. *Molecular and Cellular Biology*, *11*(4), 1846–1853. https://doi.org/10.1128/mcb.11.4.1846.

Domanitskaya, E., Anllo, L., & Schüpbach, T. (2014). Phantom, a cytochrome P450 enzyme essential for ecdysone biosynthesis, plays a critical role in the control of border cell migration in Drosophila. *Developmental Biology*, *386*(2), 408–418.

Dressel, U., Thormeyer, D., Altincicek, B., Paululat, A., Eggert, M., Schneider, S., et al. (1999). Alien, a highly conserved protein with characteristics of a corepressor for members of the nuclear hormone receptor superfamily. *Molecular and Cellular Biology*, *19*(5), 3383–3394.

Droujinine, I. A., & Perrimon, N. (2016). Interorgan communication pathways in physiology: Focus on Drosophila. *Annual Review of Genetics*, *50*, 539–570. https://doi.org/10.1146/annurev-genet-121415-122024.

Drummond, A. E., Britt, K. L., Dyson, M., Jones, M. E., Kerr, J. B., O'Donnell, L., et al. (2002). Ovarian steroid receptors and their role in ovarian function. *Molecular and Cellular Endocrinology*, *191*(1), 27–33.

Drummond-Barbosa, D. (2019). Local and physiological control of germline stem cell lineages in Drosophila melanogaster. *Genetics*, *213*(1), 9–26.

Drummond-Barbosa, D., & Spradling, A. C. (2001). Stem cells and their progeny respond to nutritional changes during drosophila oogenesis. *Developmental Biology*, *231*(1), 265–278. https://doi.org/10.1006/dbio.2000.0135.

Duhart, J. C., Parsons, T. T., & Raftery, L. A. (2017). The repertoire of epithelial morphogenesis on display: Progressive elaboration of Drosophila egg structure. *Mechanisms of Development*, *148*, 18–39.

Evans, R. M., & Mangelsdorf, D. J. (2014). Nuclear receptors, RXR, and the big bang. *Cell*, *157*(1), 255–266. https://doi.org/10.1016/j.cell.2014.03.012.
Feng, K., Palfreyman, M. T., Hasemeyer, M., Talsma, A., & Dickson, B. J. (2014). Ascending SAG neurons control sexual receptivity of Drosophila females. *Neuron*, *83*(1), 135–148. https://doi.org/10.1016/j.neuron.2014.05.017.
Forbes, A., & Lehmann, R. (1998). Nanos and Pumilio have critical roles in the development and function of Drosophila germline stem cells. *Development*, *125*(4), 679–690.
Freeman, M. R., Dobritsa, A., Gaines, P., Segraves, W. A., & Carlson, J. R. (1999). The dare gene: Steroid hormone production, olfactory behavior, and neural degeneration in Drosophila. *Development*, *126*(20), 4591–4602.
Gancz, D., Lengil, T., & Gilboa, L. (2011). Coordinated regulation of niche and stem cell precursors by hormonal signaling. *PLoS Biology*, *9*(11), e1001202.
Gao, Y., Mao, Y., Xu, R. G., Zhu, R., Zhang, M., Sun, J., et al. (2019). Defining gene networks controlling the maintenance and function of the differentiation niche by an in vivo systematic RNAi screen. *Journal of Genetics and Genomics*, *46*(1), 19–30. https://doi.org/10.1016/j.jgg.2018.10.008.
Gaskins, A. J., & Chavarro, J. E. (2018). Diet and fertility: A review. *American Journal of Obstetrics and Gynecology*, *218*(4), 379–389. https://doi.org/10.1016/j.ajog.2017.08.010.
Gauhar, Z., Sun, L. V., Hua, S., Mason, C. E., Fuchs, F., Li, T.-R., et al. (2009). Genomic mapping of binding regions for the Ecdysone receptor protein complex. *Genome Research*, *19*(6), 1006–1013.
Gilbert, L. I. (2004). Halloween genes encode P450 enzymes that mediate steroid hormone biosynthesis in Drosophila melanogaster. *Molecular and Cellular Endocrinology*, *215*(1–2), 1–10. https://doi.org/10.1016/j.mce.2003.11.003.
Gilbert, L. I., Rybczynski, R., & Warren, J. T. (2002). Control and biochemical nature of the ecdysteroidogenic pathway. *Annual Review of Entomology*, *47*(1), 883–916.
Gilbert, L. I., Serafin, R. B., Watkins, N. L., & Richard, D. S. (1998). Ecdysteroids regulate yolk protein uptake by Drosophila melanogaster oocytes. *Journal of Insect Physiology*, *44*(7–8), 637–644. https://doi.org/10.1016/s0022-1910(98)00020-1.
Gilbert, L. I., & Warren, J. T. (2005). A molecular genetic approach to the biosynthesis of the insect steroid molting hormone. *Vitamins and Hormones*, *73*, 31–57. https://doi.org/10.1016/s0083-6729(05)73002-8.
Gonsalves, S. E., Neal, S. J., Kehoe, A. S., & Westwood, J. T. (2011). Genome-wide examination of the transcriptional response to ecdysteroids 20-hydroxyecdysone and ponasterone A in Drosophila melanogaster. *BMC Genomics*, *12*(1), 475. https://doi.org/10.1186/1471-2164-12-475.
Griswold, M. D., & Hogarth, C. (2018). Beyond stem cells: Commitment of progenitor cells to meiosis. *Stem Cell Research*, *27*, 169–171. https://doi.org/10.1016/j.scr.2018.01.032.
Grive, K. J., & Freiman, R. N. (2015). The developmental origins of the mammalian ovarian reserve. *Development*, *142*(15), 2554–2563. https://doi.org/10.1242/dev.125211.
Gu, L., Liu, H., Gu, X., Boots, C., Moley, K. H., & Wang, Q. (2015). Metabolic control of oocyte development: Linking maternal nutrition and reproductive outcomes. *Cellular and Molecular Life Sciences*, *72*(2), 251–271. https://doi.org/10.1007/s00018-014-1739-4.
Hackney, J. F., Pucci, C., Naes, E., & Dobens, L. (2007). Ras signaling modulates activity of the ecdysone receptor EcR during cell migration in the Drosophila ovary. *Developmental Dynamics: An Official Publication of the American Association of the Anatomists*, *236*(5), 1213–1226.
Handler, A. M. (1982). Ecdysteroid titers during pupal and adult development in Drosophila melanogaster. *Developmental Biology*, *93*(1), 73–82.
Harshman, L. G., Loeb, A. M., & Johnson, B. A. (1999). Ecdysteroid titers in mated and unmated Drosophila melanogaster females. *Journal of Insect Physiology*, *45*(6), 571–577. https://doi.org/10.1016/s0022-1910(99)00038-4.

Hartman, T. R., Ventresca, E. M., Hopkins, A., Zinshteyn, D., Singh, T., O'Brien, J. A., et al. (2015). Novel tools for genetic manipulation of follicle stem cells in the Drosophila ovary reveal an integrin-dependent transition from quiescence to proliferation. *Genetics, 199*(4), 935–957. https://doi.org/10.1534/genetics.114.173617.

Hasemeyer, M., Yapici, N., Heberlein, U., & Dickson, B. J. (2009). Sensory neurons in the Drosophila genital tract regulate female reproductive behavior. *Neuron, 61*(4), 511–518. https://doi.org/10.1016/j.neuron.2009.01.009.

Heck, B. W., Zhang, B., Tong, X., Pan, Z., Deng, W.-M., & Tsai, C.-C. (2012). The transcriptional corepressor SMRTER influences both Notch and ecdysone signaling during Drosophila development. *Biology Open, 1*(3), 182–196.

Henrich, V. C., Szekely, A. A., Kim, S. J., Brown, N. E., Antoniewski, C., Hayden, M. A., et al. (1994). Expression and function of the ultraspiracle (usp) gene during development of Drosophila melanogaster. *Developmental Biology, 165*(1), 38–52.

Hill, R. J., Billas, I. M., Bonneton, F., Graham, L. D., & Lawrence, M. C. (2013). Ecdysone receptors: From the Ashburner model to structural biology. *Annual Review of Entomology, 58*, 251–271. https://doi.org/10.1146/annurev-ento-120811-153610.

Hinnant, T. D., Merkle, J. A., & Ables, E. T. (2020). Coordinating proliferation, polarity, and cell fate in the Drosophila female germline. *Frontiers in Cell and Development Biology, 8*, 19. https://doi.org/10.3389/fcell.2020.00019.

Hitrik, A., Popliker, M., Gancz, D., Mukamel, Z., Lifshitz, A., Schwartzman, O., et al. (2016). Combgap promotes ovarian niche development and chromatin association of EcR-binding regions in BR-C. *PLoS Genetics, 12*(11), e1006330. https://doi.org/10.1371/journal.pgen.1006330.

Hodin, J., & Riddiford, L. M. (1998). The ecdysone receptor and ultraspiracle regulate the timing and progression of ovarian morphogenesis during Drosophila metamorphosis. *Development Genes and Evolution, 208*(6), 304–317. https://doi.org/10.1007/s004270050186.

Hoffmeister, H., Grützmacher, H., & Dünnebeil, K. (1967). Studies on the structure and biochemical action of ecdysterone. *Zeitschrift fur Naturforschung. Teil B, Chemie, Biochemie, Biophysik, Biologie und Verwandte Gebiete, 22*(1), 66.

Hohos, N. M., & Skaznik-Wikiel, M. E. (2017). High-fat diet and female fertility. *Endocrinology, 158*(8), 2407–2419. https://doi.org/10.1210/en.2017-00371.

Hu, X., Cherbas, L., & Cherbas, P. (2003). Transcription activation by the ecdysone receptor (EcR/USP): Identification of activation functions. *Molecular Endocrinology, 17*(4), 716–731. https://doi.org/10.1210/me.2002-0287.

Huang, Y. C., Lu, Y. N., Wu, J. T., Chien, C. T., & Pi, H. (2014). The COP9 signalosome converts temporal hormone signaling to spatial restriction on neural competence. *PLoS Genetics, 10*(11), e1004760. https://doi.org/10.1371/journal.pgen.1004760.

Huang, P., Sahai-Hernandez, P., Bohm, R. A., Welch, W. P., Zhang, B., & Nystul, T. (2014). Enhancer-trap flippase lines for clonal analysis in the Drosophila ovary. *G3 (Bethesda), 4*(9), 1693–1699. https://doi.org/10.1534/g3.114.010710.

Igarashi, F., Ogihara, M. H., Iga, M., & Kataoka, H. (2018). Cholesterol internalization and metabolism in insect prothoracic gland, a steroidogenic organ, via lipoproteins. *Steroids, 134*, 110–116.

Jang, A. C.-C., Chang, Y.-C., Bai, J., & Montell, D. (2009). Border-cell migration requires integration of spatial and temporal signals by the BTB protein Abrupt. *Nature Cell Biology, 11*(5), 569.

Jevitt, A., Chatterjee, D., Xie, G., Wang, X. F., Otwell, T., Huang, Y. C., et al. (2020). A single-cell atlas of adult Drosophila ovary identifies transcriptional programs and somatic cell lineage regulating oogenesis. *PLoS Biology, 18*(4), e3000538. https://doi.org/10.1371/journal.pbio.3000538.

Ji, S., Li, C., Hu, L., Liu, K., Mei, J., Luo, Y., et al. (2017). Bam-dependent deubiquitinase complex can disrupt germ-line stem cell maintenance by targeting cyclin A. *Proceedings of the National Academy of Sciences of the United States of America, 114*(24), 6316–6321. https://doi.org/10.1073/pnas.1619188114.

Jiang, C., Baehrecke, E. H., & Thummel, C. S. (1997). Steroid regulated programmed cell death during Drosophila metamorphosis. *Development, 124*(22), 4673–4683.

Jin, Z., Kirilly, D., Weng, C., Kawase, E., Song, X., Smith, S., et al. (2008). Differentiation-defective stem cells outcompete normal stem cells for niche occupancy in the Drosophila ovary. *Cell Stem Cell, 2*(1), 39–49.

Jowett, T., & Postlethwait, J. H. (1980). The regulation of yolk polypeptide synthesis in Drosophila ovaries and fat body by 20-hydroxyecdysone and a juvenile hormone analog. *Developmental Biology, 80*(1), 225–234.

Kahney, E. W., Snedeker, J. C., & Chen, X. (2019). Regulation of Drosophila germline stem cells. *Current Opinion in Cell Biology, 60*, 27–35.

Karlson, P. (2003). On the hormonal control of insect metamorphosis. A historical review. *International Journal of Developmental Biology, 40*(1), 93–96.

Karpiuk, O., Najafova, Z., Kramer, F., Hennion, M., Galonska, C., König, A., et al. (2012). The histone H2B monoubiquitination regulatory pathway is required for differentiation of multipotent stem cells. *Molecular Cell, 46*(5), 705–713.

King-Jones, K., & Thummel, C. S. (2005). Nuclear receptors—A perspective from Drosophila. *Nature Reviews Genetics, 6*(4), 311–323. https://doi.org/10.1038/nrg1581.

Kirilly, D., Wang, S., & Xie, T. (2011). Self-maintained escort cells form a germline stem cell differentiation niche. *Development, 138*(23), 5087–5097. https://doi.org/10.1242/dev.067850.

Knapp, E. M., Li, W., Singh, V., & Sun, J. (2020). Nuclear receptor Ftz-f1 promotes follicle maturation and ovulation partly via bHLH/PAS transcription factor Sim. *eLife, 9*, e54568. https://doi.org/10.7554/eLife.54568.

Knapp, E., & Sun, J. (2017). Steroid signaling in mature follicles is important for Drosophila ovulation. *Proceedings of the National Academy of Sciences of the United States of America, 114*(4), 699–704.

Koelle, M. R., Talbot, W. S., Segraves, W. A., Bender, M. T., Cherbas, P., & Hogness, D. S. (1991). The Drosophila EcR gene encodes an ecdysone receptor, a new member of the steroid receptor superfamily. *Cell, 67*(1), 59–77.

König, A., & Shcherbata, H. R. (2015). Soma influences GSC progeny differentiation via the cell adhesion-mediated steroid-let-7-Wingless signaling cascade that regulates chromatin dynamics. *Biology Open, 4*(3), 285–300.

König, A., Yatsenko, A. S., Weiss, M., & Shcherbata, H. R. (2011). Ecdysteroids affect Drosophila ovarian stem cell niche formation and early germline differentiation. *The EMBO Journal, 30*(8), 1549–1562.

Kozlova, T., & Thummel, C. S. (2003). Essential roles for ecdysone signaling during *Drosophila* mid-embryonic development. *Science, 301*(5641), 1911–1914. https://doi.org/10.1126/science.1087419.

Kreher, J., Kovač, K., Bouazoune, K., Mačinković, I., Ernst, A. L., Engelen, E., et al. (2017). EcR recruits dMi-2 and increases efficiency of dMi-2-mediated remodelling to constrain transcription of hormone-regulated genes. *Nature Communications, 8*, 14806. https://doi.org/10.1038/ncomms14806.

Kugler, S. J., Gehring, E. M., Wallkamm, V., Kruger, V., & Nagel, A. C. (2011). The Putzig-NURF nucleosome remodeling complex is required for ecdysone receptor signaling and innate immunity in Drosophila melanogaster. *Genetics, 188*(1), 127–139. https://doi.org/10.1534/genetics.111.127795.

Lam, G. T., Jiang, C., & Thummel, C. S. (1997). Coordination of larval and prepupal gene expression by the DHR3 orphan receptor during Drosophila metamorphosis. *Development, 124*(9), 1757–1769.

Latcheva, N. K., Viveiros, J. M., & Marenda, D. R. (2019). The Drosophila chromodomain protein kismet activates steroid hormone receptor transcription to govern axon pruning and memory in vivo. *iScience, 16*, 79–93. https://doi.org/10.1016/j.isci.2019.05.021.

Laws, K. M., & Drummond-Barbosa, D. (2017). Control of germline stem cell lineages by diet and physiology. *Results and Problems in Cell Differentiation, 59*, 67–99. https://doi.org/10.1007/978-3-319-44820-6_3.

Li, T., & Bender, M. (2000). A conditional rescue system reveals essential functions for the ecdysone receptor (EcR) gene during molting and metamorphosis in Drosophila. *Development, 127*(13), 2897–2905.

Li, Y., Ma, Q., Cherry, C. M., & Matunis, E. L. (2014). Steroid signaling promotes stem cell maintenance in the Drosophila testis. *Developmental Biology, 394*(1), 129–141. https://doi.org/10.1016/j.ydbio.2014.07.016.

Li, Y., Minor, N. T., Park, J. K., McKearin, D. M., & Maines, J. Z. (2009). Bam and Bgcn antagonize Nanos-dependent germ-line stem cell maintenance. *Proceedings of the National Academy of Sciences of the United States of America, 106*(23), 9304–9309.

Li, T. R., & White, K. P. (2003). Tissue-specific gene expression and ecdysone-regulated genomic networks in Drosophila. *Developmental Cell, 5*(1), 59–72.

Lin, H., & Spradling, A. C. (1993). Germline stem cell division and egg chamber development in transplanted Drosophila germaria. *Developmental Biology, 159*(1), 140–152.

Liu, T., Wang, Q., Li, W., Mao, F., Yue, S., Liu, S., et al. (2017). Gcn5 determines the fate of Drosophila germline stem cells through degradation of Cyclin A. *The FASEB Journal, 31*(5), 2185–2194. https://doi.org/10.1096/fj.201601217R.

Lu, Y., Anderson, W. R., Zhang, H., Feng, S., & Pick, L. (2013). Functional conservation of Drosophila FTZ-F1 and its mammalian homologs suggests ligand-independent regulation of NR5A family transcriptional activity. *Development Genes and Evolution, 223*(3), 199–205. https://doi.org/10.1007/s00427-012-0435-3.

Luo, L., Wang, H., Fan, C., Liu, S., & Cai, Y. (2015). Wnt ligands regulate Tkv expression to constrain Dpp activity in the *Drosophila* ovarian stem cell niche. *The Journal of Cell Biology, 209*(4), 595–608. https://doi.org/10.1083/jcb.201409142.

Luzzo, K. M., Wang, Q., Purcell, S. H., Chi, M., Jimenez, P. T., Grindler, N., et al. (2012). High fat diet induced developmental defects in the mouse: Oocyte meiotic aneuploidy and fetal growth retardation/brain defects. *PLoS One*, 7(11), e49217. https://doi.org/10.1371/journal.pone.0049217.

Mahowald, A. P., & Kambysellis, M. P. (1980). Oogenesis. In M. Ashbumer, & T. R. F. Wright (Eds.), *The genetics and biology of Drosophila* (pp. 141–225). New York: Academic Press.

Mangelsdorf, D. J., Thummel, C., Beato, M., Herrlich, P., Schütz, G., Umesono, K., et al. (1995). The nuclear receptor superfamily: The second decade. *Cell, 83*(6), 835–839.

Manning, L., Sheth, J., Bridges, S., Saadin, A., Odinammadu, K., Andrew, D., et al. (2017). A hormonal cue promotes timely follicle cell migration by modulating transcription profiles. *Mechanisms of Development, 148*, 56–68. https://doi.org/10.1016/j.mod.2017.06.003.

Margolis, J., & Spradling, A. (1995). Identification and behavior of epithelial stem cells in the Drosophila ovary. *Development, 121*(11), 3797–3807.

Martinot, E., Sedes, L., Baptissart, M., Holota, H., Rouaisnel, B., Damon-Soubeyrand, C., et al. (2017). The bile acid nuclear receptor FXRalpha is a critical regulator of mouse germ cell fate. *Stem Cell Reports, 9*(1), 315–328. https://doi.org/10.1016/j.stemcr.2017.05.036.

Marvin, K. A., Reinking, J. L., Lee, A. J., Pardee, K., Krause, H. M., & Burstyn, J. N. (2009). Nuclear receptors homo sapiens Rev-erbbeta and Drosophila melanogaster E75 are thiolate-ligated heme proteins which undergo redox-mediated ligand switching and bind CO and NO. *Biochemistry*, *48*(29), 7056–7071. https://doi.org/10.1021/bi900697c.

Matsuoka, S., Armstrong, A. R., Sampson, L. L., Laws, K. M., & Drummond-Barbosa, D. (2017). Adipocyte metabolic pathways regulated by diet control the female germline stem cell lineage in drosophila melanogaster. *Genetics*, *206*(2), 953–971. https://doi.org/10.1534/genetics.117.201921.

McBrayer, Z., Ono, H., Shimell, M., Parvy, J. P., Beckstead, R. B., Warren, J. T., et al. (2007). Prothoracicotropic hormone regulates developmental timing and body size in Drosophila. *Developmental Cell*, *13*(6), 857–871. https://doi.org/10.1016/j.devcel.2007.11.003.

McDonald, S. I., Beachum, A. N., Hinnant, T. D., Blake, A. J., Bynum, T., Hickman, E. P., et al. (2019). Novel cis-regulatory regions in ecdysone responsive genes are sufficient to promote gene expression in Drosophila ovarian cells. *Gene Expression Patterns*, *34*, 119074.

McLaughlin, J. M., & Bratu, D. P. (2015). Drosophila melanogaster oogenesis: An overview. In *Drosophila oogenesis* (pp. 1–20). Springer.

Mendes, C. C., & Mirth, C. K. (2016). Stage-specific plasticity in ovary size is regulated by insulin/insulin-like growth factor and ecdysone signaling in Drosophila. *Genetics*, *202*(2), 703–719. https://doi.org/10.1534/genetics.115.179960.

Miller, W. L. (2008). Steroidogenic enzymes. In *Vol. 13. Disorders of the human adrenal cortex* (pp. 1–18). Karger Publishers.

Mirth, C. K., Nogueira Alves, A., & Piper, M. D. W. (2019). Turning food into eggs: Insights from nutritional biology and developmental physiology of Drosophila. *Current Opinion in Insect Science*, *31*, 49–57. https://doi.org/10.1016/j.cois.2018.08.006.

Morris, L. X., & Spradling, A. C. (2011). Long-term live imaging provides new insight into stem cell regulation and germline-soma coordination in the Drosophila ovary. *Development*, *138*(11), 2207–2215.

Morris, L. X., & Spradling, A. C. (2012). Steroid signaling within drosophila ovarian epithelial cells sex-specifically modulates early germ cell development and meiotic entry. *PLoS One*, 7(10), e46109. https://doi.org/10.1371/journal.pone.0046109.

Mottier-Pavie, V. I., Palacios, V., Eliazer, S., Scoggin, S., & Buszczak, M. (2016). The Wnt pathway limits BMP signaling outside of the germline stem cell niche in Drosophila ovaries. *Developmental Biology*, *417*(1), 50–62. https://doi.org/10.1016/j.ydbio.2016.06.038.

Musille, P. M., Pathak, M., Lauer, J. L., Griffin, P. R., & Ortlund, E. A. (2013). Divergent sequence tunes ligand sensitivity in phospholipid-regulated hormone receptors. *The Journal of Biological Chemistry*, *288*(28), 20702–20712. https://doi.org/10.1074/jbc.M113.472837.

Niwa, R., Matsuda, T., Yoshiyama, T., Namiki, T., Mita, K., Fujimoto, Y., et al. (2004). CYP306A1, a cytochrome P450 enzyme, is essential for ecdysteroid biosynthesis in the prothoracic glands of Bombyx and Drosophila. *Journal of Biological Chemistry*, *279*(34), 35942–35949.

Niwa, R., & Niwa, Y. S. (2014). Enzymes for ecdysteroid biosynthesis: Their biological functions in insects and beyond. *Bioscience, Biotechnology, and Biochemistry*, *78*(8), 1283–1292. https://doi.org/10.1080/09168451.2014.942250.

Ohlstein, B., & McKearin, D. (1997). Ectopic expression of the Drosophila Bam protein eliminates oogenic germline stem cells. *Development*, *124*(18), 3651–3662.

Okamoto, N., Viswanatha, R., Bittar, R., Li, Z., Haga-Yamanaka, S., Perrimon, N., et al. (2018). A membrane transporter is required for steroid hormone uptake in Drosophila. *Developmental Cell*, *47*(3), 294–305 e297. https://doi.org/10.1016/j.devcel.2018.09.012.

Ono, H., Rewitz, K. F., Shinoda, T., Itoyama, K., Petryk, A., Rybczynski, R., et al. (2006). Spook and Spookier code for stage-specific components of the ecdysone biosynthetic pathway in Diptera. *Developmental Biology, 298*(2), 555–570. https://doi.org/10.1016/j.ydbio.2006.07.023.

Oro, A. E., McKeown, M., & Evans, R. M. (1990). Relationship between the product of the Drosophila ultraspiracle locus and the vertebrate retinoid X receptor. *Nature, 347*(6290), 298–301. https://doi.org/10.1038/347298a0.

Oro, A., McKeown, M., & Evans, R. M. (1992). The Drosophila retinoid X receptor homolog ultraspiracle functions in both female reproduction and eye morphogenesis. *Development, 115*(2), 449–462.

Osterfield, M., Berg, C. A., & Shvartsman, S. Y. (2017). Epithelial patterning, morphogenesis, and evolution: Drosophila eggshell as a model. *Developmental Cell, 41*(4), 337–348. https://doi.org/10.1016/j.devcel.2017.02.018.

Palanker, L., Necakov, A. S., Sampson, H. M., Ni, R., Hu, C., Thummel, C. S., et al. (2006). Dynamic regulation of Drosophila nuclear receptor activity in vivo. *Development, 133*(18), 3549–3562. https://doi.org/10.1242/dev.02512.

Panth, N., Gavarkovs, A., Tamez, M., & Mattei, J. (2018). The influence of diet on fertility and the implications for public health nutrition in the United States. *Frontiers in Public Health, 6*, 211. https://doi.org/10.3389/fpubh.2018.00211.

Pardee, K., Necakov, A. S., & Krause, H. (2011). Nuclear receptors: Small molecule sensors that coordinate growth metabolism and reproduction. *Sub-Cellular Biochemistry, 52*, 123–153. https://doi.org/10.1007/978-90-481-9069-0_6.

Parra-Peralbo, E., & Culi, J. (2011). Drosophila lipophorin receptors mediate the uptake of neutral lipids in oocytes and imaginal disc cells by an endocytosis-independent mechanism. *PLoS Genetics*, 7(2), e1001297.

Parvy, J. P., Blais, C., Bernard, F., Warren, J. T., Petryk, A., Gilbert, L. I., et al. (2005). A role for betaFTZ-F1 in regulating ecdysteroid titers during post-embryonic development in Drosophila melanogaster. *Developmental Biology, 282*(1), 84–94. https://doi.org/10.1016/j.ydbio.2005.02.028.

Peercy, B. E., & Starz-Gaiano, M. (2020). Clustered cell migration: Modeling the model system of Drosophila border cells. *Seminars in Cell & Developmental Biology, 100*, 167–176. https://doi.org/10.1016/j.semcdb.2019.11.010.

Pepling, M. E. (2012). Follicular assembly: Mechanisms of action. *Reproduction, 143*(2), 139–149. https://doi.org/10.1530/rep-11-0299.

Perinthottathil, S., & Kim, C. (2011). Bam and Bgcn in Drosophila germline stem cell differentiation. *Vitamins & Hormones, 87*, 399–416. Elsevier.

Peterson, J. S., Timmons, A. K., Mondragon, A. A., & McCall, K. (2015). The end of the beginning: Cell death in the germline. *Current Topics in Developmental Biology, 114*, 93–119. https://doi.org/10.1016/bs.ctdb.2015.07.025.

Petryk, A., Warren, J. T., Marqués, G., Jarcho, M. P., Gilbert, L. I., Kahler, J., et al. (2003). Shade is the Drosophila P450 enzyme that mediates the hydroxylation of ecdysone to the steroid insect molting hormone 20-hydroxyecdysone. *Proceedings of the National Academy of Sciences of the United States of America, 100*(24), 13773–13778. https://doi.org/10.1073/pnas.2336088100.

Pivonello, C., Muscogiuri, G., Nardone, A., Garifalos, F., Provvisiero, D. P., Verde, N., et al. (2020). Bisphenol A: An emerging threat to female fertility. *Reproductive Biology and Endocrinology, 18*(1), 22. https://doi.org/10.1186/s12958-019-0558-8.

Port, F., Strein, C., Stricker, M., Rauscher, B., Heigwer, F., Zhou, J., et al. (2020). A large-scale resource for tissue-specific CRISPR mutagenesis in Drosophila. *eLife, 9*, e53865. https://doi.org/10.7554/eLife.53865.

Purcell, S. H., & Moley, K. H. (2011). The impact of obesity on egg quality. *Journal of Assisted Reproduction and Genetics, 28*(6), 517–524. https://doi.org/10.1007/s10815-011-9592-y.

Quinlan, M. E. (2016). Cytoplasmic streaming in the Drosophila oocyte. *Annual Review of Cell and Developmental Biology*, *32*, 173–195. https://doi.org/10.1146/annurev-cellbio-111315-125416.

Rajan, A., & Perrimon, N. (2011). Drosophila as a model for interorgan communication: Lessons from studies on energy homeostasis. *Developmental Cell*, *21*(1), 29–31. https://doi.org/10.1016/j.devcel.2011.06.034.

Reinking, J., Lam, M. M., Pardee, K., Sampson, H. M., Liu, S., Yang, P., et al. (2005). The Drosophila nuclear receptor e75 contains heme and is gas responsive. *Cell*, *122*(2), 195–207. https://doi.org/10.1016/j.cell.2005.07.005.

Rewitz, K. F., Yamanaka, N., Gilbert, L. I., & O'Connor, M. B. (2009). The insect neuropeptide PTTH activates receptor tyrosine kinase torso to initiate metamorphosis. *Science*, *326*(5958), 1403–1405. https://doi.org/10.1126/science.1176450.

Richards, G. (1981). The radioimmune assay of ecdysteroid titres in Drosophila melanogaster. *Molecular and Cellular Endocrinology*, *21*(3), 181–197.

Roa, J., & Tena-Sempere, M. (2014). Connecting metabolism and reproduction: Roles of central energy sensors and key molecular mediators. *Molecular and Cellular Endocrinology*, *397*(1–2), 4–14. https://doi.org/10.1016/j.mce.2014.09.027.

Robertson, C. W. (1936). The metamorphosis of Drosophila melanogaster, including an accurately timed account of the principal morphological changes. *Journal of Morphology*, *59*(2), 351–399.

Romani, P., Bernardi, F., Hackney, J., Dobens, L., Gargiulo, G., & Cavaliere, V. (2009). Cell survival and polarity of Drosophila follicle cells require the activity of ecdysone receptor B1 isoform. *Genetics*, *181*(1), 165–175.

Romani, P., Gargiulo, G., & Cavaliere, V. (2016). The ecdysone receptor signalling regulates microvilli formation in follicular epithelial cells. *Cellular and Molecular Life Sciences*, *73*(2), 409–425. https://doi.org/10.1007/s00018-015-1999-7.

Rust, K., & Nystul, T. (2020). Signal transduction in the early *Drosophila follicle* stem cell lineage. *Current Opinion in Insect Science*, *37*, 39–48. https://doi.org/10.1016/j.cois.2019.11.005.

Schauer, S., Callender, J., Henrich, V. C., & Spindler-Barth, M. (2011). The N-terminus of ecdysteroid receptor isoforms and ultraspiracle interacts with different ecdysteroid response elements in a sequence specific manner to modulate transcriptional activity. *The Journal of Steroid Biochemistry and Molecular Biology*, *124*(3–5), 84–92. https://doi.org/10.1016/j.jsbmb.2011.01.013.

Schonbaum, C. P., Perrino, J. J., & Mahowald, A. P. (2000). Regulation of the vitellogenin receptor during Drosophila melanogaster oogenesis. *Molecular Biology of the Cell*, *11*(2), 511–521. https://doi.org/10.1091/mbc.11.2.511.

Schubiger, M., Carré, C., Antoniewski, C., & Truman, J. W. (2005). Ligand-dependent de-repression via EcR/USP acts as a gate to coordinate the differentiation of sensory neurons in the Drosophila wing. *Development*, *132*(23), 5239–5248. https://doi.org/10.1242/dev.02093.

Schubiger, M., Tomita, S., Sung, C., Robinow, S., & Truman, J. W. (2003). Isoform specific control of gene activity in vivo by the Drosophila ecdysone receptor. *Mechanisms of Development*, *120*(8), 909–918. https://doi.org/10.1016/s0925-4773(03)00134-5.

Schubiger, M., & Truman, J. W. (2000). The RXR ortholog USP suppresses early metamorphic processes in Drosophila in the absence of ecdysteroids. *Development*, *127*(6), 1151–1159.

Schwartz, M. B., Kelly, T. J., Imberski, R. B., & Rubenstein, E. C. (1985). The effects of nutrition and methoprene treatment on ovarian ecdysteroid synthesis in Drosophila melanogaster. *Journal of Insect Physiology*, *31*(12), 947–957.

Shen, R., Weng, C., Yu, J., & Xie, T. (2009). eIF4A controls germline stem cell self-renewal by directly inhibiting BAM function in the Drosophila ovary. *Proceedings of the National Academy of Sciences of the United States of America*, *106*(28), 11623–11628.

Shlyueva, D., Stelzer, C., Gerlach, D., Yáñez-Cuna, J. O., Rath, M., Boryń, Ł. M., et al. (2014). Hormone-responsive enhancer-activity maps reveal predictive motifs, indirect repression, and targeting of closed chromatin. *Molecular Cell*, *54*(1), 180–192.

Sieber, M. H., & Spradling, A. C. (2015). Steroid signaling establishes a female metabolic state and regulates SREBP to control oocyte lipid accumulation. *Current Biology*, *25*(8), 993–1004.

Sieber, M. H., & Spradling, A. C. (2017). The role of metabolic states in development and disease. *Current Opinion in Genetics & Development*, *45*, 58–68. https://doi.org/10.1016/j.gde.2017.03.002.

Slaidina, M., Banisch, T. U., Gupta, S., & Lehmann, R. (2020). A single-cell atlas of the developing Drosophila ovary identifies follicle stem cell progenitors. *Genes & Development*, *34*(3–4), 239–249. https://doi.org/10.1101/gad.330464.119.

Soller, M., Bownes, M., & Kubli, E. (1999). Control of oocyte maturation in sexually mature Drosophila females. *Developmental Biology*, *208*(2), 337–351.

Song, X., Wong, M. D., Kawase, E., Xi, R., Ding, B. C., McCarthy, J. J., et al. (2004). Bmp signals from niche cells directly repress transcription of a differentiation-promoting gene, *bag of marbles*, in germline stem cells in the *Drosophila* ovary. *Development*, *131*(6), 1353–1364. https://doi.org/10.1242/dev.01026.

Song, X., Zhu, C.-H., Doan, C., & Xie, T. (2002). Germline stem cells anchored by adherens junctions in the Drosophila ovary niches. *Science*, *296*(5574), 1855–1857.

Stoiber, M., Celniker, S., Cherbas, L., Brown, B., & Cherbas, P. (2016). Diverse hormone response networks in 41 independent drosophila cell lines. *G3: Genes | Genomes | Genetics*, *6*(3), 683–694. https://doi.org/10.1534/g3.115.023366.

Sun, J., Smith, L., Armento, A., & Deng, W. M. (2008). Regulation of the endocycle/gene amplification switch by Notch and ecdysone signaling. *The Journal of Cell Biology*, *182*(5), 885–896. https://doi.org/10.1083/jcb.200802084.

Sun, J., & Spradling, A. C. (2012). NR5A nuclear receptor Hr39 controls three-cell secretory unit formation in Drosophila female reproductive glands. *Current Biology*, *22*(10), 862–871. https://doi.org/10.1016/j.cub.2012.03.059.

Szakmary, A., Cox, D. N., Wang, Z., & Lin, H. (2005). Regulatory relationship among piwi, pumilio, and bag-of-marbles in Drosophila germline stem cell self-renewal and differentiation. *Current Biology*, *15*(2), 171–178.

Talbot, W. S., Swyryd, E. A., & Hogness, D. S. (1993). Drosophila tissues with different metamorphic responses to ecdysone express different ecdysone receptor isoforms. *Cell*, *73*(7), 1323–1337.

Talmor, A., & Dunphy, B. (2015). Female obesity and infertility. *Best Practice & Research. Clinical Obstetrics & Gynaecology*, *29*(4), 498–506. https://doi.org/10.1016/j.bpobgyn.2014.10.014.

Terashima, J., & Bownes, M. (2006). E75A and E75B have opposite effects on the apoptosis/development choice of the Drosophila egg chamber. *Cell Death and Differentiation*, *13*(3), 454.

Terashima, J., Takaki, K., Sakurai, S., & Bownes, M. (2005). Nutritional status affects 20-hydroxyecdysone concentration and progression of oogenesis in Drosophila melanogaster. *The Journal of Endocrinology*, *187*(1), 69–79. https://doi.org/10.1677/joe.1.06220.

Thomas, H. E., Stunnenberg, H. G., & Stewart, A. F. (1993). Heterodimerization of the Drosophila ecdysone receptor with retinoid X receptor and ultraspiracle. *Nature*, *362*(6419), 471–475. https://doi.org/10.1038/362471a0.

Tiwari, M. D., Zeitler, D. M., Meister, G., & Wodarz, A. (2019). Molecular profiling of stem cell-like female germ line cells in Drosophila delineates networks important for stemness and differentiation. *Biology Open*, *8*(11), bio046789. https://doi.org/10.1242/bio.046789.

Truman, J. W. (2019). The evolution of insect metamorphosis. *Current Biology*, *29*(23), R1252–R1268. https://doi.org/10.1016/j.cub.2019.10.009.

Tu, M. P., Yin, C. M., & Tatar, M. (2002). Impaired ovarian ecdysone synthesis of Drosophila melanogaster insulin receptor mutants. *Aging Cell*, *1*(2), 158–160. https://doi.org/10.1046/j.1474-9728.2002.00016.x.

Tzolovsky, G., Deng, W.-M., Schlitt, T., & Bownes, M. (1999). The function of the broad-complex during Drosophila melanogaster oogenesis. *Genetics*, *153*(3), 1371–1383.

Uryu, O., Ameku, T., & Niwa, R. (2015). Recent progress in understanding the role of ecdysteroids in adult insects: Germline development and circadian clock in the fruit fly Drosophila melanogaster. *Zoological Letters*, *1*, 32. https://doi.org/10.1186/s40851-015-0031-2.

Uyehara, C. M., & McKay, D. J. (2019). Direct and widespread role for the nuclear receptor EcR in mediating the response to ecdysone in Drosophila. *Proceedings of the National Academy of Sciences*, *116*(20), 9893–9902.

Uyehara, C. M., Nystrom, S. L., Niederhuber, M. J., Leatham-Jensen, M., Ma, Y., Buttitta, L. A., et al. (2017). Hormone-dependent control of developmental timing through regulation of chromatin accessibility. *Genes & Development*, *31*(9), 862–875. https://doi.org/10.1101/gad.298182.117.

Wang, Z., & Lin, H. (2004). Nanos maintains germline stem cell self-renewal by preventing differentiation. *Science*, *303*(5666), 2016–2019.

Wang, X., Wang, H., Liu, L., Li, S., Emery, G., & Chen, J. (2020). Temporal coordination of collective migration and lumen formation by antagonism between two nuclear receptors. *iScience*, *23*(7), 101335. https://doi.org/10.1016/j.isci.2020.101335.

Warren, J. T., Petryk, A., Marques, G., Parvy, J. P., Shinoda, T., Itoyama, K., et al. (2004). Phantom encodes the 25-hydroxylase of Drosophila melanogaster and Bombyx mori: A P450 enzyme critical in ecdysone biosynthesis. *Insect Biochemistry and Molecular Biology*, *34*(9), 991–1010. https://doi.org/10.1016/j.ibmb.2004.06.009.

Warren, J. T., Wismar, J., Subrahmanyam, B., & Gilbert, L. I. (2001). Woc (without children) gene control of ecdysone biosynthesis in Drosophila melanogaster. *Molecular and Cellular Endocrinology*, *181*(1–2), 1–14.

Warren, J. T., Yerushalmi, Y., Shimell, M. J., O'Connor, M. B., Restifo, L. L., & Gilbert, L. I. (2006). Discrete pulses of molting hormone, 20-hydroxyecdysone, during late larval development of Drosophila melanogaster: Correlations with changes in gene activity. *Developmental Dynamics: An Official Publication of the American Association of the Anatomists*, *235*(2), 315–326.

Weaver, L. N., & Drummond-Barbosa, D. (2019). The nuclear receptor seven up functions in adipocytes and oenocytes to control distinct steps of Drosophila oogenesis. *Developmental Biology*, *456*(2), 179–189. https://doi.org/10.1016/j.ydbio.2019.08.015.

Weikum, E. R., Liu, X., & Ortlund, E. A. (2018). The nuclear receptor superfamily: A structural perspective. *Protein Science*, *27*(11), 1876–1892. https://doi.org/10.1002/pro.3496.

Wilcockson, S. G., & Ashe, H. L. (2019). Drosophila ovarian germline stem cell cytocensor projections dynamically receive and attenuate BMP signaling. *Developmental Cell*, *50*(3), 296–312 e295. https://doi.org/10.1016/j.devcel.2019.05.020.

Wilk, R., Hu, J., & Krause, H. M. (2013). Spatial profiling of nuclear receptor transcription patterns over the course of Drosophila development. *G3 (Bethesda)*, *3*(7), 1177–1189. https://doi.org/10.1534/g3.113.006023.

Wismar, J., Habtemichael, N., Warren, J. T., Dai, J. D., Gilbert, L. I., & Gateff, E. (2000). The mutation without children(rgl) causes ecdysteroid deficiency in third-instar larvae of Drosophila melanogaster. *Developmental Biology*, *226*(1), 1–17. https://doi.org/10.1006/dbio.2000.9811.

Woodard, C. T., Baehrecke, E. H., & Thummel, C. S. (1994). A molecular mechanism for the stage specificity of the Drosophila prepupal genetic response to ecdysone. *Cell*, *79*(4), 607–615. https://doi.org/10.1016/0092-8674(94)90546-0.

Xie, T., & Spradling, A. C. (1998). Decapentaplegic is essential for the maintenance and division of germline stem cells in the Drosophila ovary. *Cell*, *94*(2), 251–260.

Yamada, M., Murata, T., Hirose, S., Lavorgna, G., Suzuki, E., & Ueda, H. (2000). Temporally restricted expression of transcription factor betaFTZ-F1: Significance for embryogenesis, molting and metamorphosis in Drosophila melanogaster. *Development*, *127*(23), 5083–5092.

Yan, Y. L., & Postlethwait, J. H. (1990). Vitellogenesis in Drosophila: Sequestration of a yolk polypeptide/invertase fusion protein into developing oocytes. *Developmental Biology*, *140*(2), 281–290. https://doi.org/10.1016/0012-1606(90)90078-w.

Yao, T.-P., Forman, B. M., Jiang, Z., Cherbas, L., Chen, J. D., McKeown, M., et al. (1993). Functional ecdysone receptor is the product of EcR and Ultraspiracle genes. *Nature*, *366*(6454), 476–479. https://doi.org/10.1038/366476a0.

Yatsenko, A. S., & Shcherbata, H. R. (2018). Stereotypical architecture of the stem cell niche is spatiotemporally established by miR-125-dependent coordination of Notch and steroid signaling. *Development*, *145*(3), dev159178. https://doi.org/10.1242/dev.159178.

Yoo, J., Ko, S., Kim, H., Sampson, H., Yun, J. H., Choe, K. M., et al. (2011). Crystal structure of Fushi tarazu factor 1 ligand binding domain/Fushi tarazu peptide complex identifies new class of nuclear receptors. *The Journal of Biological Chemistry*, *286*(36), 31225–31231. https://doi.org/10.1074/jbc.M111.252916.

Zhang, C., Robinson, B. S., Xu, W., Yang, L., Yao, B., Zhao, H., et al. (2015). The ecdysone receptor coactivator Taiman links Yorkie to transcriptional control of germline stem cell factors in somatic tissue. *Developmental Cell*, *34*(2), 168–180.

Zhang, R., Wesevich, V., Chen, Z., Zhang, D., & Kallen, A. N. (2020). Emerging roles for noncoding RNAs in female sex steroids and reproductive disease. *Molecular and Cellular Endocrinology*, *518*, 110875. https://doi.org/10.1016/j.mce.2020.110875.

Ziegler, R., & Van Antwerpen, R. (2006). Lipid uptake by insect oocytes. *Insect Biochemistry and Molecular Biology*, *36*(4), 264–272.

CHAPTER TWELVE

Steroid hormones and first trimester vascular remodeling

V.A. Cañumil, E. Bogetti, F.L. de la Cruz Borthiry, M.L. Ribeiro, and J.S. Beltrame*

Center of Pharmacological and Botanical Studies (CEFyBO), School of Medicine University of Buenos Aires (UBA)—National Research Council (CONICET), Buenos Aires, Argentina

*Corresponding author: e-mail address: jbeltrame@fmed.uba.ar

Contents

Abstract

Successful implantation and placentation require neoangiogenesis and the remodeling of the uterine spiral arteries. Progesterone and estradiol control various of the placental functions, but their role in vascular remodeling remains controversial. Therefore, this chapter aims to summarize the current knowledge regarding the role of steroid hormones in the uteroplacental vascular remodeling during the first trimester of gestation.

Introduction

Embryo implantation is a critical step for the establishment and maintenance of a healthy pregnancy (Cha, Sun, & Dey, 2012; Wang & Dey, 2006). In humans, the apposition of the blastocyst in the maternal endometrium defines the onset of the implantation process. In a unique period called "window of implantation," the maternal endometrium is receptive and interacts with the active embryo. Once the blastocyst apposes and invades

Vitamins and Hormones, Volume 116
ISSN 0083-6729
https://doi.org/10.1016/bs.vh.2021.02.003

the uterine epithelium, develops stable communication to establish the maternal-fetal interface (Pijnenborg, Dixon, Robertson, & Brosens, 1980).

A successful pregnancy requires angiogenesis, spiral artery remodeling, and later placentation. In mammals, the placenta ensures the exchange in nutrients and blood flow during gestation. Deregulation of the vascular adaptations of the maternal-fetal interface has a major clinical impact resulting in several pregnancy complications such as implantation failure and preeclampsia (Plaisier et al., 2009; Zhou, Damsky, & Fisher, 1997).

The processes that take place at the maternal-fetal interface are synchronized by the master hormones that regulate the majority of the reproductive events, progesterone (P4) and estradiol (E2). P4 and E2 regulate the development of the neovasculature during early pregnancy (Clark, Pru, & Pru, 2017) and modulate uterine blood flow that follows the remodeling of the uterine arteries and placental angiogenesis (Chen, Sheehan, Brennecke, & Keogh, 2012; Maliqueo, Echiburú, & Crisosto, 2016). This chapter aims to summarize the current knowledge regarding the role of steroid hormones in the uterine-placental vascular remodeling during the first trimester of gestation.

1. Vascular remodeling at the maternal-fetal interface

Uterine blood supply consists of multiple branches that decreases in diameter while invading the endometrium and myometrium. During pregnancy, the vascular bed of the uterus changes drastically, where new vessels are formed and the existing vessels dilate. Adequate uterine blood flow as a result of angiogenesis and spiral artery remodeling is required to ensure the optimal transfer of nutrients and oxygen to the fetus.

The formation of new vessels from pre-existing ones is called neovascularization, or angiogenesis. Throughout implantation and the invasion of the stroma by trophoblast cells, an extensive angiogenesis process occurs (Pepe & Albrecht, 1995) resulting in a large increase of blood flow to the uterus. This is a unique feature. In adults, the variations in the blood flow are in response to local vasodilatation or vasoconstriction. However, the increase in uterine blood flow during pregnancy is mainly due to the formation of the placental vascular bed (Reynolds & Redmer, 2001). The classic angiogenic process begins when endothelial cells become motile and protrude as a result of a proangiogenic stimulus.

Thus, endothelial migration is one of the earlier steps in the angiogenic cascade. The independent motility cell, denominated tip cell, originates a precise configuration with leader and follower cells (Potente, Gerhardt, & Carmeliet, 2011). Tips cells follow a chemotactic gradient and commander new sprouts. Then followers' cells (stalk cells) establish a lumen and proliferate to sustain sprouting elongation. When the environmental signals decline, the sprouting process finishes and quiescence is re-established (Nowak-Sliwinska, Alitalo, Allen, et al., 2018; Potente et al., 2011).

Blood supply into the uterus is established by several branches of the uterine and ovarian arteries (Fig. 1). Afterward, small radial arteries enter the endometrium, ramified and penetrate two-thirds of the myometrium through the spiral arteries (Espinoza et al., 2006; Osol & Mandala, 2009). Human spiral arteries comprise a single endothelial layer upon an elastic lamina surrounded by vascular smooth muscle cells with the associated extracellular matrix. These strained and rigid coiled-appearance vessels are transformed into larger, low-resistance vessels that allow the significant increase in blood flow detailed before (Demir, Yaba, & Huppertz, 2010; Harris et al., 2006; Leach et al., 2006). As a result of the spiral artery remodeling, vascular resistance decreases. Particularly in humans, trophoblast differentiates and allows these fetal and maternal adaptations to pregnancy (Liu et al., 2018; Vento-Tormo et al., 2018). The precursors of the differentiated cells of the placenta are trophoblasts stem cells that give rise to both syncytiotrophoblast and cytotrophoblast (Roberts & Fisher, 2011). Then, extravillous trophoblast differentiates into endovascular trophoblast to remodel the spiral arteries. In this process, the endovascular trophoblast invades maternal spiral arteries, disrupts the endothelium interactions, and replaces endothelial and muscle cells of these vessels. These adaptations of the maternal arteries allow low-pressure intervillous flow and ensure adequate blood flow in response to the increased metabolic demands of the embryo (Demir et al., 2010).

Accumulating evidence suggests an active regulation of sex steroids in the vascular remodeling of the maternal-fetal interface during early gestation. The potential role of placental hormones as possible biomarkers to predict certain pregnancy-related pathologies has become an active focus of research. The involvement of steroid hormones in angiogenesis (Section 2) and spiral artery remodeling (Section 3) at the maternal-fetal interface will be reviewed in the next sections.

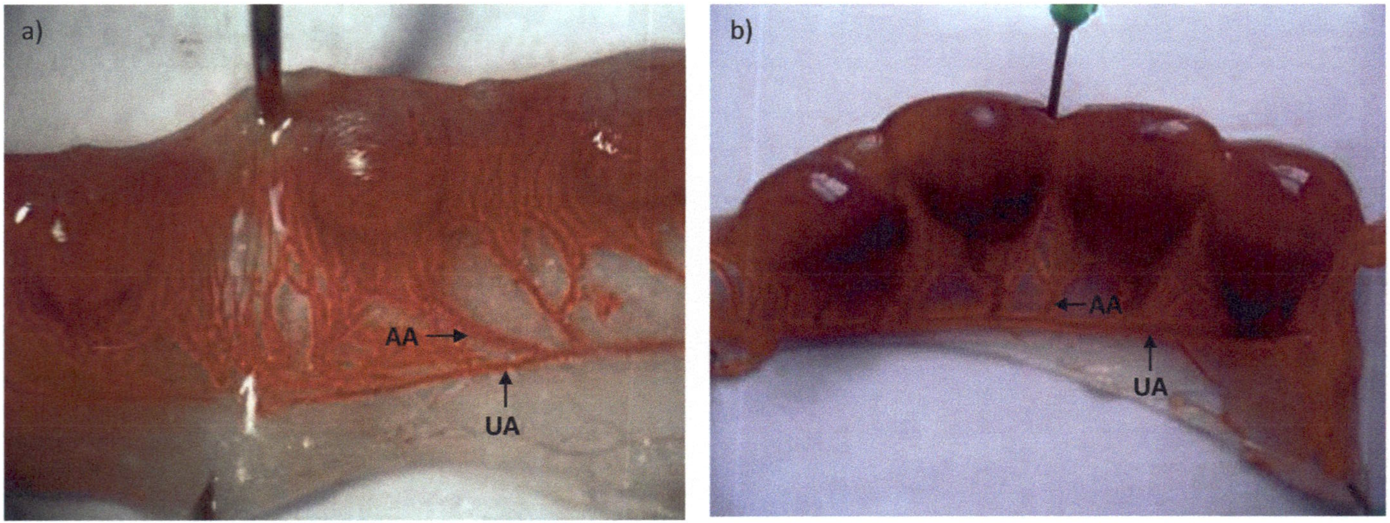

Fig. 1 Macro-vasculature of the implantation sites of the rat. Blood supply into the uterus is established by several branches of the uterine and ovarian arteries. The first branch of the uterine artery is the arcuate artery. Afterward, small radial arteries enter the myometrium, ramified and penetrate two-thirds of the endometrium through the spiral arteries. Throughout gestation, an increment in uterine blood flow and vascular remodeling is mandatory for a healthy pregnancy. (A) Rat uterus in day 8 of gestation (B) rat uterus in day 15 of gestation. UA: uterine artery. AA: arcuate artery.

2. Sex steroids and maternal uterine vasculature

As we mentioned, the extensive rearrangement in the structure and function of the uterine vasculature facilitates the increase in uteroplacental blood flow. The role of sex hormones in the onset of these changes has been supported by several studies using different experimental approaches (Mandala, 2020). However, their specific role in vascular remodeling remains controversial (Chen et al., 2012; Maliqueo et al., 2016).

Steroid hormones are present in extremely high concentrations in the maternal circulation and orchestrate most of the reproductive events. These elevated plasma concentrations found during gestation lead to different theories given to explain this new endocrine environment that may influence the maternal-fetal vascular adaptations. Following ovulation, P4 regulates the decidualization process, where stromal fibroblasts differentiate into decidual cells (Aplin, Myers, Timms, & Westwood, 2020; Leitao et al., 2010). These specialized cells are involved in the regulation of the synchronized events that prepare the maternal vessels to the arrival of the trophoblast. Different studies showed that the decidualization process can be induced in the absence of the estrogen receptor α (ERα) (Curtis, Goulding, Eddy, & Korach, 2002; Paria, Tan, Lubahn, Dey, & Das, 1999), suggesting that is mainly a progesterone-dependent response. Also, P4 augments uterine natural killer (uNK) cells by an indirect mechanism (Bulmer & Lash, 2005). uNK cells secrete a wide range of angiogenic mediators during early pregnancy (i.e., vascular endothelial growth factor) (Hanna et al., 2006; Wallace, Fraser, & Cartwright, 2012) and are localized in the decidua near the maternal vessels prior to trophoblast invasion (Smith, Dunk, Aplin, Harris, & Jones, 2009). Fraser and collaborators determine a functional role for decidual NK cells in the disruption of endothelial structures and suggest that an impairment of this process may contribute to the reduced vessel remodeling in pregnancies with a high uterine artery resistance index.

Different studies observed downregulation of the ERα in decidual cells surrounding the embryo, suggesting a role of ER on the implantation window in mice (Ma, Song, Das, Paria, & Dey, 2003; Robertshaw, Bian, & Das, 2016). In the endometrial receptivity, E2 seems to be essential for a P4 primed uterus to the receptive state (Ma et al., 2003; Paulson, 2011). During the window of implantation and in early pregnancy, the corpus luteum is the main source of hormone production. Then, from 6 to 8 weeks of gestation until labor, the placenta progressively becomes the main source of sex hormone in humans (Costa, 2016; Edey et al., 2018; Napso, Yong, Lopez-Tello, & Sferruzzi-Perri, 2018). In contrast, rodents' corpus luteum produces steroid hormones until the end of pregnancy (Goyeneche, Deis, Gibori, & Telleria, 2003).

Following blastocyst attachment, decidual and myometrial uterine vascular remodeling begins. Uteroplacental blood flow increases in pregnant women up to 450–800 mL/min, being more than 1000% higher than in nonpregnant women (baseline value of 20–50 mL/min) (Osol & Mandala, 2009; Palmer et al., 1992). To support this increase in blood flow, profound uterine vascular changes are needed, while cytotrophoblast cells invade into maternal spiral arteries and trigger spiral artery remodeling (Section 3). The group of Habara et al. (2002) observed elevated blood flow resistance in uterine arteries of women with unexplained recurrent pregnancy loss using Doppler ultrasonography. This technique became an important tool for noninvasive determinations of the uterine blood flow and nowadays is used to detect increased vascular resistance in the uterine arteries. Reconciling the available data, sex steroids seem to modulate the uterine blood flow overregulation of uterine vessels and placental angiogenesis (Maliqueo et al., 2016). Interestingly, uterine arteries express receptors for both P4 and E2 and are reactive to these steroids (de Ziegler, Bessis, & Frydman, 1991; Osol & Mandala, 2009). E2 promotes vasodilatation of uterine and placental arteries, suggesting a role in the regulation of human uteroplacental blood flow (Corcoran et al., 2014). Moreover, an increase in the percentage of the vascular area and density of vessels are observed under E2 action in placental tissue in different models (Albrecht & Pepe, 2010; Albrecht, Robb, & Pepe, 2004; Robb, Pepe, & Albrecht, 2004). Also, Jobe et al. (2010) suggest its participation in the angiogenic process, as E2 induces human umbilical vein endothelial cell line (HUVEC) proliferation through classical ERα and ERβ receptors. In this sense, E2 enhances the proliferation and migration of HUVEC cell line (Morales et al., 1995). The mechanisms involved in these angiogenic effects of E2 seems to be regulated by the vascular endothelial growth factor (VEGF) and nitric oxide (NO) (Kim, Moriarty, & Bender, 2008). Furthermore, the renin-angiotensin system, a hormone system that regulates blood pressure, is regulated by E2 (Pepe & Albrecht, 1995).

The physiological effects of steroid hormones are mediated predominately by nuclear receptors. P4 binds at least to five different isoforms of the nuclear progesterone receptor (PR), being PR-A and PR-B the most commonly expressed in different tissues (Li & O'Malley, 2003). After ligand binding, P4 receptors dimerize and activate downstream different signaling cascades. In human tissues, PR-A and PR-B are generally expressed at similar levels. During early and late gestation, PR protein is present in the placenta (Shanker & Rao, 1999) and endothelial cells of the decidual tissue (Maliqueo et al., 2016).

Studies performed in mice show implantation failure and decidualization defects in PR-A knockout females. In contrast, PR-B knockout mice seem to be reproductive-normal having other defects not related to the implantation process (Mulac-Jericevic, Lydon, DeMayo, & Conneely, 2003; Mulac-Jericevic, Mullinax, DeMayo, Lydon, & Conneely, 2000). Similar to P4, the classical E2 signaling pathway is through the nuclear receptors, ERα and ERβ (Mandala, 2020). These receptors undergo a conformational change that allows interaction with coactivators that dimerize after ligand binding and activates target genes. Different studies demonstrate that the distribution of both receptors differs. ERα has a broad expression pattern being predominantly expressed in the uterus, mammary gland, testis, pituitary, liver, kidney, heart, and skeletal muscles, whereas ERβ is expressed in the ovary, prostate, epididymis, lung, and hypothalamus. These tissue distribution and expression pattern might explain the phenotypic differences observed in knockout mice for the estrogen receptors: ERα knockout mice are infertile (Curtis et al., 2002; Lubahn et al., 1993), while ERβ knockout female mice exhibit reduced fertility (Krege et al., 1998). Besides, the human placenta expresses the ERα and ERβ protein in early gestation and at term (Bechi et al., 2006; Bukovsky et al., 2003). Particularly, ERα is present mainly in the villous cytotrophoblast, whereas ERβ is expressed by the syncytiotrophoblast (Bechi et al., 2006).

Proper reproductive functions depend also on the levels of hormone production. It has been observed that elevated hormonal states associated with fertility treatments are correlated to adverse obstetric and perinatal outcomes (Sun et al., 2019). Indeed, the low pregnancy success rate of *in vitro* fertilization related to implantation failure as a result of uterine refractoriness could be explained by high P4 and E2 levels due to ovarian hyperstimulation programs (Fatemi & Van Vaerenbergh, 2015; Ma et al., 2003). Moreover, Meyer et al. (2019) observed that female mice exposed to high E2 doses have smaller implantation sizes at day 8 of gestation, and the embryos die intrauterine at day 10 of gestation. On the other hand, hormones found in low levels in plasma are also associated with poor pregnancy outcomes. It has been observed that women with preeclampsia have lower circulating levels of E2 compared with normotensive pregnancy (Wan et al., 2018). In addition, as reported by Murphy, Smith, Giles, and Clifton (2006), low estrogen concentrations were associated with shallow trophoblast invasion and insufficient placental development, suggesting a specific role of the steroid hormones in the maintenance of the proper fetal growth through the supply of oxygen and nutrients.

All these data support the notion that a proper function of the vascular adaptations at the maternal-fetal interface depends on concentration levels, mode and timing of P4 and E2 production, as well as localization and expression of sex hormones receptors.

3. Spiral artery remodeling

Multiple factors intervene during the early stages of gestation to achieve an adequate blood flow into the uterus. In particular, as observed in Fig. 2, spiral artery remodeling includes: (1) invasion and migration of extravillous trophoblast, (2) transformation of extravillous cytotrophoblast into an endothelial-like phenotype, often referred to as endovascular trophoblast, (3) subsequent replacement of the endothelial and muscle layers, and (4) fibrinoid deposition.

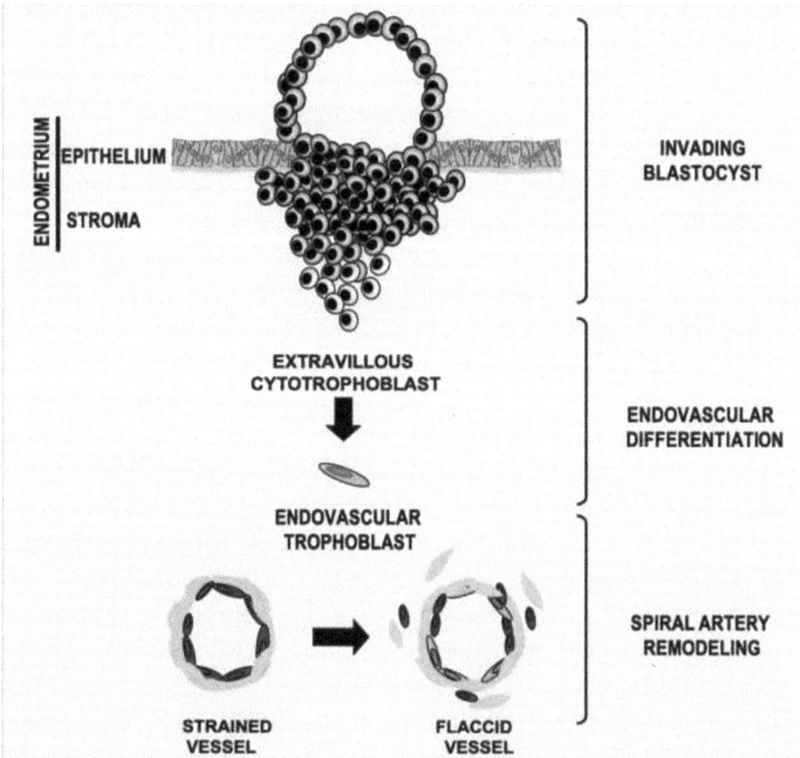

Fig. 2 Spiral artery remodeling. The strained coiled-appearance spiral arteries are transformed into larger, low-resistance vessels by a mechanism called spiral artery remodeling. The extravillous trophoblast differentiates into endovascular trophoblast to remodel these vessels. Later, the endovascular trophoblast invades maternal spiral arteries, disrupts the endothelium interactions, and replaces endothelial and muscle lining. These adaptations ensure adequate blood flow in response to the increased metabolic demands of the embryo.

After attachment of the embryo to the uterus, the main goal of trophoblast migration and invasion is to reach the maternal spiral arteries. The precursors of the differentiated cells of the placenta are a proliferative population called trophoblasts stem cells which give rise to both syncytiotrophoblast and cytotrophoblast (Roberts & Fisher, 2011). In humans, trophectoderm first undergoes expansion and differentiation into villous trophoblast and extravillous trophoblast. Cytotrophoblast cells are localized adjacent to the villous mesenchyme, while extravillous cytotrophoblast are in the cell column and invading the uterine wall (Chang, Wakeland, & Parast, 2018). The villous cytotrophoblast cells fuse and form the multinucleated syncytiotrophoblast, which has primarily secretory functions (Maltepe & Fisher, 2015). On the other hand, the extravillous trophoblast is involved in the anchoring of the fetoplacental unit to the uterine wall. Interestingly, extravillous trophoblast in the uterine wall could have two distinct fates: interstitial or endovascular trophoblast (Da Silva-Arnold, James, Al-Khan, Zamudio, & Illsley, 2015; Soncin, Natale, & Parast, 2015). The endovascular extravillous trophoblast is the main actor in spiral artery remodeling (Kam, Gardner, Loke, & King, 1999; Zhou, Fisher, et al., 1997).

Particularly, the presence of the endovascular cytotrophoblast has been detected in the spiral arteries from 8 weeks of gestation being more preponderant at 10 weeks (Pijnenborg et al., 1980). The endovascular trophoblast disrupts the endothelial cells barrier of these vessels and replaces the endothelial and muscle lining (Demir et al., 2010). The process finally ends when the extracellular matrix is replaced by fibrinoid (Whitley & Cartwright, 2009, 2010). Approximately 150 human spiral arteries acquire greater caliber and blood flow with low resistance, shifting the average luminal size from 200 μm to 2 mm (Benirschke & Kaufmann, 2000; Boyd & Hamilton, 1970; Lyall, 2005).

Remarkably, the physiological transformations of the spiral arteries seem to be more likely to happen in the center in comparison to the periphery of the placental bed (Brosens, Robertson, & Dixon, 1967). Besides the remodeling of the spiral arteries, the extravillous trophoblast also invades and remodels uterine glands, veins and lymphatic vessels (Moser et al., 2010; Moser, Weiss, Gauster, Sundl, & Huppertz, 2015; Windsperger et al., 2017).

Endothelial structures are disturbed in pregnant women with high resistance rates, an index related to the degree of the spiral artery remodeling (Fraser, Whitley, Thilaganathan, & Cartwright, 2015). Indeed, endothelial cells from preeclamptic patients are more susceptible to the arrival of the trophoblast than in normal pregnancies (Wang et al., 2004).

It is interesting to mention that in some pathologies associated with poor vascular remodeling, the presence of the interstitial cytotrophoblast but not endovascular trophoblast is observed in the spiral arteries. Also, the absence of vascular changes in the spiral arteries is a feature of some obstetric pathologies as fetal growth restriction and preeclampsia. These data suggest that failures in vascular remodeling could not be explained only with an aberrant trophoblast invasion (Pijnenborg, Vercruysse, Verbist, & Van Assche, 1998) and demonstrate that spiral artery remodeling is a multi-factor mechanism.

The main problem that arises when investigating the physiopathology of gestation is the impossibility of obtaining samples of first trimester placenta given the existing legislation in some countries. Besides, the amounts of primary trophoblast in the first trimester placenta are limited and are usually contaminated with other cell types. Also, trophoblast primary cells rapidly cease their proliferation and their transfection is difficult (Graham et al., 1993; Knöfler, 2010). Also, interspecies variability restricts the study of the role of sex hormones during spiral artery remodeling and placentation. In humans, the placenta is the main source of hormone production (Costa, 2016; Edey et al., 2018; Malassine, Frendo, & Evain-Brion, 2003). In contrast, the corpus luteum of the rodents produce steroid hormones along with pregnancy. Also, Albrecht and Pepe (1990) and Pepe and Albrecht (1995) observed that the human trophoblast secretes P4 and E2 in higher amounts than other mammals. To date, most studies have focused on the uteroplacental blood flow and angiogenesis, in part because of the impossibility mentioned above and in part due to the availability of non-invasive techniques as arteries and umbilical Doppler ultrasonography. *In vitro* models have been developed to overcome these limitations. Immortalized trophoblast cell lines obtained from first trimester elective termination of pregnancy (e.g., HTR-8/SVneo cells or SWAN-71) have been widely used as models to study the contribution of the human trophoblast in spiral artery remodeling.

3.1 Sex steroid and spiral artery remodeling

3.1.1 Trophoblast invasion and migration

Adequate trophoblast invasion and migration is required to spiral artery remodeling. Trophoblast cells differentiate into villous or extravillous trophoblast cells (Da Silva-Arnold et al., 2015). Then, extravillous trophoblast migrates and invades the maternal endometrium to reach the maternal spiral arteries. The relevance of these processes is evidenced by the fact that a shallow invasion does not allow adequate vascular transformation, as observed in preeclampsia (Moffett-King, 2002).

Trophoblast invasion and migration is regulated by extracellular matrix degradation by matrix metalloproteinases (MMPs). MMP-2 and MMP-9 are highly expressed by the placental trophoblast (Isaka et al., 2003) and trophoblast cell lines (Beltrame, Sordelli, Cañumil, Franchi, & Ribeiro, 2018; Chen, Wong, Brennecke, & Keogh, 2011).

Activation of MMPs is correlated with the invasive competence of trophoblast cells. P4 seems to act as a negative regulator of trophoblast invasion as significantly inhibited HTR-8/SVneo migration by reduced MMP-2 and -9 activity (Chen et al., 2011). Also, early first trimester trophoblast cells (6–8 weeks) have reduced invasive properties under P4 action, while late trophoblast cells (9–12 weeks) show increased invasion (Goldman & Shalev, 2006). These dual P4 actions were also observed in mice (Dai et al., 2003). The authors observed that P4 promoted the invasion of primary trophoblast giant cells at the early stage of embryo implantation, but inhibited the invasion of the secondary trophoblast giant cells.

Recent studies suggest a role of E2 in MMPs regulation. In this sense, an E2 receptor agonist (GPR30) enhances trophoblast cell invasion by upregulating MMP9 expression and the PI3K-Akt signaling pathway (Tong et al., 2016). Also, E2 promotes trophoblast invasion by activating a serum and glucocorticoid inducible kinase (SGK1). SGK1 protein levels are decreased in patients with early pregnancy loss (He et al., 2019).

3.1.2 Extravillous trophoblast differentiation into endovascular trophoblast

The transformation into endovascular trophoblast is crucial to achieving the vascular adaptations that are mandatory for a healthy pregnancy and is a requirement for spiral artery remodeling (Pijnenborg, Vercruysse, & Hanssens, 2006). HTR-8/SVneo trophoblast cell line has been used to model the acquisition of the endovascular phenotype by the invading trophoblast (Highet et al., 2015; Highet, Zhang, Heinemann, & Roberts, 2012; Hu et al., 2010) the initial step of spiral artery remodeling. This cell line was obtained from explant cultures of human first trimester placenta (8–10 weeks of gestation) and immortalized by transfection with a cDNA construct that encodes the SV40 large T antigen (Graham et al., 1993). Although these cells are highly invasive *in vitro*, they are non-tumorigenic or metastatic. HTR-8/SVneo cells exhibit various properties of extravillous cytotrophoblast including the expression of cytokeratin 7, 8, and 18, human leukocyte antigen (HLA) framework antigen W6/32, as well as an integrin profile characteristic of the invasive cytotrophoblast (Graham et al., 1993;

Irving et al., 1995). These cells secrete variable levels of hCG (Armant et al., 2006), and express cytoplasmic and membrane-associated HLA-G (Kalkunte et al., 2008; Kilburn et al., 2000).

Tube formation assay is a technique used to study the ability of extravillous trophoblast to differentiate to an endothelial-like phenotype (Fukushima et al., 2008; Highet et al., 2015, 2012; Hu et al., 2010; Lala, Girish, Cloutier-Bosworth, & Lala, 2012; Waddell, Evans, Jabbour, & Denison, 2011; Wang et al., 2018). Previous studies demonstrate that HTR-8/SVneo cells do not express endothelial markers when cultured on plastic or glass (Highet et al., 2016; Hu et al., 2010). However, when this cell line is cultured upon Matrigel or Geltrex it forms capillary-like structures and changes the protein pattern expressing vascular endothelial markers (Highet et al., 2016; Hu et al., 2010). A gene expression microarray approach identified genes associated with the tube-forming phenotype of HTR-8/SVneo and demonstrated that the endovascular trophoblast expresses a gene profile similar to endothelial cells (i.e., EDN1, TGF-β2, BMP4, EDNRA and ANGPTL4). This expression of endothelial markers supports endovascular differentiation during tube formation and strongly suggest that HTR-8/SVneo tube formation represents the acquisition of an endovascular phenotype.

Experiments carried out in HTR-8/SVneo trophoblast cells showed that this cell line constitutively expresses both P4 (PR-A and PR-B) and E2 (ER-α and ER-β) receptors (Lee, Park, & Lee, 2014; Patel, Kilburn, Imudia, Armant, & Skafar, 2015). Also, steroid hormones regulate HTR-8/SVneo endothelial-like phenotype and capillary-like response (Beltrame, Sordelli, Cañumil, Franchi, & Ribeiro, 2018). The concentrations of P4 and E2 used in these sets of experiments were selected according to the levels reported in the plasma and placental tissue in the first trimester of pregnancy (Chen et al., 2012). Interestingly, P4 and E2 increase HTR-8/SVneo tube formation in a concentration-dependent manner but only a specific combination of P4 and E2 stimulate tube formation (P4:10^{-7}M + E2:10^{-5}M).

Higher P4 levels seem to affect placenta development as are correlated with the development of preeclampsia (Walsh & Coulter, 1989). Besides, high concentrations of P4 alone or in combination with E2 have a deleterious effect on basal tube formation in HTR-8/SVneo trophoblast cells (Beltrame et al., 2018). As mentioned before, it has been observed that high hormonal states associated with fertility treatments early in gestation may lead to subsequent development of adverse pregnancy outcomes (Sun et al., 2019). Sun et al. (2019) hypothesized that this early estrogen exposure as part of the fertility treatment may drive to increased placental hormone production to maintain these pregnancies in a high hormonal state. Moreover, embryos at day 10 of gestation show impaired uterine spiral artery remodeling and embryo death in female mice exposed to high E2 doses (Meyer et al., 2019). These results reinforce the notion that there must be an adequate regulation of the steroid hormones tone during endovascular differentiation at the maternal-fetal interface, which strengthens the concept that the balance of female sex hormones is critical for pregnancy (Cha et al., 2012; Lim & Wang, 2010; Ramathal, Bagchi, Taylor, & Bagchi, 2010).

3.1.3 Replacement of the endothelial and muscle lining of spiral arteries

The multifactorial mechanisms of spiral artery remodeling involve not only invasion, migration and endovascular trophoblast differentiation, but also the replacement of muscle and endothelial cells of these vessels (Fig. 3) (Cartwright, Fraser, Leslie, Wallace, & James, 2010; Whitley & Cartwright, 2009, 2010).

Loss of endothelial cells involves migration and apoptosis, and loss of vascular smooth muscle cells occurs by dedifferentiation and apoptosis (Beltrame et al., 2019; Robson et al., 2019; Sato, Fujiwara, & Konishi, 2012; Whitley & Cartwright, 2009, 2010). These mechanisms seem to be regulated indirectly or directly by hormones. Remarkably, studies performed in endometrial arteries demonstrated that non-pregnant spiral vessels present high sensitivity to steroid hormones stimuli compared with non-hormone responsive basal arteries (Blaustein, 2013). In this sense, endothelial and

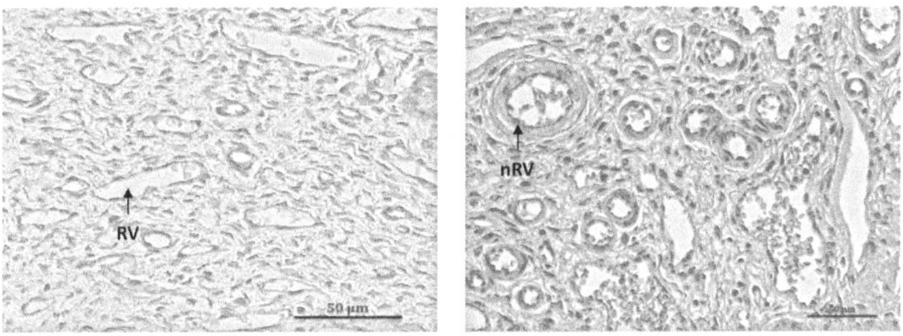

Fig. 3 Replacement of the endothelial and muscle lining of the spiral arteries. Loss of endothelial cells involves migration and apoptosis, and loss of vascular smooth muscle cells occurs by dedifferentiation and apoptosis. These mechanisms seem to be regulated by steroid hormones. Implantation sites of day 8 of gestation stained with hematoxylin and eosin (400 ×). In the non-remodeled vessels, the presence of the vascular muscle cells is evident. nRV: non-remodeled vessel. RV: remodeled vessel.

vascular smooth muscle cells expressed PR and ER (Joswig, Hach-Wunderle, Ziegler, & Nawroth, 1999; Vázquez et al., 1999). PR regulates endothelial and vascular smooth muscle cell proliferation, migration and apoptosis *in vitro* (Cutini & Massheimer, 2019; Cutini, Sellés, & Massheimer, 2009). Moreover, PR promotes vasodilation by a decrease of intracellular calcium (Pang & Thomas, 2019). Also, P4 stimulates the differentiation of a subtype of decidual cells into endothelial and smooth muscle cells (Wang et al., 2013).

While endovascular trophoblast is essential for spiral artery remodeling, some initial steps occur in the absence of trophoblast cells. This trophoblast-independent remodeling seems to be mediated by decidual cells and uNK. uNK cells represent approximately 70% of the decidual leukocyte population (Lash, Robson, & Bulmer, 2010) and are attracted to the vicinity of the vessels where they induce the destabilization of the vascular smooth muscle layers (Robson et al., 2012; Smith et al., 2009). The loss of smooth muscle cells from the spiral arteries would indeed transform them into dilated flaccid vessels (Burton, Woods, Jauniaux, & Kingdom, 2009). It has been observed that E2 influences differentiation and activation of uterine NK and may be involved in the early reorganization of vascular vessels before the arrival of the trophoblast (Gibson, Greaves, Critchley, & Saunders, 2015).

Using multiple antibodies in the immunohistochemistry technique could be a suitable strategy to study spiral artery remodeling. The use of cytokeratin or HLA-G, von Willebrand factor, and α-smooth muscle actin as markers of trophoblast, endothelium, and smooth muscle respectively acknowledge more precise documentation of the degree of transformation, or the lack of it (Labarrere et al., 2017; Robson et al., 2019). In this sense, the impaired invasion and latter replacement of vascular cells lead to retention of vascular smooth muscle cells in the spiral arteries as observed with the retain α-actin stain in these vessels (Labarrere et al., 2017). These vascular smooth muscle cells promote vasoconstriction, suggesting a decrease in the utero-placental blood flow.

4. Steroid hormones regulate vascular remodeling through lipid mediators

While uterine vasculature adaptations are well studied, the factors that induce and regulate these processes are not fully understood. The existing data suggest that numerous players act downstream steroid-regulation in vascular remodeling. In this sense, some of the most widely studied mediators of physiological responses in reproduction are the phosphorylated lipids. In particular, lysophosphatidic acid (LPA) is a bioactive lipid that regulates several female reproductive functions through G protein-coupled receptors (Hama et al., 2007; Ye et al., 2005). LPA is produced mainly by the lysophospholipase-D enzyme that has been found expressed in the human placental trophoblast (Iwasawa et al., 2009). The activity of lysophospholipase-D is increased in serum along gestation producing micromolar levels of LPA (Tokumura, Kanaya, et al., 2002), and soon after delivery, decreases to non-pregnant levels (Tokumura, Majima, et al., 2002).

The participation of LPA signaling during early gestation has been demonstrated in mice, rats, pigs, cattle, and buffalo (Aikawa et al., 2017; Jeong et al., 2016; Sadam et al., 2017; Sordelli et al., 2012; Woclawek-Potocka et al., 2009; Ye et al., 2005; Ye, Herr, Diao, Rivera, & Chun, 2011). LPA and the LPA3 receptor play major roles in the implantation process in mice demonstrated in an LPA3 knockout model (Ye, 2008, Ye et al., 2005). Besides, micromolar concentrations of LPA augment the production of crucial molecules involved in the angiogenesis and decidualization of the rat uterus (Beltrame et al., 2013; Sordelli et al., 2012) Consistent with this, a pharmacological blockade of LPA3 signaling in the rat increases embryo resorption due to a decrease in the length of the uterine and arcuate arteries that irrigate the implantation sites (Sordelli et al., 2017). Following this notion, Jeong and collaborators observed that the LPA-LPA3 system participates in the development of the trophoblast during early pregnancy in pigs. All these data reinforce the notion of LPA as a pro-implantation lipid with vascular action. Remarkably, in the LPA signaling pathway, the actions of P4 and E2 appear to be mediated also by the LPA3 receptor. The balance of P4 and E2 is disrupted in the uterus of the LPA3 knockout mice leading to delayed implantation (Diao et al., 2015) suggesting a role in modulating uterine receptivity and the mechanisms trigger by the implantation process.

The studies on the vascular action driven by LPA in different animal models reveal that it might also be of importance in humans. It has been observed that P4 and E2 enhance the formation of capillary-like structures in HTR-8/SV-neo human trophoblast cells by a mechanism that implies the participation of LPA and at least LPA3 (Fig. 4) (Beltrame, Sordelli, Cañumil, Alonso, et al., 2018). Additionally, P4 and E2 increase the mRNA levels of lysophospholipase-D in this cell line, suggesting a regulation of the LPA pathway (Beltrame, Sordelli, Cañumil, Alonso, et al., 2018). Interestingly, reduced levels of LPA3 receptor have been found in the endometrium of

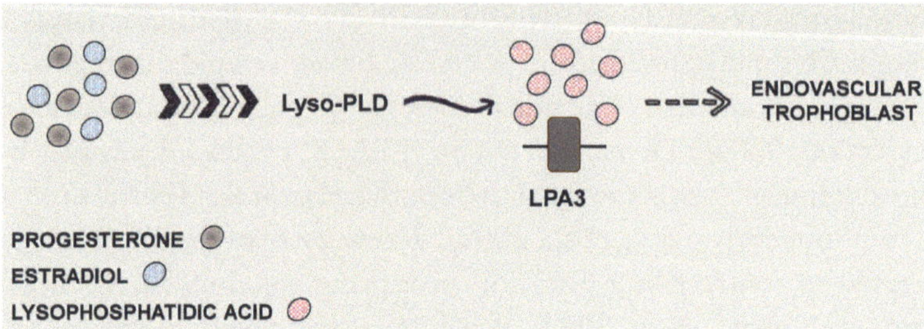

Fig. 4 A fine balance of P4 and E2 is necessary to induce trophoblast capillary-like structures. Steroid hormones regulate the acquisition of the endovascular trophoblast phenotype *in vitro*. Besides, this endovascular differentiation is mediated by the LPA3 receptor. *Adapted from Beltrame J.S., Sordelli M.S., Cañumil V.A., Alonso, C.A.I, Perez Martinez S. & Ribeiro M.L. (2018). Steroid hormones induce in vitro human first trimester trophoblast tubulogenesis by the lysophosphatidic acid pathway. Molecular and Cellular Endocrinology, 478, 126–132. https://doi.org/10.1016/j.mce.2018.08.003.*

patients with recurrent implantation failure and endometriosis (Achache, Tsafrir, Prus, Reich, & Revel, 2010; Wei, St Clair, Fu, Stratton, & Nieman, 2009).

All the evidence contributes to the hypothesis that steroid hormones present at the maternal-fetal interface could modulate directly or indirectly mechanisms associated with the LPA/LPA3 pathway in early gestation. However, it is necessary to mention that the levels of numerous members of the lipids′ pathways are altered in infertility (Sun et al., 2019). Additionally, throughout pregnancy, steroid hormones modulate the majority of the systems of the mother. In this sense, sex hormones might alter the levels of different mediators as prolactin or placental lactogens, which in turn, may contribute to the physiological changes in pregnancy (Napso et al., 2018). Moreover, prostaglandins and nitric oxide pathways, crucial modulators of the vascular tone, could also be involved while they are regulated by LPA and steroid hormones in different reproductive processes (Beltrame, Sordelli, Cañumil, Alonso, et al., 2018; Beltrame et al., 2013; Farina et al., 2007, 2004; Sordelli et al., 2012). Further work is required to outline how steroid hormones control their actions and the downstream pathways on the extensive vascular remodeling of the maternal fetal interface during early gestation.

5. Conclusions

The extensive vascular remodeling that suffers the maternal-fetal interface is essential for embryo growth and to achieve a successful gestation.

Steroid hormones contribute to the modulation of the processes that are critical for adequate fetal development, uterine blood flow, angiogenesis, and spiral artery remodeling. Understanding the role of sex hormones in the vascular dynamics of the uterus and later placentation will provide further insight into the origin of disorders associated with vascular remodeling, and would help to design new approaches to alleviate these pathologies.

References

Achache, H., Tsafrir, A., Prus, D., Reich, R., & Revel, A. (2010). Defective endometrial prostaglandin synthesis identified in patients with repeated implantation failure undergoing in vitro fertilization. *Fertility and Sterility*, *94*, 1271–1278. https://doi.org/10.1016/j.fertnstert.2009.07.1668.

Aikawa, S., Kano, K., Inoue, A., Wang, J., Saigusa, D., Nagamatsu, T., et al. (2017). Autotaxin-lysophosphatidic acid-LPA3 signaling at the embryo epithelial boundary controls decidualization pathways. *EMBO Journal*, *36*, 2146–2160.

Albrecht, E. D., & Pepe, G. J. (1990). Placental steroid hormone biosynthesis in primate pregnancy. *Endocrine Reviews*, *11*, 124–150. https://doi.org/10.1210/edrv-11-1-124.

Albrecht, E. D., & Pepe, G. J. (2010). Estrogen regulation of placental angiogenesis and fetal ovarian development during primate pregnancy. *International Journal of Developmental Biology*, *54*, 397.

Albrecht, E. D., Robb, V. A., & Pepe, G. J. (2004). Regulation of placental vascular endothelial growth/permeability factor expression and angiogenesis by estrogen during early baboon pregnancy. *Journal of Clinical Endocrinology and Metabolism*, *89*, 5803–5809.

Aplin, J. D., Myers, J. E., Timms, K., & Westwood, M. (2020). Tracking placental development in health and disease. *Nature Reviews Endocrinology*, *16*, 479–494. https://doi.org/10.1038/s41574-020-0372-6.

Armant, D. R., Kilburn, B. A., Petkova, A., Edwin, S. S., Duniec-Dmuchowski, Z. M., Edwards, H. J., et al. (2006). Human trophoblast survival at low oxygen concentrations requires metalloproteinase mediated shedding of heparin-binding EGF-like growth factor. *Development*, *133*, 751–759.

Bechi, N., Letta, F., Romagnoli, R., Focardi, S., Corsi, I., Buffi, C., et al. (2006). Estrogen-like response top-nonylphenol in human first trimester placenta and BeWo choriocarcinoma cells. *Toxicological Sciences*, *93*, 75–81. https://doi.org/10.1093/toxsci/kfl043.

Beltrame, J. S., Scotti, L., Sordelli, M. S., Cañumil, V. A., Franchi, A. M., Parborell, F., et al. (2019). Lysophosphatidic acid induces the crosstalk between the endovascular human trophoblast and endothelial cells in vitro. *Journal of Cellular Physiology*, *234*, 6274–6285. https://doi.org/10.1002/jcp.27358.

Beltrame, J. S., Sordelli, M. S., Cañumil, V. A., Alonso, C. A. I., Perez, M. S., & Ribeiro, M. L. (2018). Steroid hormones induce in vitro human first trimester trophoblast tubulogenesis by the lysophosphatidic acid pathway. *Molecular and Cellular Endocrinology*, *478*, 126–132. https://doi.org/10.1016/j.mce.2018.08.003.

Beltrame, J. S., Sordelli, M. S., Cañumil, V. A., Franchi, A. M., & Ribeiro, M. L. (2018). Lysophosphatidic acid-triggered pathways promote the acquisition of trophoblast endovascular phenotype in vitro. *Journal of Cellular Biochemistry*, *119*, 758–772. https://doi.org/10.1002/jcb.26239.

Beltrame, J. S., Sordelli, M. S., Cella, M., Perez, M. S., Franchi, A. M., & Ribeiro, M. L. (2013). Lysophosphatidic acid increases the production of pivotal mediators of decidualization and vascularization in the rat uterus. *Placenta*, *34*, 751–756.

Benirschke, K., & Kaufmann, P. (2000). *Pathology of the human placenta* (4th ed., p. 1050). Springer.

Blaustein, A. (2013). *Pathology of the female genital tract*. Springer Science & Business Media.

Boyd, J. D., & Hamilton, W. J. (1970). *The human placenta*. Cambridge: W. Heffer & Sons.

Brosens, I., Robertson, W. B., & Dixon, H. G. (1967). The physiological response of the vessels of the placental bed to normal pregnancy. *Journal of Pathology and Bacteriology, 93*, 569–579. https://doi.org/10.1002/path.1700930218.

Bukovsky, A., Cekanova, M., Caudle, M. R., Wimalasena, J., Foster, J. S., Henley, D. C., et al. (2003). Expression and localization of estrogen receptor-alpha protein in normal and abnormal term placentae and stimulation of trophoblast differentiation by estradiol. *Reproductive Biology and Endocrinology, 6*, 1–13.

Bulmer, J. N., & Lash, G. E. (2005). Human uterine natural killer cells: A reappraisal. *Molecular Immunology, 42*, 511–521. https://doi.org/10.1016/j.molimm.2004.07.035.

Burton, G. J., Woods, A. W., Jauniaux, E., & Kingdom, J. C. P. (2009). Rheological and physiological consequences of conversion of the maternal spiral arteries for utero-placental blood flow during human pregnancy. *Placenta, 30*, 473–482.

Cartwright, J. E., Fraser, R., Leslie, K., Wallace, A. E., & James, J. L. (2010). Remodelling at the maternal–fetal interface: Relevance to human pregnancy disorders. *Reproduction, 140*, 803–813.

Cha, J., Sun, X., & Dey, S. K. (2012). Mechanisms of implantation: Strategies for successful pregnancy. *Nature Medicine, 18*, 1754–1767.

Chang, C., Wakeland, A., & Parast, M. (2018). Trophoblast lineage specification, differentiation and their regulation by oxygen tension. *Journal of Endocrinology, 236*, 43–56. https://doi.org/10.1530/JOE-17-0402.

Chen, J. Z., Sheehan, P. M., Brennecke, S. P., & Keogh, R. J. (2012). Vessel remodelling, pregnancy hormones and extravillous trophoblast function. *Molecular and Cellular Endocrinology, 349*, 138–144.

Chen, J., Wong, M. H., Brennecke, S. P., & Keogh, R. J. (2011). The effects of human chorionic gonadotrophin, progesterone and oestradiol on trophoblast function. *Molecular and Cellular Endocrinology, 342*, 73–80. https://doi.org/10.1016/j.mce.2011.05.034.

Clark, N. C., Pru, C. A., & Pru, J. K. (2017). Novel regulators of hemodynamics 556 in the pregnant uterus. *Progress in Molecular Biology and Translational Science, 145*, 181–216.

Corcoran, J. J., Nicholson, C., Sweeney, M., Charnock, C. J., Robson, S. C., Westwood, M., et al. (2014). Human uterine and placental arteries exhibit tissue-specific acute responses to 17β-estradiol and estrogen-receptor-specific agonists. *Molecular Human Reproduction, 20*, 433–441. https://doi.org/10.1093/molehr/gat095.

Costa, M. (2016). The endocrine function of human placenta: An overview. *Reproductive Biomedicine Online, 32*, 14–43.

Curtis, H. S., Goulding, E. H., Eddy, E. M., & Korach, K. S. (2002). Studies using the estrogen receptor alpha knockout uterus demonstrate that implantation but not decidualization-associated signaling is estrogen dependent. *Biology of Reproduction, 67*, 1268–1277. https://doi.org/10.1095/biolreprod67.4.1268.

Cutini, P., & Massheimer, V. L. (2019). In vitro effects of progesterone and the synthetic progestin medroxyprogesterone acetate on vascular remodeling. *Molecular and Cellular Endocrinology, 498*, 110543.

Cutini, P., Sellés, J., & Massheimer, V. L. (2009). Cross-talk between rapid and long term effects of progesterone on vascular tissue. *The Journal of Steroid Biochemistry and Molecular Biology, 115*, 36–43.

Da Silva-Arnold, S., James, J. L., Al-Khan, A., Zamudio, S., & Illsley, N. P. (2015). Differentiation of first trimester cytotrophoblast to extravillous trophoblast involves an epithelial-mesenchymal transition. *Placenta, 36*, 1412–1418.

Dai, B., Cao, Y., Liu, W., Li, S., Yang, Y., Chen, S., et al. (2003). Dual roles of progesterone in embryo implantation in mouse. *Endocrinology, 21*, 123–132. https://doi.org/10.1385/ENDO:21:2:123.

de Ziegler, D., Bessis, R., & Frydman, R. (1991). Vascular resistance of uterine arteries: Physiological effects of estradiol and progesterone. *Fertility and Sterility*, *55*, 775–779.

Demir, R., Yaba, A., & Huppertz, B. (2010). Vasculogenesis and angiogenesis in the endometrium during menstrual cycle and implantation. *Acta Histochemica*, *112*, 203–214.

Diao, H., Li, R., El Zowalaty, A. E., Xiao, S., Zhao, F., Dudley, E. A., et al. (2015). Deletion of lysophosphatidic acid receptor 3 (Lpar3) disrupts fine local balance of progesterone and estrogen signaling in mouse uterus during implantation. *Biology of Reproduction*, *93*, 123. https://doi.org/10.1095/biolreprod.115.131110.

Edey, L. F., Georgiou, H., O'Dea, K. P., Mesiano, S., Herbert, B. R., Lei, K., et al. (2018). Progesterone, the maternal immune system and the onset of parturition in the mouse. *Biology of Reproduction*, *98*, 376–395. https://doi.org/10.1093/biolre/iox146.

Espinoza, J., Romero, R., Kim, Y. M., Kusanovic, J. P., Hassan, S., Erez, O., et al. (2006). Normal and abnormal transformation of the spiral arteries during pregnancy. *Journal of Perinatal Medicine*, *34*, 447–458. https://doi.org/10.1515/JPM.2006.089.

Farina, M., Billi, S., Leguizamón, G., Weissmann, C., Guadagnoli, T., Ribeiro, M. L., et al. (2007). Secretory and cytosolic phospholipase A2 activities and expression are regulated by oxytocin and estradiol during labor. *Reproduction*, *134*, 355–364.

Farina, M., Ribeiro, M. L., Weissmann, C., Estevez, A., Billi, S., Vercelli, C., et al. (2004). Biosynthesis and catabolism of prostaglandin F2alpha (PGF2alpha) are controlled by progesterone in the rat uterus during pregnancy. *The Journal of Steroid Biochemistry and Molecular Biology*, *91*, 211–218.

Fatemi, H. M., & Van Vaerenbergh, I. (2015). Significance of premature progesterone rise in IVF. *Current Opinion in Obstetrics & Gynecology*, *27*, 242–248.

Fraser, R., Whitley, G. S. J., Thilaganathan, B., & Cartwright, J. E. (2015). Decidual natural killer cells regulate vessel stability: Implications for impaired spiral artery remodeling. *Journal of Reproductive Immunology*, *110*, 54–60. https://doi.org/10.1016/j.jri.2015.04.003.

Fukushima, K., Murata, M., Hachisuga, M., Tsukimori, K., Seki, H., Takeda, S., et al. (2008). Gene expression profiles by microarray analysis during matrigel-induced tube formation in a human extravillous trophoblast cell line: Comparison with endothelial cells. *Placenta*, *29*, 898–904. https://doi.org/10.1016/j.placenta.2008.07.015.

Gibson, D. A., Greaves, E., Critchley, H. O. D., & Saunders, P. T. K. (2015). Estrogen-dependent regulation of human uterine natural killer cells promotes vascular remodelling via secretion of CCL2. *Human Reproduction*, *30*, 1290–1301. https://doi.org/10.1093/humrep/dev067.

Goldman, S., & Shalev, E. (2006). Difference in progesterone-receptor isoforms ratio between early and late first-trimester human trophoblast is associated with differential cell invasion and matrix metalloproteinase 2 expression. *Biology of Reproduction*, *74*, 13–22.

Goyeneche, A. A., Deis, R. P., Gibori, G., & Telleria, S. M. (2003). Progesterone promotes survival of the rat Corpus luteum in the absenceof cognate receptors. *Biology of Reproduction*, *68*, 151–158. https://doi.org/10.1095/biolreprod.102.007898.

Graham, C. H., Hawley, T. S., Hawley, R. G., MacDougall, J. R., Kerbel, R. S., Khoo, N., et al. (1993). Establishment and characterization of first trimester human trophoblast cells with extended lifespan. *Experimental Cell Research*, *206*, 204–211.

Habara, T., Nakatsuka, M., Konishi, H., Asagiri, K., Noguchi, S., & Kudo, T. (2002). Elevated blood flow resistance in uterine arteries of women unexplained recurrent pregnancy loss. *Human Reproduction*, *17*, 190–194.

Hama, K., Aoki, J., Inoue, A., Endo, T., Amano, T., Motoki, R., et al. (2007). Embryo spacing and implantation timing are differentially regulated by LPA3-mediated lysophosphatidic acid signaling in mice. *Biology of Reproduction*, *77*, 954–959.

Hanna, J., Goldman-Wohl, D., Hamani, Y., Avraham, I., Greenfield, C., Natanson-Yaron, S., et al. (2006). Decidual NK cell regulate key developmental processes at the human fetal–maternal interface. *Nature Medicine, 12*, 1065–1074.

Harris, L. K., Keogh, R. J., Wareing, M., Baker, P. N., Cartwright, J. E., Aplin, J. D., et al. (2006). Invasive trophoblasts stimulate vascular smooth muscle cell apoptosis by a fas ligand-dependent mechanism. *American Journal of Pathology, 169*, 1863–1874.

He, W. H., Jin, M. M., Liu, A. P., Zhou, Y., Hu, X. L., Zhu, Y. M., et al. (2019). Estradiol promotes trophoblast viability and invasion by activating SGK1. *Biomedicine & Pharmacotherapy, 117*, 109092. https://doi.org/10.1016/j.biopha.2019.109092.

Highet, A. R., Buckberry, S., Mayne, B. T., Khoda, S. M., Bianco-Miotto, T., & Roberts, C. T. (2016). First trimester trophoblasts forming endothelial-like tubes in vitro emulate a "blood vessel development" gene expression profile. *Gene Expression Patterns, 21*, 103–110.

Highet, A. R., Khoda, S. M., Buckberry, S., Leemaqz, S., Miotto, T. N., Harrington, E., et al. (2015). Hypoxia induced HIF-1/HIF-2 activity alters trophoblast transcriptional regulation and promotes invasion. *European Journal of Cell Biology, 94*, 589–602. https://doi.org/10.1016/j.ejcb.2015.10.004.

Highet, A. R., Zhang, V. J., Heinemann, G. K., & Roberts, C. T. (2012). Use of Matrigel in culture affects cell phenotype and gene expression in the first trimester trophoblast cell line HTR8/SVneo. *Placenta, 33*, 586–588. https://doi.org/10.1016/j.placenta.2012.04.003.

Hu, Y., Eastabrook, G., Tan, R., MacCalman, C. D., Dutz, J. P., & von Dadelszen, P. (2010). Decidual NK cell-derived conditioned medium enhances capillary tube and network organization in an extravillous cytotrophoblast cell line. *Placenta, 31*, 213–221.

Irving, J. A., Lysiak, J. J., Graham, C. H., Hearn, S., Han, V. K. M., & Lala, P. K. (1995). Characteristics of trophoblast cells migrating from first trimester chorionic villus explants and propagated in culture. *Placenta, 16*, 413–433.

Isaka, K., Usuda, S., Ito, H., Sagawa, Y., Nakamura, H., Nishi, H., et al. (2003). Expression and activity of matrix metalloproteinase 2 and 9 in human trophoblasts. *Placenta, 24*, 53–64.

Iwasawa, Y., Fujii, T., Nagamatsu, T., Kawana, K., Okudaira, S., Miura, S., et al. (2009). Expression of autotaxin, an ectoenzyme that produces lysophosphatidic acid, in human placenta. *American Journal of Reproductive Immunology, 62*, 90–95.

Jeong, W., Seo, H., Sung, Y., Ka, H., Song, G., & Kim, J. (2016). Lysophosphatidic acid (LPA) receptor 3-mediated LPA signal transduction pathways: A possible relationship with early development of Peri-implantation porcine conceptus. *Biology of Reproduction, 94*, 104. https://doi.org/10.1095/biolreprod.115.137174.

Jobe, S. O., Ramadoss, J., Koch, J. M., Jiang, Y., Zheng, J., & Magness, R. R. (2010). Estradiol-17beta and its cytochrome P450- and catechol-O-methyltransferase-derived metabolites stimulate proliferation in uterine artery endothelial cells: Role of estrogen receptor-alpha versus estrogen receptor-beta. *Hypertension, 55*, 1005–1011. https://doi.org/10.1161/HYPERTENSIONAHA.109.146399.

Joswig, M., Hach-Wunderle, V., Ziegler, R., & Nawroth, P. P. (1999). Postmenopausal hormone replacement therapy and the vascular wall: Mechanisms of 17 beta-estradiol's effects on vascular biology. *Experimental and Clinical Endocrinology & Diabetes, 107*, 477–487. https://doi.org/10.1055/s-0029-1232556.

Kalkunte, S., Lai, Z., Tewari, N., Chichester, C., Romero, R., Padbury, J., et al. (2008). In vitro and in vivo evidence for lack of endovascular remodeling by third trimester trophoblasts. *Placenta, 29*, 871–878.

Kam, E. P., Gardner, L., Loke, Y. W., & King, A. (1999). The role of trophoblast in the physiological change in decidual spiral arteries. *Human Reproduction, 14*, 2131–2138.

Kilburn, B. A., Wang, J., Duniec-Dmuchowski, Z. M., Leach, R. E., Romero, R., & Armant, D. R. (2000). Extracellular matrix composition and hypoxia regulate the expression of HLAG and integrins in a human trophoblast cell line. *Biology of Reproduction*, *62*, 739–747.

Kim, K. H., Moriarty, K., & Bender, J. R. (2008). Vascular cell signaling by membrane estrogen receptors. *Steroids*, *73*, 864–869.

Knöfler, M. (2010). Critical growth factors and signalling pathways controlling human trophoblast invasion. *International Journal of Developmental Biology*, *54*, 269–280.

Krege, J. H., Hodgin, J. B., Couse, J. F., Enmark, E., Warner, M., Mahler, J. F., et al. (1998). Generation and reproductive phenotypes of mice lacking estrogen receptor beta. *Proceedings of the National Academy of Sciences of the United States of America*, *95*, 15677–15682.

Labarrere, C. A., DiCarlo, H. L., Bammerlin, E., Hardin, J. W., Kim, Y. M., Chaemsaithong, P., et al. (2017). Failure of physiologic transformation of spiral arteries, endothelial and trophoblast cell activation, and acute atherosis in the basal plate of the placenta. *American Journal of Obstetrics and Gynecology*, *21*, 287.e1–287.e16. https://doi.org/10.1016/j.ajog.2016.12.029.

Lala, N., Girish, G. V., Cloutier-Bosworth, A., & Lala, P. K. (2012). Mechanisms in decorin regulation of vascular endothelial growth factor-induced human trophoblast migration and acquisition of endothelial phenotype. *Biology of Reproduction*, *87*, 1–14.

Lash, G. E., Robson, S. C., & Bulmer, J. N. (2010). Functional role of uterine natural killer (uNK) cells in human early pregnancy decidua. *Placenta*, *31*, 87–92.

Leach, L., Badet, J., Brownbill, P., Harris, L., Keogh, R., Kalionis, B., et al. (2006). Endothelium, blood vessels and angiogenesis. A workshop report. *Placenta*, *27*, 26–29.

Lee, B. H., Park, T. C., & Lee, H. J. (2014). Influence of caffeine on the expression of human chorionic gonadotropin and progesterone receptors in human trophoblast cell lines. *Acta Obstetricia et Gynecologica Scandinavica*, *93*, 1334–1335.

Leitao, B., Jones, M. C., Fusi, L., Higham, J., Lee, Y., Takano, M., et al. (2010). Silencing of the JNK pathway maintains progesterone receptor activity in decidualizing human endometrial stromal cells exposed to oxidative stress signals. *The FASEB Journal*, *24*, 1541–1551.

Li, X., & O'Malley, B. W. (2003). Unfolding the action of progesterone receptors. *Journal of Biological Chemistry*, *278*, 39261–39264. https://doi.org/10.1074/jbc.R300024200.

Lim, H. J., & Wang, H. (2010). Uterine disorders and pregnancy complications: Insights from mouse models. *Journal of Clinical Investigation*, *120*, 1004–1015.

Liu, Y., Fan, X., Wang, R., Lu, X., Dang, Y. L., Wang, H., et al. (2018). Single-cell RNA-seq reveals the diversity of trophoblast subtypes and patterns of differentiation in the human placenta. *Cell Research*, *28*, 819–832. https://doi.org/10.1038/s41422-018-0066-y.

Lubahn, D. B., Moyer, J. S., Golding, T. S., Couse, J. F., Korach, K. S., & Smithies, O. (1993). Alteration of reproductive function but not prenatal sexual development after insertional disruption of the mouse estrogen receptor gene. *Proceedings of the National Academy of Sciences of the United States of America*, *90*, 11162–11166. https://doi.org/10.1073/pnas.90.23.11162.

Lyall, F. (2005). Priming and remodelling of human placental bed spiral arteries during pregnancy: A review. *Placenta*, *26*, 31–S36.

Ma, W. G., Song, H., Das, S. K., Paria, B. C., & Dey, S. K. (2003). Estrogen is a critical determinant that specifies the duration of the window of uterine receptivity for implantation. *Proceedings of the National Academy of Sciences of the United States of America*, *100*, 2963–2968.

Malassine, A., Frendo, J. L., & Evain-Brion, D. (2003). A comparison of placental development and endocrine functions between the human and mouse model. *Human Reproduction Update*, *9*, 531–539. https://doi.org/10.1093/humupd/dmg043.

Maliqueo, M., Echiburú, B., & Crisosto, N. (2016). Sex steroids modulate uterine-placental vasculature: Implications for obstetrics and neonatal outcomes. *Frontiers in Physiology*, *26*(7), 152.

Maltepe, E., & Fisher, S. J. (2015). Placenta: The forgotten organ. *Annual Review of Cell and Developmental Biology*, *31*, 523–552.

Mandala, M. (2020). Influence of estrogens on uterine vascular adaptation in normal and pre-eclamptic pregnancies. *International Journal of Molecular Sciences*, *21*, 2592. https://doi.org/10.3390/ijms21072592.

Meyer, N., Santamaria, C. G., Müller, J. E., Schumacher, A., Rodriguez, H. A., & Zenclussen, A. C. (2019). Exposure to 17α-ethinyl estradiol during early pregnancy affects fetal growth and survival in mice. *Enviromental Pollution*, *251*, 493–501. https://doi.org/10.1016/j.envpol.2019.04.144.

Moffett-King, A. (2002). Natural killer cells and pregnancy. *Nature Reviews Immunology*, *2*, 656–663. https://doi.org/10.1038/nri886.

Morales, D. E., McGowan, K. A., Grant, D. S., Maheshwari, S., Bhartiya, D., Cid, M. C., et al. (1995). Estrogen promotes angiogenic activity in human umbilical vein endothelial cells in vitro and in a murine model. *Circulation*, *91*, 755–763. https://doi.org/10.1161/01.cir.91.3.75.

Moser, G., Gauster, M., Orendi, K., Glasner, A., Theuerkauf, R., & Huppertz, B. (2010). Endoglandular trophoblast, an alternative route of trophoblast invasion? Analysis with novel confrontation co-culture models. *Human Reproduction*, *25*, 1127–1136.

Moser, G., Weiss, G., Gauster, M., Sundl, M., & Huppertz, B. (2015). Evidence from the very beginning: Endoglandular trophoblasts penetrate and replace uterine glands in situ and in vitro. *Human Reproduction*, *30*, 2747–2757.

Mulac-Jericevic, B., Lydon, J. P., DeMayo, F. J., & Conneely, O. M. (2003). Defective mammary gland morphogenesis in micelacking the progesterone receptor B isoform. *Proceedings of the National Academy of Sciences of the United States of America*, *100*, 9744–9749. https://doi.org/10.1073/pnas.1732707100.

Mulac-Jericevic, B., Mullinax, R. A., DeMayo, F. J., Lydon, J. P., & Conneely, O. M. (2000). Subgroup of reproductive functions of progesterone mediated by progesterone receptor-B isoform. *Science*, *289*, 1751–1754. https://doi.org/10.1126/science.289.5485.1751.

Murphy, V. E., Smith, R., Giles, W. B., & Clifton, V. L. (2006). Endocrine regulation of human fetal growth: The role of the mother, placenta, and fetus. *Endocrine Reviews*, *27*, 141–169. https://doi.org/10.1210/er.2005-0011.

Napso, T., Yong, H. E. J., Lopez-Tello, J., & Sferruzzi-Perri, A. M. (2018). The role of placental hormones in mediating maternal adaptations to support pregnancy and lactation. *Frontiers in Physiology*, *9*, 1091. https://doi.org/10.3389/fphys.2018.01091.

Nowak-Sliwinska, P., Alitalo, K., Allen, E., et al. (2018). Consensus guidelines for the use and interpretation of angiogenesis assays. *Angiogenesis*, *21*, 425–532. https://doi.org/10.1007/s10456-018-9613-x.

Osol, G., & Mandala, M. (2009). Maternal uterine vascular remodeling during pregnancy. *Physiology*, *24*, 58–71. https://doi.org/10.1152/physiol.00033.2008.

Palmer, S. K., Zamudio, S., Coffin, C., Parker, S., Stamm, E., & Moore, L. G. (1992). Quantitative estimation of human uterine artery blood flow and pelvic blood flow redistribution in pregnancy. *Obstetrics and Gynecology*, *80*, 1000–1006.

Pang, Y., & Thomas, P. (2019). Role of mPRα (PAQR7) in progesterone-induced Ca2+ decrease in human vascular smooth muscle cells. *Journal of Molecular Endocrinology*, *63*, 199–213. https://doi.org/10.1530/JME-19-0019. 31416049.

Paria, B. C., Tan, J., Lubahn, D. B., Dey, S. K., & Das, S. K. (1999). Uterine decidual response occurs in estrogen receptor-alpha-deficient mice. *Endocrinology*, *140*, 2704–2710.

Patel, S., Kilburn, B., Imudia, A., Armant, D. R., & Skafar, D. F. (2015). Estradiol elicits proapoptotic and antiproliferative effects in human trophoblast cells. *Biology of Reproduction, 93*, 1–10.

Paulson, R. J. (2011). Hormonal induction of endometrial receptivity. *Fertility and Sterility, 96*, 3. https://doi.org/10.1016/j.fertnstert.2011.07.1097.

Pepe, G. J., & Albrecht, E. D. (1995). Actions of placental and fetal adrenal steroid hormones in primate pregnancy. *Endocrine Reviews, 16*, 608–648. https://doi.org/10.1210/edrv-16-5-608.

Pijnenborg, R., Dixon, G., Robertson, W. B., & Brosens, I. (1980). Trophoblastic invasion of human decidua from 8 to 18 weeks of pregnancy. *Placenta, 1*, 3e19.

Pijnenborg, R., Vercruysse, L., & Hanssens, M. (2006). The uterine spiral arteries in human pregnancy: Facts and controversies. *Placenta, 27*, 939e58.

Pijnenborg, R., Vercruysse, L., Verbist, L., & Van Assche, F. A. (1998). Interaction of interstitial trophoblast with placental bed capillaries and venules of normotensive and pre-eclamptic pregnancies. *Placenta, 19*, 569–575.

Plaisier, M., Dennert, I., Rost, E., Koolwijk, P., van Hinsbergh, V. W., & Helmerhorst, F. M. (2009). Decidual vascularization and the expression of angiogenic growth factors and proteases in first trimester spontaneous abortions. *Human Reproduction, 24*, 185–197.

Potente, M., Gerhardt, H., & Carmeliet, P. (2011). Basic and therapeutic aspects of angiogenesis. *Cell, 146*, 873–887.

Ramathal, C. Y., Bagchi, I. C., Taylor, R. N., & Bagchi, M. K. (2010). Endometrial decidualization: Of mice and men. *Seminars in Reproductive Medicine, 28*, 17–26.

Reynolds, L. P., & Redmer, D. A. (2001). Angiogenesis in the placenta. *Biology of Reproduction, 64*, 1033–1040. https://doi.org/10.1095/biolreprod64.4.1033.

Robb, V. A., Pepe, G. J., & Albrecht, E. D. (2004). Acute temporal regulation of placental vascular endothelial growth/permeability factor expression in baboons by estrogen. *Biology of Reproduction, 71*, 1694–1698.

Roberts, R. M., & Fisher, S. J. (2011). Trophoblast stem cells. *Biology of Reproduction, 84*, 412–421. https://doi.org/10.1095/biolreprod.110.088724.

Robertshaw, I., Bian, F., & Das, S. K. (2016). Mechanisms of uterine estrogen signaling during early pregnancy in mice: An update. *Journal of Molecular Endocrinology, 56*, 127–138. https://doi.org/10.1530/JME-15-0300.

Robson, A., Harris, L. K., Innes, B. A., Lash, G. E., Aljunaidy, M. M., Aplin, J. D., et al. (2012). Uterine natural killer cells initiate spiral artery remodeling in human pregnancy. *The FASEB Journal, 26*, 4876–4885. https://doi.org/10.1096/fj.12-210310.

Robson, A., Lash, G. E., Innes, B. A., Zhang, J. Y., Robson, S. C., & Bulmer, J. N. (2019). Uterine spiral artery muscle dedifferentiation. *Human Reproduction, 34*, 1428–1438. https://doi.org/10.1093/humrep/dez124.

Sadam, A., Parida, S., Padol, A. R., Verma, A. D., Baba, N. A., Khuman, W. M., et al. (2017). Study of lysophosphatidic acid receptors (LPARs) in buffalo uterus demonstrated upregulation of LPAR1 and LPAR6 in early pregnancy. *Theriogenology, 99*, 90–97. https://doi.org/10.1016/j.theriogenology.2017.05.019.

Sato, Y., Fujiwara, H., & Konishi, I. (2012). Mechanism of maternal vascular remodeling during human pregnancy. *Reproductive Medicine and Biology, 11*, 27–36. https://doi.org/10.1007/s12522-011-0102-9.

Shanker, Y. G., & Rao, A. J. (1999). Progesterone receptor expression in the human placenta. *Molecular Human Reproduction, 5*, 481–486. https://doi.org/10.1093/molehr/5.5.481.

Smith, S. D., Dunk, C. E., Aplin, J. D., Harris, L. K., & Jones, R. L. (2009). Evidence for immune cell involvement in decidual spiral arteriole remodeling in early human pregnancy. *American Journal of Pathology, 174*, 1959–1971. https://doi.org/10.2353/ajpath.2009.080995.

Soncin, F., Natale, D., & Parast, M. M. (2015). Signaling pathways in mouse and human trophoblast differentiation: A comparative review. *Cellular and Molecular Life Sciences*, *72*, 1291–1302.

Sordelli, M. S., Beltrame, J. S., Cella, M., Gervasi, M. G., Perez, M. S., Burdet, J., et al. (2012). Interaction between lysophosphatidic acid, prostaglandins and the endocannabinoid system during the window of implantation in the rat uterus. *PLoS One*, *7*, e46059.

Sordelli, M. S., Beltrame, J. S., Zotta, E., Gomez, N., Dmytrenko, G., Sales, M. E., et al. (2017). Endogenous lysophosphatidic acid participates in vascularisation and decidualization at the maternal-fetal interface in the rat. *Reproduction, Fertility, and Development*, *29*, 2112–2126. https://doi.org/10.1071/RD16235.

Sun, T., Lee, B., Kinchen, J., Wang, E. T., Gonzalez, T. L., Chan, J. L., et al. (2019). Differences in first-trimester maternal metabolomic profiles in pregnancies conceived from fertility treatments. *Journal of Clinical Endocrinology and Metabolism*, *104*, 1005–1019. https://doi.org/10.1210/jc.2018-01118J.

Tokumura, A., Kanaya, Y., Miyake, M., Yamano, S., Irahara, M., & Fukuzawa, K. (2002). Increased production of bioactive lysophosphatidic acid by serum lysophospholipase D in human pregnancy. *Biology of Reproduction*, *67*, 1386–1392.

Tokumura, A., Majima, E., Kariya, Y., Tominaga, K., Kogure, K., Yasuda, K., et al. (2002). Identification of human plasma lysophospholipase D, a lysophosphatidic acid-producing enzyme, as autotaxin, a multifunctional phosphodiesterase. *Journal of Biological Chemistry*, *277*(42), 39436–39442.

Tong, C., Feng, X., Chen, J., Qi, X., Zhou, L., Shi, S., et al. (2016). G protein-coupled receptor 30 regulates trophoblast invasion and its deficiency is associated with preeclampsia. *Journal of Hypertension*, *34*, 710–718. https://doi.org/10.1097/HJH.0000000000000844.

Vázquez, F., Rodríguez-Manzaneque, J. C., Lydon, J. P., Edwards, D. P., O'Malley, B. W., & Iruela-Arispe, M. L. (1999). Progesterone regulates proliferation of endothelial cells. *Journal of Biological Chemistry*, *274*, 2185–2192. https://doi.org/10.1074/jbc.274.4.2185.

Vento-Tormo, R., Efremova, M., Botting, R. A., Turco, M. Y., Vento-Tormo, M., Meyer, K. B., et al. (2018). Single-cell reconstruction of the early maternal-fetal interface in humans. *Nature*, *563*, 347–353. https://doi.org/10.1038/s41586-018-0698-6.

Waddell, J. M., Evans, J., Jabbour, H. N., & Denison, F. C. (2011). CTGF expression is up-regulated by PROK1 in early pregnancy and influences HTR-8/Svneo cell adhesion and network formation. *Human Reproduction*, *26*, 67–75. https://doi.org/10.1093/humrep/deq294.

Wallace, A. E., Fraser, R., & Cartwright, J. E. (2012). Extravillous trophoblast and decidual natural killer cells: A remodelling partnership. *Human Reproduction Update*, *18*, 458–471. https://doi.org/10.1093/humupd/dms015.

Walsh, S. W., & Coulter, S. (1989). Increased placental progesterone may cause decreased placental prostacyclin production in preeclampsia. *American Journal of Obstetrics and Gynecology*, *161*, 1586–1592.

Wan, J., Hub, Z., Zenga, K., Yinc, Y., Zhaoc, M., Chend, M., et al. (2018). The reduction in circulating levels of estrogen and progesterone in women with preeclampsia. *Pregnancy Hypertension*, *11*, 18–25.

Wang, F., Cao, G., Liu, Q., Li, X., Song, M., & Zhang, Z. (2018). Retinol-binding protein 4 regulates the biological functions and molecular mechanisms of JEG-3 cells. *International Journal of Clinical and Experimental Pathology*, *11*, 5877–5884.

Wang, H., & Dey, S. K. (2006). Roadmap to embryo implantation: Clues from mouse models. *Nature Reviews. Genetics*, 7, 185–199.

Wang, Y., Lewis, D. F., Gu, Y., Zhang, Y., Alexander, J. S., & Granger, D. N. (2004). Placental trophoblast-derived factors diminish endothelial barrier function. *Journal of Clinical Endocrinology and Metabolism*, *89*, 2421–2428.

Wang, Q., Shen, L., Huang, W., Song, Y., Xiao, L., Xu, W., et al. (2013). Vasculogenesis of decidua side population cells of first-trimester pregnancy. *Stem Cell Research & Therapy*, *4*, 50.

Wei, Q., St Clair, J. B., Fu, T., Stratton, P., & Nieman, L. K. (2009). Reduced expression of biomarkers associated with the implantation window in women with endometriosis. *Fertility and Sterility*, *91*, 1686–1691.

Whitley, G. S., & Cartwright, J. E. (2009). Trophoblast-mediated spiral artery remodelling: A role for apoptosis. *Journal of Anatomy*, *215*, 21–26. https://doi.org/10.1111/j.1469-7580.2008.01039.x.

Whitley, G. S., & Cartwright, J. E. (2010). Cellular and molecular regulation of spiral artery remodelling: Lessons from the cardiovascular field. *Placenta*, *31*, 465–474. https://doi.org/10.1016/j.placenta.2010.03.002.

Windsperger, K., Dekan, S., Pils, S., Golletz, C., Kunihs, V., Fiala, C., et al. (2017). Extravillous trophoblast invasion of venous as well as lymphatic vessels is altered in idiopathic, recurrent, spontaneous abortions. *Human Reproduction*, *32*, 1208–1217. https://doi.org/10.1093/humrep/dex058.

Woclawek-Potocka, I., Komiyama, J., Saulnier-Blache, J. S., Brzezicka, E., Bah, M. M., Okuda, K., et al. (2009). Lysophosphatic acid modulates prostaglandin secretion in the bovine uterus. *Reproduction*, *137*, 95–105.

Ye, X. (2008). Lysophospholipid signaling in the function and pathology of the reproductive system. *Human Reproduction Update*, *14*, 519–536.

Ye, X., Hama, K., Contos, J. J., Anliker, B., Inoue, A., Skinner, M. K., et al. (2005). LPA3-mediated lysophosphatidic acid signalling in embryo implantation and spacing. *Nature*, *435*, 104–108.

Ye, X., Herr, D. R., Diao, H., Rivera, R., & Chun, J. (2011). Unique uterine localization and regulation may differentiate LPA3 from other lysophospholipid receptors for its role in embryo implantation. *Fertility and Sterility*, *95*, 2107–2113.

Zhou, Y., Damsky, C. H., & Fisher, S. J. (1997). Preeclampsia is associated with failure of human cytotrophoblasts to mimic a vascular adhesion phenotype. One cause of defective endovascular invasion in this syndrome? *Journal of Clinical Investigation*, *99*, 2152–2164.

Zhou, Y., Fisher, S. J., Janatpour, M., Genbacev, O., Dejana, E., Wheelock, M., et al. (1997). Human cytotrophoblasts adopt a vascular phenotype as they differentiate. A strategy for successful endovascular invasion? *Journal of Clinical Investigation*, *99*, 2139–2151.

CHAPTER THIRTEEN

Nuclear receptors and differentiation of oligodendrocyte precursor cells

Vito Antonio Baldassarro[a,*], Alessandra Flagelli[a], Michele Sannia[a], and Laura Calzà[b,c,d]

[a]Interdepartmental Center for Industrial Research in Health Sciences and Technologies, University of Bologna, Bologna, Italy
[b]Montecatone Rehabilitation Institute, Imola, Bologna, Italy
[c]IRET Foundation, Ozzano Emilia, Bologna, Italy
[d]Department of Pharmacy and Biotechnology, University of Bologna, Bologna, Italy
*Corresponding author: e-mail address: vito.baldassarro2@unibo.it

Contents

Abstract

Oligodendrocytes are the cells responsible for myelin formation during development and in adulthood, both for normal myelin turnover and myelin repair. These highly specialized cells derive from the oligodendrocyte precursor cells (OPCs), through a complex differentiation process involving genetic and epigenetic regulation mechanisms, which switch the phenotype from a migratory and replicative precursor to a mature post-mitotic cell. The process is regulated by a plethora of molecules, involving neurotransmitters, growth factors, hormones and other small molecules, and is mainly driven by nuclear receptors (NRs). NRs are transcription factors with heterogeneous ligand-dependent and independent actions which differ for the cell target, the responsive gene and the formation of NR homo- or heterodimers.

This chapter highlights the role of NRs in regulating OPC differentiation, also in view of drug discovery strategies aimed at targeting pathological conditions which interfere with both developmental myelination and remyelination in adulthood.

Vitamins and Hormones, Volume 116
ISSN 0083-6729
https://doi.org/10.1016/bs.vh.2021.02.002

1. Nuclear receptor biology

The nuclear receptor (NR) superfamily is composed of seven subfamilies of transcription factors (48 members total in humans) playing fundamental roles in different biological processes. NRs present a common structure, consisting of a variable N-terminal domain (NTD), a conserved DNA-binding domain (DBD) containing two zinc-finger motifs (domain C), a variable and flexible linker known as a hinge region (domain D), and a 12-helical ligand-binding domain (LBD; domain E). Some NRs also contain a variable C-terminal (domain F) (Rastinejad, Huang, Chandra, & Khorasanizadeh, 2013) (Fig. 1A).

NRs are activated by a specific ligand-receptor interaction, which produces a conformational change leading to the binding of the NR to

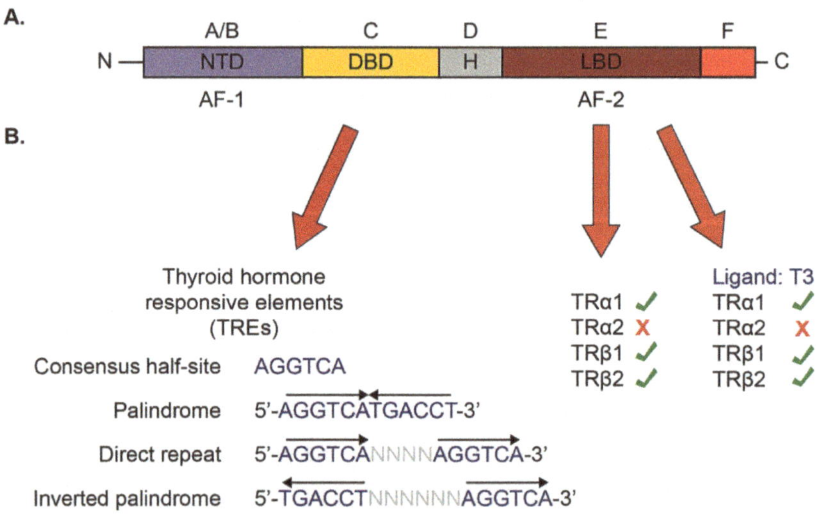

Fig. 1 Nuclear receptors. (A) Nuclear receptors (NRs) consist of an N-terminal domain (NTD; domain A/B), a conserved DNA-binding domain (DBD; domain C), a variable hinge region (H, domain D), a ligand-binding domain (LBD, domain E), and in some cases a variable C-terminal domain (F). The NTD domain contains the activation function-1 (AF-1) involved in ligand-independent transactivation, while the LBD contains activation function-2 (AF-2), the ligand-dependent coregulator binding site, and the hormone/ligand binding pocket (LBP). All domains undergo post-translational modifications, are involved in protein-protein interactions, and are involved in dimerization (De Bosscher et al., 2020). (B) Thyroid hormone receptors bind specific DNA sequences called thyroid hormone responsive elements (TREs) identified by specific palindrome, direct repeat and inverted palindrome sequences. Of the four TR isoforms, TRα2 is the only one which does not contain AF-2 and is unable to bind triiodothyronine (T3).

specific DNA sequences, and the recruitment of co-regulators, chromatin remodelers, and the transcription machinery. Although small lipophilic ligands have been identified for the different NRs (steroids, retinoids, phospholipids), some receptors remain "orphan," with no ligands yet identified (Weikum, Liu, & Ortlund, 2018).

NRs form a wide variety of dimeric states, binding a huge number of DNA response elements and, depending on the recruited complexes, may act as transcription repressor or promoter. Classically, it has been defined that in the absence of the ligand, NRs recruit co-repressors such as N-CoR1 and SMRT, which in turn recruit histone deacetylases. When the receptor is activated by the ligand, the change in the conformation of the protein leads to the recruitment of positive co-regulators of DNA transcription, usually harboring histone acetyltransferase activity (Rastinejad et al., 2013).

Depending on their structure, NRs can act as monomers, homodimers (identified for most steroid receptors), and heterodimers. RXRs (retinoic X receptors), for example, are a typical component in heterodimers, binding different partners: LXR (liver X receptor), PPAR (peroxisome proliferation activated receptor), PXR (pregnane X receptor), RAR (retinoic acid receptor), TR (thyroid receptor), VDR (vitamin D receptor), and other orphan receptors (De Bosscher, Desmet, Clarisse, Estébanez-Perpiña, & Brunsveld, 2020).

The importance of dimerization was highlighted by the isolation of the orphan receptor RXR and the identification of its endogenous ligand (9-*cis* retinoic acid), defining the first "adopted" orphan receptor, and leading to what was termed the "big bang" of NR biology. The expanding wave of discovery included the ability of RXRs to heterodimerize, the characterization of dietary lipid metabolites as ligands for RXR-partnered orphan receptors, and the elucidation of endocrine and paracrine signaling mediated by heterodimers and their role in developmental and metabolic pathways (Evans & Mangelsdorf, 2014).

In recent years, further insight into NRs has been gained by the *in vitro* identification of atypical heterodimers not involving RXRs, potentially implicated in different physiological and pathological processes (De Bosscher et al., 2020).

This complexity has led to the identification of at least 10 different mechanisms of direct and indirect crosstalk between the nuclear receptors at DNA level, also showing context-specific differences (e.g., tissue or cell type) (De Bosscher et al., 2020).

Since NR action is regulated by ligands, it is possible to act pharmacologically on their biology: indeed they show a druggable nature accounting for the 3% of all human drug targets, and are responsible for the therapeutic effect of 16% of all drugs on the market (Santos et al., 2016).

2. From neural precursor to mature oligodendrocyte: Riding the waves

NRs play a fundamental role in the development and biology of stem cells, both in embryonic/fetal and adult stem cells, being involved in stemness and differentiation, adipogenesis, myogenesis, chondrogenesis, osteoblastogenesis, and neural stem cell proliferation and differentiation (Jeong & Mangelsdorf, 2009). NRs are also key players in neuroectodermal differentiation, leading to oligodendrocyte generation.

Oligodendrocytes (OLs) are the myelin forming cells in the central nervous system (CNS), differentiating from the glial precursor known as the oligodendrocyte precursor cell (OPC), first discovered by Raff et al. in 1983 (Raff, Miller, & Noble, 1983) and recognized as the NG2 expressing cell for NG2 proteoglycan expression on the plasma membrane.

OPCs arise from multipotent neural precursor/stem cells (NSCs), which also produce neurons and astrocytes, and give rise to OPCs in three generation waves separated spatially and temporally. At mid-gestation in rodents, OPCs are detectable in the ventricular germinal zones (VZs) in the brain and spinal cord: in particular, the radial glia in ventral VZs give rise to OPCs at E12.5 in mice, E14 in rats and E45 in humans, which rapidly migrate in all directions and proliferate, colonizing the entire developing CNS at E15 in mice.

In the telencephalon, OPCs first appear in the ventral VZ of the medial ganglionic eminence (MGE) at E12.5 in mice, and precursor generation starts soon after from the lateral ganglionic eminence (LGE) in a second wave of precursors which migrate in a lateral-to medial direction from E16. After birth, a third-generation wave arises from the cortical VZ, and the previously produced MGE-derived OPCs are gradually eliminated.

In the spinal cord, two waves occur, the first on the ventral side, generated by the same precursors which produce the motor neurons (pMN progenitors). Soon after this event, around E15.5 in mice, a second wave of OPCs starts on the dorsal side of the spinal cord, which transdifferentiate from the radial glia. Most of these precursors differentiate into astrocytes, but a subset starts to express OPC markers while keeping its endfeet attached to the pial undersurface, in a process which lowers their migratory capability. These poorly migratory OPCs remain on the dorsal side where they gradually displace the previously ventrally derived precursors, reaching 20% of the total OPCs in the mouse spinal cord, while the other 80% is produced by the first wave of pMNs. During development, part of the generated OPCs will maturate as OLs, wrapping the cell membrane around the nude axon and producing the myelin sheath; another part remains undifferentiated in a relatively quiescent state in both the white and gray matter.

The complex differentiation/maturation mechanism is orchestrated by a plethora of factors leading to a tight control of transcriptional regulation (Emery & Lu, 2015), occurring as early as the network which regulates the fate of the NSCs. Multiple gene expression and epigenetic regulation occur throughout the entire process, which also defines different stages of the process, from undifferentiated precursor, through the pre-OL state to mature/myelinating OL. Different markers appear and disappear depending on the differentiation state, shading from one stage to the next and in parallel with the changes in cell morphology.

Different NRs have been identified as fundamental regulators of overall CNS development, including TRs (thyroid hormone receptors), RXRs (retinoid X receptors), RARs (retinoic acid receptors), PPARs (peroxisome proliferator-activated receptors), LXRs (liver X receptors), VDR (vitamin D receptor) and COUP-TF (chicken ovalbumin upstream promoter transcription factor) (Elbaz & Popko, 2019; Olivares, Moreno-Ramos, & Haider, 2015). Other nuclear receptors binding progesterone (PR), androgen (AR) and estrogen (ER) also appear to be involved in the process, regulating general homeostasis, cell proliferation, differentiation, and myelination (Ghoumari, Abi Ghanem, Asbelaoui, Schumacher, & Hussain, 2020).

Specific NRs participate in the control of NSC homeostasis, affecting the ability of the multipotent stem cell to proliferate or differentiate into one of the three lineages, and the TRs emerge as fundamental regulators of both NSC homeostasis and OPC differentiation in this context[a]. The action of these NRs is tightly controlled in target tissues and organs by a local regulation of their ligand, i.e., TH. The prohormone tetraiodothyronine (L-thyroxine, T4), secreted by the thyroid gland, is converted to the active form L-triiodothyronine (T3) in target tissues by specific enzymes and

[a] Given the topic of the chapter, only genomic affects will be considered. THs exert a number of biological functions through non-genomic mechanisms, part of which are related to the mitochondria, and part to plasma membrane binding sites. These topics are covered by excellent reviews (Davis, Goglia, & Leonard, 2016).

transporters, and the local concentrations are regulated by the synthesis and inactivation of deiodinase enzymes (Calzà et al., 2018).

The TRs include two different members, TRα (NR1S1) and TRβ (NR1A2), activated by ligand interaction (thyroid hormones, THs) and binding specific DNA regions indicated as thyroid hormone response elements (TREs) (Fig. 1B). These NRs belong to Class II, meaning that they predominantly form heterodimers with RXRs (Evans & Mangelsdorf, 2014), an NR widely studied for its role in remyelination, but they also act as heterodimers with other NRs, such as VDR and PPAR (De La Fuente et al., 2015).

Various TR-responsive genes have been identified in the CNS (e.g., *Dpb*, albumin site D-binding protein and *Klf9*, kruppel-like factor 9), some of which overlap with T3 target genes in amphibian metamorphosis (e.g., *Ezh2*, enhancer of the zaste 2 polycomb repressive complex 2 subunit) (Chatonnet, Flamant, & Morte, 2015). Through the same genomic action, T3 also regulates its own production, secretion and local metabolism, working as a transcription suppressor through the binding to TRβ2 for TSH (thyroid stimulating hormone), the main regulator of the hypothalamic-pituitary-thyroid axis (Hirahara et al., 2020), and for TRH (thyrotropin releasing hormone) through TRβ (Dupré et al., 2004) and the ligand-independent TRa2 action (Guissouma et al., 2014), allowing the T3-mediated negative feedback of the system which regulates circulating TH levels. At tissue level, it directly activates the expression of *Dio3* (iodothyronine deiodinases 3), the enzyme responsible for the inactivation of both T3 and T4 (Gil-Ibáñez, Bernal, & Morte, 2014).

THs are involved in neuronal differentiation and migration, synaptogenesis and myelination, as shown by the severe cognitive deficits and locomotor disorders in individuals with cretinism, a condition caused by untreated TH deficiency during pregnancy, and which leads to underdevelopment of the brain. It has also been shown that the absence of TRα leads to dwarfism, metabolic disorders and brain underdevelopment, while a TRβ deficiency leads to learning, memory and development deficiencies (Ortiga-Carvalho, Sidhaye, & Wondisford, 2014). In OPC biology, T3 acts as a cell cycle regulator, pushing to exit the cell cycle, mediated by a downregulation of TRα and upregulation of TRβ1 expression, the isoform that drives the final differentiation. This process leads to the morphological and functional maturation of the OL, characterized by the expression of specific proteins such as MBP (myelin basic protein) and PLP (proteolipid protein) (Kim & Petratos, 2019) (Fig. 2). In mature OLs, the promoter region of MBP contains a TR binding site (Farsetti, Mitsuhashi, Desvergne, Robbins, & Nikodem, 1991).

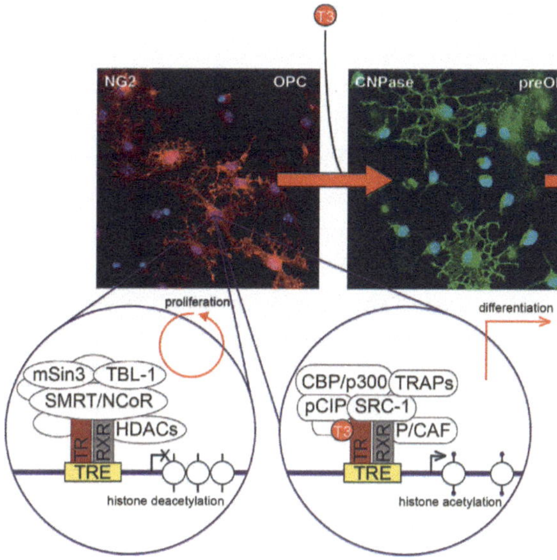

Fig. 2 Oligodendrocyte precursor cell differentiation. Thyroid hormone receptor (TR) works mainly as a heterodimer with retinoic X receptor (RXR), binding specific thyroid hormone responsive elements (TREs) in the DNA. When triiodothyronine (T3) is not present, corepressors are recruited (e.g., mSin3, TBL1 and SMRT/NCoR) together with histone deacetylases (HDAC), blocking the transcription of target genes and leaving the oligodendrocyte precursor cell (OPC) in a proliferative state. The binding of T3 generates a conformational change in the complex, allowing the recruitment of co-activators (e.g., CBP/p300, TRAPs, pCIP, SRC-1, P/CAF) and histone acetylation, activating the transcription of target genes. Thus, T3 pushes the OPC out of the cell cycle, starting the differentiation process, from precursor to pre-oligodendrocyte (pre-OL) and finally reaching the full maturation stage (mature oligodendrocytes). The figure shows original pictures of neural stem cell-derived OPC cultures, using specific markers to identify the differentiation/maturation stages of the OL lineage: NG2 for OPCs, CNPase for pre-OLs and double positivity for CNPase (green) and MBP (red) for mature OLs. Abbreviations: HDAC, histone deacetylases; mSin3, Paired amphipathic helix protein; SMRT/NCoR, Silencing mediator of retinoic acid and thyroid hormone receptor/nuclear receptor corepressor; TBL-1, transducin β-like protein-1; TRAPs, triiodothyronine receptor auxiliary proteins; CBP/p300, CREB binding protein/histone acetyltransferase; pCIP, p300/CBP-interacting protein, histone acetyltransferase; SRC-1, steroid receptor coactivator-1; P/CAF, p300/CBP-associated factor, histone acetyltransferase; NG2, neural/glial antigen 2, Chondroitin sulfate proteoglycan; CNPase, 2′,3′-Cyclic-nucleotide 3′-phosphodiesterase; MBP, myelin basic protein.

TRs also regulate expression in absence of the ligand, usually acting as repressors. When un-ligated, TRs recruit a multiprotein complex which includes corepressors, such as NCoR1 or SMRT, mediating the transcriptional repression by recruiting other factors responsible for histone deacetylation. The conformational change of the receptors which occurs in the case of T3 binding causes the dismissal of corepressors and recruitment of coregulators (e.g., SRC family and p300/CBP) and leads to histone acetylation and transcriptional activation (Fig. 2). From this classical description of the T3-mediated mechanism, it is clear that T3-sensitivity is not only linked to T3 intracellular concentration, but also to cellular levels of both TR-specific corepressors and coactivators (Mendoza & Hollenberg, 2017). However, the role of TRs and co-regulators is not fully understood, since they also show a ligand-independent gene activation (Astapova et al., 2008; Joseph-Bravo, Jaimes-Hoy, & Charli, 2015).

Another fundamental NR ligand in the regulation of OPC homeostasis and differentiation during development is retinoic acid (RA), a molecule synthesized from dietary vitamin A (retinol) after two oxidative steps (Napoli, 2012). The role of RA in OPC differentiation, along with that of THs and glucocorticoids, was proven in a seminal *in vitro* study by Raff's group in 1994 (Barres, Lazar, & Raff, 1994). Its activity in myelin development has recently been proven by experiments knocking down the expression of one of the two enzymes involved in RA production, the retinaldehyde dehydrogenase 2 (RALDH2), which leads to an impairment in OPC number and differentiation (Morrison, Smith, & Huang, 2020).

While TH and RA emerge as the main players as NR ligands in the regulation of OPC differentiation machinery, other NRs also seem to play an important role in developmental myelination. LXR activity is crucial for the homeostasis of cholesterol, the major constituent of the myelin sheath, and different LXR isoforms seem to control myelination in different brain areas, such as the corpus callosum, cortex and cerebellum. Studies involving the cerebellum, a highly myelinated CNS structure, have proved that LXRβ expression increases postnatally with a peak at P21, which overlaps with the myelination period of this area (Meffre et al., 2015). LXRs interact with Wnt/β-Catenin signaling, regulating the transcription of MPZ (myelin protein zero) and PMP22 (peripheral myelin protein 22), two fundamental myelin proteins, as demonstrated by the observation of fewer layers in myelin sheaths (Makoukji et al., 2011), and hypomyelination of the corpus callosum and optic nerve (Xu et al., 2014) in LXRs −/− mice.

COUP-TFI gene depletion also results in delayed myelination and hypomyelination, due to a delay in OPC differentiation, while not affecting cell migration, proliferation or death (Yamaguchi et al., 2004).

Interestingly, most of the single NR knock-out animals are viable and show little or no evident deficits in myelination and myelin functions, while TR double knock-out mice show minor CNS abnormalities (Calzà, Forrest, Vennström, & Hökfelt, 2000). This is probably due to the redundancy of

NRs and of compensation mechanisms, which are missing in *in vitro* conditions, where strong effects on OPC differentiation and maturation can be observed, such as a depletion of NR expression, or inhibition of specific activity of these receptors (see Section 4).

3. Remyelination in adulthood: A developmental memory?

Part of the OPCs produced during the three generation waves that colonize the developing white and gray matter remains undifferentiated and quiescent, and gives rise to the pool of precursors which remain throughout adulthood (Hughes, Kang, Fukaya, & Bergles, 2013). Resident OPCs account for 5%–8% of total CNS cells, and represent the major proliferating population of the CNS (Fernandez-Castaneda & Gaultier, 2016). OPCs also guarantee normal myelin turnover in the healthy adult brain, where new myelin is continuously generated and remodeled (Yeung et al., 2014), and contribute to CNS plasticity (Sampaio-Baptista & Johansen-Berg, 2017).

OPCs are therefore not only responsible for developmental myelination and normal myelin turnover, but also for myelin repair (remyelination) during adulthood (Bergles & Richardson, 2016; Yeung et al., 2014). In fact, resident and newly generated OPCs from NSCs of the SVZ, when driven by specific stimuli including demyelination insults (Fukushima et al., 2015; Gonzalez-Perez, 2014), can proliferate, migrate to the nude axons, and differentiate into mature OLs, leading to complete anatomical and functional restoration (Crawford, Chambers, & Franklin, 2013; Trotter, Karram, & Nishiyama, 2010). Remyelination is currently the only known true regenerative capability of the CNS (Crawford et al., 2013).

The ability of OPCs to correctly establish or repair myelin sheath, however, is severely impaired during certain pathological conditions, both during development and adulthood, in which OPCs and OLs seem to be the most vulnerable cells in the CNS (Bradl & Lassmann, 2010; Lee et al., 2012).

Given that OPCs share the same biological objective during development and adulthood, it has been suggested that myelin repair in adulthood may resemble the developmental myelination mechanism (Franklin & Hinks, 1999), and this hypothesis drove research into remyelination-enhancing therapies for many years. Various pieces of evidence, however, suggest an alternative hypothesis, based on a wide range of differences in

OPC biology, proposing an age-dependent view (Baldassarro, Marchesini, Giardino, & Calzà, 2017; Miron, Kuhlmann, & Antel, 2011).

In addition, the vulnerability of cells belonging to the oligodendroglial lineage appears to be dependent on developmental and maturation stage (Amaral, Hadera, Tavares, Kotter, & Sonnewald, 2016). To further complicate this picture, the decrease in OL number with age should also be taken into account (Hill, Li, & Grutzendler, 2018), a phenomenon which possibly reflects a reduced capacity of the remaining OPCs to form new myelinating cells, leading to a decline in myelin content, in parallel with a physiological age-dependent reduction in white matter (Giorgio et al., 2010).

The different age-related properties of OPCs are already evident at the gestational stages, when human fetal OPCs exhibit differences in myelination profile (Cui et al., 2012): these differences are even more evident in human OPCs derived from the fetal brain during the period of maximum oligogenesis, in contrast to adult OPCs derived from the subcortical white matter (Windrem et al., 2004). In both cases, adult-derived OPCs myelinate more readily and are able to ensheath a higher number of axons compared to OPCs derived from the fetal brain. In addition, *in vitro* experiments demonstrate a difference in metabolism between OPCs derived from fetal and from adult NSCs, with fetal cells mostly glycolytic, reflecting a different vulnerability to noxious stimuli (Baldassarro, Marchesini, Giardino, & Calzà, 2019).

In a similar way to their involvement in embryonic development, NR ligands play a fundamental role in the mechanism of adult NSC differentiation toward an OL lineage. Systemic hypothyroidism leads to tissue and cellular hypothyroidism and subsequent dysfunctions in T3-dependent cellular activities, including OPC differentiation/maturation (Fernandez, Pirondi, Manservigi, Giardino, & Calzà, 2004). While even fetal and adult OPCs differ in their response to differentiating stimuli, the main mechanism inducing the differentiation remains unaltered, and the intracellular levels of T3 lead to proper remyelination capability, as occurs during embryonic development, and pathologies where this equilibrium is altered are clear evidence of this mechanism (Sandoval-Hernández, Contreras, Jaramillo, & Arboleda, 2016).

OPC differentiation in pathological conditions has been thoroughly investigated, in particular with regard to multiple sclerosis (MS), the most common demyelinating disease, and in related animal models. The prevalent clinical form of MS (remitting-relapsing) is characterized by the establishment of an inflammatory environment in small areas throughout the entire

CNS, which periodically worsen to cause demyelination, axonal degeneration, and brain atrophy over the years, which affects both the gray and white matter, and which is responsible for cognitive symptoms (Grassiot, Desgranges, Eustache, & Defer, 2009). While remyelination may occur efficiently throughout resident and NSCs-derived OPCs in the early stages of the disease (Rodriguez & Scheithauer, 1994), this repair process is progressively impaired, leading to remyelination failure. Given that numerous OPCs are present in those termed "fresh" demyelination plaques (Prineas et al., 1989), progressive remyelination failure is mainly attributed to the OPC differentiation block (Kuhlmann et al., 2008) mediated by inflammatory cytokines (Fernández, Baldassarro, Sivilia, Giardino, & Calzà, 2016). In this context, our laboratory provided *in vitro* and *in vivo* evidence of a direct involvement of local TH metabolism as a major cause of OPC differentiation block and the consequent failure of axonal remyelination (Calzà et al., 2018; Calzà, Fernandez, & Giardino, 2010; Calzà, Fernández, & Giardino, 2015): indeed, inflammation leads to the increase of Dio3 expression and activity in the tissue. Since Dio3 is the enzyme which deactivates both T3 and T4, this leads to a decrease in intracellular T3 concentration. When Dio3 activity is pharmacologically blocked, OPC differentiation is restored, thus proving a direct correlation between intracellular T3 content and OPC differentiation (Fernández et al., 2016). Notably, different animal models of MS have shown that systemic and local administration of THs increases local T3 availability and helps to combat remyelination failure (Calzà et al., 2018), discoveries which have led to safety and dose-finding clinical trials for the use of TH for MS (ClinicalTrials.gov Identifier: NCT02760056).

Other NR ligands which ameliorate MS-like symptoms in animal models are 9-*cis*-retinoic acid and 15-deoxy-12,14-prostaglandin J2 which specifically bind RXRs (Huang et al., 2011). Of the three types of RXR, all of which are differentially expressed in remyelination, RXRγ shows the greater differential expression, being highly expressed in OLs during remyelination and in OLs in patients with MS. In fact, in 2011 it was demonstrated for the first time that this NR is involved in OPC differentiation and promotes remyelination *in vivo* (Huang et al., 2011). Interestingly, RXRγ −/− mice are viable and normally developed, while adult animals show TH resistance and dysregulation of CNS homeostasis, leading to changes in metabolic activity and depressive behavior (Brown et al., 2000; Haugen et al., 2004; Krezel et al., 1996; Krzyzosiak et al., 2010). Moreover, the loss of RXRγ seems to be compensated when remyelination occurs in models of focal chemical demyelination obtained by lysolecithin injection (Huang et al., 2011). The RXR-VDR heterodimer, described as fundamental for OPC differentiation during development, also seems to play a key role in remyelination, as also highlighted by the increase of VDR expression in OL lineage in patients with MS (De La Fuente et al., 2015).

In addition to RXRs, the RA signaling pathway also promotes remyelination, driving OPC differentiation through RARα activity in different and complex pathological conditions, such as spinal cord injury (Goncalves et al., 2019).

ER ligand treatment also exerts protective effects on EAE susceptibility, revealing a possible role of these NRs in the remyelination mechanism (Crawford et al., 2010), which may be indirectly regulated by other NR ligands. This is the case of PPARα and PPARγ agonists which ameliorate EAE symptoms, inhibiting the microglia and astrocytes and preventing the release of pro-inflammatory cytokines (Drew, Xu, & Racke, 2008; Gocke et al., 2009).

4. *Ex vivo*: Evidence from the dish

In vitro cultures of primary OPCs from neonatal rodents are classically isolated from mixed cortical glial cultures by different shaking steps (O'Meara, Ryan, Colognato, & Kothary, 2011), or differentiated from fetal and adult NSCs (Chen et al., 2007). Both *in vitro* systems are powerful tools for investigating the mechanisms driving the differentiation and maturation of these cells.

Since the beginning of research on OPCs, *in vitro* models have been used to study the molecules involved in differentiation control. In the original experiments by Raff and collaborators on OPCs isolated from the rat optic nerve, it was described that OPC survival and proliferation requires the presence of astrocytes in the same culture; alternatively these may be substituted by adding PDGF to the medium (Raff, Abney, & Fok-Seang, 1985; Raff, Lillien, Richardson, Burne, & Noble, 1988). In particular, the proliferation/differentiation balance was described as being promptly regulated by the balance between PDGF (platelet derived growth factor), which keeps the cells proliferating, and T3, which pushes the OPCs out of the cell cycle toward differentiation into mature OLs, a process mediated by TRα1 (Billon, Tokumoto, Forrest, & Raff, 2001). However, further experiments in

OPCs derived from the cerebral cortex of rat pups have demonstrated that GC-1, a selective TRβ ligand, is sufficient to trigger the differentiation mechanism (Baxi et al., 2014).

In vitro models have demonstrated that T3 drives the OPC out of the cell cycle following a defined number of cell replications (Perez-Juste & Aranda, 1999) and directly regulates the expression level of different regulatory proteins (i.e., p18 and p27 Cdk) (Puzianowska-Kuznicka, Pietrzak, Turowska, & Nauman, 2006) (Tokumoto, Apperly, Gao, & Raff, 2002). Currently, the protocols driving OPC differentiation in a dish, from embryonic stem cells to mature OLs, are well defined, such as where T3 exerts its role in the differentiation pipeline (Goldman & Kuypers, 2015).

In the classical *in vitro* experiment on purified OPCs from the optic nerve, however, all tested molecules, i.e., T3, all-trans-RA and glucocorticoids, were able to activate the effector mechanisms of the OPC cell cycle exit and differentiation induction (Barres et al., 1994), suggesting an overlap or interactions between these nuclear receptor ligands. A number of authors have also described that activation of different RARs (NRs specific to all-trans-RA ligands) drives NSC differentiation, with RARα and RARγ leading to astrocytes and OL lineage specification, and RARβ inducing neuronal differentiation (Goncalves et al., 2009, 2005).

NRs also play specific roles in fetal NSC differentiation (Gkikas, Tsampoula, & Politis, 2017). For example, the *Trb* gene is highly expressed in NSCs isolated from fetal forebrain compared to adult SVZ-derived NSCs, matching with a lower expression of *Rxrg* (Baldassarro, Krężel, et al., 2019). A different NR expression profile may drive a different capability of OPCs to differentiate or respond to noxious stimuli, corroborating the hypothesis that OPC intrinsic factors are responsible for the differences between fetal and adult cells. In fact, differences in NR expression at the multipotent stage reflect a reduced capability of adult NSC-derived OPCs to give rise to mature OLs, pointing to a probable TRβ/RXRγ heterodimer equilibrium as being responsible for this different age-dependent yield.

TR/RXR dimers are seen as fundamental regulators of the process, even if RXRγ −/− animals are viable, as described above, through unknown compensation mechanisms. Since adult NSCs and NSC-derived OPCs show a lower expression in *Rxrg* mRNA, RXRγ −/− animals are an extreme phenotype of this expression profile. In fact, when translating OPC differentiation in a dish, precursors derived from both fetal and adult RXRγ −/− NSCs are unable to properly mature (Baldassarro, Krężel, et al., 2019), proving the fundamental contribution of this NR to T3-mediated differentiation.

5. Conclusion

NRs are the main drivers of the OPC differentiation and maturation machinery, although the complex network of the NR-mediated regulation is not yet entirely clear. Their druggable nature makes these transcription factors fundamental targets for developing therapeutic strategies aimed at treating pathological conditions which interfere with both myelination and remyelination.

References

Amaral, A. I., Hadera, M. G., Tavares, J. M., Kotter, M. R. N., & Sonnewald, U. (2016). Characterization of glucose-related metabolic pathways in differentiated rat oligodendrocyte lineage cells. *Glia*, *64*(1), 21–34. https://doi.org/10.1002/glia.22900.

Astapova, I., Lee, L. J., Morales, C., Tauber, S., Bilban, M., & Hollenberg, A. N. (2008). The nuclear corepressor, NCoR, regulates thyroid hormone action in vivo. *Proceedings of the National Academy of Sciences of the United States of America*, *105*(49), 19544–19549. https://doi.org/10.1073/pnas.0804604105.

Baldassarro, V. A., Krężel, W., Fernández, M., Schuhbaur, B., Giardino, L., & Calzà, L. (2019). The role of nuclear receptors in the differentiation of oligodendrocyte precursor cells derived from fetal and adult neural stem cells. *Stem Cell Research*, *37*, 101443. https://doi.org/10.1016/j.scr.2019.101443.

Baldassarro, V. A., Marchesini, A., Giardino, L., & Calzà, L. (2017). PARP activity and inhibition in fetal and adult oligodendrocyte precursor cells: Effect on cell survival and differentiation. *Stem Cell Research*, *22*, 54–60. https://doi.org/10.1016/J.SCR.2017.05.011.

Baldassarro, V. A., Marchesini, A., Giardino, L., & Calzà, L. (2019). Differential effects of glucose deprivation on the survival of fetal versus adult neural stem cells-derived oligodendrocyte precursor cells. *Glia*, *68*, 898–917. https://doi.org/10.1002/glia.23750.

Barres, B. A., Lazar, M. A., & Raff, M. C. (1994). A novel role for thyroid hormone, glucocorticoids and retinoic acid in timing oligodendrocyte development. *Development*, *120*(5), 1097–1108.

Baxi, E. G., Schott, J. T., Fairchild, A. N., Kirby, L. A., Karani, R., Uapinyoying, P., et al. (2014). A selective thyroid hormone β receptor agonist enhances human and rodent oligodendrocyte differentiation. *Glia*, *62*(9), 1513–1529. https://doi.org/10.1002/glia.22697.

Bergles, D. E., & Richardson, W. D. (2016). Oligodendrocyte development and plasticity. *Cold Spring Harbor Perspectives in Biology*, *8*(2). https://doi.org/10.1101/cshperspect.a020453.

Billon, N., Tokumoto, Y., Forrest, D., & Raff, M. (2001). Role of thyroid hormone receptors in timing oligodendrocyte differentiation. *Developmental Biology*, *235*(1), 110–120. https://doi.org/10.1006/dbio.2001.0293.

Bradl, M., & Lassmann, H. (2010). Oligodendrocytes: Biology and pathology. *Acta Neuropathologica*, *119*, 37–53. https://doi.org/10.1007/s00401-009-0601-5.

Brown, N. S., Smart, A., Sharma, V., Brinkmeier, M. L., Greenlee, L., Camper, S. A., et al. (2000). Thyroid hormone resistance and increased metabolic rate in the RXR-γ-deficient mouse. *Journal of Clinical Investigation*, *106*(1), 73–79. https://doi.org/10.1172/JCI9422.

Calzà, L., Baldassarro, V. A., Fernandez, M., Giuliani, A., Lorenzini, L., & Giardino, L. (2018). Thyroid hormone and the white matter of the central nervous system: From development to repair. *Vitamins and Hormones*, *106*, 253–281. https://doi.org/10.1016/BS.VH.2017.04.003.

Calzà, L., Fernandez, M., & Giardino, L. (2010). Cellular approaches to central nervous system remyelination stimulation: Thyroid hormone to promote myelin repair via endogenous stem and precursor cells. *Journal of Molecular Endocrinology*, *44*, 13–23. https://doi.org/10.1677/JME-09-0067.

Calzà, L., Fernández, M., & Giardino, L. (2015). Role of the thyroid system in myelination and neural connectivity. *Comprehensive Physiology*, *5*(3), 1405–1421. https://doi.org/10.1002/cphy.c140035.

Calzà, L., Forrest, D., Vennström, B., & Hökfelt, T. (2000). Expression of peptides and other neurochemical markers in hypothalamus and olfactory bulb of mice devoid of all known thyroid hormone receptors. *Neuroscience*, *101*(4), 1001–1012. https://doi.org/10.1016/S0306-4522(00)00420-6.

Chatonnet, F., Flamant, F., & Morte, B. (2015). A temporary compendium of thyroid hormone target genes in brain. *Biochimica et Biophysica Acta - Gene Regulatory Mechanisms*, *1849*, 122–129. Elsevier. https://doi.org/10.1016/j.bbagrm.2014.05.023.

Chen, Y., Balasubramaniyan, V., Peng, J., Hurlock, E. C., Tallquist, M., Li, J., et al. (2007). Isolation and culture of rat and mouse oligodendrocyte precursor cells. *Nature Protocols*, *2*, 1044–1051. https://doi.org/10.1038/nprot.2007.149.

Crawford, A. H., Chambers, C., & Franklin, R. J. M. (2013). Remyelination: The true regeneration of the central nervous system. *Journal of Comparative Pathology*, *149*, 242–254. https://doi.org/10.1016/j.jcpa.2013.05.004.

Crawford, D. K., Mangiardi, M., Song, B., Patel, R., Du, S., Sofroniew, M. V., et al. (2010). Oestrogen receptor β ligand: A novel treatment to enhance endogenous functional remyelination. *Brain*, *133*(10), 2999–3016. https://doi.org/10.1093/brain/awq237.

Cui, Q. L., D'Abate, L., Fang, J., Leong, S. Y., Ludwin, S., Kennedy, T. E., et al. (2012). Human fetal oligodendrocyte progenitor cells from different gestational stages exhibit substantially different potential to myelinate. *Stem Cells and Development*, *21*(11), 1831–1837. https://doi.org/10.1089/scd.2011.0494.

Davis, P. J., Goglia, F., & Leonard, J. L. (2016). Nongenomic actions of thyroid hormone. *Nature Reviews Endocrinology*, *12*, 111–121. Nature Publishing Group. https://doi.org/10.1038/nrendo.2015.205.

De Bosscher, K., Desmet, S. J., Clarisse, D., Estébanez-Perpiña, E., & Brunsveld, L. (2020). Nuclear receptor crosstalk — Defining the mechanisms for therapeutic innovation. *Nature Reviews Endocrinology*, *16*, 363–377. https://doi.org/10.1038/s41574-020-0349-5.

De La Fuente, A. G., Errea, O., Van Wijngaarden, P., Gonzalez, G. A., Kerninon, C., Jarjour, A. A., et al. (2015). Vitamin D receptor-retinoid X receptor heterodimer signaling regulates oligodendrocyte progenitor cell differentiation. *Journal of Cell Biology*, *211*(5), 975–985. https://doi.org/10.1083/jcb.201505119.

Drew, P. D., Xu, J., & Racke, M. K. (2008). PPAR-γ: Therapeutic potential for multiple sclerosis. *PPAR Research*, *2008*. https://doi.org/10.1155/2008/627463.

Dupré, S. M., Guissouma, H., Flamant, F., Seugnet, I., Scanlan, T. S., Baxter, J. D., et al. (2004). Both thyroid hormone receptor (TR)β1 and TRβ2 isoforms contribute to the regulation of hypothalamic thyrotropin-releasing hormone. *Endocrinology*, *145*(5), 2337–2345. https://doi.org/10.1210/en.2003-1209.

Elbaz, B., & Popko, B. (2019). Molecular control of oligodendrocyte development. *Trends in Neurosciences*, *42*, 263–277. Elsevier Ltd. https://doi.org/10.1016/j.tins.2019.01.002.

Emery, B., & Lu, Q. R. (2015). Transcriptional and epigenetic regulation of oligodendrocyte development and myelination in the central nervous system. *Cold Spring Harbor Perspectives in Biology*, 7(9). https://doi.org/10.1101/cshperspect.a020461.

Evans, R. M., & Mangelsdorf, D. J. (2014). Nuclear receptors, RXR, and the big bang. *Cell, 157*, 255–266. Cell Press. https://doi.org/10.1016/j.cell.2014.03.012.

Farsetti, A., Mitsuhashi, T., Desvergne, B., Robbins, J., & Nikodem, V. M. (1991). Molecular basis of thyroid hormone regulation of myelin basic protein gene expression in rodent brain. *Journal of Biological Chemistry, 266*(34), 23226–23232. Retrieved from https://pubmed.ncbi.nlm.nih.gov/1720778/.

Fernández, M., Baldassarro, V. A., Sivilia, S., Giardino, L., & Calzà, L. (2016). Inflammation severely alters thyroid hormone signaling in the central nervous system during experimental allergic encephalomyelitis in rat: Direct impact on OPCs differentiation failure. *Glia, 64*(9), 1573–1589. https://doi.org/10.1002/glia.23025.

Fernandez, M., Pirondi, S., Manservigi, M., Giardino, L., & Calzà, L. (2004). Thyroid hormone participates in the regulation of neural stem cells and oligodendrocyte precursor cells in the central nervous system of adult rat. *European Journal of Neuroscience, 20*(8), 2059–2070. https://doi.org/10.1111/j.1460-9568.2004.03664.x.

Fernandez-Castaneda, A., & Gaultier, A. (2016). Adult oligodendrocyte progenitor cells – Multifaceted regulators of the CNS in health and disease. In *Brain, behavior, and immunity*. Cambridge, MA, USA: Academic Press Inc. https://doi.org/10.1016/j.bbi.2016.01.005.

Franklin, R. J. M., & Hinks, G. L. (1999). Understanding CNS remyelination: Clues from developmental and regeneration biology. *Journal of Neuroscience Research, 58*(2), 207–213. https://doi.org/10.1002/(SICI)1097-4547(19991015)58:2<207::AID-JNR1>3.0.CO;2-1.

Fukushima, S., Nishikawa, E., Furube, E., Muneoka, S., Ono, K., Takebayashi, H., et al. (2015). Oligodendrogenesis in the fornix of adult mouse brain; the effect of LPS-induced inflammatory stimulation. *Brain Research, 1627*, 52–69. https://doi.org/10.1016/j.brainres.2015.09.011.

Ghoumari, A. M., Abi Ghanem, C., Asbelaoui, N., Schumacher, M., & Hussain, R. (2020). Roles of progesterone, testosterone and their nuclear receptors in central nervous system myelination and Remyelination. *International Journal of Molecular Sciences, 21*(9), 3163. https://doi.org/10.3390/ijms21093163.

Gil-Ibáñez, P., Bernal, J., & Morte, B. (2014). Thyroid hormone regulation of gene expression in primary cerebrocortical cells: Role of thyroid hormone receptor subtypes and interactions with retinoic acid and glucocorticoids. *PLoS One, 9*(3), 91692. https://doi.org/10.1371/journal.pone.0091692.

Giorgio, A., Santelli, L., Tomassini, V., Bosnell, R., Smith, S., De Stefano, N., et al. (2010). Age-related changes in grey and white matter structure throughout adulthood. *NeuroImage, 51*(3), 943–951. https://doi.org/10.1016/j.neuroimage.2010.03.004.

Gkikas, D., Tsampoula, M., & Politis, P. K. (2017). Nuclear receptors in neural stem/progenitor cell homeostasis. *Cellular and Molecular Life Sciences, 74*, 4097–4120. Birkhauser Verlag AG. https://doi.org/10.1007/s00018-017-2571-4.

Gocke, A. R., Hussain, R. Z., Yang, Y., Peng, H., Weiner, J., Ben, L.-H., et al. (2009). Transcriptional modulation of the immune response by peroxisome proliferator-activated receptor-α agonists in autoimmune disease. *The Journal of Immunology, 182*(7), 4479–4487. https://doi.org/10.4049/jimmunol.0713927.

Goldman, S. A., & Kuypers, N. J. (2015). How to make an oligodendrocyte. *Development (Cambridge), 142*(23), 3983–3995. https://doi.org/10.1242/dev.126409.

Goncalves, M. B. C. V., Agudo, M., Connor, S., McMahon, S., Minger, S. L., Maden, M., et al. (2009). Sequential RARβ and α signalling in vivo can induce adult forebrain neural progenitor cells to differentiate into neurons through Shh and FGF signalling pathways. *Developmental Biology, 326*(2), 305–313. https://doi.org/10.1016/j.ydbio.2008.11.018.

Goncalves, M. B. C. V., Boyle, J., Webber, D. J., Hall, S., Minger, S. L., & Corcoran, J. P. T. (2005). Timing of the retinoid-signalling pathway determines the expression of neuronal markers in neural progenitor cells. *Developmental Biology, 278*(1), 60–70. https://doi.org/10.1016/j.ydbio.2004.10.015.

Goncalves, M. B., Wu, Y., Clarke, E., Grist, J., Hobbs, C., Trigo, D., et al. (2019). Regulation of myelination by exosome associated retinoic acid release from NG2-positive cells. *Journal of Neuroscience, 39*(16), 3013–3027. https://doi.org/10.1523/JNEUROSCI.2922-18.2019.

Gonzalez-Perez, O. (2014). The ventricular-subventricular zone: A source of oligodendrocytes in the adult brain. *Frontiers in Cellular Neuroscience, 8*, 137–138. Frontiers Research Foundation. https://doi.org/10.3389/fncel.2014.00137.

Grassiot, B., Desgranges, B., Eustache, F., & Defer, G. (2009). Quantification and clinical relevance of brain atrophy in multiple sclerosis: A review. *Journal of Neurology, 256*, 1397–1412. https://doi.org/10.1007/s00415-009-5108-4.

Guissouma, H., Ghaddab-Zroud, R., Seugnet, I., Decherf, S., Demeneix, B., & Clerget-Froidevaux, M. S. (2014). TR alpha 2 exerts dominant negative effects on hypothalamic Trh transcription in vivo. *PLoS One, 9*(4). https://doi.org/10.1371/journal.pone.0095064.

Haugen, B. R., Jensen, D. R., Sharma, V., Pulawa, L. K., Hays, W. R., Krezel, W., et al. (2004). Retinoid X receptor γ-deficient mice have increased skeletal muscle lipoprotein lipase activity and less weight gain when fed a high-fat diet. *Endocrinology, 145*(8), 3679–3685. https://doi.org/10.1210/en.2003-1401.

Hill, R. A., Li, A. M., & Grutzendler, J. (2018). Lifelong cortical myelin plasticity and age-related degeneration in the live mammalian brain. *Nature Neuroscience, 21*(5), 683–695. https://doi.org/10.1038/s41593-018-0120-6.

Hirahara, N., Nakamura, H. M., Sasaki, S., Matsushita, A., Ohba, K., Kuroda, G., et al. (2020). Liganded T3 receptor β2 inhibits the positive feedback autoregulation of the gene for GATA2, a transcription factor critical for thyrotropin production. *PLoS One, 15*(1), e0227646. https://doi.org/10.1371/journal.pone.0227646.

Huang, J. K., Jarjour, A. A., Oumesmar, B. N., Kerninon, C., Williams, A., Krezel, W., et al. (2011). Retinoid X receptor gamma signaling accelerates CNS remyelination. *Nature Neuroscience, 14*(1), 45–55. https://doi.org/10.1038/nn.2702.

Hughes, E. G., Kang, S. H., Fukaya, M., & Bergles, D. E. (2013). Oligodendrocyte progenitors balance growth with self-repulsion to achieve homeostasis in the adult brain. *Nature Neuroscience, 16*(6), 668–676. https://doi.org/10.1038/nn.3390.

Jeong, Y., & Mangelsdorf, D. J. (2009). Nuclear receptor regulation of stemness and stem cell differentiation. *Experimental and Molecular Medicine, 41*, 525–537. Nature Publishing Group. https://doi.org/10.3858/emm.2009.41.8.091.

Joseph-Bravo, P., Jaimes-Hoy, L., & Charli, J. L. (2015). Regulation of TRH neurons and energy homeostasis-related signals under stress. *Journal of Endocrinology, 224*, R139–R159. BioScientifica Ltd. https://doi.org/10.1530/JOE-14-0593.

Kim, M. J., & Petratos, S. (2019). Oligodendroglial lineage cells in thyroid hormone-deprived conditions. *Stem Cells International*. https://doi.org/10.1155/2019/5496891.

Krezel, W., Dupé, V., Mark, M., Dierich, A., Kastner, P., & Chambon, P. (1996). RXRγ null mice are apparently normal and compound RXRα+/−/RXRβ−/−/RXRγ−/− mutant mice are viable. *Proceedings of the National Academy of Sciences of the United States of America, 93*(17), 9010–9014. https://doi.org/10.1073/pnas.93.17.9010.

Krzyzosiak, A., Szyszka-Niagolov, M., Wietrzych, M., Gobaille, S., & Muramatsu, S. ichi, & Krezel, W. (2010). Retinoid X receptor gamma control of affective behaviors involves dopaminergic signaling in mice. *Neuron, 66*(6), 908–920. https://doi.org/10.1016/j.neuron.2010.05.004.

Kuhlmann, T., Miron, V., Cuo, Q., Wegner, C., Antel, J., & Bruck, W. (2008). Differentiation block of oligodendroglial progenitor cells as a cause for remyelination failure in chronic multiple sclerosis. *Brain, 131*(7), 1749–1758. https://doi.org/10.1093/brain/awn096.

Lee, Y., Morrison, B. M., Li, Y., Lengacher, S., Farah, M. H., Hoffman, P. N., et al. (2012). Oligodendroglia metabolically support axons and contribute to neurodegeneration. *Nature, 487*(7408), 443–448. https://doi.org/10.1038/nature11314.

Makoukji, J., Shackleford, G., Meffre, D., Grenier, J., Liere, P., Lobaccaro, J. M. A., et al. (2011). Interplay between LXR and Wnt/β-catenin signaling in the negative regulation of peripheral myelin genes by oxysterols. *Journal of Neuroscience, 31*(26), 9620–9629. https://doi.org/10.1523/JNEUROSCI.0761-11.2011.

Meffre, D., Shackleford, G., Hichor, M., Gorgievski, V., Tzavara, E. T., Trousson, A., et al. (2015). Liver X receptors alpha and beta promote myelination and remyelination in the cerebellum. *PNAS, 112*(24), 7587–7592. https://doi.org/10.1073/pnas.1424951112.

Mendoza, A., & Hollenberg, A. N. (2017). New insights into thyroid hormone action. *Pharmacology and Therapeutics, 173*, 135–145. Elsevier Inc. https://doi.org/10.1016/j.pharmthera.2017.02.012.

Miron, V. E., Kuhlmann, T., & Antel, J. P. (2011). Cells of the oligodendroglial lineage, myelination, and remyelination. *BBA - Molecular Basis of Disease, 1812*, 184–193. https://doi.org/10.1016/j.bbadis.2010.09.010.

Morrison, V. E., Smith, V. N., & Huang, J. K. (2020). Retinoic acid is required for oligodendrocyte precursor cell production and differentiation in the postnatal mouse corpus callosum. *ENeuro*, 7(1). https://doi.org/10.1523/ENEURO.0270-19.2019.

Napoli, J. L. (2012). Physiological insights into all-trans-retinoic acid biosynthesis. *Biochimica et Biophysica Acta - Molecular and Cell Biology of Lipids, 1821*, 152–167. https://doi.org/10.1016/j.bbalip.2011.05.004.

Olivares, A. M., Moreno-Ramos, O. A., & Haider, N. B. (2015). Role of nuclear receptors in central nervous system development and associated diseases. *Journal of Experimental Neuroscience, 9*(Suppl. 2), 93–121. SAGE Publications Ltd. https://doi.org/10.4137/JEN.S25480.

O'Meara, R. W., Ryan, S. D., Colognato, H., & Kothary, R. (2011). Derivation of enriched oligodendrocyte cultures and oligodendrocyte/neuron myelinating co-cultures from post-natal murine tissues. *Journal of Visualized Experiments, 54*, e3324. https://doi.org/10.3791/3324.

Ortiga-Carvalho, T. M., Sidhaye, A. R., & Wondisford, F. E. (2014). Thyroid hormone receptors and resistance to thyroid hormone disorders. *Nature Reviews Endocrinology, 10*, 582–591. Nature Publishing Group. https://doi.org/10.1038/nrendo.2014.143.

Perez-Juste, G., & Aranda, A. (1999). The cyclin-dependent kinase inhibitor p27(Kip1) is involved in thyroid hormone-mediated neuronal differentiation. *Journal of Biological Chemistry, 274*(8), 5026–5031. https://doi.org/10.1074/jbc.274.8.5026.

Prineas, J. W., Kwon, E. E., Goldenberg, P. Z., Ilyas, A. A., Quarles, R. H., Benjamins, J. A., et al. (1989). Multiple sclerosis. Oligodendrocyte proliferation and differentiation in fresh lesions. *Laboratory Investigation, 61*(5), 489–503. Retrieved from https://pubmed.ncbi.nlm.nih.gov/2811298/.

Puzianowska-Kuznicka, M., Pietrzak, M., Turowska, O., & Nauman, A. (2006). *Thyroid hormones and their receptors in the regulation of cell proliferation.* Retrieved from www.actabp.pl.

Raff, M. C., Abney, E. R., & Fok-Seang, J. (1985). Reconstitution of a developmental clock in vitro: A critical role for astrocytes in the timing of oligodendrocyte differentiation. *Cell, 42*(1), 61–69. https://doi.org/10.1016/S0092-8674(85)80101-X.

Raff, M. C., Lillien, L. E., Richardson, W. D., Burne, J. F., & Noble, M. D. (1988). Platelet-derived growth factor from astrocytes drives the clock that times oligodendrocyte development in culture. *Nature, 333*(6173), 562–565. https://doi.org/10.1038/333562a0.

Raff, M. C., Miller, R. H., & Noble, M. (1983). A glial progenitor cell that develops in vitro into an astrocyte or an oligodendrocyte depending on culture medium. *Nature, 303*(5916), 390–396. https://doi.org/10.1038/303390a0.

Rastinejad, F., Huang, P., Chandra, V., & Khorasanizadeh, S. (2013). Understanding nuclear receptor form and function using structural biology. *Journal of Molecular Endocrinology, 51*, T1–T21. https://doi.org/10.1530/JME-13-0173.

Rodriguez, M., & Scheithauer, B. (1994). Ultrastructure of multiple sclerosis. *Ultrastructural Pathology*, *18*(1–2), 3–13. https://doi.org/10.3109/01913129409016267.
Sampaio-Baptista, C., & Johansen-Berg, H. (2017). White matter plasticity in the adult brain. *Neuron*, *96*, 1239–1251. Cell Press. https://doi.org/10.1016/j.neuron.2017.11.026.
Sandoval-Hernández, A., Contreras, M. J., Jaramillo, J., & Arboleda, G. (2016). Regulation of oligodendrocyte differentiation and myelination by nuclear receptors: Role in neurodegenerative disorders. *Advances in Experimental Medicine and Biology*, *949*, 287–310. https://doi.org/10.1007/978-3-319-40764-7_14.
Santos, R., Ursu, O., Gaulton, A., Bento, A. P., Donadi, R. S., Bologa, C. G., et al. (2016). A comprehensive map of molecular drug targets. *Nature Reviews Drug Discovery*, *16*(1), 19–34. https://doi.org/10.1038/nrd.2016.230.
Tokumoto, Y. M., Apperly, J. A., Gao, F. B., & Raff, M. C. (2002). Posttranscriptional regulation of p18 and p27 Cdk inhibitor proteins and the timing of oligodendrocyte differentiation. *Developmental Biology*, *245*(1), 224–234. https://doi.org/10.1006/dbio.2002.0626.
Trotter, J., Karram, K., & Nishiyama, A. (2010). NG2 cells: Properties, progeny and origin. *Brain Research Reviews*, *63*(1–2), 72–82. https://doi.org/10.1016/j.brainresrev.2009.12.006.
Weikum, E. R., Liu, X., & Ortlund, E. A. (2018). The nuclear receptor superfamily: A structural perspective. *Protein Science*, *27*, 1876–1892. Blackwell Publishing Ltd. https://doi.org/10.1002/pro.3496.
Windrem, M. S., Nunes, M. C., Rashbaum, W. K., Schwartz, T. H., Goodman, R. A., McKhann, G., et al. (2004). Fetal and adult human oligodendrocyte progenitor cell isolates myelinate the congenitally dysmyelinated brain. *Nature Medicine*, *10*(1), 93–97. https://doi.org/10.1038/nm974.
Xu, P., Xu, H., Tang, X., Xu, L., Wang, Y., Guo, L., et al. (2014). Liver X receptor β is essential for the differentiation of radial glial cells to oligodendrocytes in the dorsal cortex. *Molecular Psychiatry*, *19*(8), 947–957. https://doi.org/10.1038/mp.2014.60.
Yamaguchi, H., Zhou, C., Lin, S. C., Durand, B., Tsai, S. Y., & Tsai, M. J. (2004). The nuclear orphan receptor COUP-TFI is important for differentiation of oligodendrocytes. *Developmental Biology*, *266*(2), 238–251. https://doi.org/10.1016/j.ydbio.2003.10.038.
Yeung, M. S. Y., Zdunek, S., Bergmann, O., Bernard, S., Salehpour, M., Alkass, K., et al. (2014). Dynamics of oligodendrocyte generation and myelination in the human brain. *Cell*, *159*(4), 766–774. https://doi.org/10.1016/j.cell.2014.10.011.

CHAPTER FOURTEEN

Oxysterols and mesenchymal stem cell biology

Cadiele Oliana Reichert, Fabio Alessandro de Freitas, Débora Levy, and Sérgio Paulo Bydlowski*
Lipids, Oxidation and Cell Biology Team, Laboratory of Immunology (LIM19), Heart Institute (InCor), School of Medicine, University of São Paulo, São Paulo, Brazil
*Corresponding author: e-mail address: spbydlow@usp.br

Contents

Abstract

Mesenchymal stem cells have the ability to differentiate into several cell types when exposed to determined substances, including oxysterols. Oxysterols are cholesterol products derived from its auto-oxidation by reactive species or from enzymatic action. They are present in the body in low quantities under physiological conditions and exhibit several physiological and pharmacological actions according to both the types of oxysterol and tissue. Some of them are cytotoxic while others have been shown to promote cell differentiation through the action on several different receptors, such as nuclear LXR receptors and Smoothened receptor ligands. Here, we review the main pathways by which oxysterols have been associated with cell differentiation and death of mesenchymal stem cells.

1. Introduction

Mesenchymal stem cells (MSC) are multipotent cells with high capacity of proliferation, found in several tissues as bone marrow and adipose tissue. They are characterized by their capacity to differentiate *in vitro* in adipocytes, chondrocytes and osteoblasts under proper stimuli,

Vitamins and Hormones, Volume 116
ISSN 0083-6729
https://doi.org/10.1016/bs.vh.2021.02.004

while exhibiting or lacking several membrane markers. Besides, they have immunomodulatory and anti-inflammatory properties, and can promote tissue repair, maintaining homeostasis (Gao, Xu, Asada, & Frenette, 2018). Because of these properties, MSCs have been increasingly studied in regenerative medicine (Di Trapani et al., 2016; Vizoso, Eiro, Cid, Schneider, & Perez-Fernandez, 2017).

Oxysterols are compounds derived from cholesterol, which have several functions. Oxysterols regulate cholesterol biosynthesis and absorption, acting on several transcription factors and receptors, such as sterol regulatory element binding proteins (SREBPs) and liver X receptor (LXR) (Du & Yang, 2013; Levy, de Melo, Ruiz, & Bydlowski, 2017). Recent studies have described the action of some oxysterols and analogs as inducers of osteogenic differentiation, acting on signaling processes such as activation of Hedgehog pathway, which has been associated with stem cell renewal and bone cell differentiation (Lee, Kim, Han, Kang, & Heo, 2017). In addition, some oxysterols are involved in cell death, including apoptosis, autophagy, and oxyapoptophagy, a particular type of cell death associated with oxidative stress promoted by oxysterols (Paz et al., 2019). This chapter will address some recent aspects of mesenchymal stem cells death and differentiation promoted by oxysterols.

2. Mesenchymal stem cells

Mesenchymal stem cells were first described in 1970 by Friedenstein et al. In bone marrow, two distinct cell populations: hematopoietic and stromal cells, were described to form a cooperative system (Afanasyev, Elstner, & Zander, 2009; Dimmeler & Losordo, 2011; Friedenstein, Chailakhjan, & Lalykina, 1970; Maria Ruiz, Levy, & Bydlowski, 2014; Williams & Hare, 2011). Friedenstein was also the first to demonstrate that these cells had the ability to differentiate in mesodermal cell lines and identified their importance in hematopoietic niche control (Afanasyev et al., 2009; Williams & Hare, 2011). Mesenchymal stem cells in bone marrow are a relatively rare cell population, present in a proportion of 1:10,000 nucleated cells (Williams & Hare, 2011). These cells were considered as fibroblast colony forming cells (CFU-F), by their fibroblastoid aspects and capacity of forming colonies in culture (Afanasyev et al., 2009; Bajada, Mazakova, Richardson, & Ashammakhi, 2008; Charbord, 2010; Lv, Tuan, Cheung, & Leung, 2014; Ohishi & Schipani, 2010; Schwindt, Barnabé, & Mello, 2005; Sylvester & Longaker, 2004; Williams & Hare, 2011).

Mesenchymal stem cells have self-renewal and differentiation abilities, widely distributed in the body (Kolf, Cho, & Tuan, 2007). Although bone marrow is the best known and most studied source of mesenchymal stem cells, they can be found in most tissues: tendons (Bajada et al., 2008; Ohishi & Schipani, 2010), periodontal ligaments (Bajada et al., 2008), synovial membrane, skeletal muscles, adipose tissue (Lv et al., 2014; Maria Ruiz et al., 2014; Ohishi & Schipani, 2010), peripheral blood (in tiny quantities and with distinct surface markers) (Bajada et al., 2008; Lv et al., 2014; Maria Ruiz et al., 2014), dental pulp, endometrium, umbilical cord blood, umbilical cord (Lv et al., 2014; Maria Ruiz et al., 2014; Ohishi & Schipani, 2010), placenta (Lee, 2008; Lv et al., 2014), amniotic fluid and amniotic membrane (Bajada et al., 2008; Lv et al., 2014; Maria Ruiz et al., 2014; Ohishi & Schipani, 2010).

To establish a standard method for mesenchymal stem cell characterization, the International Society of Cell Therapy (ISCT) proposed minimum criteria: (a) plastic adhesion of cells in standardized culture conditions; (b) presence of surface cell marker expression of CD73, CD90 and CD105, and absence of CD34, CD45, HLA-DR, CD14 or CD11b, CD79a, or CD19; (c) under *in vitro* standardized culture conditions they must have the ability to differentiate in osteoblasts, adipocytes and chondrocytes (Lee, 2008; Maria Ruiz et al., 2014; Ohishi & Schipani, 2010; Sylvester & Longaker, 2004; Williams & Hare, 2011). Other cell surface markers CD13, CD29, CD44 and CD10 can also be expressed by mesenchymal stem cells (Kolf et al., 2007; Lv et al., 2014).

All of these criteria can be used for human mesenchymal stem cell standardization. However, they may not be applied to other animal species. For example, murine mesenchymal stem cells have different surface cell markers. Although they have similar behavior to human mesenchymal stem cells (Peister et al., 2004). It is also possible that during mesenchymal stem cell isolation and expansion, some cell surface markers may cease to be expressed while others appear and so a species-specific characterization of mesenchymal stem cells is a challenge (Jones et al., 2002).

Although there are criteria for classifying a cell as a mesenchymal stem cell, it has been observed that not all mesenchymal stem cells are equal in their potentiality, indicating that they are heterogeneous cell populations of different concomitant lineages. The different protein and cytokine expression profiles found in mesenchymal stem cells obtained from different sources can justify this idea, and this suggests that mesenchymal stem cells are more likely to give rise to tissues related to their origin (Lv et al., 2014).

Niches are structural and functional units that regulate the proliferation, differentiation, and maintenance of stem cells (Gao et al., 2018). In this way, niche regulation is the key role of mesenchymal stem cells in different tissues, and this function is attracting the attention of the scientific community (Williams & Hare, 2011). Although the hematopoietic niche is the best known and most studied, there are other niches where mesenchymal stem cells can be found such as the hair follicle bulb niche (Pasolli, 2011), and the intestinal epithelium niche (Yeung, Chia, Kosinski, & Kuo, 2011). In addition, Scaffidi and Misteli (2008) suggested that mesenchymal stem cells can be found in tissues involved with progeria (Hutchinson-Gilford's syndrome) and may be also be involved with age regulation (Scaffidi & Misteli, 2008). Therefore, there are hypotheses that suggest the participation of mesenchymal stem cells as precursors of certain cell lineages, as regulators of the function of other cells, as participants in niche regulation and in aging control processes.

Other characteristics attributed to mesenchymal stem cells are their immunomodulatory properties both *in vivo* and *in vitro*. These cells can evade the immune system, recognizing and actively inhibiting an immunologic response. For these characteristics, the mesenchymal stem cells are considered to be immunoprivileged, being capable of surviving and differentiating in immunocompatible-incompatible transplants in both allogenic and xenogenic cases. Another characteristic attributed to the immunologic properties of mesenchymal stem cells is their capacity to induce tolerance in allotransplants in immunocompetent receptors, or even in xenotransplants. Some of these mechanisms for evading the immune response can be attributed to the hypoimmunogenicity of mesenchymal stem cells, modulation of T cell phenotype and immunosuppression of the local environment (Atoui & Chiu, 2012).

Mesenchymal stem cells express moderate levels of Human Leukocyte Antigen (HLA) ABC and do not express HLA-DR (Devine & Hoffman, 2000; Majumdar et al., 2003), T-cell costimulatory molecules B7 (Devine & Hoffman, 2000) and CD40 ligand (Tse, Pendleton, Beyer, Egalka, & Guinan, 2003). In addition, these characteristics of mesenchymal stem cells, added to its immunosuppressive activity *via* cell-cell contact with target immune cells and secretion of soluble factors such as indolamine 2,3-dioxigenase, heme oxygenate-1 and nitric oxide, can explain how mesenchymal stem cells are able to support allogenic tolerance (Chabannes et al., 2007; Krampera et al., 2006; Ren et al., 2008; Sato et al., 2007). This immunomodulatory effect is obtained by the interaction of mesenchymal stem cells and immune cells of both the innate and adaptative systems (Williams & Hare, 2011).

Due to their immunomodulatory properties, mesenchymal stem cell infusion has been demonstrating valuable results in graft-*versus*-host diseases in hematopoietic stem cell transplantation, mainly in steroid-refractory cases (Brooke et al., 2007). Mesenchymal stem cells can secrete growth factors and cytokines that modulate the immune response in graft-*versus*-host diseases and can contribute to the expansion and differentiation of host cells in a hematopoietic stem cell transplant (Paz et al., 2019). In addition, the use of mesenchymal stem cells is being studied in autoimmune disease treatment due to their immunomodulatory properties and anti-inflammatory cytokines production, mainly Interleukine-10 (IL-10) (Choi et al., 2008).

Tissue engineering is a new goal in medical science and mesenchymal stem cells can be used in this way for properties such as self-renewal, differentiation into different cell lineages, and because of their ubiquitous localization (Abdallah & Kassem, 2008). Mesenchymal stem cells can be used in regenerative medicine due to their ability to home on sites of injury or disease, and for their secretoma, which contain bioactive factors and promote immunomodulatory and trophic roles implicated in their regenerative properties (Caplan, 2017). Vizoso et al. (2017) suggested that the secretoma of mesenchymal stem cells can be used in regenerative medicine with more advantages than using the cells themselves. Di Trapani et al. (2016) further suggest that the mesenchymal stem cell extracellular vesicles can be standardized and used as a cell-free way of treatment for inflammatory diseases.

The clinical use of mesenchymal stem cells has also been studied. For example, they have been used in the treatment of type I diabetes, decreasing the glucose levels (Katuchova et al., 2015); in the prevention of bronchopulmonary dysplasia (BPD) (Namba, 2019); in hepatic diseases, using MSC extracellular vesicles (Borrelli et al., 2018). Importantly, studies in rats suggested that mesenchymal stem cells can lose their immunomodulatory privileges when they undergo a differentiation process (Hatzistergos et al., 2010). The source of mesenchymal stem cells is another factor that has proved to be very important, because it has been demonstrated that when these cells are present in elderly people or individuals with diseases such as kidney diseases, ischemic cardiac disease, diabetes or others, these cells have a very important reduction in their proliferative and differentiation capacity, in addition to growth of apoptosis both *in vivo* than *in vitro* (Nayan et al., 2011; Zhuo et al., 2010).

Multipotent adult progenitor cells (MAPC) is a type of MSC similar to bone marrow mesenchymal stem cells (Jiang et al., 2002). These cells have a great capacity for expansion and can differentiate into tissues of mesenchymal lineage, in addition to endothelial and endoderm tissue. *In vitro* these cells, from mice bone marrow, can differentiate into cells of the three germinal leaflets (Jiang et al., 2002).

3. Oxysterols

3.1 Origin and formation of oxysterols

Oxysterols are sterols derived from the oxidation of the cholesterol molecule. Cholesterol ($C_{27}H_{46}O$) is one of the main constituents of plasma membranes. Furthermore, cholesterol is involved in physiological processes of signaling, proliferation, cell death and cell resistance phenotype. It has a hydrophobic tetracyclic structure, a sterol nucleus containing a β-hydroxyl group in position 3, and a branched isooctyl side chain. Due to its high body distribution and their three-dimensional structure, cholesterol molecule can be oxidized, forming oxysterols, by reactive oxygen species and free radicals (non-enzymatic formation), or biotransformation by enzymes (enzymatic formation), such as: oxidoreductases, hydrolases and reductases (Du & Yang, 2013; Levy et al., 2017). Oxysterols are intermediates in the pathway of bile acid formation and steroidal hormone biosynthesis, or as auto-oxidation products in some tissues, under high oxidative stress conditions. Oxysterols have in their structure one or more oxygen atoms linked to the ring structure or the side chain, which may be a hydroxyl, carbonyl, epoxy, hydroperoxide or carboxyl group (Du & Yang, 2013; Griffiths & Jörnvall, 2014). Table 1 shows the main oxysterols including the origin of their formation.

The main oxysterols derived from enzymatic action are: side-chain oxysterol and 4β-hydroxycholesterol (4β-OHC), which can be used as endogenous markers for CYP3A4 and CYP3A5 activity; 7α-hydroxycholesterol (7α-OHC), an intermediate in the biosynthesis of bile acids, metabolized by 7α-hydroxylase (CYP7A1), which is a liver enzyme (Massey & Pownall, 2006).

Non-enzymatic oxidation takes place in the unsaturation present in the cholesterol B ring, as well as in the allylic positions and the tertiary carbons of the side chain. These positions are vulnerable to attack by reactive oxygen species, transition metals, free radicals, molecular chlorine, atomic oxygen, and ozone (Griffiths & Jörnvall, 2014; Patel, Diczfalusy, Dzeletovic, Wilson, & Darley-Usmar, 1996). The main oxisterols of non-enzymatic oxidation origin are: 7-ketocholesterol (7-KC), 7α-hydroxycholesterol, 7β-hydroxycholesterol, 5α,6α-epoxycholesterol, 5β,6β-epoxycolesterol,

Table 1 Oxysterols and their biological effects in *in vitro* and *in vivo* experiments.

Oxysterol	Formation	Effects	Model	References
7β-hydroxycholesterol (7β-OHC)	Auto-oxidation: reactive oxygen species	Cytotoxic effects: cell death; Oxiapoptophagy	158 N murine oligodendrocytes	Nury et al. (2020)
7α-hydroxycholesterol (7α-OHC)	CYP7A1 enzyme	Cytotoxic effects: cell death	158 N murine oligodendrocytes	Nury et al. (2020)
7-ketocholesterol (7-KC)	Auto-oxidation: reactive oxygen species CYP7A1 enzyme	Cytotoxic effects: cell death; Oxiapoptophagy	MSC derived from human bone marrow	Paz et al. (2019)
5β, 6β-epoxy cholesterol (5β, 6β-epoxy)	Auto-oxidation: reactive oxygen species	Cytotoxic effects: cell death	MSC derived from human adipose tissue	Silva et al. (2017)
3α-5β-6α-cholestane-3,6-diol (diol)	Auto-oxidation: reactive oxygen species	Cytotoxic effects: cell death	MSC derived from human adipose tissue	Silva et al. (2017)
cholestan-3β, 5α, 6β triol (triol)	Auto-oxidation: reactive oxygen species	Cytotoxic effects: cell death	MSC derived from human adipose tissue	Silva et al. (2017)
20(*S*)-hydroxycholesterol (20S)	CYP7A1 and CYP11A1 enzymes	Differentiation osteogenesis	M2-10B4 cells and	Lee et al. (2017)
22(*R*)-hydroxycholesterol (22R)	CYP11A1 enzyme	Differentiation: chondrogenesis and osteogenesis	MSC derived from human adipose tissues and M2-10B4 cells	Kim et al. (2020) and Lee et al. (2017)
22(*S*)-hydroxycholesterol (22S)	CYP11A1 enzyme	Differentiation osteogenesis	M2-10B4 cells and MSC derived from human adipose tissue	Lee et al. (2017)

Continued

Table 1 Oxysterols and their biological effects in *in vitro* and *in vivo* experiments.—cont'd

Oxysterol	Formation	Effects	Model	References
24(S)-hydroxycholesterol (24S)	CYP46A1 and CYP39A1 enzymes	Cytotoxic effects: cell death; Oxiapoptophagy	Monocytic cells	Nury et al. (2020)
24(S), 25-epoxycholesterol	CYP46A1 enzyme	Differentiation neurogenesis	Mouse	Theofilopoulos et al. (2019)
Oxy34, Oxy49 and Oxy133	Synthetic analogue	Differentiation osteogenesis	M2-10B4 cells and rabbit MSC from bone marrow	Montgomery et al. (2014)

cholestane-3β,5α,6β-triol (α-triol), 3β,5α-dihydroxy-colestane-6-ona, 20α-hydroxycolesterol and 3β-hydroxy-5α-colestane-6-ona. The addition of polar groups makes the hydroperoxide cholesterol and oxysterols more water-soluble than cholesterol, facilitating spontaneous diffusion between fat-soluble membranes. Endogenous and exogenous oxysterols are located on the plasma membrane in microdomains rich in cholesterol, sphingomyelin and phosphatidylcholine (Levy et al., 2017; Massey & Pownall, 2006; Vurusaner, Leonarduzzi, Gamba, Poli, & Basaga, 2016).

The interest in studying the action of oxysterols in different cell types comes from discoveries, *in vitro* and *in vivo*, of the biological action in the development of some diseases, such as atherosclerosis, neurodegenerative diseases and in some types of cancer. Currently, it is known that oxysterol plays an important role in modulating the gene expression of transcription factors that act in lipid homeostasis; reverse cholesterol transport; in signaling pathways related to both apoptotic death and cell proliferation; alteration in mitochondrial membrane potential; immunomodulatory function, as well as in mesenchymal stem cells differentiation (Brown, Sharpe, & Rogers, 2020; Dias et al., 2019; Griffiths & Jörnvall, 2014; Levy et al., 2019; Nury et al., 2020). We are going to focus on the cell signaling pathways that are associated with cell differentiation in mesenchymal stem cells with oxysterol.

3.2 Oxysterols and intracellular signaling pathways

3.2.1 Nuclear receptors

Nuclear receptors (NR) are composed of 48 receptors, which have structural homology and domains and function. They are known as transcription factors inducible by lipophilic ligands. Nuclear receptors present five domains: variable N-terminal region, DNA binding region, variable region, ligand binding domain and variable C-terminal region (Ma & Nelson, 2019). Biological functions of NR activation, synthesis and intra and extracellular lipid traffic, are regulated by cholesterol, oxysterols, sphingolipids, phospholipids, phosphatidylcholine, glucocorticoid, fatty acids, retinoic acid and vitamin D, among other ligands (Ma & Nelson, 2019; Vurusaner et al., 2016).

3.2.2 Oxysterols and liver X receptors (LXRs)

LXR receptors were described between 1994 and 1996 as heterodimers of retinoid X receptors (RXR). Initially known as orphan receptors, they were relocated to nuclear receptor class with the discovery that 22 (*R*)- hydroxycholesterol, 24(*S*)-hydroxycholesterol and 24(*S*), 25-epoxycholesterol are *in vivo* LXR ligands, acting as sterol sensors. Currently, it is known that there are two LXR isoforms, LXR-α and LXR-β (encoded by Nr1h3 and Nr1h2 genes, respectively), with structural homology of approximately 77% (Willy et al., 1995). Inducible LXR-α is highly expressed in metabolically active tissues, such as: liver, adrenal gland, intestine, adipose tissue, macrophages, lung and kidney, while LXR-β is expressed in all types of tissue or cells (Traversari, Sozzani, Steffensen, & Russo, 2014). The presence of hydroxyl group in the side chain in oxysterol structure is necessary for the binding and high affinity activation of LXR. Moreover, the introduction of an oxygen into the sterol B ring results in a ligand with selectivity to the LXRα (Janowski et al., 1999). In nucleus, LXR receptors remain heterodimerized with RXR receptors. LXR-RXR complex is linked to the responsive element LXR (LXRE) in DNA, a region of direct repetition 4 (DR4), recognized as the consensus sequence (5′-AGGTCA-NNNN-AGGTCA-3′) (Willy et al., 1995). At baseline, LXR-RXR complex is associated with co-repressors (complexes containing NCoR or SMRT) that suppress genetic expression. On the other hand, binding of oxysterols to the LXR-RXR complex generates a signaling cascade, with recruitment of coactivators such as P300; CBP and SRC-1, resulting in gene transcription of proteins and factors associated with the regulation of lipid metabolism, including cholesterol (Levy et al., 2017) (Fig. 1).

In addition, oxysterols control the cholesterol absorption (LXR-RXR pathway) by increasing the expression of ATP-binding cassette (ABC) proteins, mainly the subfamily ABC-A (A1) and ABC-G (G1, G5 and G8). ABC-A1 transporter mediates the efflux of cholesterol and phospholipids to apolipoproteins ApoA1 and ApoE low in lipid, forming nascent high-density lipoproteins (HDL). The ABC-G subfamily is responsible for cholesterol efflux and influx in macrophages and other cells. Another way that LXR receptors regulate cholesterol transport is by inhibiting the LDL receptor (LDLR) pathway through transcriptional induction of IDOL (Inducible Degrader Of the LDLR), an E3 ubiquitin ligase that triggers ubiquitination of the LDLR in its cytoplasmic domain, thereby targeting it for degradation (Chawla et al., 2001; Khovidhunkit, Moser, Shigenaga, Grunfeld, & Feingold, 2003; Yvan-charvet, Wang, & Tall, 2010).

3.2.3 Sterol regulatory element-binding protein (SREBP)

SREBPs are composed of two proteins: SREBP1 and SREBP2. These proteins are structurally related and control lipogenesis by upregulation of genes that are sterol-regulated. They belong to a large class of transcription factors

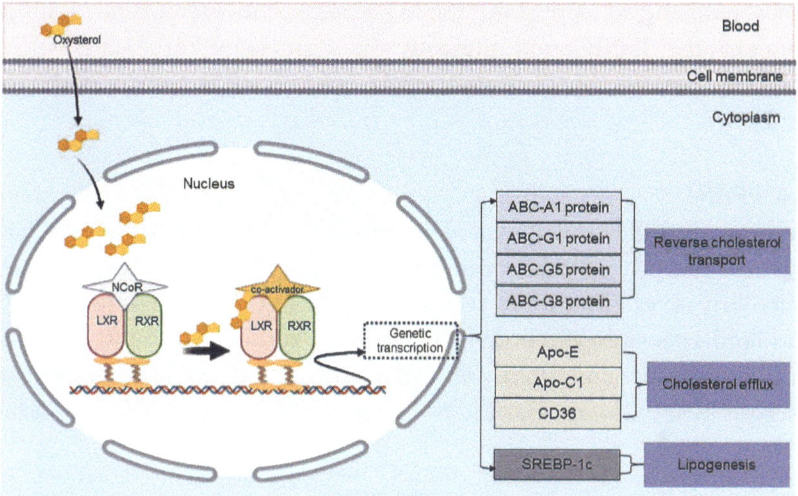

Fig. 1 Oxysterols activate liver X receptors (LXR). LXR form heterodimers with retinoid X receptors (RXR), in nucleus. LXR-RXR complex is linked to the responsive element LXR (LXRE) in DNA. In the basal (or inactivated) form, the LXR-RXR complex is associated with repression complexes such as NCoR (nuclear receptor co-repressor), which prevent the gene transition. The oxysterols bind to the LXR receptors activating a signaling cascade and recruiting co-activators such as SRC-1 (steroid receptor coactivator-1), P300, CBP (CREB-binding protein), initiating the genetic transcription of factors and proteins associated with reverse cholesterol transport, cholesterol efflux and influx and lipogenesis.

containing a basic helix-loop-helix-leucine zipper (bHLH-Zip) motif. The SREBP1 protein has two isoforms: SREBP1a and SREBP1c, being expressed in variable between tissues; SREB2 is ubiquitously expressed. Both SREBP1 and SREB2 are associated and form a complex with another endoplasmatic reticum (ER) membrane protein, the SREBP-cleavage activating protein (SCAP) (Sato, 2010). Wang, Sato, Brown, Hua, and Goldstein (1994) discovered that the membrane-bound precursor in sterol-depleted cells is cleaved in order to generate a soluble N-terminal fragment (apparent molecular mass, 68 kD) which translocates to the nucleus. The fragment, including the bHLH-ZIP domain, activates transcription of genes for the LDL receptor and HMG-CoA synthase. SCAP protein is associated with INSIG in the ER membrane. For SREBP cleavage by the S1P protease to occur, it is necessary for the SREBP-SCAP complex to move from the ER to the Golgi complex (Janowski, Shan, & Russell, 2001; Radhakrishnan, Ikeda, Hyock, Brown, & Goldstein, 2007). The release of the SREBP-SCAP complex occurs with the dissociation of INSIG in the ER membrane. Some oxysterols inhibit the cleavage of SREBP1, and the 68-kD nuclear form is rapidly catabolized, thereby reducing transcription. The 25-hydroxycholesterol binds with high affinity to the INSIG and prevents the dissociation of SREBP-SCAP-INSIG complex in ER (Janowski et al., 2001; Radhakrishnan et al., 2007) (Fig. 2).

3.2.4 PPARs

Peroxisome proliferator-activated receptors (PPARs) are part of the family of nuclear receptors. PPARs can be activated by fatty acids and their metabolic derivatives. The three receptors: PPAR-α, PPAR-ɣ and PPAR-δ, act as lipid sensors in the body. PPAR-α is expressed in the liver; PPAR-ɣ in adipose tissues and PPAR-δ in all tissues, although it has less expression in hepatic tissue (Evans, Barish, & Wang, 2004). The PPAR-γ is induced

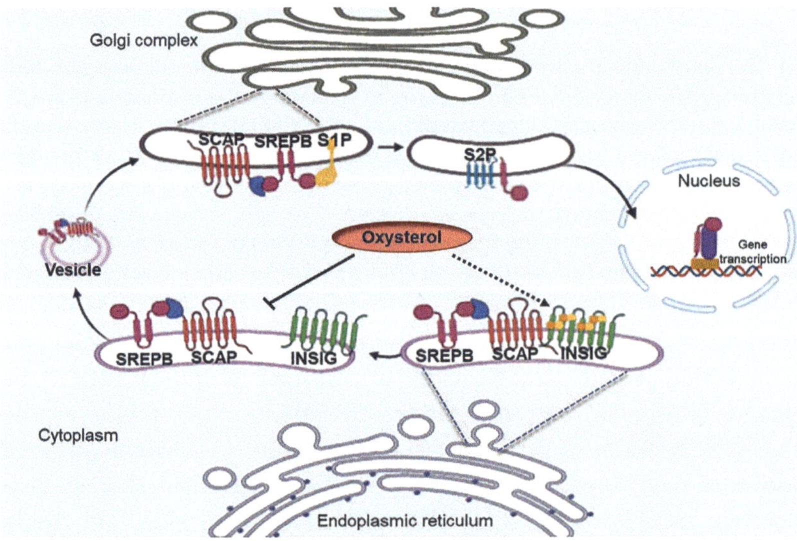

Fig. 2 Oxysterols inhibit Sterol regulatory element-binding protein (SREBP) pathway. SREBP transcription factors (SREBP 1 and SREBP2) control lipid homeostasis. They are activated in response to low levels of cellular cholesterol. SREBPs are proteins present in the membrane in the endoplasmic reticulum (ER), and form a complex with another ER membrane protein, SCAP (SREBP cleavage-activating protein). In basal or inactivated form, SREBP-SCAP complex is retained in the ER associated with the INSIG protein. The SREBP is activated by the SCAP-INSIG cleavage. Then, the SREBP-SCAP complex translocates from vesicles to the Golgi complex. In the Golgi membrane, SREBP is cleaved by the proteins S1P (site-1 protease) and S2P (site-2 protease). The fragment with bHLH-ZIP domain of SREBP protein translocates to the nucleus. There it binds to the steroid response elements (SRE) in the promoter regions of genes that encode enzymes involved in cholesterol synthesis, including squalene synthase and HMG-CoA reductase, and the LDLR gene. Some oxysterols, such as 25-hydroxycholesterol, bind with high affinity to the INSIG and prevent the dissociation of SREBP-SCAP-INSIG complex in ER.

during adipocyte differentiation and forced expression of PPAR-γ in non-adipogenic cell lines, and it effectively converts them into mature adipocytes. In addition, PPAR-γ knockout mice fail to develop adipose tissue (Chawla et al., 2001; Evans et al., 2004). Studies *in vitro* suggest that PPAR-γ is the final effector of adipogenesis in a transcriptional cascade that also involves members of the C/EBP family of transcription factors. Moreover, PPAR-γ regulates a cholesterol efflux pathway in macrophages by increasing the expression of ABC-A1 through LXRα (Chawla et al., 2001; Evans et al., 2004). The oxysterol 27-Hydroxycholesterol promotes cholesterol transport by upregulating the expression of PPAR-γ, LXR-α, ABCA1 and ApoE in C6 cells (An et al., 2017). On the other hand, 7-ketocholesterol binds in CD1d and results in PPARγ activation and lipid accumulation in HEK 293 cells (Rosales, Tang, & Geng, 2015).

3.2.5 Hedgehog pathway

Hedgehog (Hh) signaling pathway is responsible for organization, development, proliferation, and cell differentiation in the embryonic period. Mammalian homologs of Hh include three proteins: Sonic hedgehog (Shh), Indian hedgehog (Ihh) and desert hedgehog (Dhh). Shh signaling remains quiescent in adult mammalian tissues. However, in multiple adult tissues it becomes active during differentiation, proliferation and maintenance of stem cells (Jeng, Chang, & Lin, 2020). Hh signaling depends on the primary cilium, an organelle that forms microtubules protruding out of the plasma membrane. The components of the Hh pathway move and accumulate in the cilium at different stages during pathway signaling (Bangs & Anderson, 2017; Gorojankina, 2016). Hh signaling begins by connecting the Hh ligand to the Patched1 receptor (Ptch1), inhibiting Ptch1. At baseline, Ptch1 suppresses Smoothened (SMO) expression. When Ptch1 is inhibited, SMO activation occurs. Next, a signaling cascade begins, with the recruitment and activation of glioma-associated-oncogene (GLI) transcription factors. Suppressor of fused (Sufu) is one of the main inhibitors of the Hh pathway, which binds proteins Gli1, Gli2, Gli3, inactivating them in proteasomes. Active SMO inhibits the processing of Gli (Bangs & Anderson, 2017; Gorojankina, 2016).

The SMO protein, a G-protein-coupled receptor (GPCR) of the Frizzled-class (class-F), contains two distinct ligand-binding sites: transmembrane domain (TMD) and cysteine-rich domain (CRD). Both are separated by intermediate link domain, which stabilizes the SMO inactive state (Byrne et al., 2016). The CRD is the binding site for cholesterol and other sterols, such as some oxysterols (Byrne et al., 2016). Moreover, inactivation of PTCH1 by Hedgehog allows transmembrane sterols to access the 7-transmembrane site that leads to SMO activation. In human cells, 24(*S*),25-epoxycholesterol is an endogenous ligand of PTCH1, which can stimulate Hedgehog signaling and initiate SMO signaling pathway (Deshpande et al., 2019; Qi et al., 2019). In addition, the conserved extracellular CRD is the site of action for oxysterols on SMO, including 20(*S*)-OHC, 7-keto-25- and 7-keto-27-hydroxycholesterol, involving residues structurally analogous to those contacting the Wnt lipid adduct in the homologous Frizzled CRD. Activation of SMO by oxysterols can be independent of modulation by cyclopamine and regulation mediated by Ptch1 in Shh pathway (Janda, Waghray, Levin, Thomas, & Garcia, 2012; Myers et al., 2013). The Shh signaling pathway and SMO are associated with both Wnt and RUNX pathways for bone differentiation (Alman, 2015; Yang, Andre, Ye, & Yang, 2015) (Fig. 3).

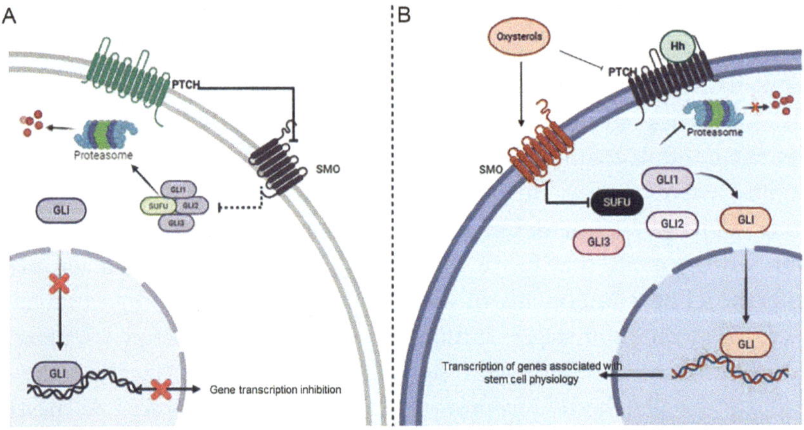

Fig. 3 Oxysterols bind to the SMO and activate Hedgehog pathway: The Hedgehog signaling pathway is essential for the maintenance of stem cells. In the basal state (A) the pathway remains inactive. The transmembrane protein PTCH1 (Patched1) inhibits the activation of the SMO receptor. Thus, the family of transcription factors Gli (GLI1, GLI2 and GLI3) are degraded in proteasomes, after ubiquitination by the SUFU (Suppressor of fused) complex. The degradation of GLI proteins inhibits gene transcription. The Hh signaling pathway is activated (B) when the Hh linker binds to the PTCH1 receptor and inhibits PTCH1 protein. Inhibition of PTCH1 activates SMO. Some oxysterols, such as 24(S),25-epoxycholesterol, 20(*S*)-hydrocholesterol, 7-keto-25- and 7-keto-27-hydroxycholesterol, can activate Hedgehog pathway, inhibiting PTCH1, and/or binding directly to the SMO in CRD domain. When SMO is activated, SUFU complex is inactivated. Then, GLI transcription factors translocate to the nucleus activating genes associated with stem cell biology, such as BMPs proteins, FGF4, VEGF, Myf5, Ptch1, Ptch2, Nkx gene, Rab34, Pax family (Pax6, Pax7, Pax9), Jagged1and Runx2 (Levy et al., 2017).

The transcription factors RUNX1, RUNX2 and RUNX3, are composed of an α subunit, which binds to DNA through a high-affinity Runt domain, and β subunit, encoded by the CBFB gene (Carbonare, Innamorati, & Valenti, 2012). RUNX transcription factors are important for bone formation, osteoblast differentiation, growth plate cartilage hypertrophy, cell migration and bone vascular invasion, among others (Carbonare et al., 2012). *In vitro*, Ptch1-deficient osteoblast precursor cells differentiate into osteoblasts at an accelerated rate in response to the transcription factor RUNX2 and reduced generation of the Gli3 repressor and, *in vivo*, Ptch1-deficient ($Ptch1^{+/-}$) mice demonstrated an increase in bone mass (Ohba et al., 2008).

3.2.6 Wnt signaling

The Wnt signaling pathway regulates several phenomena during the embryonic period, such as organogenesis, ontogenesis and morphogenesis. In adult tissue, changes in the Wnt pathway lead to carcinogenesis. However, it is known that the Wnt pathway is essential for the renewal of stem cells. The Wnt signaling pathway can be classified in canonical and non-canonical pathways. The non-canonical can be subdivided into Wnt/Ca^{+} and Planar Cell Polarity (PCP) (Carreira-Barbosa & Nunes, 2020). Both pathways are active after binding the Wnt protein to the receptor Frizzled (Fz), in the extracellular N-terminal domain rich in cysteine. Then, there is a signal transduction that activates the protein kinase Disheveled (Dsh) (Carreira-Barbosa & Nunes, 2020). In the canonical pathway, Dsh activation inhibits the GSK3 enzyme, and β-catenin present in the cytoplasm translocates to the nucleus, where it acts as a co-activator of several genes and transcription factors, among which are genes responsible for the differentiation of osteoblasts (Valenta, Hausmann, & Basler, 2012; Zeng, Wang, Dong, Ma, & De Liu, 2020).

4. Oxysterol and mesenchymal stem cells

Cellular differentiation by oxysterols was demonstrated by several reports, for example, 7-ketocholesterol inducing monocyte differentiation (Hayden et al., 2002), and oxysterols 22(*R*)- and 25-hydroxycholesterol inducing human keratinocytes differentiation (Hanley et al., 2000), both *in vitro*. These studies demonstrated the cell differentiation potential of oxysterols (Fig. 4).

One of the first studies that suggested that oxidized lipids play a role in bone cell function was treatment with certain oxidized lipids and

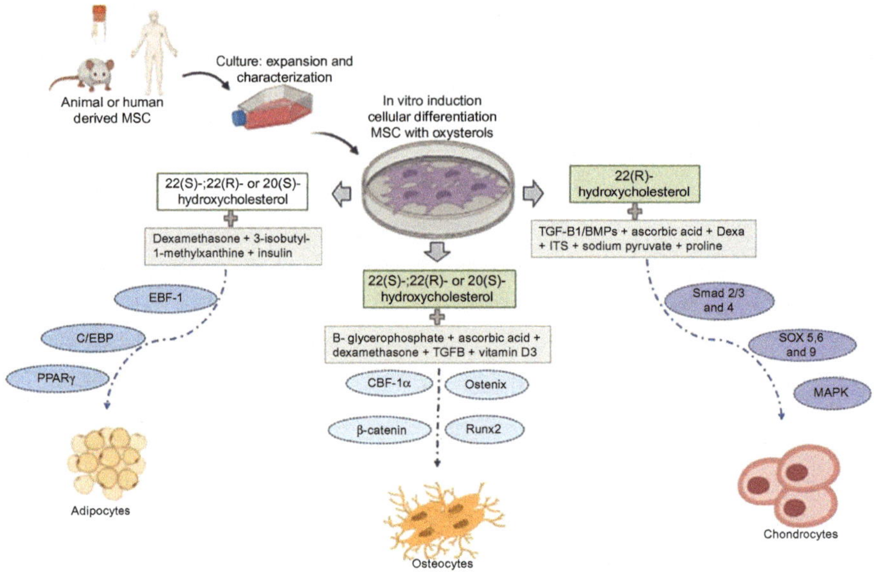

Fig. 4 Cell differentiation of mesenchymal stem cells (MSC) by oxysterols. MSCs isolated from humans and/or animals should be characterized according to the recommendations of the International Society of Cell Therapy (ISCT). *In vitro*, under suitable conditions and proper culture medium for each differentiation, MSC can be differentiated into adipocytes, osteocytes and chondrocytes. Adipogenic differentiation involves the activation of the transcription factors associated with PPARɣ, C/EBP, EBF-1. However, oxysterols such 22(S), 22(R) or 20(S) hydroxycholesterol inhibit adipogen differentiation. On the other hand, these oxysterols induce osteogenic differentiation through the signaling pathways that involve β-catenin and Runx-2. In addition, 22(*R*)-hydroxycholesterol activates MSC differentiation in chondrocytes through Smad, SOX and MAPK pathways.

lipoproteins that inhibit osteoblastic differentiation (Parhami, 2003). Parhami and colleagues studied how osteoblastic differentiation was affected by minimally oxidized low-density lipoprotein (MM-LDL) in murine marrow stromal cells, M2–10B4. MM-LDL inhibited osteoblastic differentiation and increased adipogenic differentiation demonstrated by inhibition of collagen I processing, alkaline phosphatase activity and mineralization of the cells. A similar effect was demonstrated in MC3T3-E1 proosteoblastic cells that inhibited osteoblast differentiation when treated with oxidized lipids. In addition, marrow stromal cells from C57BL/6 mice fed with a high fat atherogenic diet did not undergo osteogenic differentiation *in vitro* (Parhami et al., 1999).

Subsequently, it was demonstrated that the cholesterol and its derivative biosynthetic pathway are important for the proper development of MSCs in functional osteoblastic cells capable of forming a mineralized matrix.

Parhami and colleagues used a M2-10B4 cell line to examine the effect of two inhibitors of cellular cholesterol biosynthesis, mevastatin (compactin) and mevinolin (lovastatin), on the regulation of osteoblastic differentiation. They found inhibition of calcium mineral deposits and a decreased activity and expression of alkaline phosphatase in MSC cultures, but the expression of osteocalcin was not affected by the treatment. The authors suggested that products from the cholesterol biosynthetic pathway are essential for the maturation and mineralization of MSCs and for the proper development of MSCs into functional osteoblastic cells (Parhami et al., 2002).

Studies in humans and animals have shown that hypercholesterolemia, and the resulting deposition and oxidation of lipids in tissues, could lead to osteoporotic bone loss (Parhami, 2003). MSCs are the most frequent progenitors for both osteoblasts and adipocytes. It has been suggested that bone marrow adipocytes can be diverted from an osteogenic to an adipogenic program. This happens as a result of changes in pluripotent mesenchymal cells within the bone marrow stroma (Chan & Duque, 2002). Bone turnover is regulated by osteoblasts and osteoclasts that produce and reabsorb bone, respectively (Parhami, Garfinkel, & Demer, 2000). Most of the therapeutic interventions that are presently approved for osteoporosis block osteoclastic bone resorption, but anabolic agents that target osteoblasts can also be effective in reversing or preventing osteoporotic bone loss (Parhami, 2003).

Kha and colleagues raised the hypothesis that the oxysterols generated by osteoprogenitor cells, as well as from exogenous sources, may be involved in osteogenic differentiation. They provided the first evidence of osteogenic activity of oxysterols. They used the oxysterols 22(*R*)-hydroxycholesterol (22*R*), 20(*S*)-hydroxycholesterol (20*S*), and 22(*S*)-hydroxycholesterol (22S), individually or in combination, in M2–10B4 cell line. It has been shown that an increase in early and late markers of osteogenic differentiation, including osteocalcin mRNA expression and mineralization, increased alkaline phosphatase activity and decreased markers of adipogenic differentiation, proving pro-osteogenic and anti-adipogenic effects of oxysterols. Mesenchymal cell differentiation into osteoblasts was regulated by the activity of cyclooxygenase (COX) and by ERK-dependent mechanisms (Kha et al., 2004). Dwyer and colleagues studies on 20(*S*)- and 22(*S*)-hydroxycholesterol combination in the M2–10B4 cell line, demonstrated that the combination of oxysterols stimulates hedgehog pathway activity by indirectly activating the pathway component Smoothened, and by this pathway the oxysterols exert their osteoinductive effects (Dwyer et al., 2007).

In order to deepen the knowledge of the pathways involved in oxysterol-induced osteoblastic differentiation of 20(*S*), 22(*S*) and 22(*R*)-hydroxycholesterol in the M2-10B4 cell line, Richardson et al. (2007) demonstrated that the differentiation is mediated through a protein kinase C (PKC)- and protein kinase A (PKA)-dependent mechanism. This mechanism increases the Runx2 activity and regulates other markers of osteoblastic differentiation, including osteocalcin mRNA expression and alkaline phosphatase (ALP) activity. The increase of osteogenic differentiation markers selectively induced expression of Wnt target genes in the absence of β-catenin accumulation or TCF/Lef activation, suggesting the involvement of a non-canonical Wnt pathway in pro-osteogenic differentiation events (Amantea, Kim, Meliton, Tetradis, & Parhami, 2008).

The effect of 20(*S*)-hydroxycholesterol on the M2-10B4 (M2) cell line also induces the expression of the Notch target genes HES-1, HEY-1, and HEY-2; moreover, the induction was both through activation of Hedgehog signaling and *via* a pathway independent of canonical Notch signaling (Kim, Meliton, Bourquard, Hahn, & Parhami, 2010; Kim, Meliton, Tetradis, et al., 2010). On the other hand, the molecular mechanisms by which 20(*S*)-hydroxycholesterol inhibits adipogenic differentiation of mouse MSCs was investigated by Kim and colleagues. It was shown that the inhibition of adipogenesis by 20(*S*)-hydroxycholesterol is mediated predominantly through a hedgehog pathway-dependent mechanism that inhibited PPARγ mRNA expression and adipogenic differentiation of these cells.

All the previous studies were performed using M2-10B4 bone marrow stromal cell line. Woltje et al. (2015) studied whether 20(*S*)- and 22(*S*)-hydroxycholesterol combination was able to promote cell differentiation of bone marrow from Lewis rats and demonstrated that the oxysterol combination accelerates osteogenic differentiation. They found that there was an increase in the amount of mineral deposition when compared with classical osteogenic differentiation media. There was also enhanced alkaline phosphatase activity, and increased expression of osteopontin, osteocalcin and calcium sedimentation.

Animal experiments were done to demonstrate the ability of oxysterols to stimulate osteogenic differentiation using a critical-sized rat calvarial defect model with a scaffold coated with low-dose and high-dose mixture of 22 (*S*)-hydroxycholesterol with 20(*S*)-hydroxycholesterol. The oxysterols significantly but modestly induced bone formation when compared to uncoated control scaffolds, irrespective of oxysterol dose (Aghaloo et al., 2007).

Recently, another formulation was studied *in vitro* and *in vivo*, 20(*S*)-hydroxycholesterol, alone or in combination with simvastatin (Huang et al., 2019). This study was conducted in bone marrow stromal cells (BMSCs) isolated from bone marrow aspirates from Wistar rats and in critical-sized bone defects in rabbits. This study showed that 20(*S*)-hydroxycholesterol synergized with simvastatin and enhanced calcium sedimentation, ALP activity, and expression of osteogenesis-related genes ALP, osteocalcin and BMP-2. The oxysterol combination activated Raf/MEK/ERK signaling in BMSCs. The combined treatment was able to enhance bone regeneration in a rabbit calvarial defect model.

In the *in vitro* osteoinductive effects, oxysterols were used at micromolar doses that *in vivo* had only modest effects in inducing bone formation. Analogues of 20(*S*)-hydroxycholesterol were designed to improve *in vitro* and *in vivo* osteoinductive activity. At that time, three semi-synthetic formulations, named Oxy34, Oxy49 and Oxy133, were formulated and published. These new molecules were able to induce the osteogenic differentiation of mouse MSC and were shown to effectively stimulate new bone formation in rat and rabbit models (Hokugo et al., 2013; Johnson et al., 2011; Montgomery et al., 2014). Another protocol for osteogenic differentiation was the use of 20(*S*)-hydroxycholesterol with stearylamine in a liposome formulation in hydrogels. It was demonstrated that these formulations induced osteogenic differentiation *in vitro* in mouse bone marrow stromal cell lines by the presence of ALP and mineralization staining, and the expression of Runx2, osteocalcin and ALP. The mechanism involved was the upregulation of the Hh signaling activity. The formulation was also able to enhance calvarial healing without delivery of additional therapeutic agents, indicating their intrinsic bone-forming potential (Cui et al., 2017).

All studies conducted using the mouse MSC cell line and animal model demonstrated that some oxysterols stimulate osteogenic differentiation of MSCs through various signaling pathways, including Hedgehog, Wnt, PKC, PKA, Notch and Raf/MEK/ERK. Multiple pathways are necessary for osteogenic differentiation to occur.

Few studies of cell differentiation and oxysterols have been conducted on human MSCs and the results are controversial. Plaisant et al. (2009) used human mesenchymal stem cell lines derived from adipose tissue and mesenchymal stem cells from bone marrow. They demonstrated that osteoblast differentiation of human MSC is associated with a decrease in Hh signaling when used in a combination of 20(*S*) and 22(*S*)-hydroxycholesterol.

Activation of Hh signaling inhibits human osteoblast differentiation, in contrast to what has been observed in rodent cell lines.

Results described by Yalom and colleagues were the opposite from those of Plaisant and colleagues. They found that 20-hydroxycholesterol can lead to osteogenic differentiation in human adipose-derived stem cells, and that the process is mediated at least in part by the hedgehog signaling pathway (Yalom et al., 2014). These results were also supported by Bakshi and col. (2019), who studied human immortalized bone marrow mesenchymal stromal cells in a combination of 20(*S*) and 22(*S*)-hydroxycholesterol. They described potent osteoinductive activity of oxysterols with increased expression of osteocalcin, osteopontin, and alkaline phosphatase activity.

The mechanisms involved in the use of a hydroxycholesterol combination were studied on a rodent alveolar cleft model. Bakshi et al. (2019) showed that implantation of a collagen sponge incorporated with hydroxycholesterols was able to regenerate bone without any inflammatory cells, indicating a robust production of bone mass without adverse effects in the defect with involvement of hedgehog signaling.

Liu, Yuan, Xu, Wang, and Zhang (2005) used human periodontal ligament stem cell (PDLSC) and experimentally created alveolar bone defects in Sprague–Dawley rat models to verify the ability of the 20(*S*)-hydroxycholesterol and 22(*S*)-hydroxycholesterol combination to enhance the osteogenesis and provide periodontium regeneration. This combination of oxysterols increased the osteogenic activity of PDLSCs by increasing intracellular calcium levels, alkaline phosphatase activity, matrix mineralization, and mRNA expression of osteogenic markers, and increased protein levels of liver X receptor (LXR) α and β, hedgehog (Hh) receptor SMO and the transcription factor Gli1. *In vivo*, the oxysterols enhanced bone formation in the defect sites, showing equivalent efficacy to recombinant human bone morphogenetic protein-2. These results demonstrated that oxysterol combinations could be agents for regenerating alveolar bone and periodontal tissues in clinical trials.

Other oxysterols were also evaluated for cell differentiation. Liu et al. (2005) showed that cholestane-3b,5a,6b-triol (triol) on primary rat bone MSCs inhibited osteoblastic differentiation as demonstrated by inhibition of matrix mineralization, alkaline phosphatase activity and osteocalcin secretion. Furthermore, triol was found to induce apoptosis in MSCs.

Oxysterols were also studied in their capacity to influence other physiologic function such as protein prenylation, platelet aggregation, and cholesterol homeostasis regulation (Silva et al., 2017). A particularly interesting effect of several oxysterols is the promotion of cell death. This effect has been observed in several cell lines and occurs by several mechanisms, including apoptosis and autophagic induction and regulation of gene expression (Levy et al., 2018). Levy and colleagues have also studied the effect of 7-ketocholesterol (7-KC) on MSCs derived from human adipose tissue. They showed that 7-KC was cytotoxic to the cells, although with an IC50 higher than that found in normal cancer cell lines. The death was in part by apoptosis with mitochondrial hyperpolarization and changes in actin organization (Levy et al., 2014). Murdolo et al. (2016) also studied the effect of 7-KC and 5,6-secosterol on MSCs derived from human adipose tissue and found a time- and concentration-dependent decrease of cell viability. Both oxysterols activated the canonical WNT and the MAPKs pathways in a time sensitive and concerted manner.

Cell death promoted by other oxysterols has been also studied on MSCs derived from human adipose tissue. Cholestane-3b,5a,6b-triol (triol) and (3α-5β-6α)-cholestane-3,6-diol (diol) led to a complex mode of cell death, depending on the type of oxysterol and its concentration. Triol was described as the most effective oxysterol in promoting cell death, followed by diol (Silva et al., 2017). Both oxysterols led to the activation of caspase-3/7 and MSCs apoptosis at higher doses. Autophagy was also observed in these cells. Mitochondrial alterations were observed only with triol treatment. 5β-6β-epoxy-cholesterol was demonstrated to be cytotoxic only at high doses, and 3,5-cholestan-7-one and 7-oxocholest-5-en-3-β-yl acetate had no effect on human MSC (Silva et al., 2017).

7-Ketocholesterol was also studied in MSCs from patients with acute myeloid leukemia (AML) (Paz et al., 2019). Defects in AML mesenchymal stem cells (LMSCs) were demonstrated to lead to dysfunctional crosstalk between the bone marrow niche and hematopoietic stem cell, potentially resulting in AML initiation or propagation. 7-KC led to cell death, but less efficiently than MSC from healthy patients. 7-KC exposure triggered both the extrinsic pathway of apoptosis and intrinsic mitochondrial pathway, with increased ROS generation by LMSCs, with the presence of autophagy but absence of necroptosis. 7-KC is able to downregulate the SHh protein in LMSCs, but without modification of SMO expression.

In conclusion, oxysterols have shown to promote several biological activities on mesenchymal stem cells, mainly cytotoxicity and cell differentiation. More studies are needed to deeply understand the precise mechanisms and roles of different oxysterols in mesenchymal stem cell biology in health and disease, as well as the therapeutical potential of oxysterols, mainly in regenerative medicine.

Acknowledgment

The figures in this chapter were made at Biorender.

References

Abdallah, B. M., & Kassem, M. (2008). Human mesenchymal stem cells: From basic biology to clinical applications. *Gene Therapy*, *15*(2), 109–116. https://doi.org/10.1038/sj.gt.3303067.

Afanasyev, B., Elstner, E., & Zander, A. (2009). A. J. Friedenstein, founder of the mesenchymal stem cell concept. *Cellular Therapy and Transplantation*, *1*, 35–38. https://doi.org/10.3205/ctt-2009-en-000029.01.

Aghaloo, T. L., Amantea, C. M., Cowan, C. M., Richardson, J. A., Wu, B. M., Parhami, F., et al. (2007). Oxysterols enhance osteoblast differentiation in vitro and bone healing in vivo. *Journal of Orthopaedic Research*, *25*(11), 1488–1497.

Alman, B. A. (2015). The role of hedgehog signalling in skeletal health and disease. *Nature Reviews Rheumatology*, *11*(9), 552–560. https://doi.org/10.1038/nrrheum.2015.84.

Amantea, C. M., Kim, W. K., Meliton, V., Tetradis, S., & Parhami, F. (2008). Oxysterol-induced osteogenic differentiation of marrow stromal cells is regulated by Dkk-1 inhibitable and PI3-kinase mediated signaling. *Journal of Cellular Biochemistry*, *105*(2), 424–436.

An, Y., Zhang, D., Di, Y., L, H., Ma, W. W., Lu, Y. H., et al. (2017). 27-hydroxycholesterol regulates cholesterol synthesis and transport in C6 glioma cells. *Neurotoxicology*, *59*, 88–97. https://doi.org/10.1016/j.neuro.2017.02.001.

Atoui, R., & Chiu, R. C. (2012). Concise review: Immunomodulatory properties of mesenchymal stem cells in cellular transplantation: Update, controversies, and unknowns. *Stem Cells Translational Medicine*, *1*(3), 200–205. https://doi.org/10.5966/sctm.2011-0012.

Bajada, S., Mazakova, I., Richardson, J. B., & Ashammakhi, N. (2008). Updates on stem cells and their applications in regenerative medicine. *Journal of Tissue Engineering and Regenerative Medicine*, *2*(4), 169–183. https://doi.org/10.1002/term.83.

Bakshi, R., Hokugo, A., Zhou, S., Zhang, Z., Wang, L., Rezzadeh, K., et al. (2019). Application of hydroxycholesterols for alveolar cleft osteoplasty in a rodent model. *Plastic and Reconstructive Surgery*, *143*(5), 1385–1395.

Bangs, F., & Anderson, K. V. (2017). Primary cilia and mammalian hedgehog signaling. *Cold Spring Harbor Perspectives in Biology*, *9*(5), 1–21. https://doi.org/10.1101/cshperspect.a028175.

Borrelli, D. A., Yankson, K., Shukla, N., Vilanilam, G., Ticer, T., & Wolfram, J. (2018). Extracellular vesicle therapeutics for liver disease. *Journal of Controlled Release*, *273*, 86–98. https://doi.org/10.1016/j.jconrel.2018.01.022.

Brooke, G., Cook, M., Blair, C., Han, R., Heazlewood, C., Jones, B., et al. (2007). Therapeutic applications of mesenchymal stromal cells. *Seminars in Cell & Developmental Biology*, *18*(6), 846–858. https://doi.org/10.1016/j.semcdb.2007.09.012.

Brown, A. J., Sharpe, L. J., & Rogers, M. J. (2020). Oxysterols: From physiological tuners to pharmacological opportunities. *British Journal of Pharmacology*, 1–15. March https://doi.org/10.1111/bph.15073.

Byrne, E. F. X., Sircar, R., Miller, P. S., Hedger, G., Luchetti, G., Nachtergaele, S., et al. (2016). Structural basis of smoothened regulation by its extracellular domains. *Nature*, *535*(7613), 517–522. https://doi.org/10.1038/nature18934.

Caplan, A. I. (2017). Mesenchymal stem cells: Time to change the name! *Stem Cells Translational Medicine*, *6*(6), 1445–1451. https://doi.org/10.1002/sctm.17-0051.

Carbonare, L. D., Innamorati, G., & Valenti, M. T. (2012). Transcription factor Runx2 and its application to bone tissue engineering. *Stem Cell Reviews and Reports*, *8*(3), 891–897. https://doi.org/10.1007/s12015-011-9337-4.

Carreira-Barbosa, F., & Nunes, S. C. (2020). Wnt signaling: Paths for cancer progression. *Advances in Experimental Medicine and Biology, 1219*, 189–202. https://doi.org/10.1007/978-3-030-34025-4_10.

Chabannes, D., Hill, M., Merieau, E., Rossignol, J., Brion, R., Soulillou, J. P., et al. (2007). A role for heme oxygenase-1 in the immunosuppressive effect of adult rat and human mesenchymal stem cells. *Blood, 110*(10), 3691–3694. https://doi.org/10.1182/blood-2007-02-075481.

Chan, G. K., & Duque, G. (2002). Age-related bone loss: Old bone, new facts. *Gerontology, 48*(2), 62–71.

Charbord, P. (2010). Bone marrow mesenchymal stem cells: Historical overview and concepts. *Human Gene Therapy, 21*(9), 1045–1056. https://doi.org/10.1089/hum.2010.115.

Chawla, A., Boisvert, W. A., Lee, C.-H., Laffitte, B. A., Barak, Y., Joseph, S. B., et al. (2001). A PPARγ -LXR-ABCA1 pathway in macrophages is involved in cholesterol efflux and atherogenesis we provide evidence here for convergence of PPARγ. *Molecular Cell, 7*, 161–171.

Choi, J. J., Yoo, S. A., Park, S. J., Kang, Y. J., Kim, W. U., Oh, I. H., et al. (2008). Mesenchymal stem cells overexpressing interleukin-10 attenuate collagen-induced arthritis in mice. *Clinical and Experimental Immunology, 153*(2), 269–276. https://doi.org/10.1111/j.1365-2249.2008.03683.x.

Cui, Z. K., Kim, S., Baljon, J. J., Doroudgar, M., Lafleur, M., Wu, B. M., et al. (2017). Design and characterization of a therapeutic non-phospholipid liposomal nanocarrier with osteoinductive characteristics to promote bone formation. *ACS Nano, 11*(8), 8055–8063.

Deshpande, I., Liang, J., Hedeen, D., Roberts, K. J., Zhang, Y., Ha, B., et al. (2019). Smoothened stimulation by membrane sterols drives hedgehog pathway activity. *Nature, 571*(7764), 284–288. https://doi.org/10.1038/s41586-019-1355-4.

Devine, S. M., & Hoffman, R. (2000). Role of mesenchymal stem cells in hematopoietic stem cell transplantation. *Current Opinion in Hematology*, 7(6), 358–363. https://doi.org/10.1097/00062752-200011000-00007.

Di Trapani, M., Bassi, G., Midolo, M., Gatti, A., Kamga, P. T., Cassaro, A., et al. (2016). Differential and transferable modulatory effects of mesenchymal stromal cell-derived extracellular vesicles on T, B and NK cell functions. *Scientific Reports, 6*, 24120. https://doi.org/10.1038/srep24120.

Dias, I. H., Borah, K., Amin, B., Griffiths, H. R., Sassi, K., Lizard, G., et al. (2019). Localisation of oxysterols at the sub-cellular level and in biological fluids. *Journal of Steroid Biochemistry and Molecular Biology, 193*(June), 105426. https://doi.org/10.1016/j.jsbmb.2019.105426.

Dimmeler, S., & Losordo, D. (2011). Stem cells review series: An introduction. *Circulation Research, 109*(8), 907–909. https://doi.org/10.1161/circresaha.111.255570.

Du, X., & Yang, H. (2013). Endosomal cholesterol trafficking: Protein factors at a glance Niemann-pick type C proteins. *Acta Biochimica et Biophysica Sinica, 45*(1), 11–17. https://doi.org/10.1093/abbs/gms095.Advance.

Dwyer, J. R., Sever, N., Carlson, M., Nelson, S. F., Beachy, P. A., & Parhami, F. (2007). Oxysterols are novel activators of the hedgehog signaling pathway in pluripotent mesenchymal cells. *The Journal of Biological Chemistry, 282*(12), 8959–8968.

Evans, R. M., Barish, G. D., & Wang, Y. X. (2004). PPARs and the complex journey to obesity. *Nature Medicine, 10*(4), 355–361. https://doi.org/10.1038/nm1025.

Friedenstein, A. J., Chailakhjan, R. K., & Lalykina, K. S. (1970). The development of fibroblast colonies in monolayer cultures of Guinea-pig bone marrow and spleen cells. *Cell and Tissue Kinetics, 3*(4), 393–403. https://doi.org/10.1111/j.1365-2184.1970.tb00347.x.

Gao, X., Xu, C., Asada, N., & Frenette, P. S. (2018). The hematopoietic stem cell niche: From embryo to adult. *Development, 145*(2), 1–12. https://doi.org/10.1242/dev.139691.

Gorojankina, T. (2016). Hedgehog signaling pathway: A novel model and molecular mechanisms of signal transduction. *Cellular and Molecular Life Sciences, 73*(7), 1317–1332. https://doi.org/10.1007/s00018-015-2127-4.
Griffiths, W., & Jörnvall, H. (2014). Oxysterols. *Biochemical and Biophysical Research Communications, 446*(3), 645–646. https://doi.org/10.1016/j.bbrc.2014.04.001.
Hanley, K., Ng, D. C., He, S. S., Lau, P., Min, K., Elias, P. M., et al. (2000). Oxysterols induce differentiation in human keratinocytes and increase Ap-1-dependent involucrin transcription. *The Journal of Investigative Dermatology, 114*(3), 545–553.
Hatzistergos, K. E., Quevedo, H., Oskouei, B. N., Hu, Q., Feigenbaum, G. S., Margitich, I. S., et al. (2010). Bone marrow mesenchymal stem cells stimulate cardiac stem cell proliferation and differentiation. *Circulation Research, 107*(7), 913–922. https://doi.org/10.1161/circresaha.110.222703.
Hayden, J. M., Brachova, L., Higgins, K., Obermiller, L., Sevanian, A., Khandrika, S., et al. (2002). Induction of monocyte differentiation and foam cell formation in vitro by 7-ketocholesterol. *Journal of Lipid Research, 43*(1), 26–35.
Hokugo, A., Sorice, S., Yalom, A., Lee, J. C., Li, A., Zuk, P., et al. (2013). In vitro study of a novel oxysterol for osteogenic differentiation on rabbit bone marrow stromal cells. *Plastic and Reconstructive Surgery, 132*(1), 70e–80e.
Huang, Y., Lin, Y., Rong, M., Liu, W., He, J., & Zhou, L. (2019). 20(S)-hydroxycholesterol and simvastatin synergistically enhance osteogenic differentiation of marrow stromal cells and bone regeneration by initiation of Raf/MEK/ERK signaling. *Journal of Materials Science. Materials in Medicine, 30*(8), 87.
Janda, C. Y., Waghray, D., Levin, A. M., Thomas, C., & Garcia, K. C. (2012). Structural basis of Wnt. *Science, 337*(July), 59–64. https://doi.org/10.1126/science.1222879. Structural.
Janowski, B. A., Grogan, M. J., Jones, S. A., Wisely, G. B., Kliewer, S. A., Corey, E. J., et al. (1999). Structural requirements of ligands for the oxysterol liver X receptors LXRa and LXRβ. *Proceedings of the National Academy of Sciences of the United States of America, 96*(1), 266–271. https://doi.org/10.1073/pnas.96.1.266.
Janowski, B. A., Shan, B., & Russell, D. W. (2001). The hypocholesterolemic agent LY295427 reverses suppression of sterol regulatory element-binding protein processing mediated by oxysterols. *Journal of Biological Chemistry, 276*(48), 45408–45416. https://doi.org/10.1074/jbc.M108348200.
Jeng, K. S., Chang, C. F., & Lin, S. S. (2020). Sonic hedgehog signaling in organogenesis, tumors, and tumor microenvironments. *International Journal of Molecular Sciences, 21*(3), 758. https://doi.org/10.3390/ijms21030758.
Jiang, Y., Jahagirdar, B. N., Reinhardt, R. L., Schwartz, R. E., Keene, C. D., Ortiz-Gonzalez, X. R., et al. (2002). Pluripotency of mesenchymal stem cells derived from adult marrow. *Nature, 418*(6893), 41–49. https://doi.org/10.1038/nature00870.
Johnson, J. S., Meliton, V., Kim, W. K., Lee, K. B., Wang, J. C., Nguyen, K., et al. (2011). Novel oxysterols have pro-osteogenic and anti-adipogenic effects in vitro and induce spinal fusion in vivo. *Journal of Cellular Biochemistry, 112*(6), 1673–1684.
Jones, E. A., Kinsey, S. E., English, A., Jones, R. A., Straszynski, L., Meredith, D. M., et al. (2002). Isolation and characterization of bone marrow multipotential mesenchymal progenitor cells. *Arthritis and Rheumatism, 46*(12), 3349–3360. https://doi.org/10.1002/art.10696.
Katuchova, J., Harvanova, D., Spakova, T., Kalanin, R., Farkas, D., Durny, P., et al. (2015). Mesenchymal stem cells in the treatment of type 1 diabetes mellitus. *Endocrine Pathology, 26*(2), 95–103. https://doi.org/10.1007/s12022-015-9362-y.
Kha, H. T., Basseri, B., Shouhed, D., Richardson, J., Tetradis, S., Hahn, T. J., et al. (2004). Oxysterols regulate differentiation of mesenchymal stem cells: Pro-bone and anti-fat. *Journal of Bone and Mineral Research, 19*(5), 830–840.

Khovidhunkit, W., Moser, A. H., Shigenaga, J. K., Grunfeld, C., & Feingold, K. R. (2003). Endotoxin down-regulates ABCG5 and ABCG8 in mouse liver and ABCA1 and ABCG1 in J774 murine macrophages: Differential role of LXR. *Journal of Lipid Research*, *44*(9), 1728–1736. https://doi.org/10.1194/jlr.M300100-JLR200.

Kim, M. W., An, B. M., Wang, L., Lee, K. Y., Bang, J., & Joo, B. S. (2020). Effects of oxysterols on chondrogenesis of human adipose-derived stem cells. *Annals of Clinical and Laboratory Science*, *50*(2), 190–198.

Kim, W. K., Meliton, V., Bourquard, N., Hahn, T. J., & Parhami, F. (2010). Hedgehog signaling and osteogenic differentiation in multipotent bone marrow stromal cells are inhibited by oxidative stress. *Journal of Cellular Biochemistry*, *111*(5), 1199–1209.

Kim, W. K., Meliton, V., Tetradis, S., Weinmaster, G., Hahn, T. J., Carlson, M., et al. (2010). Osteogenic oxysterol, 20(S)-hydroxycholesterol, induces notch target gene expression in bone marrow stromal cells. *Journal of Bone and Mineral Research*, *25*(4), 782–795.

Kolf, C. M., Cho, E., & Tuan, R. S. (2007). Mesenchymal stromal cells. Biology of adult mesenchymal stem cells: Regulation of niche, self-renewal and differentiation. *Arthritis Research & Therapy*, *9*(1), 204. https://doi.org/10.1186/ar2116.

Krampera, M., Cosmi, L., Angeli, R., Pasini, A., Liotta, F., Andreini, A., et al. (2006). Role for interferon-gamma in the immunomodulatory activity of human bone marrow mesenchymal stem cells. *Stem Cells*, *24*(2), 386–398. https://doi.org/10.1634/stemcells.2005-0008.

Lee, K. D. (2008). Applications of mesenchymal stem cells: An updated review. *Chang Gung Medical Journal*, *31*(3), 228–236.

Lee, J. S., Kim, E., Han, S., Kang, K. L., & Heo, J. S. (2017). Evaluating the oxysterol combination of 22(S)-hydroxycholesterol and 20(S)-hydroxycholesterol in periodontal regeneration using periodontal ligament stem cells and alveolar bone healing models. *Stem Cell Research & Therapy*, *8*(1), 276.

Levy, D., Correa de Melo, T., Ohira, B. Y., Fidelis, M. L., Ruiz, J. L. M., Rodrigues, A., et al. (2018). Oxysterols selectively promote short-term apoptosis in tumor cell lines. *Biochemical and Biophysical Research Communications*, *505*(4), 1043–1049.

Levy, D., de Melo, T. C., Oliveira, B. A., Paz, J. L., de Freitas, F. A., Reichert, C. O., et al. (2019). 7-Ketocholesterol and cholestane-triol increase expression of SMO and LXRα signaling pathways in a human breast cancer cell line. *Biochemistry and Biophysics Reports*, *19*, 100604. https://doi.org/10.1016/j.bbrep.2018.12.008.

Levy, D., de Melo, T. C., Ruiz, J. L. M., & Bydlowski, S. P. (2017). Oxysterols and mesenchymal stem cell biology. *Chemistry and Physics of Lipids*, *207*, 223–230. https://doi.org/10.1016/j.chemphyslip.2017.06.009.

Levy, D., Ruiz, J. L., Celestino, A. T., Silva, S. F., Ferreira, A. K., Isaac, C., et al. (2014). Short-term effects of 7-ketocholesterol on human adipose tissue mesenchymal stem cells in vitro. *Biochemical and Biophysical Research Communications*, *446*(3), 720–725.

Liu, H., Yuan, L., Xu, S., Wang, K., & Zhang, T. (2005). Cholestane-3beta,5alpha,6beta-triol inhibits osteoblastic differentiation and promotes apoptosis of rat bone marrow stromal cells. *Journal of Cellular Biochemistry*, *96*(1), 198–208.

Lv, F. J., Tuan, R. S., Cheung, K. M., & Leung, V. Y. (2014). Concise review: The surface markers and identity of human mesenchymal stem cells. *Stem Cells*, *32*(6), 1408–1419. https://doi.org/10.1002/stem.1681.

Ma, L., & Nelson, E. R. (2019). Oxysterols and nuclear receptors. *Molecular and Cellular Endocrinology*, *484*(January), 42–51. https://doi.org/10.1016/j.mce.2019.01.016.

Majumdar, M. K., Keane-Moore, M., Buyaner, D., Hardy, W. B., Moorman, M. A., McIntosh, K. R., et al. (2003). Characterization and functionality of cell surface molecules on human mesenchymal stem cells. *Journal of Biomedical Science*, *10*(2), 228–241. https://doi.org/10.1007/bf02256058.

Maria Ruiz, J. L., Levy, D., & Bydlowski, S. P. (2014). Mesencchymal stem cells: Sources and properties in regenerative medicine. *J Hematol Transfus*, *2*(3), 1029.
Massey, J. B., & Pownall, H. J. (2006). Structures of biologically active oxysterols determine their differential effects on phospholipid membranes. *Biochemistry*, *45*(35), 10747–10758. https://doi.org/10.1021/bi060540u.
Montgomery, S. R., Nargizyan, T., Meliton, V., Nachtergaele, S., Rohatgi, R., Stappenbeck, F., et al. (2014). A novel osteogenic oxysterol compound for therapeutic development to promote bone growth: Activation of hedgehog signaling and osteogenesis through smoothened binding. *Journal of Bone and Mineral Research*, *29*(8), 1872–1885.
Murdolo, G., Piroddi, M., Tortoioli, C., Bartolini, D., Schmelz, M., Luchetti, F., et al. (2016). Free radical-derived oxysterols: Novel adipokines modulating adipogenic differentiation of adipose precursor cells. *The Journal of Clinical Endocrinology and Metabolism*, *101*(12), 4974–4983.
Myers, B. R., Sever, N., Chong, Y. C., Kim, J., Belani, J. D., Rychnovsky, S., et al. (2013). Hedgehog pathway modulation by multiple lipid binding sites on the smoothened effector of signal response. *Developmental Cell*, *26*(4), 346–357. https://doi.org/10.1016/j.devcel.2013.07.015.
Namba, F. (2019). Mesenchymal stem cells for the prevention of bronchopulmonary dysplasia. *Pediatrics International*, *61*(10), 945–950. https://doi.org/10.1111/ped.14001.
Nayan, M., Paul, A., Chen, G., Chiu, R. C., Prakash, S., & Shum-Tim, D. (2011). Superior therapeutic potential of young bone marrow mesenchymal stem cells by direct intramyocardial delivery in aged recipients with acute myocardial infarction: In vitro and in vivo investigation. *Journal of Tissue Engineering*, *2011*, 741213. https://doi.org/10.4061/2011/741213.
Nury, T., Zarrouk, A., Yammine, A., Mackrill, J. J., Vejux, A., & Lizard, G. (2020). Oxiapoptophagy: A type of cell death induced by some oxysterols. *British Journal of Pharmacology*, *1*(March), 1–9. https://doi.org/10.1111/bph.15173.
Ohba, S., Kawaguchi, H., Kugimiya, F., Ogasawara, T., Kawamura, N., Saito, T., et al. (2008). Patched1 haploinsufficiency increases adult bone mass and modulates Gli3 repressor activity. *Developmental Cell*, *14*(5), 689–699. https://doi.org/10.1016/j.devcel.2008.03.007.
Ohishi, M., & Schipani, E. (2010). Bone marrow mesenchymal stem cells. *Journal of Cellular Biochemistry*, *109*(2), 277–282. https://doi.org/10.1002/jcb.22399.
Parhami, F. (2003). Possible role of oxidized lipids in osteoporosis: Could hyperlipidemia be a risk factor? *Prostaglandins, Leukotrienes, and Essential Fatty Acids*, *68*(6), 373–378.
Parhami, F., Garfinkel, A., & Demer, L. L. (2000). Role of lipids in osteoporosis. *Arteriosclerosis, Thrombosis, and Vascular Biology*, *20*(11), 2346–2348.
Parhami, F., Jackson, S. M., Tintut, Y., Le, V., Balucan, J. P., Territo, M., et al. (1999). Atherogenic diet and minimally oxidized low density lipoprotein inhibit osteogenic and promote adipogenic differentiation of marrow stromal cells. *Journal of Bone and Mineral Research*, *14*(12), 2067–2078.
Parhami, F., Mody, N., Gharavi, N., Ballard, A. J., Tintut, Y., & Demer, L. L. (2002). Role of the cholesterol biosynthetic pathway in osteoblastic differentiation of marrow stromal cells. *Journal of Bone and Mineral Research*, *17*(11), 1997–2003.
Pasolli, H. A. (2011). The hair follicle bulge: A niche for adult stem cells. *Microscopy and Microanalysis*, *17*(4), 513–519. https://doi.org/10.1017/s1431927611000419.
Patel, R. P., Diczfalusy, U., Dzeletovic, S., Wilson, M. T., & Darley-Usmar, V. M. (1996). Formation of oxysterols during oxidation of low density lipoprotein by peroxynitrite, myoglobin, and copper. *Journal of Lipid Research*, *37*(11), 2361–2371.
Paz, J. L., Levy, D., Oliveira, B. A., de Melo, T. C., de Freitas, F. A., Reichert, C. O., et al. (2019). 7-ketocholesterol promotes oxiapoptophagy in bone marrow mesenchymal stem cell from patients with acute myeloid leukemia. *Cell*, *8*(5), 482. https://doi.org/10.3390/cells8050482.

Peister, A., Mellad, J. A., Larson, B. L., Hall, B. M., Gibson, L. F., & Prockop, D. J. (2004). Adult stem cells from bone marrow (MSCs) isolated from different strains of inbred mice vary in surface epitopes, rates of proliferation, and differentiation potential. *Blood*, *103*(5), 1662–1668. https://doi.org/10.1182/blood-2003-09-3070.

Plaisant, M., Fontaine, C., Cousin, W., Rochet, N., Dani, C., & Peraldi, P. (2009). Activation of hedgehog signaling inhibits osteoblast differentiation of human mesenchymal stem cells. *Stem Cells*, *27*(3), 703–713.

Qi, X., Liu, H., Thompson, B., McDonald, J., Zhang, C., & Li, X. (2019). Cryo-EM structure of oxysterol-bound human smoothened coupled to a heterotrimeric Gi. *Nature*, *571*(7764), 279–283. https://doi.org/10.1038/s41586-019-1286-0.

Radhakrishnan, A., Ikeda, Y., Hyock, J. K., Brown, M. S., & Goldstein, J. L. (2007). Sterol-regulated transport of SREBPs from endoplasmic reticulum to Golgi: Oxysterols block transport by binding to Insig. *Proceedings of the National Academy of Sciences of the United States of America*, *104*(16), 6511–6518. https://doi.org/10.1073/pnas.0700899104.

Ren, G., Zhang, L., Zhao, X., Xu, G., Zhang, Y., Roberts, A. I., et al. (2008). Mesenchymal stem cell-mediated immunosuppression occurs via concerted action of chemokines and nitric oxide. *Cell Stem Cell*, *2*(2), 141–150. https://doi.org/10.1016/j.stem.2007.11.014.

Richardson, J. A., Amantea, C. M., Kianmahd, B., Tetradis, S., Lieberman, J. R., Hahn, T. J., et al. (2007). Oxysterol-induced osteoblastic differentiation of pluripotent mesenchymal cells is mediated through a PKC- and PKA-dependent pathway. *Journal of Cellular Biochemistry*, *100*(5), 1131–1145.

Rosales, C., Tang, D., & Geng, Y. J. (2015). CD1d serves as a surface receptor for oxidized cholesterol induction of peroxisome proliferator-activated receptor-γ. *Atherosclerosis*, *239*(1), 224–231. https://doi.org/10.1016/j.atherosclerosis.2015.01.004.

Sato, R. (2010). Sterol metabolism and SREBP activation. *Archives of Biochemistry and Biophysics*, *501*(2), 177–181. https://doi.org/10.1016/j.abb.2010.06.004.

Sato, K., Ozaki, K., Oh, I., Meguro, A., Hatanaka, K., Nagai, T., et al. (2007). Nitric oxide plays a critical role in suppression of T-cell proliferation by mesenchymal stem cells. *Blood*, *109*(1), 228–234. https://doi.org/10.1182/blood-2006-02-002246.

Scaffidi, P., & Misteli, T. (2008). Lamin A-dependent misregulation of adult stem cells associated with accelerated ageing. *Nature Cell Biology*, *10*(4), 452–459. https://doi.org/10.1038/ncb1708.

Schwindt, T. T., Barnabé, G. F., & Mello, L. E. A. M. (2005). Proliferar ou diferenciar? Perspectivas de destino das células-tronco. *Jornal Brasileiro de Neurocirurgia*, *16*(1), 13–19. https://doi.org/10.22290/jbnc.v16i1.505.

Silva, S. F., Levy, D., Ruiz, J. L. M., de Melo, T. C., Isaac, C., Fidelis, M. L., et al. (2017). Oxysterols in adipose tissue-derived mesenchymal stem cell proliferation and death. *The Journal of Steroid Biochemistry and Molecular Biology*, *169*, 164–175.

Sylvester, K. G., & Longaker, M. T. (2004). Stem cells: Review and update. *Archives of Surgery*, *139*(1), 93–99. https://doi.org/10.1001/archsurg.139.1.93.

Theofilopoulos, S., Abreu de Oliveira, W. A., Yang, S., Yutuc, E., Saeed, A., Abdel-Khalik, J., et al. (2019). 24(*S*),25-Epoxycholesterol and *cholesterol 24S-hydroxylase* (*CYP46A1*) overexpression promote midbrain dopaminergic neurogenesis *in vivo*. *The Journal of Biological Chemistry*, *294*(11), 4169–4176. https://doi.org/10.1074/jbc.RA118.005639.

Traversari, C., Sozzani, S., Steffensen, K. R., & Russo, V. (2014). LXR-dependent and -independent effects of oxysterols on immunity and tumor growth. *European Journal of Immunology*, *44*(7), 1896–1903. https://doi.org/10.1002/eji.201344292.

Tse, W. T., Pendleton, J. D., Beyer, W. M., Egalka, M. C., & Guinan, E. C. (2003). Suppression of allogeneic T-cell proliferation by human marrow stromal cells: Implications in transplantation. *Transplantation*, *75*(3), 389–397. https://doi.org/10.1097/01.Tp.0000045055.63901.A9.

Valenta, T., Hausmann, G., & Basler, K. (2012). The many faces and functions ofÎ 2-catenin. *EMBO Journal, 31*(12), 2714–2736. https://doi.org/10.1038/emboj.2012.150.

Vizoso, F. J., Eiro, N., Cid, S., Schneider, J., & Perez-Fernandez, R. (2017). Mesenchymal stem cell secretome: Toward cell-free therapeutic strategies in regenerative medicine. *International Journal of Molecular Sciences, 18*(9). https://doi.org/10.3390/ijms18091852.

Vurusaner, B., Leonarduzzi, G., Gamba, P., Poli, G., & Basaga, H. (2016). Oxysterols and mechanisms of survival signaling. *Molecular Aspects of Medicine, 49*, 8–22. https://doi.org/10.1016/j.mam.2016.02.004.

Wang, X., Sato, R., Brown, M. S., Hua, X., & Goldstein, J. L. (1994). SREBP-1, a membrane-bound transcription factor released by sterol-regulated proteolysis. *Cell*, 77(1), 53–62. https://doi.org/10.1016/0092-8674(94)90234-8.

Williams, A. R., & Hare, J. M. (2011). Mesenchymal stem cells: Biology, pathophysiology, translational findings, and therapeutic implications for cardiac disease. *Circulation Research, 109*(8), 923–940. https://doi.org/10.1161/circresaha.111.243147.

Willy, P. J., Umesono, K., Ong, E. S., Evans, R. M., Heyman, R. A., & Mangelsdorf, D. J. (1995). LXR, a nuclear receptor that defines a distinct retinoid response pathway. *Genes and Development, 9*(9), 1033–1045. https://doi.org/10.1101/gad.9.9.1033.

Woltje, M., Bobel, M., Heiland, M., Beck-Broichsitter, B., Al-Dam, A., Grobe, A., et al. (2015). Purmorphamine and oxysterols accelerate and promote osteogenic differentiation of mesenchymal stem cells in vitro. *In Vivo, 29*(2), 247–254.

Yalom, A., Hokugo, A., Sorice, S., Li, A., Segovia Aguilar, L. A., Zuk, P., et al. (2014). In vitro osteoinductive effects of hydroxycholesterol on human adipose-derived stem cells are mediated through the hedgehog signaling pathway. *Plastic and Reconstructive Surgery, 134*(5), 960–968.

Yang, J., Andre, P., Ye, L., & Yang, Y. Z. (2015). The hedgehog signalling pathway in bone formation. *International Journal of Oral Science*, 7(2), 73–79. https://doi.org/10.1038/ijos.2015.14.

Yeung, T. M., Chia, L. A., Kosinski, C. M., & Kuo, C. J. (2011). Regulation of self-renewal and differentiation by the intestinal stem cell niche. *Cellular and Molecular Life Sciences, 68*(15), 2513–2523. https://doi.org/10.1007/s00018-011-0687-5.

Yvan-charvet, L., Wang, N., & Tall, A. R. (2010). The role of HDL, ABCA1 and ABCG1 transporters in cholesterol. *Arteriosclerosis, Thrombosis, and Vascular Biology, 30*(2), 139–143. https://doi.org/10.1161/ATVBAHA.108.179283.The.

Zeng, X., Wang, Y., Dong, Q., Ma, M. X., & De Liu, X. (2020). DLX2 activates Wnt1 transcription and mediates Wnt/β-catenin signal to promote osteogenic differentiation of hBMSCs. *Gene, 744*(March), 144564. https://doi.org/10.1016/j.gene.2020.144564.

Zhuo, Y., Li, S. H., Chen, M. S., Wu, J., Kinkaid, H. Y., Fazel, S., et al. (2010). Aging impairs the angiogenic response to ischemic injury and the activity of implanted cells: Combined consequences for cell therapy in older recipients. *The Journal of Thoracic and Cardiovascular Surgery, 139*(5), 1286–1294. 1294.e1281-1282 https://doi.org/10.1016/j.jtcvs.2009.08.052.

FSC
www.fsc.org
MIX
Papier | Fördert gute Waldnutzung
FSC® C083411

Kolibri 360 GmbH
Ferdinand-Jühlke-Straße 7
99095 Erfurt, Deutschland
produktsicherheit@kolibri360.de